Food
The Chemistry of its Components
5th Edition

for

Isaac Leon

Food
The Chemistry of its Components
5th Edition

T.P. Coultate
formerly of London South Bank University

RSCPublishing

ISBN: 978-0-85404-111-4

A catalogue record for this book is available from the British Library

Published by The Royal Society of Chemistry,
Thomas Graham House, Science Park, Milton Road,
Cambridge CB4 0WF, UK

Registered Charity Number 207890

For further information see our web site at www.rsc.org

Foreword

Like science, cuisine thrives on curiosity.

In 1999 I began visiting the flavour and fragrance company Firmenich in Geneva, to exchange ideas and explore culinary science. On one occasion I was in the office of Dr Alan Parker, at the wonderfully named Laboratory for the Physics of Soft Edible Matter, and noticed a shelf full of RSC books with titles that held a lot of appeal for a chef with a growing interest in the molecular properties of food: *Basic Principles of Colloid Science*; *Carbohydrate Chemistry*; *Fatty Acids* and, of course, *Food: The Chemistry of Its Components*.

As soon as I got back to Bray, I looked up the RSC's publications list and ordered lots of their books. Tom's turned out to be particularly useful. Lucid, accessible and clearly expressed, it was exactly the kind of science primer I was looking for. I read it from cover to cover and then invited Tom over to the Fat Duck to find out more.

In person, Tom is just as entertaining and passionate about his subject as he is on the page. It made for a truly stimulating discussion, during which he talked, among other things, about gellan gum – a versatile gelling agent with superb flavour release that has since become a vital part of the kitchen's resources and a key ingredient in a number of the dishes I've created, including amazingly fresh, smooth purées, salmon poached in liquorice, and a cup of tea that appears to be hot and cold at the same time!

Food: The Chemistry of Its Components is full of similarly valuable, thought-provoking information, guaranteed to spark off all kinds of insights and ideas. It deserves to be on the bookshelf not just of every chef and scientist, but of anyone interested in food or how the world around us works.

Heston Blumenthal, Bray, 2008

Preface to the Fifth Edition

When the Royal Society of Chemistry originally invited me to write this book in 1980 the proposed readership was clearly delineated, namely teachers of GCE Advanced Level Chemistry in school sixth forms. Apparently they were badly in need of background information to help them prepare their students for the new topic of Food Chemistry which had appeared in some of their syllabuses. By the time the first edition actually appeared a few years later the need for an up to date textbook of introductory food chemistry to support my own BSc Food Science students, at what was then the Polytechnic of the South Bank, London, was also becoming obvious. The favourable response to the first edition told me that for once two birds had been killed with one stone! As new editions have appeared over the years the original objective has, per-haps inevitably, suffered from what I believe is known in military cir-cles as "mission creep". This edition is over twice the size of the first, as interesting new topics have demanded attention and chapters devoted to water, minerals and "undesirables" have been added.

My determination to ensure that chemical formulae are included for virtually every food substance mentioned in the text may make the contents look a little intimidating. However, I continue to try to make the actual words as approachable as possible, lending themselves to being read as well as "looked up". Despite appearances, readers with only the vaguest recollection of school chemistry should find most of it within their grasp, especially in the later chapters. The separate inclusion of more advanced material begun in the fourth edition has now been further developed with the Special Topics included at the end of most chapters.

Food: The Chemistry of its Components, Fifth edition
By T.P. Coultate
© T.P. Coultate, 2009
Published by the Royal Society of Chemistry, www.rsc.org

As in previous editions, a selection of Further Reading is also included with each chapter, wherever possible restricted to 21st century material. Extensive references to research literature are not provided, but each chapter ends with a selection of articles, headed Recent Reviews. These will be available to most university students and can provide a route to advanced literature. The temptation to include addresses of relevant and authoritative websites has been resisted since, unlike the traditional scientific literature, they tend to be ephemeral and are unlikely to survive through the lifetime of a textbook.

The internet can provide a wealth of information—but it must be used with considerable caution. It has always been the rule that one should not necessarily believe something just because it has appeared in a book, but such scepticism is absolutely essential when exploring websites. For example, an internet search using "choline" and "vitamin" as keywords will locate the official statement that choline is not a vitamin (*see* Chapter 8), followed by innumerable commercial sites which claim the exact opposite, as a means of boosting sales of choline as a "dietary supplement".

Information on nutrient consumption, exposure to pesticides and similar material is largely based on British sources, a limitation for which I apologise to readers elsewhere in the world. Similarly, where legislation is mentioned, for instance regarding what additives are or are not permitted, the situation in Britain and the European Community is reported for illustrative purposes. The appropriate authorities such as the UK Food Standards Agency must always be consulted for an authoritative view of the legal position.

A BRIEF NOTE ABOUT CONCENTRATIONS

The concentrations of chemical components are expressed in a number of different styles in this book, depending on the context and the concentrations concerned. Readers may find the following helpful.

(a) However they are expressed, concentrations always imply the amount *contained*, rather than the amount *added*. Thus "5 g of X per 100 g of foodstuff" implies that 100 g of the foodstuff contains 95 g of substance(s) other than X.

(b) The abbreviation ppm means parts per million, *i.e.* grams per million grams, or more usually milligrams per kilogram. One ppb, or part per billion, corresponds to one microgram per kilogram.

(c) Amounts contained in 100 g (or 100 cm^3) are often expressed as simple percentages. Where necessary the terms w/w, v/v or w/v are added to indicate whether volumes or weights or both are involved. Thus "5% w/v" means that $100\,cm^3$ of a liquid contains 5 g of a solid, either dissolved or in suspension. Note that millilitres (ml) and litres (l) are no longer considered acceptable. Although the replacement for the ml, the cubic centimetre (cm^3), is widely recognised, there is little sign that the cubic decimetre, or dm^3, has yet taken over from the litre—except in teaching laboratories, which of course must always endeavour to toe the party line.

(d) Very often a strictly mathematical style is adopted, with "per" expressing the power of minus one. Since mathematically:

$$x^{-1} = 1/x$$

$5\,\mu g\,kg^{-1}$ becomes a convenient way of writing 5 micrograms per kilogram. This brief but mathematically rigorous style comes into its own when the rates of intake of substances such as toxins have to be related to the size of the animal consuming them, as in "5 milligrams per day per kilogram body weight", which abbreviates to:

$$5\,mg\,kg^{-1}\ body\ weight\ per\ day$$

A quantity, say 10 mg, per cubic centimetre, cm^3, would be written: $10\,mg\,cm^{-3}$.

Tom Coultate, 2008

Contents

The Author		xvii
Acknowledgements		xviii
Chapter 1	**Introduction**	
	Further Reading	5
Chapter 2	**Sugars**	
	Monosaccharides	7
	Oligosaccharides	20
	Sugars as Solids	26
	Sugars in Solution	27
	Decomposition	30
	The Maillard Reaction	33
	Special Topics	41
	1. Reducing Group Reactions	41
	2. Sugar Cane, Sugar Beet and Tequila	42
	3. Monosaccharide Conformations	43
	4. Acrylamide	44
	Further Reading	48
	Recent Reviews	48
Chapter 3	**Polysaccharides**	
	Starch	51
	Pectins	60
	Seaweed Polysaccharides	67
	Cellulose, Hemicelluloses and Fibre	72

Food: The Chemistry of its Components, Fifth edition
By T.P. Coultate
© T.P. Coultate, 2009
Published by the Royal Society of Chemistry, www.rsc.org

Gums   80
Special Topics   85
   1   Chemically Modified Starches   85
   2   Syrups from Starch   87
   3   Glycaemic Index   89
   4   Further Details of Pectin Structure   92
Further Reading   95
Recent Reviews   95

Chapter 4   Lipids

Fatty Acids: Structure and Distribution   97
   Conjugated Linoleic Acids   105
Essential Fatty Acids   106
   Fatty Acids and Coronary Heart Disease   108
Reactions of Unsaturated Fatty Acids   112
   Hydrogenation, Margarine and *trans* Fatty Acids   113
   Rancidity   115
   Antioxidants   123
Triglycerides   127
   Melting and Crystallisation   130
   Cocoa Butter and Chocolate   133
   Fractionation   135
   Interesterification   137
Polar Lipids   139
   Milk Fat, Cream and Butter   147
   Synthetic Emulsifiers   148
   Phytosteroids   150
Special Topics   152
   1   Hydrogenation in Detail   152
   2   Singlet and Triplet Oxygen   153
   3   Triglyceride Crystals   154
Further Reading   157
Recent Reviews   158

Chapter 5   Proteins

Amino Acids   160
Protein Structure   163
Essential Amino Acids and Protein Quality   168
Analysis   173
Food Protein Systems   176
   Milk   176
   Cheese   182
   Egg   184

Meat 187
Bread 198
Special Topics 208
1 Myoglobin and Free Radicals 208
2 Wheat Genes and Chromosomes 210
3 Further Details of Gluten Proteins 211
Further Reading 212
Recent Reviews 213

Chapter 6 Colours

Chlorophylls 214
Carotenoids 218
Anthocyanins 227
Betalaines 234
Melanins 236
Tea 239
Turmeric and Cochineal 243
Artificial Food Colorants 244
Inorganic Food Colorants 249
Restrictions on the Use of Colours in Foodstuffs 249
The Molecular Basis of Colour 251
Special Topics 255
1 Flavonoids, Tannins and Health 255
2 Colour Measurement 258
Further Reading 263
Recent Reviews 263

Chapter 7 Flavours

Taste 268
Sweetness 268
Bitterness 278
Saltiness 282
Sourness 282
Astringency 284
Pungency 285
Meatiness 290
Odour 294
Meat 295
Fruit 297
Vegetables 301
Herbs and Spices 305
Synthetic Flavourings 306
Special Topics 309

1   Off-Flavours in Meat   309
2   Taints   310
Further Reading   310
Recent Reviews   311

Chapter 8   Vitamins

Thiamin (Vitamin B_1, Aneurine)   315
Riboflavin (Vitamin B_2)   318
Pyridoxine (Vitamin B_6, Pyridoxol)   321
Niacin (Nicotinic Acid, Nicotinamide)   323
Cobalamin (Cyanocobalamin, Vitamin B_{12})   326
Folic Acid (Folacin)   328
Biotin and Pantothenic Acid   330
Ascorbic Acid (Vitamin C)   332
Retinol (Vitamin A)   338
Cholecalciferol (Vitamin D, Calciferol)   343
Vitamin E (α-Tocopherol)   346
Vitamin K (Phylloquinone, Menaquinones)   350
Special Topics   352
    1.   Details of the Coenzyme Functions of Thiamine
         and Pyridoxine   352
    2.   Further Details of the Role of Ascorbic Acid   353
    3.   Non-Vitamins   355
Further Reading   357
Recent Reviews   358

Chapter 9   Preservatives

Sodium Chloride   361
Nitrites   362
Smoke   366
Sulfur Dioxide   367
Benzoates   370
Other Organic Acids   372
Nisin and Natamycin   373
Irradiation   375
Further Reading   380
Recent Review   380

Chapter 10 Undesirables

Endogenous Toxins in Foods Derived from Plants   382
Endogenous Toxins in Foods of Animal Origin   394
Mycotoxins   398

Bacterial Toxins   404
Allergens   407
Toxic Agricultural Residues   412
Toxic Metal Residues   421
    Lead   421
    Mercury   424
    Arsenic   426
    Cadmium   426
    Tin and Aluminium   427
Toxins Generated During Heat Treatment of Food   428
Packaging Residues   430
Environmental Pollutants   432
Special Topics   434
    1. Favism   434
    2. Total Intake of Undesirable Metals   436
Further Reading   437
Recent Reviews   438

Chapter 11 Minerals

The Bulk Minerals   440
    Sodium   440
    Potassium   441
    Magnesium   442
    Calcium   442
    Phosphorus   444
The Trace Minerals   446
    Iron   446
    Copper   449
    Zinc   450
    Selenium   450
    Iodine   451
    Other Trace Minerals   453
Further Reading   454
Recent Reviews   454

Chapter 12 Water

Water Structure   456
Interactions of Water with Food Components   461
Interactions of Water with Food Materials   466
Water Binding   468
Water Determination   470
Further Reading   474
Recent Reviews   474

Appendix

Nutritional Requirements and Dietary Sources 475
Further Reading 478

Subject Index 480

The Author

Until he retired from full-time teaching Tom Coultate was Principal Lecturer in Food Biochemistry at London South Bank University, on courses ranging from HNC to MSc and particularly the BSc in Food Science. He has had a professional interest in food ever since leaving school to join Unilever's Colworth laboratories, where he continued with his studies part-time gaining an ONC (in Chemistry), an HNC (in Applied Biology) and an MIBiol (in Biochemistry). Before taking up teaching he gained a PhD at the University of Leicester for studies on the biochemistry of thermophilic bacteria. Dr Coultate is a Fellow of the Institute of Food Science and Technology. From time to time he contributes articles to food magazines and enjoys giving talks to local societies on food topics.

Acknowledgements

In preparing this fifth edition I have been as dependent as ever on the contributions of my family, colleagues, students and readers of previous editions.

My wife Ann has continued to provide the same invaluable encouragement and support she first contributed over 25 years ago. My sons Edward and Ben, keen amateur photographers, have graduated from helping to spend the royalties to contributing valuable artistic advice on the design of the cover.

Ever since I started work on the first edition I have been dependent on the support of colleagues and friends. They have been an indispensable source of encouragement and ideas, details of the latest research (including their own) and expert guidance through controversial or unfamiliar topics. This list of names is intended to include contributors to all five editions and I sincerely apologise to anyone whose name should be here but isn't:

Jenny Ames, Peter Barnes, Alan Beeby, Martin Chaplin, Barbara Crook, Jill Davies, Peter Ellis, Ailbhe Fallon, Paul Gillard, (the late) Pat Hastilow, John Henley, Nick Henson, Mike Hibbs, (the late) Mike Hill, David Ledward, Dominic Man, Neil Morgan, Bryan Reuben, Sibel Roller, Dave Rosie, Tom Sanders, David Shuker, Ken Spears, Melvyn Stevens, Sam Sumar, Graham Sworn, Geoff Talbot, Jan van Mechelen, Dave Walsh and Robin Wyers.

Of course the opinions expressed in this book, and any errors that remain, must be placed at my door, not theirs. Several organisations have provided essential services, notably the libraries of London South Bank University, the Open University at Milton Keynes and King's College, London University. The regular on-line bulletins from Reading Scientific Services, Ltd., have also kept me aware of the latest developments. The staff of the UK Food Standards Agency have been a constant source of up to date information on legislative issues. What is recorded in this book was correct at the time of writing, but readers should always consult the FSA or other authorities directly when authoritative information is required about what is or is not currently permitted.

Writing a book like this would not be the rewarding task it is without the support, and often blunt criticism, of my students, for many years at London South Bank University and most recently at King's College, London University. Perhaps they will never improve on the mnemonic for the hexose sugars but they still spot errors, especially in diagrams and formulae, with ruthless efficiency.

As with all previous editions I have been able to rely on my friends at the Royal Society of Chemistry, particularly Janet Freshwater, Katrina Harding and Caroline Wain, for their support and encouragement. Special thanks must go to my editor, Don Sanders. He has done great work repairing the infelicities of my English and made me rethink a number of passages that simply did not actually say what I thought they did!

Finally I come to my friend Heston Blumenthal. Much of my readership has always consisted of students and teachers, of chemistry, food science, nutrition, *etc.*, together with scientists and technicians involved in the food industry and its associated organisations. Until I met Heston it had never occurred to me that what I had to say might be of interest to that other food world, of fine dining and internationally renowned chefs. However when Heston introduced me to "molecular gastronomy" my ideas changed, to say the least. That he has consented to write the Foreword will ensure that this book is not entirely without merit and is for me a great honour for which I am exceedingly grateful.

Tom Coultate, 2008

CHAPTER 1

Introduction

For the chemists of the 18th and 19th centuries an understanding of the chemical nature of our food was a major objective. They realised that this knowledge was essential if dietary standards, and with them health and prosperity, were to improve. Inevitably it was the food components present in large amounts, the carbohydrates, fats and proteins, that were the first nutrients to be described in chemical terms. However, it was also widely recognised that much of the food, and the drink, on sale to the general public was very likely to have been adulterated. The chemists of the day took the blame for some of this:

> "There is in this city [London] a certain fraternity of chemical operators who work underground in holes, caverns and dark retirements ... They can squeeze Bordeaux from the sloe and draw champagne from an apple."
>
> The Tatler, 1710.

But by the middle of the 19th century the chemists were deeply involved in exposing the malpractices of food suppliers. Chemistry was brought to bear on the detection of dangerous colourings in confectionery, additional water in milk, beer, wines and spirits, and many other unexpected food ingredients. A major incentive for the development of chemical analysis was financial. The British government's activities were largely funded by excise duties on alcohol and tea and large numbers of chemists were employed to protect this revenue.

Food: The Chemistry of its Components, Fifth edition
By T.P. Coultate
© T.P. Coultate, 2009
Published by the Royal Society of Chemistry, www.rsc.org

As physiologists and physicians began to relate their findings to the chemical knowledge of foodstuffs the need for reliable analytical techniques increased. Twentieth century laboratory techniques were essential for the study of vitamins and the other components that occur in similarly small amounts, including the natural and artificial colouring and flavouring compounds.

Until the use of gas chromatography (GC) became widespread in the 1960s the classical techniques of "wet chemistry" were the rule, but since that time increasingly sophisticated instrumental techniques have taken over. The latest methods are now so sensitive that many food components can be detected and quantified at such low levels (parts per billion, 1 microgram per kilogram are commonplace) that one can have serious doubts as to whether the presence of a particular pesticide residue or environmental toxin at the detection limits of the analysis can really have any biological or health significance. It is perhaps some consolation for the older generation of food chemists that some of the methods of proximate analysis[i] which they, like the author, struggled with in the 1960s are still in use today, albeit in automated apparatus rather than by using extravagant examples of the glassblower's art.

By the time of World War II it appeared that most of the questions being asked of food chemists by nutritionists, agriculturalists and others had been answered. This was certainly true as far as questions of the variety, "What is this substance and how much is present?" were concerned. However, as reflected in this book, over the past few decades new questions have been asked and many answers are still awaited. Chemists are nowadays expected to quantify the innumerable undesirable substances, both natural and man-made, which are to be found in our food. The greatest challenge food chemists now face is to explain the *behaviour* of food components. What happens when food is processed, stored, cooked, chewed, digested and absorbed? Much of the stimulus for this type of enquiry has come from the food manufacturing industry. For example, for many of us the observation that the starch in a dessert product provides a certain amount of energy has been overtaken in importance by the need to know which type of starch will give just the right degree of thickening, and what is the molecular basis for the differences between one starch and another. And furthermore, that a dessert product must have a long shelf life, look as pretty and taste nearly as good as the one served in an expensive restaurant.

In recent years these apparently rather superficial aspects of food have begun to take on a wider significance. Nutritionists, physiologists and

---

[i] The determination of total fat, protein, carbohydrate, ash (as a measure of metals) and water.

other scientists now recognise what consumers have always known—
that there is more to the business of feeding people than compiling a list
of nutrients in the correct proportions. This is just as true if one is
engaged in famine relief as it is in a five-star restaurant. To satisfy a
nutritional need a foodstuff must be acceptable, and to be acceptable it
must first look and then taste "right".

We have also become increasingly conscious of two other aspects of
our food. As some of us have become more affluent, our food intake is
no longer limited by our income and we have begun to suffer from
the Western "disease" of overnutrition. Our parents are appalled when
our children follow sound dietetic advice and discard the calorie-laden
fat from around their sliced ham, and food scientists are called upon
to devise butter substitutes with minimal fat content. Closely associated with this issue is the intense public interest in the "chemicals" in
our food. The author's very first chemistry teacher, Mr Crosland,
refused to accept the word "chemical" as a noun. It is a great pity that
his point of view never found wider acceptance, and the general public's
appreciation of the language of chemistry continues to leave much to be
desired. For example, it is unlikely that many people would buy coleslaw
from the delicatessen if the label actually listed its active ingredients in
detail:

ethanoic acid,
α-D-glucopyranosyl-(1,2)-β-D-fructofuranose,
p-hydroxybenzyl and indoylmethyl glucosinolates,
S-propenyl and other S-alkyl cysteine sulfoxides,[ii]
β-carotene (and other carotenoids),
phosphatidylcholine.

The issue of chemicals in food is closely linked to the pursuit of
"naturalness" as a guarantee of "healthiness". The enormous diversity
of the diets consumed by *Homo sapiens* as a colonist of this planet makes
it impossible to define the ideal diet. Diet-related disease, including
starvation, is a major cause of death but it appears that while choice of
diet can certainly influence the manner of our passing, diet has no
influence on its inevitability. As chemists work together with nutritionists, doctors, epidemiologists and other scientists to understand
what it is we are eating and what it will do to us, we will come to
understand the compromises essential to the human diet. After all, our

---

[ii] The replacement of *ph* with *f* in sulphur compounds in this book is in deference to the rules of the
International Union of Pure and Applied Chemistry (IUPAC).

success on this planet is to some extent owed to our extraordinary ability to adapt our eating habits to what is available in the immediate environment. Whether that environment is an arctic waste, a tropical rain forest, or a hamburger-infested inner city, humans actually cope rather well.

This book sets out to introduce the chemistry of our diet. Chapters 2 to 5 cover food's macro-components and are overtly chemical in character. These are the substances whose chemical properties exert the major influence on the obvious physical characteristics of foodstuffs. If we are to understand the properties of food gels we are going to need a firm grasp of the chemical properties of polysaccharides. Similarly, we will not understand the unique properties that cocoa butter gives to chocolate without becoming involved in the crystallography of triglycerides.

Chapters 6–11 are devoted to substances drawn together by the nature of their contribution to food, for example as colours, vitamins, *etc.*, rather than by their chemical classification. Although this means that less attention can be devoted to the chemical behaviour of these substances individually, there is still much that chemists can contribute to our understanding of flavour, appearance and nutritional value. Science is rarely as tidy as one might wish and it is inevitable that some food components have found their way into chapters where they do not really belong. For example, some of the flavonoids mentioned in Chapter 6 make no contribution to the colour of food. However, in terms of chemical structure they are closely related to the flavonoid anthocyanin pigments and Chapter 6 may be as good a location for them as any. The final chapter, Chapter 12, is devoted to water. Apparently the simplest food component, and certainly the most abundant, it remains one of the most poorly understood. It could be placed at the front of this book, as reviewers of the earlier editions have sometimes suggested, but it is feared that much of water's chemistry would be so intimidating as to prevent many readers persevering to the tastier chapters.

This book does not set out to be a textbook of nutrition; its author is in no way qualified to make it one. Nevertheless food chemists cannot ignore nutritional issues and wherever possible the links between the subtleties of the chemical nature of food components, and nutritional and health issues have been pointed out. Similarly, students of nutrition should find a greater insight into the chemical background of the subject valuable. It is also to be hoped that health pundits who campaign for the reduction of this or that component of our diet will gain a better appreciation of exactly what it is they are demanding, and what the knock-on effects might be.

Food chemists should never overlook the fact that the object of their study is not just another, albeit fascinating, aspect of applied science. It is all about what we eat, not just to provide nutrients for the benefit of our bodies but also to give pleasure and satisfaction to the senses. Not even the driest old scientist compliments the cook on the vitamin content of the food on his plate—it's the texture and flavour, and the company around the table, that wins every time.

FURTHER READING

McCance and Widdowson's The Composition of Foods, Food Standards Agency and Royal Society of Chemistry, Cambridge, 6th edn, 2002.

H. McGee, *On Food and Cooking*, Unwin Hyman, London, 1984.

J. Burnett, *Plenty and Want: A Social History of Diet in England from 1815 to the Present Day*, Routledge, London, 3rd edn, 1989.

M. Toussaint-Samat, *History of Food*, Blackwell, Cambridge (USA), 1992.

R. S. Kirk, R. Sawyer and H. Egan, *Pearson's Composition and Analysis of Foods*, Longman, London, 9th edn, 1991.

Human Nutrition and Dietetics, ed. J. S. Jarrow, W. P. T. James and A. Ralph, Churchill Livingstone, Edinburgh, 10th edn, 1999.

CHAPTER 2

Sugars

Sugars such as sucrose and glucose, together with polysaccharides such as starch and cellulose, are the principal components of the class of substances we call carbohydrates. In this chapter we will be concerned with the sugars and some of their derivatives; other sugar derivatives such as the polysaccharides will be considered in Chapter 3. Although chemists never seem to have the slightest difficulty in deciding whether or not a particular substance should be classified as a carbohydrate, a concise formal definition has proved elusive. The empirical formulae (*i.e.* the ratio of each type of atom in the molecule) of most of the carbohydrates we encounter in foodstuffs approximate to $(CH_2O)_n$, hence the name. It is simpler to regard them as aliphatic, as opposed to aromatic, polyhydroxy compounds which carry a carbonyl group (and, of course, the derivatives of such compounds). The special place of sugars in our everyday diet will be apparent from the data presented in Table 2.1.

Sugars are automatically associated in most people's minds with sweetness, and it is this property which normally reveals to us their presence in a food. In fact there are a great many foods, in particular elaborately processed products, where the sugar content may be less obvious. Fortunately it is now common to find information about the total sugar content (as in: "Total carbohydrates ....., of which sugars ....") on the labels of packaged food products, but this tells us little about which sugars are present. In fact, only a few different sugars are common in foods, as shown in Figure 2.1.

Food: The Chemistry of its Components, Fifth edition
By T.P. Coultate
© T.P. Coultate, 2009
Published by the Royal Society of Chemistry, www.rsc.org

Table 2.1 The total sugar contents of a variety of foods and beverages. The figures are taken from *McCance and Widdowson* (Chapter 1, Further Reading) and in all cases refer to the edible portion. They should be regarded as typical rather than absolute values for the particular food concerned. The levels of sugars in eggs, cured or fresh meat, poultry, game and fish, and fats such as butter and margarine, are nutritionally insignificant although even traces can sometimes be revealed by browning at high temperatures, caused by the Maillard reaction (*see* page 33).

Food	Total sugars (%)	Food	Total sugars (%)
White bread	2.6	Cabbage (raw)	4.0
Corn flakes	8.2	Beetroot (raw)	7.0
Sugar coated cornflakes	41.9	Onions (raw)	5.6
Digestive biscuits	13.6	Cooking apples (raw)	8.9
Gingernuts	35.8	Eating apples (raw)	11.8
Rich fruit cake	48.4	Bananas	20.9
Cow's milk (whole)	4.8	Grapes	15.4
Human milk	7.2	Oranges	8.5
Cheese (hard)	0.1	Raisins	69.3
Cheese (processed)	0.9	Peanuts	6.2
Yoghurt (plain)	7.8	Honey	76.4
Yoghurt (fruit)	15.7	Jam	69.0
Ice cream (dairy)	22.1	Plain chocolate	59.5
Lemon sorbet	34.2	Cola	10.5
Cheesecake (frozen)	22.2	Beer (bitter)	2.3
Beef sausages	1.8	Lager	1.5
New potatoes	1.3	Red wine	0.3
Canned baked beans	5.9	White wine (medium sweet)	3.4
Frozen peas	2.7	Port	12.0

MONOSACCHARIDES

The monosaccharides constitute the simplest group of carbohydrates and, as we shall come to see, most of them can be referred to as sugars. Monosaccharides have a backbone of three to eight carbon atoms, but only those with five or six carbon atoms are commonly encountered.

Monosaccharides which contain a carbonyl group have the suffix "-ose", and in the absence of any other identification the number of carbon atoms is indicated by terms such as triose, tetrose and pentose. The chain of carbon atoms is always straight, never branched (2.1):

Yes:

C—C—C—C—C—C—C

No:

C—C—C
       C—C

(2.1)

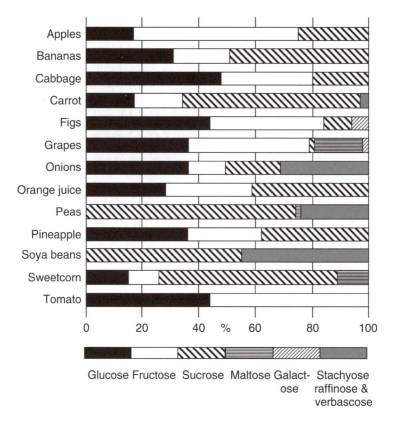

Figure 2.1   Representative percentage distribution of sugars in the total sugar content
of various foods derived from plants. There is considerable variation in
these proportions, depending on the variety, season, *etc*. The contribution
of sugars to the total weight of plant material also varies widely between
different species.

All but one of the carbon atoms carries a hydroxyl group, the
exception being that forming the carbonyl group. It is the presence of a
carbonyl group which confers reducing (*i.e.* readily oxidised) properties
on monosaccharides and many other sugars, and the carbonyl is often
referred to as the "reducing group" when considering sugar structure.
The prefixes "aldo-" and "keto-" show whether the carbonyl carbon is
the first or a subsequent carbon atom, *i.e.* whether the sugar is an
aldehyde or a ketone. Thus we refer to sugars, for example, as aldohexoses or ketopentoses. To complicate matters further, the two triose
monosaccharides are hardly ever named in this way but are referred to
as glyceraldehyde (2,3-dihydroxypropanal, 2.2) and dihydroxyacetone
(dihydroxypropanone, 2.3).

or drawn in less detail:

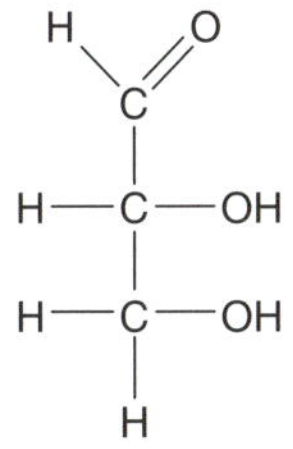

$$\begin{array}{c} H \diagdown \diagup O \\ C \\ | \\ H - C - OH \qquad (2.2) \\ | \\ H - C - OH \\ | \\ H \end{array} \qquad \begin{array}{c} CHO \\ | \\ CHOH \\ | \\ CH_2OH \end{array}$$

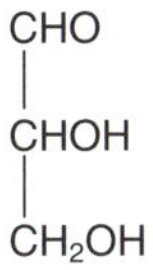

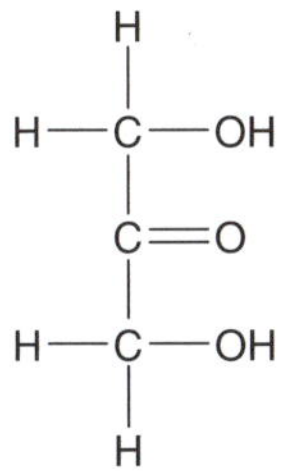

$$\begin{array}{c} H \\ | \\ H - C - OH \\ | \\ C = O \qquad (2.3) \\ | \\ H - C - OH \\ | \\ H \end{array} \qquad \begin{array}{c} CH_2OH \\ | \\ CO \\ | \\ CH_2OH \end{array}$$

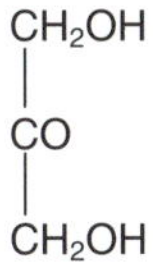

Of greatest concern to us are the aldoses [(2.4) and (2.6)], the ketoses [(2.5) and (2.7)], pentoses [(2.4) and (2.5)] and hexoses [(2.6) and (2.7)], shown here with conventional numbering of the carbon atoms:

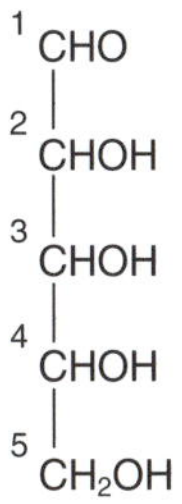

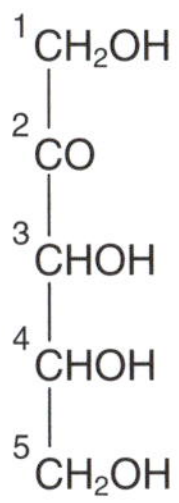

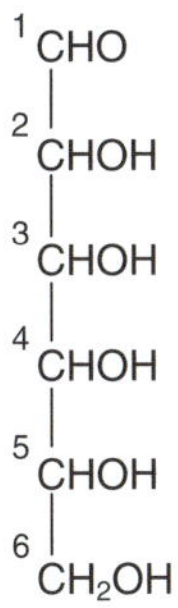

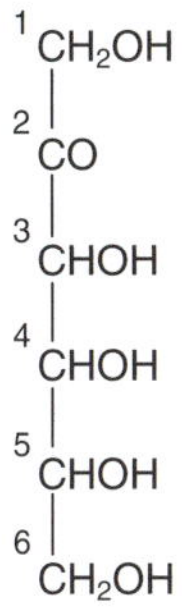

$$\begin{array}{c} ^1CHO \\ | \\ ^2CHOH \\ | \\ ^3CHOH \\ | \\ ^4CHOH \\ | \\ ^5CH_2OH \\ (2.4) \end{array} \quad \begin{array}{c} ^1CH_2OH \\ | \\ ^2CO \\ | \\ ^3CHOH \\ | \\ ^4CHOH \\ | \\ ^5CH_2OH \\ (2.5) \end{array} \quad \begin{array}{c} ^1CHO \\ | \\ ^2CHOH \\ | \\ ^3CHOH \\ | \\ ^4CHOH \\ | \\ ^5CHOH \\ | \\ ^6CH_2OH \\ (2.6) \end{array} \quad \begin{array}{c} ^1CH_2OH \\ | \\ ^2CO \\ | \\ ^3CHOH \\ | \\ ^4CHOH \\ | \\ ^5CHOH \\ | \\ ^6CH_2OH \\ (2.7) \end{array}$$

The carbon atom in each CHOH unit carries four different groups and is therefore asymmetrically substituted, *i.e.* the molecule has no plane of symmetry. The effect of this on the simplest monosaccharide, glyceraldehyde, is shown in Figure 2.2.

These two molecules appear at first sight to be identical, in that all the elements of the structure are present in each and all of these are in the same spatial relationship. In fact the two molecules are mirror images of each other. However, there is no way in which one of these could be rotated so that the two structures could be superimposed upon each other. Having four different groups attached to the central carbon atom means that the molecule as a whole has no plane of symmetry, in other

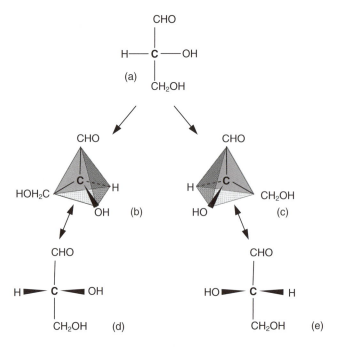

Figure 2.2 Glyceraldehyde. The simple two-dimensional structural formula (a) is compared with the two possible three-dimensional structures, (b) and (c). By convention, bonds shown as simple straight lines (*e.g.* C—CHO) lie in the plane of the page; bonds shown as a solid wedge (*e.g.* C ◄ OH) project towards the reader in front of the plane of the page; and bonds shown as a dashed line (e.g. C----H) project behind the plane of the page. The four groups attached to the central carbon atom lie at the vertices of a tetrahedron. The tetrahedral portrayal is simplified in (d) and (e). Formulae (b) and (d) show D-glyceraldehyde, (c) and (e) show L-glyceraldehyde.

words it is *asymmetric*, and its mirror images cannot be superimposed on each other. There is an obvious analogy in the relationship between one's left and right hands, and this gives rise to such isomers[i] being referred to as *chiral*, from the Greek word for hand. Mirror image pairs of asymmetric substances like this are known as *enantiomers,* from the Greek word for opposite.

All substances which contain asymmetric, *i.e.* chiral, centres are classified as members of either a D-series or L-series, depending on whether their chemical structure is derived from D-glyceraldehyde or L-glyceraldehyde, respectively. The prefixes D- and L- are derived from, but are not abbreviations of, the words *dextrorotatory* and *laevorotatory*. These terms refer

---

[i] Isomers: each of two or more substances having the same formula, but a different arrangement of atoms in the molecule and different properties.

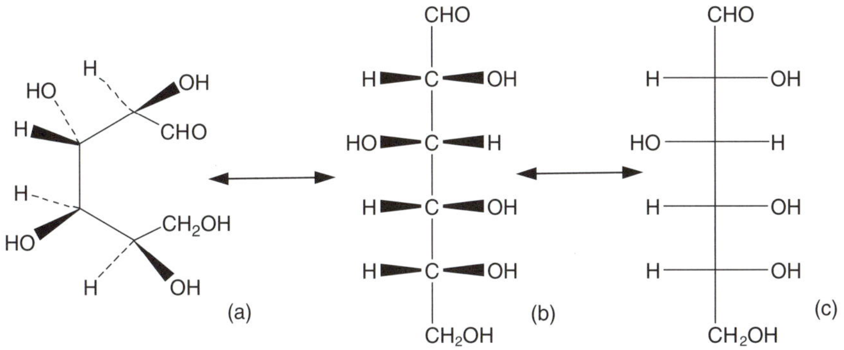

Figure 2.3 Projections of the structural formula of glucose. Versions (a) and (b) are comparable to glyceraldehydes, as shown in Figure 2.2. However it is the Fischer projection (c) which is most commonly used by carbohydrate and food chemists. Bonds drawn horizontally project forwards from the plane of the paper, and bonds drawn vertically are either in, or project behind, the plane of the paper. Asymmetric carbon atoms are not shown explicitly but are implied by the intersection of the vertical and horizontal lines.

to the phenomenon of optical activity possessed by asymmetric molecules. Both crystals and solutions of optically active substances rotate the plane of polarised light[ii] when it passes through their crystals or solutions. Enantiomeric pairs of substances, such as D- and L-glyceraldehyde, rotate polarised light in opposite directions, in this particular case to the right (clockwise) and left (anticlockwise), respectively. Measurements of optical rotation, using an instrument known as a polarimeter, made valuable contributions to the analysis of sugars in foods in the years before sophisticated methods such as high performance liquid chromatography (HPLC) took over in food analytical laboratories.

Unfortunately one cannot automatically equate membership of the D-series with rotation to the right, so the symbols (+) and (−) are used to denote rotation to the right and left. The old-fashioned names, dextrose and laevulose, are derived from the dextrorotatory and laevorotatory properties of D(+)-glucose and D(−)-fructose, respectively. Figure 2.3 shows how the formulae in Figure 2.2 are converted to the more convenient Fischer projection used in Figure 2.4 and elsewhere.

When monosaccharides contain more than one asymmetric carbon atom their inclusion in the L- or D-series is based on the configuration of

---

[ii] Polarised light is light which has been passed through a polarising filter. Unpolarised light consists of waves vibrating in all directions perpendicular to the direction of the light, whereas the wave motion of polarised light is restricted to a single direction. Polaroid™ sunglasses are a familiar examples of polarising filters. Readers requiring a more detailed treatment of this topic, which is outside the scope of this book, may wish to consult an advanced textbook of organic chemistry.

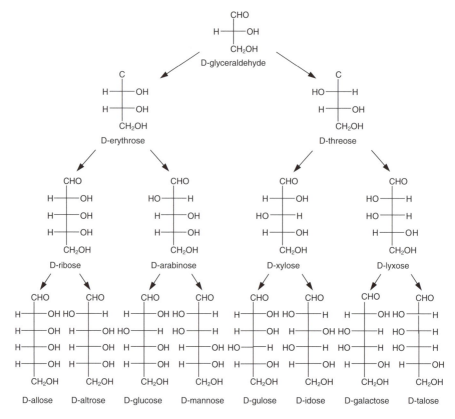

Figure 2.4 The configurations of the D-aldoses. For anyone who wishes to remember the structures or names of the aldoses the sequence, reading downwards and from left to right, may be remembered by use of the traditional, and dreadful, mnemonics: "Get Raxl!" and "All altruists gladly make gum in gallon tanks". (The author, on behalf of innumerable students, would be delighted to hear of improved versions of these mnemonics.)

their highest numbered asymmetric carbon atom (*e.g.* carbon 4 in a pentose or carbon 5 in a hexose), the one furthest from the carbonyl group. Almost all naturally occurring monosaccharides belong to the D-series.

Although only a few of them are relevant to food or even common in nature it is helpful to understand the structural relationships within the full set of aldose monosaccharides, as shown in Figure 2.4. A corresponding table of ketose sugars may be drawn up, but with the obvious exception of D-fructose (2.8) none of the ketoses are of much significance in food. The names of ketose sugars are derived by changing the "-ose" of the corresponding aldose sugar to "-ulose", as in D-xylulose (2.9). Dihydroxyacetone and fructose are exceptions to this convention.

It is important to remember that the monosaccharides of the L-series are related to L-glyceraldehyde (*see* Figure 2.2) and have configurations which are mirror images of the corresponding D-series sugars. Thus, L-glucose is (2.11) rather than (2.10), which is in fact L-idose.

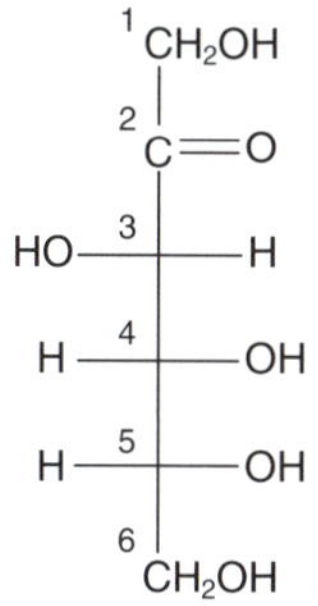

(2.8) 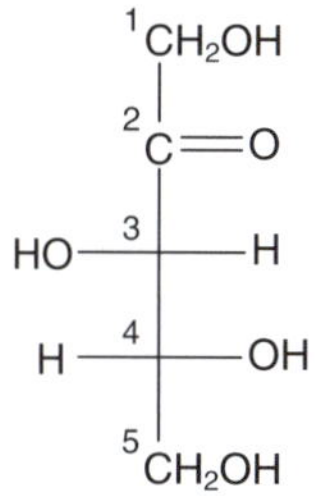   (2.9) 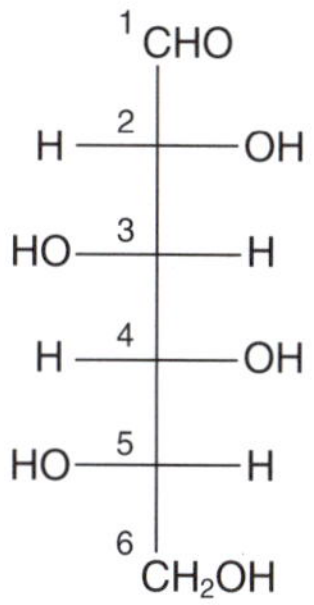   (2.10) 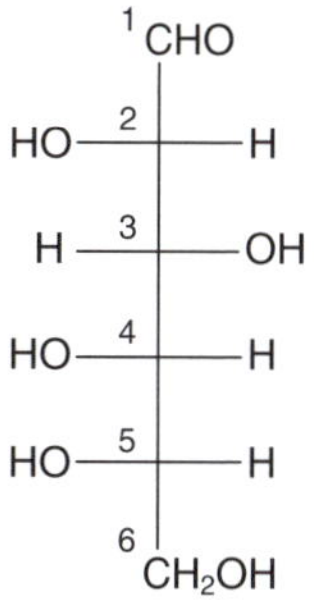   (2.11)

The straight-chain structural formulae used so far in this chapter do not account satisfactorily for some monosaccharide properties. In particular their reactions, while indicating the presence of a carbonyl group, are not entirely typical of carbonyl compounds generally. The differences are explained by the formation of a ring structure in which the carbonyl group has reacted reversibly with one or other of the hydroxyl groups at the far end of the chain, giving the structure known as a hemiacetal (Figure 2.5).

Figure 2.6 shows how a number of different structural isomers can be generated. Rings containing six atoms (five carbon and one oxygen) result when the carbonyl group of an aldose sugar (*i.e.* at C-1) reacts with a hydroxyl on C-5. The same size of ring results when the carbonyl

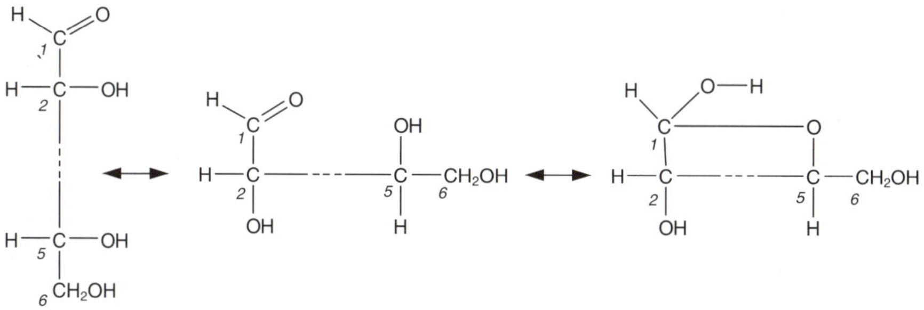

Figure 2.5 Mutarotation. The natural flexibility of the monosaccharide backbone allows a hydroxyl group at the tail of the molecule to react with the reducing group to form a ring.

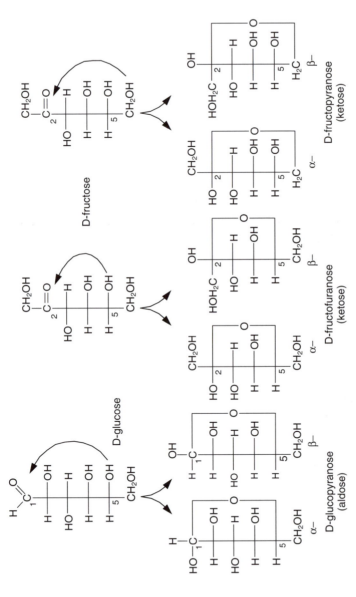

Figure 2.6 The formation of pyranose and furanose ring forms of D-glucose. Only the pyranose forms of D-glucose occur in significant amounts. This representation of the ring form has its shortcomings: C—O bonds are actually about the same length as C—C bonds and certainly do not turn through 90° halfway along. Translating the formulae of other monosaccharides from the open chains of Figure 2.2 into the more realistic Haworth rings shown in Figure 2.4 is best achieved by comparison with glucose, rather than by resorting to the three-dimensional mental gymnastics recommended by some classical organic chemistry texts.

of a ketohexose sugar (*i.e.* at C-2) reacts with a hydroxyl on C-6 (*i.e.* the –CH$_2$OH, hydroxymethyl group). Five-membered rings result when the carbonyl of a ketose sugar reacts with a hydroxyl on C-5. Six- and five-membered rings are referred to respectively as pyranose and furanose, from the structures of pyran (2.12) and furan (2.13).

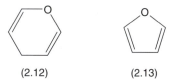

(2.12)   (2.13)

Another result of hemiacetal formation, also shown in Figures 2.5 and 2.6, is that what was previously the carbonyl carbon now has four different substituents, *i.e.* it has become a new asymmetric centre. This means that the ring form of a monosaccharide will occur as a pair of optical isomers, its α- and β-anomers.

The α- and β-anomers of a particular monosaccharide differ in optical rotation. For example, α-D-glucopyranose, which is the form in which D-glucose crystallises out of aqueous solution, has a specific rotation[iii] of + 112°, whereas that of β-D-glucopyranose, obtained by crystallisation from pyridine solutions of D-glucose, is + 19°.

Elaborate rules are used to decide which of a pair of anomers is to be designated α and which β. However, in practice we observe that α-anomers have their reducing group (*i.e.* the C-1 hydroxyl of aldose sugars and the C-2 hydroxyl of ketose sugars) on the opposite face of the ring to C-6; β-anomers have their reducing group on the same face as C-6. Careful inspection of the formulae in Figures 2.6 and 2.7 will show how this works out in the potentially tricky L-sugars and in furanose rings.

When crystals of either α- or β-D-glucose are dissolved in water the specific rotation is observed to change until, regardless of the form it started with, the solution finally gives a value of + 52°. This phenomenon is known as *mutarotation*. The transition of one anomer to another proceeds through the open-chain aldehydo form, and it is clear that it is

---

[iii] The specific rotation of a substance at 20 °C using light of the D-line of the sodium spectrum is given by the expression:

$$[\alpha]_D^{20} = \frac{100\alpha}{l \times c}$$

where α is the rotation observed in a polarimeter tube of length l dm and at a concentration of c g per100 cm^3.

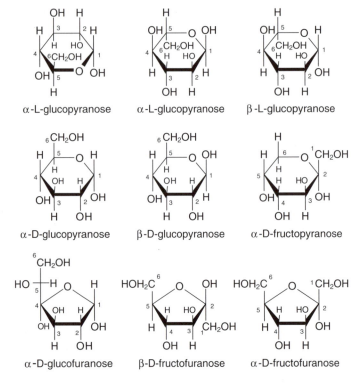

Figure 2.7   Isomers of glucose and fructose. The obvious shortcomings of the formulae
used so far in this chapter to represent molecular structure can be partially
resolved by the adoption of the Haworth convention for ring structures.
The ring is treated as planar and drawn to appear perpendicular to the plane
of the page. These structures may be compared with those in Figure 2.6. The
numbering of carbon atoms can be correlated with that in the open chain
structures (2.6) and (2.7). It should be remembered that only a few of the
isomers shown actually occur naturally in significant amounts, and also that
not all the possible isomers have been included.

this isomer which is involved in the sugar reactions which are typical of
carbonyl compounds, even though in aqueous solutions only 0.02% of
D-glucose molecules are in this form.

The structural relationships between L- and D-isomers, α- and β-
anomers, and pyranose and furanose rings in both aldose and ketose
sugars, are by no means easily mastered. If available, the use of a set of
molecular models will help to clarify the issues, but the structural formulae
set out in Figure 2.7 illustrate the essential features of the terminology in
this area.

Over the years chemists have synthesised innumerable monosaccharide derivatives, but only a few of these occur naturally or have

particular significance in food. Oxidation of the carbonyl group of aldose sugars leads to the formation of the "-onic" series of sugar acids and, depending on the exact conditions, various other products. Additional aspects of the chemical reactions of the reducing group are dealt with in the first of the Special Topics later in this chapter. The enzyme glucose oxidase catalyses the oxidation of the α-anomer of D-glucose to D-gluconolactone (2.14), which hydrolyses spontaneously to D-gluconic acid (2.15), with hydrogen peroxide as a by-product. In the popular analytical technique (*see* Figure 2.8) a second enzyme, peroxidase, is included to measure the quantity of hydrogen peroxide and thus the concentration of glucose. Besides its obvious applications in food analysis a similar reaction is used in the indicator strips used by diabetics for the measurement of blood glucose concentration. Glucose oxidase is also used in a process for removing traces of glucose from the bulk liquid

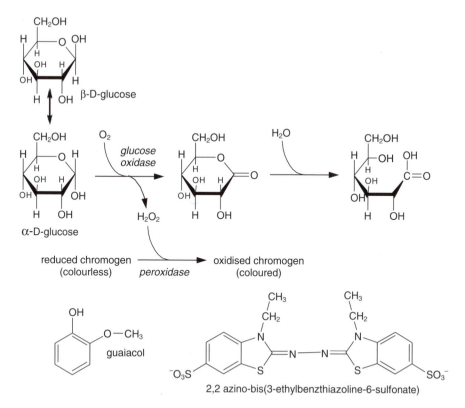

Figure 2.8   The enzymic determination of glucose. The original colour reagent (chromogen) used with this method was guaiacol, but nowadays 2,2-azino-bis(3-ethylbenzthiazoline-6-sulfonate), more commonly known as Perid™, which gives a beautiful emerald green colour, is the most popular.

egg used in commercial bakeries and elsewhere, to prevent the Maillard reaction taking place (*see* page 33).

(2.14)   (2.15)

The "-uronic" series of sugar acids are aldohexoses with a carboxyl group at C-6, such as D-galacturonic acid (2.16) and L-guluronic acid (2.17). These are important as constituents of polysaccharides such as pectins and alginates, respectively, but are of little interest in their own right.

(2.16)   (2.17)

Reduction of the carbonyl group to a hydroxyl gives sugar alcohols such as xylitol (2.18) and sorbitol (2.19). Small amounts of sorbitol occur widely in fruit (around 4% and 12% of the total sugars, respectively, in apple and pear juices), but otherwise they are extensively synthesised on the industrial scale from the corresponding aldose sugars by reduction with hydrogen. Sorbitol and other hydrogenated sugars are also used as low-calorie, *i.e.* non-metabolised, sweeteners and are considered in more detail in Chapter 7.

(2.18)   (2.19)

Sorbitol is particularly important as the starting point for the industrial synthesis of ascorbic acid (vitamin C, *see* page 337). Sugar alcohols are used to replace sugars in diabetic and other calorie-reduced food products. Although sweet, these are not absorbed in the small intestine and do not reach the bloodstream. Unfortunately their eventual arrival in the large intestine can, if large amounts have been consumed, provoke "osmotic diarrhoea", when the normal transfer of water from the colon's contents is impaired. Reduction at other positions in the sugar molecule gives deoxy sugars such as L-rhamnose (6-deoxy-L-mannose) (2.20), an important minor constituent of pectins, and 2-deoxy-D-ribose (2.21), the sugar component of DNA.

(2.20)   (2.21)

The most important derivatives of monosaccharides are those in which the hemiacetal, or reducing group, forms a glycosidic link (more strictly known as an acetal) with a hydroxyl group of another organic compound (2.22).

(2.22)

Glycosidic links are stable under ordinary conditions but are readily hydrolysed under acidic conditions or in the presence of appropriate hydrolytic enzymes. The formation of the glycosidic link has the effect of fixing the hemiacetal structure in either the α or β configuration and, of course, abolishing mutarotation. As we will see later in the description of oligosaccharides and other compounds having glycosidic links, it will be necessary when describing a particular structure to specify whether the link in a compound is in the α or β configuration. The glycoside shown (2.22) is in the α configuration.

Although any compound containing a glycosidic link is strictly speaking a glycoside, the term is usually reserved for a class of compounds which occur naturally in plants. These particular glycosides have a sugar component linked to a non-sugar component termed the aglycone. Flavonoids, found widely in plants, normally occur as glycosides with a sugar linked to one of their phenolic hydroxyl groups. The anthocyanin pigments of plants, considered in Chapter 6, are good examples. Many other well known naturally occurring glycosides are plant toxins (*see* Chapter 10), such as solanine, the cyanogenic glycosides such as amygdalin (2.23) (found in bitter almonds) and the glucoside of α-hydroxyisobutyronitrile (in cassava, *see* Chapter 10). The phenolic hydroxyl groups of flavonoids can react spontaneously with proteins through the hydroxyl groups of tyrosine residues. Blocking the phenolic hydroxyls with sugar residues maintains their contribution to the biological value of the flavonoid structure while avoiding their potentially destructive effect on the activity of enzymes, *etc.* Similarly, the presence of the sugar prevents the cyanogenic glycosides from releasing their cyanide until the circumstances are appropriate.

(2.23)

OLIGOSACCHARIDES

When a glycosidic link connects the reducing group of one monosaccharide to a hydroxyl group of another, the result is a disaccharide. Further such linkages will give rise to trisaccharides, tetrasaccharides, *etc.*, the oligosaccharides,[iv] and ultimately polysaccharides. The polysaccharides, in which hundreds or even thousands of monosaccharide units may be combined in a single molecule, are considered in Chapter 3. In the present section we will be considering the di-, tri- and tetrasaccharides which commonly occur in our food.

Even when a disaccharide is composed of two identical monosaccharide units there are numerous possible structures. This is illustrated in Figure 2.9, which shows five of the more important glucose–glucose

---

[iv] Chemists do not recognise a fixed numerical boundary between "oligo-" and "poly-". The Greek origins of these prefixes are usually translated as "few" and "many", respectively, and chemists apply the prefixes on this basis.

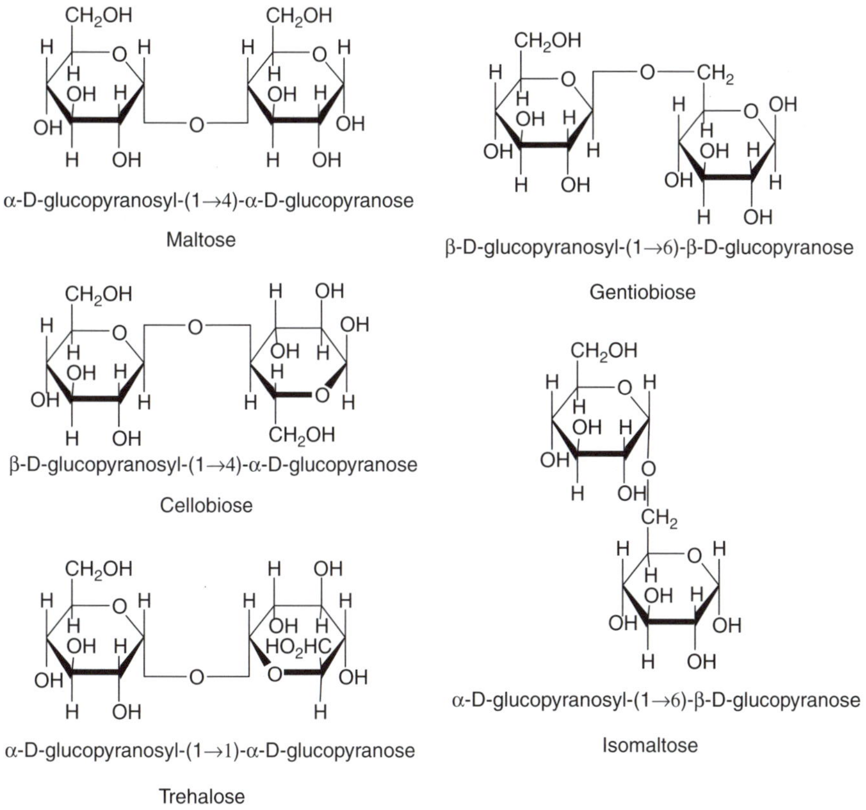

α-D-glucopyranosyl-(1→4)-α-D-glucopyranose

Maltose

β-D-glucopyranosyl-(1→6)-β-D-glucopyranose

Gentiobiose

β-D-glucopyranosyl-(1→4)-α-D-glucopyranose

Cellobiose

α-D-glucopyranosyl-(1→6)-β-D-glucopyranose

Isomaltose

α-D-glucopyranosyl-(1→1)-α-D-glucopyranose

Trehalose

Figure 2.9 Disaccharides of glucose. Both trivial and systematic names are given. In the case of the reducing oligosaccharides, maltose, cellobiose, isomaltose and gentiobiose, only one of the two possible anomers is shown.

disaccharides. Although the configuration of the hemiacetal involved in the link is no longer free to become reversed, it should not be overlooked that the uninvolved hemiacetal (*i.e.* that in the right-hand ring of the reducing sugars as they are portrayed in Figure 2.9) is still subject to mutarotation in aqueous solution. Hence the first four of these sugars occur as pairs of α- and β-anomers. Furthermore these four sugars show reducing properties similar to those of monosaccharides. In these cases it is the carbonyl group of the right-hand ring, as drawn in Figure 2.9, which provides the reducing properties. The best known of these reducing properties is the ability to reduce Cu^{2+} ions in alkaline solutions, such as Fehling's solution, to give a reddish-brown precipitate of cupric oxide. This provides a useful laboratory test to distinguish these reducing sugars from non-reducing sugars—notably trehalose and, as we shall see, sucrose.

Not all of the five disaccharides shown in Figure 2.9 occur in nature. Maltose, together with a small proportion of isomaltose, occurs in syrups obtained by the partial hydrolysis of starch, in which similar patterns of glycosidic links are found. As shown in Figure 2.1 maltose also occurs in some fruits, notably grapes. Similarly, cellobiose is formed during the acid hydrolysis of cellulose. Gentiobiose commonly occurs as the sugar component of glycosides, and trehalose is found in yeast.

The scientific names given to oligosaccharides are inevitably rather clumsy. The formally correct procedure is to name a disaccharide as a substituted monosaccharide. Thus lactose (2.24), which consists of a glucose unit carrying a D-galactose unit (in the β configuration) on its C-4 hydroxyl group, is 4-*O*-β-D-galactopyranosyl-D-glucopyranose. This system becomes very cumbersome when applied to higher oligosaccharides (those with three or more monosaccharide units) and most food scientists prefer to place the details of the link between the monosaccharides where it belongs, *i.e.* between them. Lactose is in this case, and in practice more frequently, described as β-D-galactopyranosyl-(1→4)-D-glucopyranose. In sucrose (2.25), the glycosidic link occurs between the reducing groups of *both* monosaccharide components, glucose and fructose, so that the usual formal name, α-D-glucopyranosyl-(1→2)-β-D-fructofuranose could be reversed to β-D-fructofuranosyl-(2→1)-α-D-glucopyranose, but no one ever bothers.

(2.24)   (2.25)

Lactose and sucrose are two of the most important food sugars. Lactose is the sugar present in milk[v] (approximately 5% w/v in cow's milk) and is of course a reducing sugar. Human milk, with 6.2–7.5% w/v, has one of the highest lactose concentrations of any mammal's milk. Sucrose is the familiar "sugar" of the kitchen and commerce, and occurs widely in plant materials. The sucrose we buy as "sugar" has been extracted from sugar cane or sugar beet, but sucrose is also abundant in most plant materials, particularly fruit. In terms of their

---

[v] As secreted by female mammals for feeding their young; on the other hand neither soya milk nor coconut milk contains lactose.

nutritional value, biology and most of their chemistry the two are completely indistinguishable. However, one of the Special Topics later in this chapter explains how they can be distinguished and the potential value of doing so. Although glucose in the form of its 6-phosphate derivative is the initial product of photosynthesis, it is normally converted into sucrose before transport from the leaves to the remainder of the plant. Most plants convert the sucrose to starch in the form of insoluble granules for long-term storage as an energy reserve but some plants, such as sugar beet and sugar cane, retain sucrose in solution. The sucrose and other sugars (*see* Figure 2.1) in fruit are resynthesised from starch as the fruit ripens.

Since the glucose and fructose units are joined through their hemiacetal groups, sucrose is not a reducing sugar. Under mildly acid conditions, or by the action of the enzyme invertase, sucrose is readily hydrolysed to its component monosaccharides. This phenomenon is termed *inversion* and the resulting mixture *invert* sugar, due to the effect of the hydrolysis on its optical rotation properties in solution. The specific rotation values for sucrose, glucose, and fructose are $+66.5°$, $+52.7°$, and $-92.4°$, respectively, so we can see that a dextrorotatory solution of sucrose will give a laevorotatory solution of invert sugar with specific rotation $-39.7°$.

Traditionally, invert syrup is manufactured by acid hydrolysis of sucrose using citric acid. The speed of the hydrolysis is highly dependent on both the pH and the temperature. A reduction of the pH from 4 to 3, or an increase in temperature of $20\,°C$, will in either case increase the rate by a factor of 6. The resulting syrup will not crystallise, even if manufactured with a solids content as high as 80%. Inversion can also be carried out using the enzyme invertase, which occurs in the cell walls of yeast. Boiled sweets and other confectionery products depend upon the non-crystallising property of invert syrups. Sucrose is dissolved in invert syrup and the mixture is then boiled to reduce the total water content. The result is a glass structure, a supercooled liquid which is malleable when hot but sets to form a glass-like solid as it cools. In this state the solute molecules do not form the stable ordered associations characteristic of a crystal but remain totally disordered. However, the water content is too low for the system to remain liquid, although it remains in strict chemical terms a solution. In the manufacture of boiled sweets, flavours and colours are blended in as it cools and the mass is kneaded by machine before being rolled out and moulded into the finished shape.

In modern confectionery some or all of the invert sugar is often replaced by glucose syrups, otherwise known as corn syrups, since corn (*i.e.* maize) starch is the raw material normally used in their manu-facture. In times past glucose syrups were made by the acid hydrolysis of starch. Maize starch, or other starchy raw materials such as potatoes, were heated with dilute sulfuric acid to hydrolyse the $\alpha 1 \rightarrow 4$ and $\alpha 1 \rightarrow 6$ glycosidic links between the glucose units (*see* Chapter 3). One might expect complete hydrolysis to glucose to result but in fact the reaction does not go to completion. As the reaction proceeds, the concentration of sugars becomes very high and there is a tendency for the reverse reaction to occur. However the reverse reaction is in no way constrained to re-form the $\alpha 1 \rightarrow 4$ and $\alpha 1 \rightarrow 6$ glycosidic links present in the original starch. In consequence an accumulation of oligosaccharides takes place, based on the full range of possible linkages, α or β, $1 \rightarrow 1$, $1 \rightarrow 2$, $1 \rightarrow 3$, *etc.,* since many of these are much more resistant to acid hydrolysis. These oligosaccharides are unlikely to be utilised in the human digestive process but their presence does ensure that the syrups, known com-mercially as liquid glucose,[vi] do not crystallise. Modern commercial cake recipes also take advantage of the refusal of glucose syrups to crystallise. They help the crumb of the cake to retain moisture and thereby remain soft and palatable for longer.

In the 19th century it was common for crude syrups (sold as "brewing sugar"), manufactured by the hydrolysis of potato starch with sulfuric acid, to be used as adjuncts in beer making. Adjuncts were, and remain, cheap sources of fermentable carbohydrates which reduce the brewer's dependence on expensive malted barley. This practice could have unfortunate consequences. In 1990 in the midlands and the north of England there was an incidence of over 6000 cases of arsenic poisoning among beer drinkers, including some 70 deaths. This was traced to a Salford brewery which had used contaminated brewing sugar. The crude sulfuric acid used to hydrolyse the starch had been manufactured from iron pyrites (ferrous sulfide) which was contaminated with arsenic.

Nowadays enzymic hydrolysis using thermostable α-amylases and other enzymes has superseded acid hydrolysis, making syrups available with highly controllable compositions, ranging from almost pure glu-cose to mixtures dominated by particular oligosaccharides. These enzymic methods, described in greater detail in Chapter 3, can also be used to produce syrups resembling invert syrup, known as "isosyrups". These are manufactured by using the enzyme glucose isomerase to convert glucose to an equilibrium mixture of approximately equal parts glucose and fructose.

[vi] In the food industry the terms "glucose" and "liquid glucose" are employed to refer to the mixed products of starch hydrolysis. The otherwise obsolete term "dextrose" is used in the food industry to denote pure D-glucopyranose in the crystalline state.

The inversion of sucrose is also an essential part of the process of making jam and other preserves. In making jam the fruit, together with a large amount of sugar, is boiled for a prolonged period. During this period the fruit is softened and its pectin (*see* Chapter 3), plus any extra added pectin, is brought into solution. At the same time a considerable proportion of the fruit's original water content is removed. The mild acidity leads to the hydrolysis of up to half of the added sucrose to glucose and fructose. This has a number of beneficial effects. Firstly, the total number of sugar molecules is increased by up to 50%. This gives a modest increase in sweetness (*see* Chapter 7), but most importantly it increases the proportion of water which is bound to sugar molecules and therefore unavailable to support the growth of micro-organisms (*see* Chapter 12). The invert sugar is much more soluble in water than the original sucrose, and the presence of invert sugar actually represses the crystallisation of sucrose. It is therefore easy to reach the very high solids levels (and correspondingly low water activity levels) which are needed for long term stability.

The prolonged high temperatures, reaching 105 °C during boiling, can have a drastic effect on the colour and flavour of the fruit. Commercial jam makers often reduce this effect by carrying out the boil at reduced pressure, driving off the water at a lower temperature. It is also common to replace some of the sucrose with glucose syrups prepared from starch (*see* Chapter 3). These contain significant proportions of glucose oligosaccharides, such as maltose, maltotriose (*i.e.* three glucose units linked $\alpha 1 \rightarrow 4$), isomaltose and maltotetrose. The presence of these oligosaccharides represses glucose crystallisation and enables high total sugar contents to be reached without excessive heating.

One group of higher oligosaccharides deserves particular attention. These are the galactose derivatives of sucrose:

- raffinose:
  α-D-galactopyranosyl-(1 → 6)-α-D-glucopyranosyl-(1 → 2)-
  β-D-fructofuranose
- stachyose:
  α-D-galactopyranosyl-(1 → 6)-α-D-galactopyranosyl-(1 → 6)-
  α-D-glucopyranosyl-(1 → 2)-β-D-fructofuranose
- verbascose:
  α-D-galactopyranosyl-(1 → 6)[-α-D-galactopyranosyl-(1 → 6)]$_2$-
  α-D-glucopyranosyl-(1 → 2)-β-D-fructofuranose.

These are well known for their occurrence in seeds of legumes such as peas, beans and some other vegetables (*see* Figure 2.1), and present particular problems in the utilisation of soya beans. These oligo-saccharides are neither hydrolysed nor absorbed by the human digestive system, so a meal containing large quantities of beans, for example, becomes a feast for the bacteria in the large intestine, which are able to utilise them. These bacteria produce large quantities of hydrogen and some carbon dioxide as by-products of the metabolism of the sugars. Much of the carbon dioxide and hydrogen diffuses through the walls of the colon into the bloodstream and ultimately leaves *via* the lungs, but a substantial amount remains to cause flatulence.

A similar fate befalls the lactose in milk consumed by individuals who are not able to secrete the enzyme lactase (β-galactosidase) in their small intestine. Humans, like other mammals, possess this enzyme as infants, but the majority of East Asians and many Negroes do not secrete the enzyme after weaning[vii] and are therefore unable to consume milk without risk of an upset stomach. The absence of dairy products from Chinese cuisine should therefore cause no surprise. The absence of lactase is the cause of the disorder known in the medical profession as lactose intolerance. Domestic cats have traditionally been regarded as partial to a saucer of cow's milk, but in fact adult cats are often unable to digest its lactose. To avoid the inevitable gastrointestinal upsets, so-called "cat milk" is now sold, in which the lactose has been broken down to galactose and glucose, both of which are readily absorbed from the feline small intestine.

SUGARS AS SOLIDS

Sugars in their crystalline state make important contributions to the appearance and texture of many food products, particularly con-fectionery, biscuits and cakes. The supermarket shelf usually offers the customer a choice of crystal sizes in granulated (400–600 μm) and caster sugar (200–450 μm), and a powdered form for icing (average 10–15 μm).

The relative proportions of undissolved sugar crystals and sugar solution (syrup) control the texture of the soft cream centres of choco-lates, but the special problem is that of sealing the soft centre inside the chocolate coating. The answer is the use of the enzyme, invertase. This is

---

[vii] Although never explicitly stated, the underlying implication has generally been that during the course of human evolution these ethnic groups have *lost* the ability to utilise lactose as adults. Two recent investigations have tended to confirm the opposite hypothesis. It now appears that a number of different mutations conferring lactase persistence (*i.e.* persisting from infancy into adulthood) have arisen spontaneously in Europe and Africa, mostly around 6000–8000 years ago but possibly as recently as 2000 years ago (*see* S. A. Tishkoff, *et al., Nat. Genet.*, 2006, **39**, p. 31; J. Burger, *et al., Proc. Natl. Acad. Sci. U. S. A.*, 2007, **104**, p. 3736.) Such a mutation would inevitably confer massive advantages on humans who were already engaged in cattle farming.

an enzyme obtained commercially from yeast (*Saccharomyces cerevisiae*) which catalyses the hydrolysis of sucrose to glucose and fructose. In the manufacture of soft-centred chocolates the mixture used for the centre contains a high proportion of finely milled sucrose suspended in just sufficient glucose syrup (*see* Chapter 3, page 88), together with appropriate colourings and flavourings, to produce a very stiff mouldable paste. Also included is a small amount of invertase. After the paste has been stamped out into suitable shapes it is coated with melted chocolate. During the ensuing weeks, and before the chocolates are sold to the public, the invertase hydrolyses a proportion of the sucrose. The glucose and fructose are considerably more soluble in water than their parent sucrose and as they dissolve they alter the solid–liquid balance of the centre, giving it its familiar soft creamy texture.

The normal crystalline forms of most sugars are anhydrous and consist of only one anomer. However, at temperatures below 50 °C glucose crystallises from aqueous solutions as the monohydrate of the α-pyranose anomer. Likewise, D-lactose crystallises as the monohydrate of the α-anomer. Although quite soluble (about 20 g per 100 cm^3 at room temperature), α-D-lactose monohydrate crystals are slow to dissolve and occasionally form a gritty deposit in evaporated milk. The texture of sweetened condensed milk is also highly dependent on the size of the lactose monohydrate crystals.

These considerations apart, the food chemist is primarily interested in sugars in aqueous solution.

SUGARS IN SOLUTION

Until a few decades ago our knowledge was dominated by the chemical reactions of sugars and by the ability of chemists to create a bewildering array of chemical derivatives, few of which had much relevance outside the chemical laboratory. There is now much more interest in the behaviour of the sugars themselves in the apparently simple environment of an aqueous solution. Studies using nuclear magnetic resonance (NMR) spectroscopy, and predictions based on thermodynamic calculations, have now revealed a great deal. The Haworth representation of the configuration of sugar rings has many advantages, notably that they are fairly easy to draw freehand and can demonstrate optical relationships very clearly. However, for many purposes formulae based on the cyclohexane chair structure are needed, as shown in Figure 2.10.

In Figure 2.10 many of the hydrogen atoms or hydroxyl groups in the chair structures are circled. These are those in the so-called axial position, *i.e.* their C–H bonds lie parallel to the axis of the ring. The uncircled

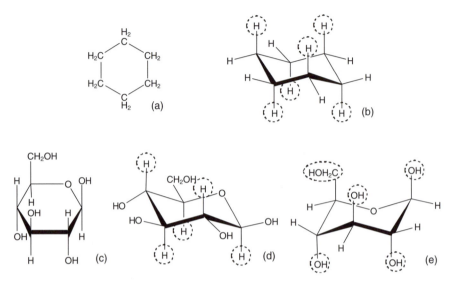

Figure 2.10 Chair structures for cyclohexane and glucose. Formula (a) is a simple
2-dimensional representation of cyclohexane. With the carbon atoms
shown with tetrahedrally distributed bonds (b), the result is the well-
known chair structure. A similar relationship exists between the familiar
Haworth structure of α-D-glucopyranose (c) and its corresponding chair
structures, (d) and (e). Here the square seat of the chair is formed by
carbons 2, 3 and 5 and the ring oxygen, with C-4 as the headrest.

hydrogen atoms and hydroxyl groups are equatorial, *i.e.* their C–H
bonds project outwards towards the equator of the ring.

In fact, as shown in Figure 2.10, there are two possible chair structures, known as *conformations,* for any given anomer. In Figure 2.10 (d)
and (e) we have the two conformations for β-D-glucose. The crucial
difference between them is that (d), designated $C1$,[viii] has all its hydroxyl
groups and the –CH$_2$OH group in equatorial positions. In contrast, the
so-called $1C$ conformation (e) has these groups in axial positions. In
spite of the very different appearance of these two molecules, none of the
asymmetric centres differs in configuration.

Theoretical studies have shown that the favoured conformation will
be the one which has the greater number of bulky substituents such
as –OH or –CH$_2$OH in equatorial positions, and where neighbouring
hydroxyls are as far apart as possible. This has been borne out by

viii The designations $C1$ or $1C$ were introduced by Reeves as part of a system for naming different
conformations. For most purposes the designation of a particular structure as $C1$ or $1C$ is most
easily achieved by comparison with the two structures shown above. If the sugar ring is portrayed
as usual with the ring oxygen towards the right and "at the back", then the conformation is $C1$
when carbon-1 in the ring is below the plane of the "seat" and $1C$ when carbon-1 lies above this
plane.

experimental work and it immediately explains why less than 1% of D-glucose molecules in solution are in the 1*C* conformation, and why the β-anomer is preferred for glucose. The correct description of a furanose ring is more difficult. There are two most probable shapes, described as the envelope (2.26) and twist (2.27) forms.

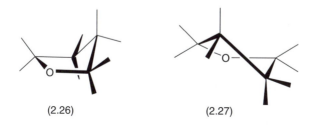

(2.26)   (2.27)

It seems best to assume that in solution furanoses are present as an equilibrium mixture of many rapidly interconverting forms, such as those above, which do not differ greatly from each other and in which no one form is dominant. In fact D-fructose occurs predominantly in the 1*C* conformation of the β-pyranose form (2.28), which contains the bulky hydroxymethyl group (–CH$_2$OH) in the equatorial position.

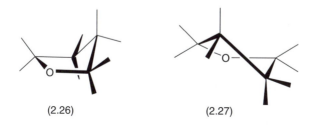

(2.28)

Of the eight D-aldohexoses, only D-glucose has a possible conformation in which all of its hydroxyl groups and the hydroxymethyl group are equatorial. This is almost certainly the explanation for the dominant role of glucose in living systems, both as a metabolic intermediate and as a structural element. The apparent untidiness of its configuration when displayed in either straight-chain or Haworth formulae is obviously irrelevant to its actual state in living systems. This issue is given further consideration in the third of the Special Topics towards the end of this chapter.

The conformation and anomer proportions of a particular sugar in solution are not only a function of the intramolecular non-bonding interactions between atoms which have been mentioned so far, they are also dependent on interactions between the sugar molecule and the solvent, in foodstuffs usually water. The fact that there is 37% of the α-anomer of D-glucose at 25 °C in aqueous solution, but 45% when the solvent is pyridine, is just one indication of the existence of strong interactions between water and sugar molecules. As we shall see, the binding of water to sugars and polysaccharides is an important contributor to the properties of many foodstuffs.

The most obvious indication of the binding of water by sugars is that concentrated solutions of sugars do not obey Raoult's Law (which states: "The relative lowering of the vapour pressure of the solvent is equal to the molecular fraction of the solute in solution"). The observed vapour pressure can be as much as 10% below the expected figure. Comparison of the observed with the expected values permits a fairly straightforward calculation of the number of water molecules that are no longer effectively part of the bulk of the water of the system, *i.e.* those which are bound to the sugar molecules directly by hydrogen bonds. The resulting *hydration numbers* approach 2 and 5 molecules of water per molecule of sugar for glucose and sucrose, respectively.

In recent years NMR and dielectric relaxation techniques have given us alternative tools for examining water binding. They have revealed a primary hydration layer containing an average number of water molecules rather higher than that deduced by departures from Raoult's Law. For example, ribose gives a value of 2.5, glucose 3.7, maltose 5.0, and sucrose 6.6. The water involved in hydration is not necessarily part of a permanent structure. NMR techniques are able to reveal water molecules as being bound if they are hydrogen-bonded for as little as one microsecond.

Further aspects of the binding of water by sugars and other food components are considered in Chapter 12. The sweetness which sugars impart to our food is discussed in Chapter 7.

DECOMPOSITION

It is not only intact sugar molecules themselves which make a positive contribution to the properties of our diet: the products of their thermal decomposition are also important food components in their own right. When sugars are heated to temperatures above 100 °C a complex series of reactions ensues, known as caramelisation, which gives rise to a wide range of flavour compounds in addition to the brown pigments we particularly associate with caramel. A related series of reactions, referred to as the Maillard reaction, involves amino compounds as well as sugars and are considered on pages 33 to 41. Since both of these groups of reactions

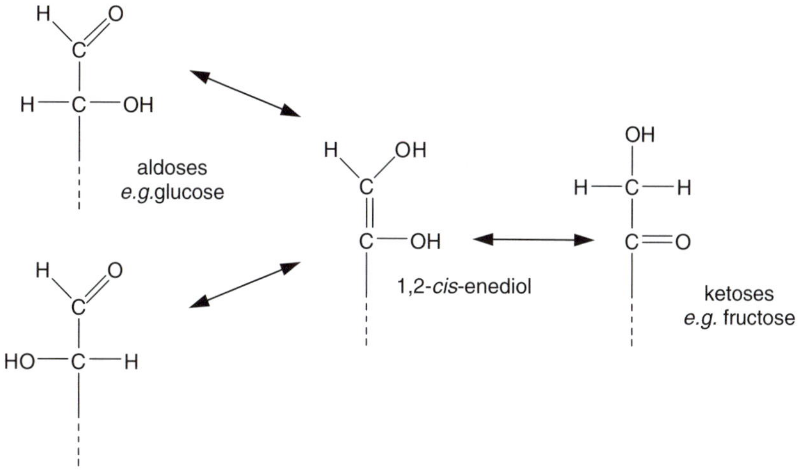

Figure 2.11   The Lobry de Bruyn – Alberda van Eckerstein transformation.

are particularly associated with the generation of brown pigments in food they are often collectively known as non-enzymic browning reactions. This distinguishes them from the enzymic browning reactions encountered in fruit and vegetables, dealt with in Chapter 6.

The first stage of the high-temperature breakdown is a reversible isomerisation (memorably known as the Lobry de Bruyn – Alberda van Eckerstein transformation) of an aldose or ketose sugar, *via* their open-chain forms, to a 1,2-*cis*-enediol intermediate (Figure 2.11). If glucose is gently heated in the presence of dilute alkali this reaction leads to the formation of a mixture of glucose, mannose and fructose.

The enediol is readily dehydrated in a neutral or acidic environment *via* the sequence of reactions shown in Figure 2.12. Hydroxymethyl-furfural (HMF) is by no means the only product of this pathway. Highly reactive α-dicarbonyl compounds (*i.e.* compounds such as 3-deoxy-aldoketose, in which pairs of carbonyl oxygen atoms are carried by neighbouring carbon atoms) occur as intermediates in these sequences and can give rise to a range of cyclic products, as well as reacting with amino compounds in the important Strecker degradation reactions discussed below.

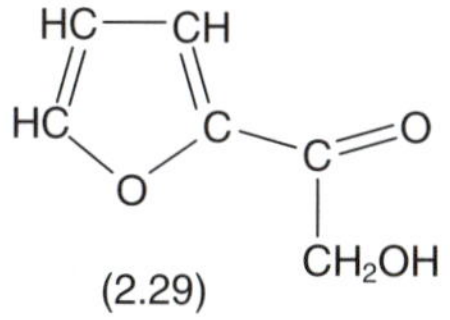

(2.29)

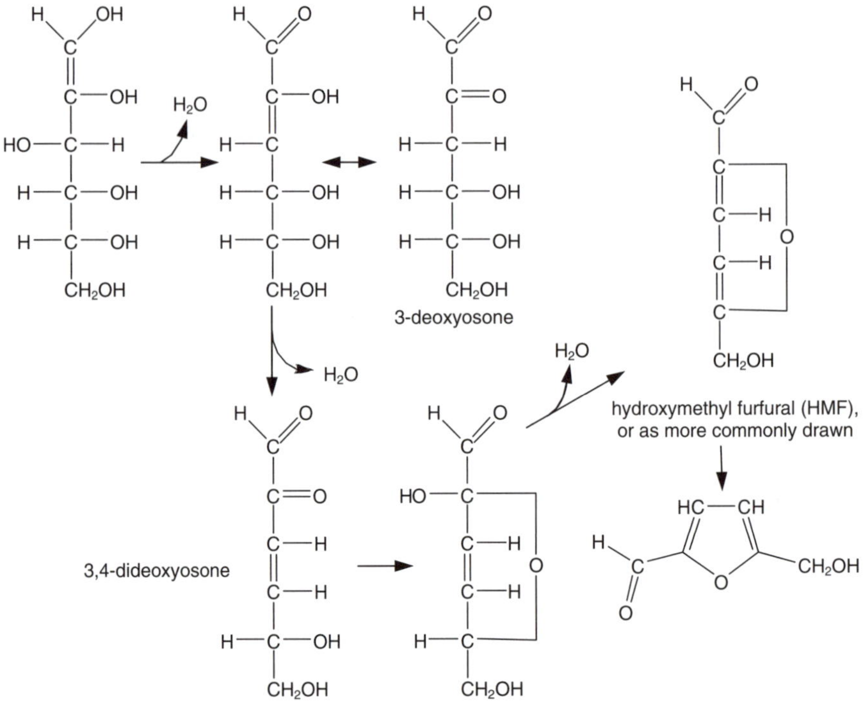

Figure 2.12 The formation of hydroxymethyl furfural by the sequential dehydration of a hexose enediol. Variation in the sites at which dehydration occurs is the source of related products such as hydroxyacetyl furan (2.29).

HMF can be readily detected in sugar-based food products which have been heated, such as boiled sweets, honey that has been adulterated with invert syrup, and "Golden Syrup".[ix] The brown pigments which characterise caramel and some other food colours arise from a poorly defined group of polymerisation reactions. These involve both HMF and its precursors. The pigment molecules have very high molecular masses and in consequence are apparently not absorbed from the intestine. Burnt sugar has always been a popular brown colouring, because it can be easily made in the home and therefore has a "natural" image. Commercially manufactured caramels for use as food colours are made in systems based on the Maillard reaction, which will be considered shortly.

[ix] Commercial invert syrup is produced by acid hydrolysis of sucrose at high temperature and is marketed as "Golden Syrup."

Small amounts of a vast range of other breakdown products give burnt sugar its characteristic acrid smell: acrolein (propenal, 2.30), pyruvaldehyde (2-oxopropanal, 2.31), and glyoxal (ethanedial, 2.32) are typical. The caramel flavour itself is reportedly due to two particular cyclic compounds, acetylformoin (2.33) and 4-hydroxy-2,5-dimethyl-3-furanone (2.34).

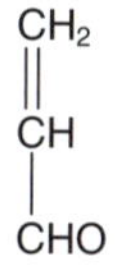

$$
\begin{array}{ccc}
CH_2 & CH_3 & CHO \\
\| & | & | \\
CH & C{=}O & CHO \\
| & | & \\
CHO & CHO & (2.32) \\
(2.30) & (2.31) &
\end{array}
$$

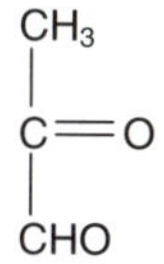

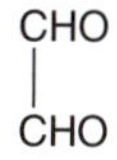

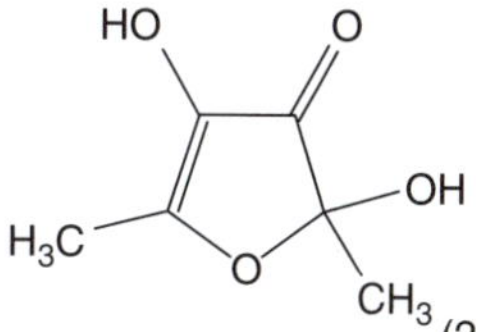

(2.33)

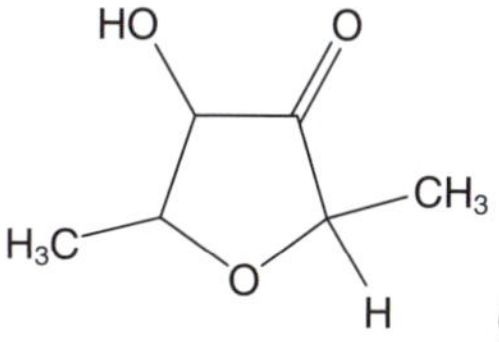

(2.34)

THE MAILLARD REACTION

In the presence of amino compounds, reducing sugars react much more readily to produce brown compounds. The amino groups which take part in the reaction may belong to free amino acids (*see* Chapter 5), the side-chains of amino acid components of protein molecules, or other amino compounds such as creatine (2.35), found in muscle. Although the formation of coloured compounds in reactions between glucose and amino compounds was first reported in 1912 by the French chemist Louis-Camille Maillard, it was not until 1953 that the essential stages of the reaction were outlined by J. E. Hodge, in a scheme which is still sometimes referred to today.

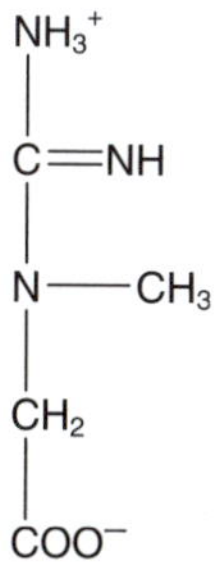

$$
\begin{array}{c}
NH_3^{+} \\
| \\
C{=}NH \\
| \\
N{-}CH_3 \\
| \\
CH_2 \\
| \\
COO^{-}
\end{array}
$$

(2.35)

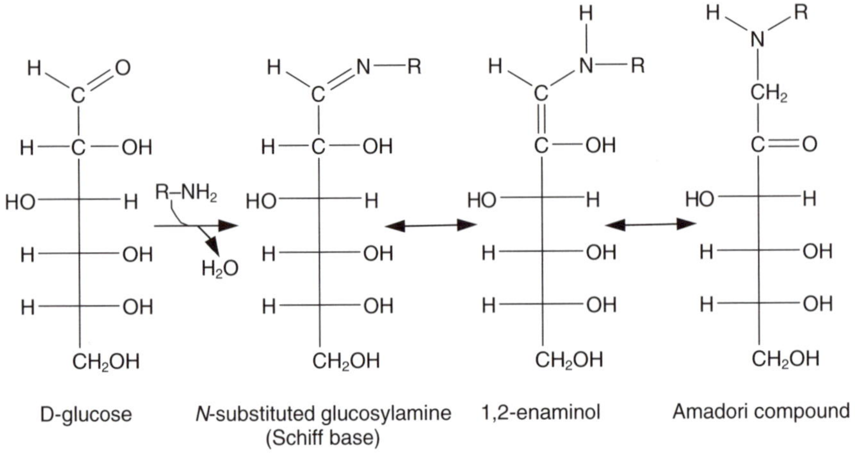

D-glucose   N-substituted glucosylamine   1,2-enaminol   Amadori compound
                     (Schiff base)

Figure 2.13   The reactions of glucose in the first stage of the Maillard reaction. R–NH$_2$ can be any compound with a free amino group, but in food systems this will normally be either a free amino acid or the side-chain amino group of an amino acid such as lysine, incorporated in a protein.

In the first stage, shown in Figure 2.13, the reducing sugar condenses with the amino compound to give an *N*-substituted glycosylamine (technically a Schiff base). This then spontaneously rearranges *via* the 1,2-enaminol to form the so-called Amadori compound, otherwise known as a 1-amino-1-deoxy-2-ketose.

The next phase as currently understood is shown in Figure 2.14. At pH values below 5, as in most food situations, the amino compound decomposes to leave a 3-deoxyaldoketose (often referred to as a 3-deoxyhexulose or 3-deoxyosone). This then loses water to give hydroxymethylfurfural, by exactly the same sequence of reactions as we encountered earlier in the case of caramelisation. If the pH is above 7 the Amadori compound breaks down by a slightly different route, which can result in a range of end-products, notably maltol and isomaltol, and various other α-dicarbonyl compounds (*i.e.* compounds having two carbonyl groups involving adjacent carbon atoms).

Of course in a real food situation a clear distinction between these two pathways would never occur and a considerable variety of ring compounds are likely to appear, some of which, such as acetylformoin (2.33), have already been mentioned in the context of caramelisation. One particular compound, sotolon, 4,5-dimethyl-3-hydroxyfuranone (2.36), has been identified as the key compound in the aroma of raw cane sugar.

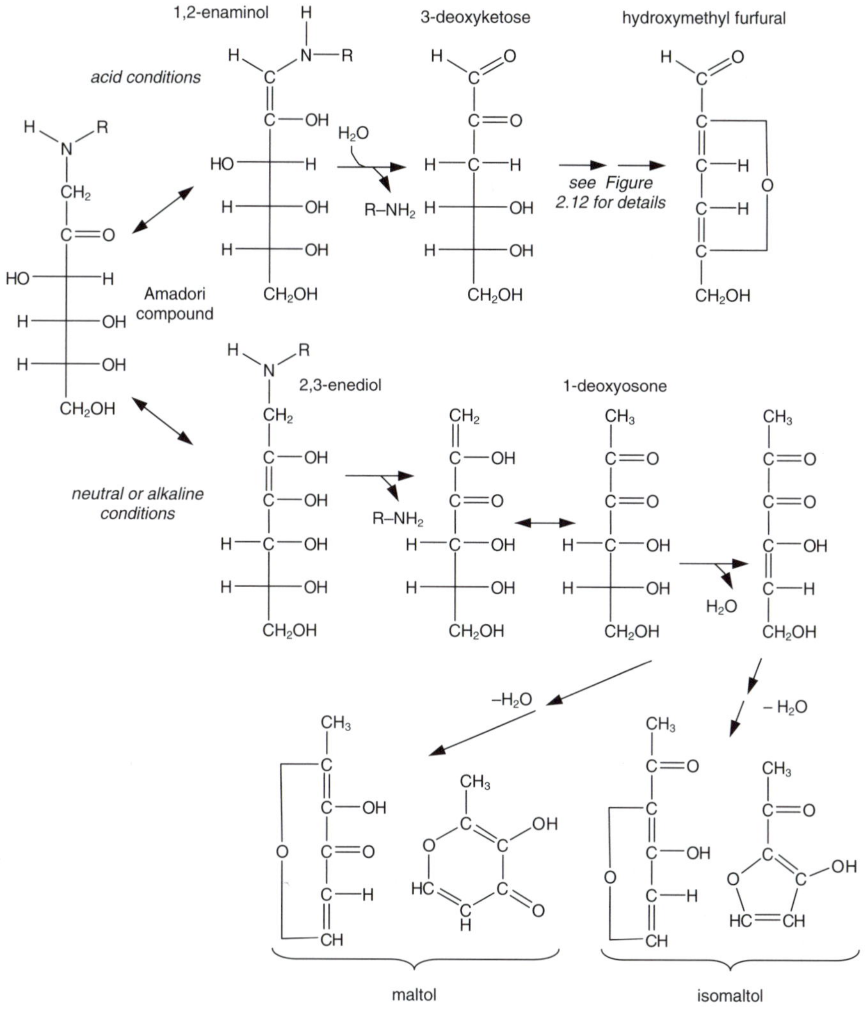

Figure 2.14 The breakdown of the Amadori compound in the Maillard reaction. In the formula of hydroxymethyl furfural shown, the carbon chain is arranged in a straight line. This has the virtue of showing the structural relationship of HMF to the original glucose but should be compared with the version shown in Figure 2.10. Similarly, both versions of the end-products, maltol and isomaltol, are shown here.

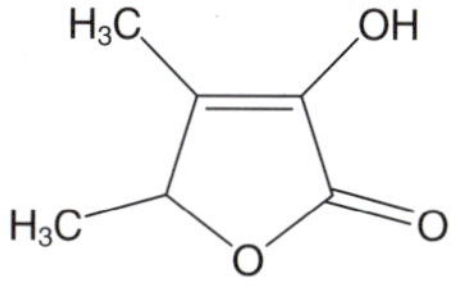

(2.36)

Although sotolon's burnt sugar aroma, detectable at levels of parts per billion, is regarded as highly desirable in many foods, particularly confectionery, it can be responsible for off-flavours in others, especially at higher levels. Maltol and isomaltol are contributors to the flavour of many cooked foodstuffs.

Disintegration of the α-dicarbonyl molecules also contributes to the flavour of foodstuffs through the formation of substances such as diacetyl (butanedione, 2.37) and acetol (hydroxypropanone, 2.38), as well as pyruvaldehyde and glyoxal, which have already been mentioned.

$$
\begin{array}{c}
CH_3 \\
| \\
C{=}O \\
| \\
C{=}O \\
| \\
CH_3
\end{array}
\qquad\qquad
\begin{array}{c}
CH_3 \\
| \\
C{=}O \\
| \\
CH_2OH
\end{array}
$$

(2.37)   (2.38)

A most important class of volatile flavour compounds arises by an interaction at elevated temperatures of α-dicarbonyl compounds with α-amino acids, known as the Strecker degradation, shown in Figure 2.15. The volatile aldehydes derived from amino acids in this reaction make a major contribution to the attractive odour during bread, cake and biscuit production. The pyrazines and other heterocyclic compounds which are often the other end-products of the Strecker degradation also make important contributions to food flavours, *e.g.* those of chocolate and roast meat (*see* Figure 7.9).

It is now recognised that the process of sealing pieces of meat before they are casseroled is another application exploiting the Maillard reaction. The authors of cookery books and television chefs have always recommended that meat should be briefly fried before being added to a casserole or similar dish where subsequent cooking is in an aqueous matrix. The cooked fatty surface has been assumed to seal in the flavours of the meat. In reality what is happening is that high temperatures in the frying pan are leading to large-scale flavour generation *via* the Maillard reaction and Strecker degradation. Whether the various flavour compounds stay inside the meat or end up in the fatty meat juice mixture in the pan is irrelevant, a good cook will ensure the entire contents of the pan all end up in the casserole.

Complex reaction sequences starting with 2,3-enaminols and involving a second amino acid in a Strecker degradation sequence have been suggested as the source of the range of pyrroles and pyridines that can

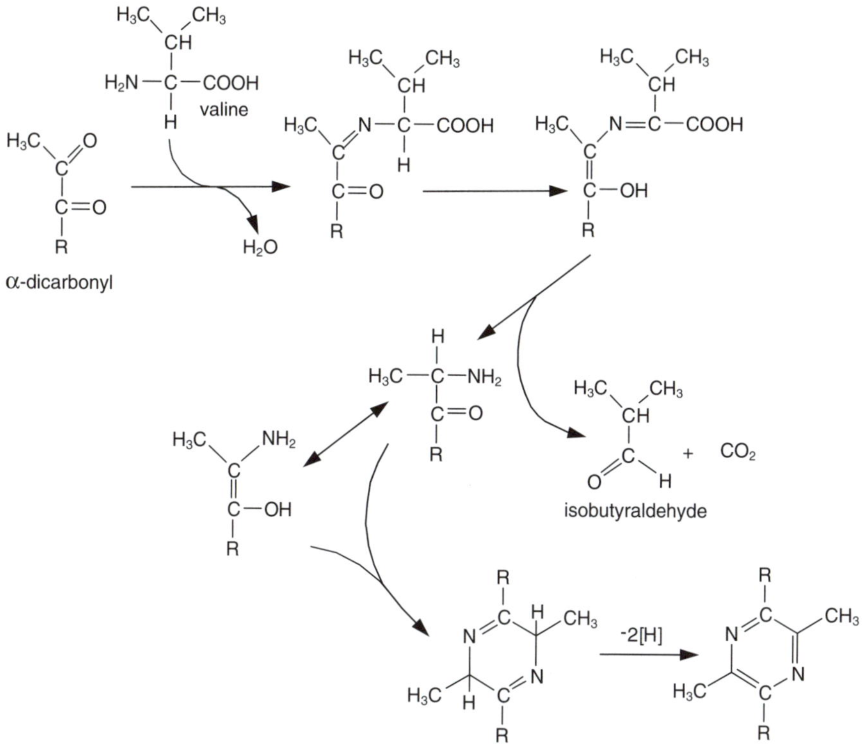

Figure 2.15 The Strecker degradation. A free amino acid (in this case illustrated by valine) reacts with an α-dicarbonyl compound generated elsewhere in the Maillard reaction. The reaction bears a superficial resemblance to transamination (see Figure 8.10. After rearrangement, the carbon skeleton of the amino acid is liberated as a volatile aldehyde. After two ex-sugar fragments (not necessarily identical, as shown here) condense, the product is readily oxidised to give the pyrazine derivative.

also arise in the Maillard reaction. These are important as they are believed to form the monomers of the polymerised brown pigments (melanoidins) which characterise the Maillard reaction. Actual food situations are far more complex than the model systems using purified starting materials in current experiments, but the structures shown in Figure 2.16 may well turn out to be a close approximation to the brown pigments we find in our food.

Low water concentrations, in which the reactants are more concentrated, favour the Maillard reaction, and it is implicated in the browning of bread crust and the less welcome discoloration that occurs from time to time in the production of powdered egg and milk. These foodstuffs are instances in which sugars are heated in the presence of

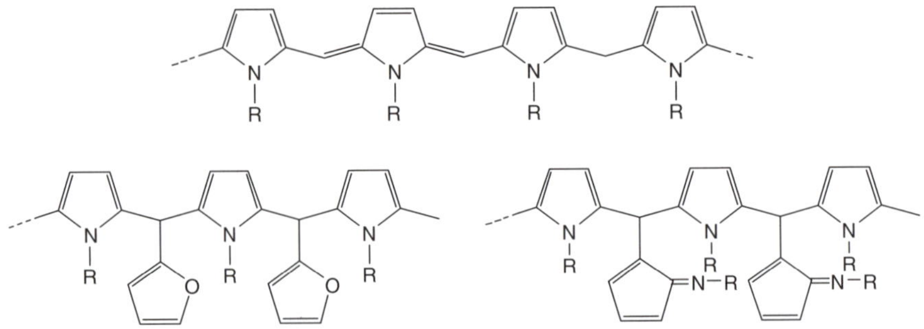

Figure 2.16   Structures postulated for the brown melanoidin pigments formed in the
Maillard reaction. In these structures R represents any of the possible
sugar fragments, such as $-CH_3$ or $-(CHOH)_2CH_2OH$, which arise during
the course of the reaction. All the structures shown here are likely to be
found in a single melanoidin molecule. It has been estimated that around
95% of molecules initially participating in the Maillard reaction eventually end up in melanoidin polymers. (Based on the proposals of
R. Tressl, *et al.*, in *The Maillard Reaction in Foods and Medicine*,
pp. 69–75, *see* Further Reading.)

protein, the source of amino groups. The amino groups most often
involved are those in the side-chains of lysine and histidine, although the
α-amino groups in any free amino acids will suffice.

While most of the effects of the Maillard reaction can be regarded as
favourable, there can be adverse nutritional consequences. When food
materials containing protein and reducing sugars are exposed to only
slightly elevated temperatures, possibly during prolonged storage, the
reaction may not go all the way to coloured end-products and may be
undetected by the consumer. However, the resulting glycosylated amino
acids are not metabolised in humans. If a sufficient proportion of the
protein surface is blocked by glycosylation the entire molecule may
become unavailable, because the proteolytic digestive enzymes cannot
gain access to the peptide bonds (*see* page 163). Lysine is the most
reactive of the vulnerable amino acids, followed by arginine, and then by
tryptophan and histidine.

Heating to reduce or eliminate its normal population of micro-
organisms is an almost universal method of extending the useful life of
milk. A side-effect of heat treatment is the Maillard reaction, although it
does not normally progress beyond the formation of a stable Amadori
compound such as α-deoxylactulosyl-lysine, shown in Figure 2.17.
Pasteurisation, or UHT treatment, of milk causes modest losses of lysine
(up to 2%) but the more aggressive heating involved in some milk
processes, such as sterilisation, evaporation or roller drying, can result in
as much as 25% of the lysine becoming unavailable to human digestion.

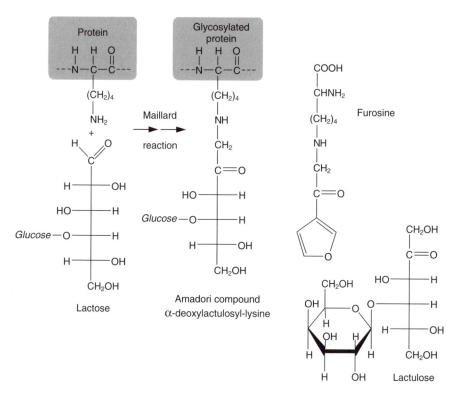

Figure 2.17 Lactose reactions during milk processing. The extent of glycosylation in a food protein can be estimated from the furosine content of the hydrolysate after it has been subjected to acid hydrolysis. Also, at high temperatures small amounts of lactose undergo the Lobry de Bruyn – Alberda van Eckerstein transformation (*see* Figure 2.11) to give lactulose, *i.e.* the glucose component of the disaccharide is converted from glucose to fructose, shown here for clarity in the open-chain structure. Measurements of furosine and lactulose levels are becoming important in establishing the processing history of commercial milk products.

A mixture of glucose and casein (the principal protein in milk) has been shown to lose as much as 70% of its lysine over a period of five days at temperatures as low as 37 °C. Conditions such as these may well occur during the transport of powdered milk for famine relief in the tropics. The fact that lysine, histidine and tryptophan must all be present in the diet, *i.e.* they are all *essential* amino acids (*see* Chapter 5), and that poor vegetarian diets are often deficient in lysine, makes this such an important issue.

The disease *diabetes mellitus* provides a very important example of protein glycosylation taking place at a low temperature, *i.e.* blood heat, and this is sufficiently food-related to be mentioned in this book.

Semi-automatic devices based on the enzymes, glucose oxidase and peroxidase (*see* Figure 2.8), provide an instant snap-shot of a patient's blood glucose concentration at the time of the test but do not give a long-term overall view. This is better provided by measuring the proportion of the patient's haemoglobin that has become glycosylated by the Maillard reaction with glucose in the blood. Normally some 3–7% of haemoglobin is in the glycosylated form (known as HbA1C), but where diabetes has resulted in persistently elevated levels of blood glucose higher percentages of HbA1C are found. Many of the complications suffered by long term victims of this disease, including the degeneration of the peripheral vascular system, sight loss (retinopathy) and kidney failure, are now known to be the result of crucial proteins in these tissues becoming glycosylated.

The Maillard reaction can pose other health problems not strictly related to nutrition. More than 20 complex heterocyclic amines, undoubtedly the products of Maillard-type reactions, have been identified in cooked meat, fish and poultry, particularly when grilled or fried, as well as in model systems. Many of these amines have been shown using the well known Ames test to be mutagenic in tests with bacteria, and must be regarded as potential, if not actual, carcinogens. While the concentration of these in grilled food is frequently very low (less than 1 ppb), one of them, 2-amino-1-methyl-6-phenylimidazo[4,5-*b*]pyridine (PhIP, 2.39) has been detected at 500 ppb in flame-cooked chicken. High temperatures (200–300 °C) and fairly long cooking times are required for the formation of heterocyclic amines; they are not produced during normal stewing, roasting or baking. It is very difficult to estimate the risk of human cancer from these compounds in real food, as opposed to model systems, but one current risk estimate is 1 in 10 000 over a normal lifetime and for typical consumption of food cooked at high temperature. Although the humble burger is grilled at high temperatures, the short cooking time make it less of a hazard than might be expected.

(2.39)

In recent years food chemists have become aware of another potentially hazardous food contaminant, acrylamide, which has now been shown to arise in the Maillard reaction. This is the subject of the last of the Special Topics later in this chapter.

Caramel is produced commercially for use as a food colour and is used in a vast range of food products, apart from the "gravy browning" found in the domestic kitchen. Around 10 000 tons are used annually in Britain, accounting for all but 5% or so of the total artificial colours used in food and drinks. Caramels are manufactured commercially by heating sucrose with catalytic amounts (a few percent) of acid or alkali (Class I caramel), sulfite (Class II), ammonia (Class III), or ammonia plus sulfite (Class IV). The various classes of caramel differ slightly in properties and consequently find different applications, especially in drinks. For example, Class I and Class II caramels are used in spirits such as whisky, Class III in beer and Class IV in soft drinks, notably the cola type.

SPECIAL TOPICS

1.   Reducing Group Reactions

Textbooks of organic chemistry give details of numerous chemical reactions of sugars, particularly those involving the carbonyl group. Many of these have little or no relevance to modern food chemistry, even though our present understanding of the structure of sugars was initially dependent upon their study. Until the advent of sophisticated chromatographic methods, notably high performance liquid chromatography (HPLC), the most important application of these reactions was in food analysis for identifying and quantifying the sugars present in a particular food material.

The carbonyl group of monosaccharides is readily oxidised, hence the reference to *reducing* sugars, *i.e.* sugars which can reduce *oxidising* agents, or oxidants. The reaction between reducing sugars and Cu^{2+} ions complexed in alkali (*e.g.* Fehling's solution or Benedict's reagent) forms the basis of the most popular chemical methods:

$$-\overset{H}{\underset{O}{C}} + 2\,Cu^{2+} + 4\,OH^- \longrightarrow -\overset{OH}{\underset{O}{C}} + Cu_2O + 2\,H_2O$$

(precipitated red copper oxide)

Although in the reaction above the product is shown as the corresponding "-onic" acid, this is not in fact the only product. Depending on the exact conditions, a number of other shorter-chain carboxylic acids are also produced. Ketose sugars such as fructose would not be expected to be oxidised but in fact they are, in reactions involving the neighbouring 3-amino-5-nitrosalicylic acid hydroxyl group. The variability in the outcome of these oxidation reactions means that they are non-stoichiometric and are difficult to apply to quantitative sugar analysis. The Lane and Eynon (titrimetric) and Munson and Walker (gravimetric) methods, using Fehling's solution, give data that have to be related to published tables to obtain results, necessarily expressed as "reducing sugars as glucose". The reaction of reducing sugars with 3,5-dinitrosalicylic acid (2.40) gives a reddish-brown product which provides the basis of a simple spectrophotometric assay, sufficiently accurate for use in college laboratory experiments.

$$\text{(2.40)} \quad + \text{ reducing sugar} \longrightarrow \text{3-amino-5-nitrosalicylate} \quad + \text{ oxidised sugar}$$

2.   Sugar Cane, Sugar Beet and Tequila

The fundamental reaction of photosynthesis in all green plants is the reaction, catalysed by the enzyme ribulose-1,5-biphosphate carboxylase, of carbon dioxide with ribulose-1,5-biphosphate to form two molecules of 3-phosphoglycerate. After a complex series of reactions these are converted to fructose-6-phosphate, and eventually to glucose and sucrose. Although this part of the process, known as the Calvin cycle, is common to all plants, two groups of plants have evolved specialised sequences of reactions which carry out the initial trapping of carbon dioxide from the atmosphere before it is transferred to the carboxylase. One such group is the so-called tropical grasses, which includes maize, *Zea mays,* ("corn" in North America) and sugar cane, *Saccharum officinarum*, the traditional source of sucrose. These use the Hatch and Slack pathway for initial carbon dioxide fixation. Another pathway, known as crassulacean acid metabolism (CAM) is used by the succulents, of which only two species, the pineapple, *Ananas comosus*, and the blue agave, *Agave tequilana*, are of interest. The leaves of the agave are the source of the sugary syrup which is fermented and then distilled to give tequila, the famous Mexican spirit. Both these variants of normal photosynthesis give improved efficiency under tropical conditions.

Table 2.2 The proportion of the stable carbon isotope ^{13}C in the total carbon of plant materials, compared with an arbitrary international standard of fossil calcium carbonate (belmenite).

	International standard	*Tropical grasses, including sugar cane and maize*	*Crassulaceae, including agave and pineapple*	*All other plants, including sugar beet and grapes*
^{13}C (%)	1.1112	1.0991	1.0975	1.0837

In contrast, sugar beet, a variety of the common beet, *Beta vulgaris*, uses the normal Calvin cycle. This grows in a wide range of temperate climates and has been an alternative source of sugar, especially in continental Europe, since the middle of the 18th century.

In terms of chemistry, biochemistry, human nutrition and health, there is no difference between cane and beet sugars once they have been purified to the colourless crystalline materials we recognise as granulated or caster sugar. Although in isolation the distinction between cane and beet sugar is purely academic, it is important from the commercial and regulatory points of view to be able to discriminate between them. The method depends upon the 1% of all naturally occurring carbon atoms (in organic compounds, atmospheric carbon dioxide and minerals such as calcium carbonate) which occur as the non-radioactive isotope ^{13}C. Almost all the rest is the ^{12}C isotope. Most chemical and biological systems, including human metabolism, cannot distinguish between the two isotopes, but the enzymes in the Calvin cycle do. As a result plant materials contain very slightly different proportions of ^{12}C and ^{13}C, depending on the photosynthetic pathway involved in their synthesis. These proportions, shown in Table 2.2, can be measured with sufficient precision by a mass spectrometric technique, stable isotope ratio analysis (SIRA), to discriminate between cane and beet sugar and to detect whether cane sugar has been dishonestly used to supplement the natural sugar content of a fruit product.

This method has been successful in demonstrating the illegal addition of cane sugar to grape juice prior to the fermentation, and in detecting the addition of corn syrups and cane sugar to maple syrup and citrus juices. Unfortunately cane sugar which finds its way into pineapple juice or into tequila fermentation is much harder to detect.

3. Monosaccharide Conformations

An analysis of the spatial distribution of the hydroxyl groups of the D-hexoses when in the *C*1 conformation is given in Table 2.3, along with

Table 2.3 The conformation of the aldohexoses. The percentages of each anomer and ring size in aqueous solution at 40 °C are shown, based on the data of S. J. Angyal, *Angew. Chem., Int. Edn. Engl.*, 1969, **8**, p. 157. The positions of the hydroxyls at C-2, C-3, and C-4 are also shown for the *C*1 conformation. The CH$_2$OH group (C-6) in a D-aldohexose is always equatorial in the *C*1 conformation, as is the anomeric hydroxyl (on C-1) of the β-configuration.

	Hydroxyl at			*% Furanose*		*% Pyranose*	
	C-2	*C-3*	*C-4*	α	β	α	β
D-Allose	eq	ax	eq	5	7	18	70
D-Altrose	ax	ax	eq	20	13	28	39
D-Glucose	eq	eq	eq	< 1	< 1	36	64
D-Mannose	ax	eq	eq	< 1	< 1	67	33
D-Gulose	eq	ax	ax	< 1	< 1	21	79
D-Idose	ax	ax	ax	16	16	31	37
D-Galactose	eq	eq	ax	< 1	< 1	27	73
D-Talose	ax	eq	ax	< 1	< 1	58	42
	C-3	*C-4*	*C-5*				
1*C*-D-Fructose	eq	eq	ax	< 1	25	8	67

eq: equatorial; ax: axial

data showing the proportions of the different anomers and ring types. The *C*1 (pyranose) ring is the only significant conformation except for D-idose and D-altrose. It is clear that with these two sugars no one conformation offers a predominance of equatorial hydroxyl groups. The same explanation can account for the behaviour of aqueous solutions of D-ribose, which have been shown to contain a broadly similar distribution of anomers and ring sizes to those of fructose, except that both the α- and β-pyranose forms exist in roughly equal proportions of the *C*1 and 1*C* conformers.

Two factors appear to control the relative proportions of α- and β-anomers in a particular sugar. The normal tendency of a hydroxyl to seek an equatorial position has to be balanced against the unfavourable character of the interaction that can occur between the ring oxygen and the anomeric hydroxyl in this position.

4. Acrylamide

In 1992 construction started on the Hallandsås railway tunnel on the west coast of Sweden. Progress had already been severely delayed by geological problems, when in 1997 it was discovered that local farm animals were dying and construction workers were being taken ill. The cause was traced to a sealing compound, polyacrylamide, used during the construction. Unfortunately this had not been polymerised properly and the monomer, acrylamide (2.41), had leached out into local water courses.

$$H_2C = C(H) - C(=O) - NH_2 \quad (2.41)$$

As part of their investigation into the poisoning Swedish scientists discovered that ordinary non-smoking[x] members of the public, examined to provide baseline data, also showed unexpectedly high levels of acrylamide exposure, although nothing like the levels found amongst tunnel workers. By 2002 this was traced to foodstuffs which had been heated to high temperatures as part of their normal cooking procedures, in particular starchy foods which had been fried.

The toxic effects of acrylamide are well established. Exposure at high levels, associated with industrial accidents, causes neurological damage. No neurotoxic effects are to be expected from the acrylamide levels found in food. Prolonged exposure at lower levels has been shown to induce tumours in laboratory animals, but there is no convincing evidence that acrylamide can cause cancer in humans. In 2007 a large study in the USA showed that the incidence of at least one common cancer, that of the breast, was unaffected by the relative amounts of acrylamide-rich foods consumed over a long period. The UK Food Standards Agency (Food Survey Information Sheet 71/05, 2005) estimates that current dietary exposure levels in the UK are at least 1000 times lower than the levels reported to cause cancer in laboratory rats.

Since 2002 worldwide investigations have provided a mass of data on the occurrence of acrylamide (*see* Table 2.4) which clearly bears out the original incrimination of starchy foods cooked at high temperature. It is unwise to attempt to draw any more than the broadest conclusions from the differences in acrylamide levels between foodstuffs. The values given in Table 2.4 can only be regarded as typical of the data gathered by laboratories around the world. Individual samples are frequently found to differ considerably from these figures. A further complication is that different countries have different views of how products such as potato crisps, chips, *etc.,* are defined.

---

[x] Acrylamide is known to occur in tobacco smoke.

Table 2.4   Typical levels of acrylamide found since 2002 in surveys of raw and
cooked foodstuffs in Great Britain and elsewhere. There are wide
variations between individual surveys reflecting, among other factors, the influence of processing conditions. References to Swedish
and Australian data are given here as examples: K. Svensson, *et al.,*
Food Chem. Toxicol., 2003, **41**, p. 158; M. Croft, *et al., Food Addit.*
Contam., 2004, **21**, p. 721. The websites of organisations such as the
Food Standards Agency (UK), the FDA (USA) and the WHO can
provide further detailed data.

Acrylamide (μg/kg)					
Potato products:		Cereal products:		Others	
raw or boiled	< 30	wheat flour	< 30	coffee (as drunk)	420
baked	190	bread	30	dry roasted peanuts	30
chips	310	toast	75	chicken nuggets	23
French fries	480	breakfast cereals	180	hamburgers	15
crisps	1050	popcorn	400		
		rye crispbread	2000		

In spite of the prevailing confidence of responsible authorities around
the world that the level of acrylamide in human diets does not constitute
a hazard, considerable attention has been devoted to discovering how it
is formed and what modifications to current food processing procedures
might reduce it. The most likely mechanism is the Maillard reaction,
hence the inclusion of the topic in this chapter. It is now generally
accepted that the amino acid, asparagine, is the source of acrylamide in a
variation of the Strecker degradation, as shown in Figure 2.18. The
involvement of an alternative mechanism involving acrolein (2.30),
which can arise in the breakdown of fats as well as sugars, has now been
discounted.

Unfortunately for those seeking to reduce the levels of acrylamide
formation, free asparagine, *i.e.* when it is not a component of proteins, is
particularly abundant in potatoes. On a weight basis asparagine can
constitute over 30% of the total free amino acids in potatoes, 0.2–0.6 g
per 100 g of their fresh weight. There is little doubt that the determined
efforts of plant breeders, combined with control of growing, storage and
processing conditions, could bring these figures down considerably, but
only at a cost. Asparagine is recognised as a major contributor to the
flavour of potatoes and potato products. The high temperatures
involved in frying (over 120 °C) which are critical to acrylamide generation are also essential for the desirable flavours and textures of our
crisps and chips.

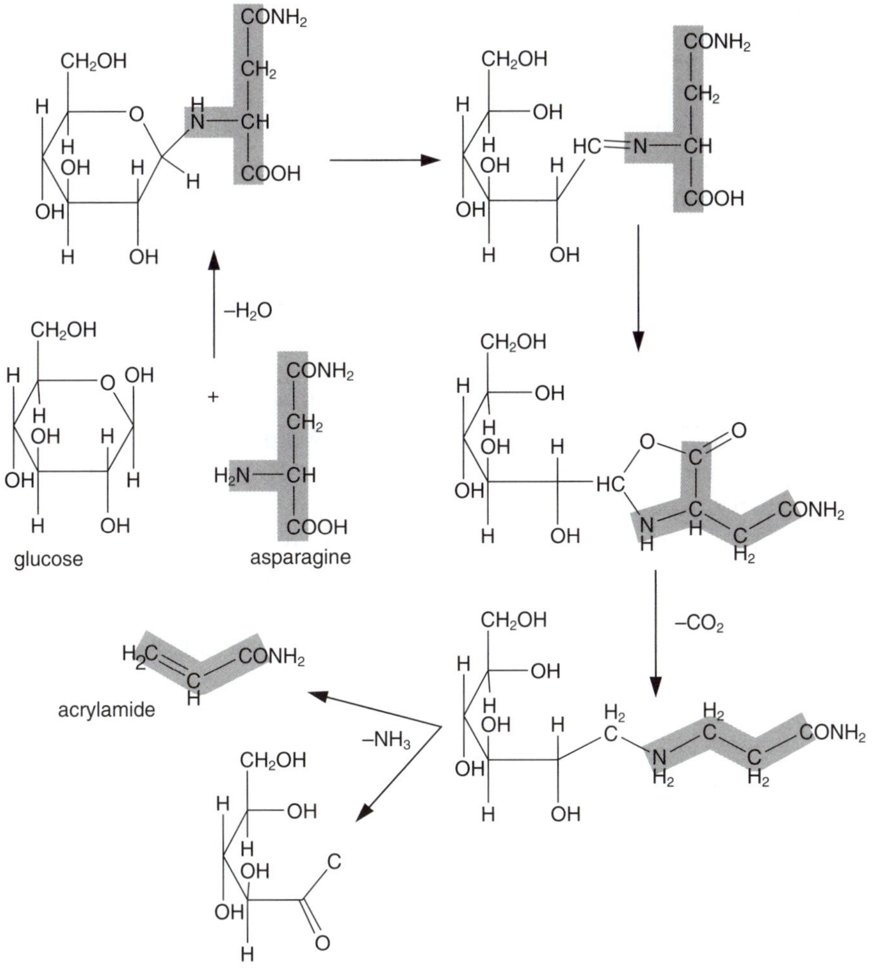

Figure 2.18 The generally accepted route for the formation of acrylamide by reaction with reducing sugars. The backbone carbon and nitrogen atoms of asparagine have been shaded to make their progress through the reactions easier to follow.

Another potential technique for reducing asparagine levels depends on the enzyme asparaginase. This enzyme, obtained commercially from the fungus *Aspergillus niger*, catalyses the removal of the side-chain amino group, converting asparagine to aspartate. It can be incorporated into dough, and it could therefore prove valuable in the manufacture of some baked products. Otherwise, acrylamide is a potential carcinogen we may have to put up with for many years to come.

FURTHER READING

S. W. Cui, *Food Carbohydrates: Chemistry, Physical Properties and Applications*, CRC Press, Boca Raton, Florida, 2005.

Chemical and Functional Properties of Food Saccharides, ed. P. Tomasik, CRC Press, Boca Raton, Florida, 2003.

D. A. Southgate, *Determination of Food Carbohydrates*, 2nd edn, Elsevier, London, 1991.

W. P. Edwards, *The Science of Sugar Confectionery*, Royal Society of Chemistry, Cambridge, 2000.

R. V. Stick, *Carbohydrates: The Sweet Molecules of Life*, Academic Press, San Diego, California, 2001.

V. S. R. Rao, P. K. Qasba, P. V. Balaji and R. Chandrasekaran, *Conformation of Carbohydrates*, Harwood Academic, Australia, 1998.

S. E. Fayle and J. Gerrard, *Maillard Reaction*, Royal Society of Chemistry, London, 2002.

H. E. Nursten, *Maillard Reaction: Chemistry, Biochemistry and Implications*, Royal Society of Chemistry, Cambridge, 2005.

The Maillard Reaction in Foods and Medicine, eds. J. O'Brien, H. E. Nursten, M. J. C. Crabbe and J. M. Ames, Royal Society of Chemistry, Cambridge, 1998.

RECENT REVIEWS

P.-A. Finot, *Historical Perspective of the Maillard Reaction in Food Science; Ann. N. Y. Acad. Sci.*, 2005, **1043**, p. 1.

I. Blank, *Current Status of Acrylamide Research in Food: Measurement, Safety Assessment, and Formation; Ann. N. Y. Acad. Sci.*, 2005, **1043**, p. 30.

D. Taeymans, *et al., A Review of Acrylamide: An Industry Perspective on Research, Analysis, Formation and Control; Crit. Rev. Food Sci. Nutr.*, 2004, **44**, p. 323.

S. D. Kelly, *Using Stable Isotope Ratio Mass Spectrometry (IRMS) in Food Authentication and Traceability, in Food Authenticity and Traceability*, ed. M. Lees, Woodhead, Cambridge, 2003.

CHAPTER 3
Polysaccharides

In Chapter 2 we considered the properties of the mono- and oligo-saccharides. We can now turn to the high molecular weight polymers of the monosaccharides: the polysaccharides.

Polysaccharide structures show considerable diversity and are often classified according to features of their chemical structure, for example whether the polymer chains are branched or linear and whether more than one type of monosaccharide is present. Only recently have we begun to discern a general basis for the relationship between the chemical structure of a polysaccharide and its physical properties. In the course of this chapter the nature of this relationship will be explored.

The one common feature of the polysaccharides which is important to food scientists is that they all occur in plants. Polysaccharides play a relatively minor role in the physiology of animals. Glycogen, which is structurally very similar to amylopectin, acts as an energy and carbohydrate reserve in the liver and muscles.[i] The other animal polysaccharides are chitin, a polymer of *N*-acetylglucosamine (3.1), the fibrous component of arthropod exoskeletons, and the proteoglycans and mucopolysaccharides, which have specialised roles in connective tissues. The molecules of these animal polysaccharides include considerable proportions of protein.

---

[i] Muscle tissue loses its glycogen in the ongoing *postmortem* metabolism before it reaches the plate as meat, but fresh liver still contains 1–2% glycogen at the time it is sold.

Food: The Chemistry of its Components, Fifth edition
By T.P. Coultate
© T.P. Coultate, 2009
Published by the Royal Society of Chemistry, www.rsc.org

$$CH_2OH$$

(3.1)

Polysaccharides have two major roles to play in plants. The first of these, as an energy and carbohydrate reserve in the tissues of seeds and tubers, is almost always provided by starch. On the other hand, the data in Table 3.1 show that leafy plant foods, as exemplified by broccoli, contain only traces of starch. The second role of polysaccharides is to provide the structural skeleton of both individual plant cells and the plant as a whole. This role is filled by a wide range of different structural types of polysaccharides, which offer a correspondingly wide range of physical characteristics.

The contribution of polysaccharides to foods derived from plants is no less diverse. From a strictly nutritional point of view starch is the only major plant polysaccharide that is readily digested in the human intestine and directly utilised as a source of energy. In fact, a high

Table 3.1   The starch contents of a variety of foods. Most of these figures are taken from *McCance and Widdowson* (*see* Chapter 1, Further Reading) and in all cases refer to the edible portion. They should be regarded as typical values rather than absolute figures for the particular food indicated.

Food	Starch (%)	Food	Starch (%)
Wheat flour (wholemeal)	61.8	Peas (petits pois)	trace
White bread	46.7	Broad beans	10.0
White rice (uncooked)	73.8	Beetroot	0.6
Spaghetti (cooked)	70.8	Broccoli	0.1
Maize (corn) flour	92.0	Carrots	0.2
Corn flakes	77.7	Chestnuts	29.6
Soya beans (raw)	4.8	Peanuts (raw)	6.3
Cassava (dry)	22.0	Walnuts	0.7
Potatoes (boiled, new)	16.7	Apples	trace
Peas (canned, processed)	14.7	Bananas	2.3

proportion of the human requirement for energy is met by the starch of seeds such as wheat, rice and maize, and tubers such as potatoes and cassava. However, it is now widely recognised that polysaccharides make a much greater contribution to human health than the mere provision of energy. Cellulose, pectins and hemicelluloses, amongst others, are commonly referred to as "fibre" but are more correctly termed "non-starch polysaccharides" (NSP). They play an essential role in the healthy function of the large intestine and in moderating the absorption of nutrients from the small intestine.

In spite of their physiological importance, it is through their influence on the texture of food that polysaccharides often make the most immediate impact on the consumer. Furthermore, the relationship between a food's texture and the underlying molecular structure of its components is revealed most clearly in the case of polysaccharides, and this relationship is a central theme in the account of the food polysaccharides which follows.

STARCH

Reflecting its role as a carbohydrate reserve, starch is found in greatest abundance in plant tissue such as tubers and the endosperm of seeds. It occurs in the form of granules, which are usually an irregular rounded shape, ranging in size from 2 to 100 µm. Both the shapes and sizes of the granules are characteristic of the species of plant and can help to identify the origin of a starch or flour.

Starch consists of two types of glucose polymer: amylose, which is essentially linear, and amylopectin, which is highly branched. These occur together in the granules, but amylose can readily be separated from starch solutions since it is much less soluble in organic solvents such as butanol. Most starches are 20–25% amylose, but there are exceptions: pea starch is around 60% amylose and the so-called waxy varieties of maize and other cereals contain little or no amylose at all.

Amylose consists of long chains of α-D-glucopyranosyl units linked, as in maltose, between their 1- and 4-positions (*see* Figure 3.1A). There is no certainty about the length of the chains, but they are generally believed to contain many thousands of glucose units, with typical average molecular weights between 2×10^5 and 2×10^6. It is becoming clear that amylose chains are not entirely linear, and they do contain a very small amount of the branching characteristic of amylopectin.

Amylopectin is a much larger molecule, comprising a million or so glucose units per molecule. As in amylose, the glucose units are joined by 1→4 glycosidic links, but some 4–5% of the glucose units are

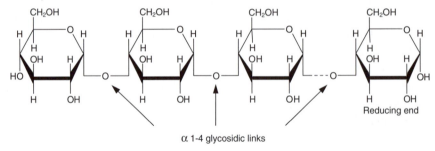

Figure 3.1   The chemical structure of amylose and amylopectin.

also involved in 1→6 links, creating the branch points shown in Figure 3.1B. This proportion of branch points results in an average chain length of 20–25 units. Over the years a great many hypotheses have been put forward to describe the arrangement of the branching of the chains in amylopectin, and for some time the tree-like arrangement proposed by Meyer and Bernfeld in 1940 was accepted. Since then new insights have been provided by the dissection of the molecule into its individual component chains using the enzyme pullulanase, which is specific for 1→6 linkages. Although the chains do indeed have the expected overall average chain length the molecule has been found to consist of two types of chain, the more abundant type having around 15 glucose units and the other about 40, referred to as the A and B chains, respectively. The so-called "cluster model" based on this data, and other supporting evidence, was proposed by Robin in 1974 and is now widely accepted (Figure 3.2A). Its essential feature is a skeleton of mainly singly branched A chains carrying clusters of B chains. In each molecule there

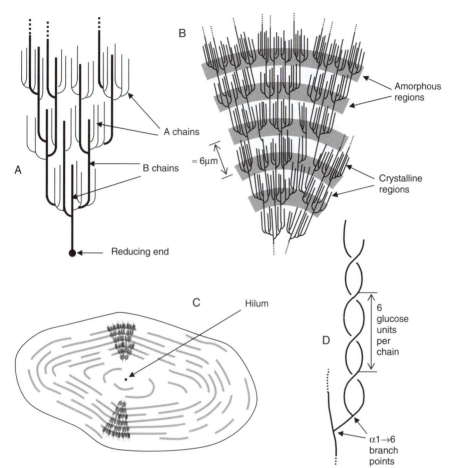

Figure 3.2 Amylopectin and the starch granule. (A) The essential features of the cluster model first proposed by Robin in 1974. (B) The organisation of the amorphous and crystalline regions (or domains) of the structure generating the concentric layers which contribute to the "growth rings" visible by light microscopy. (C) The orientation of the amylopectin molecules in a cross-section of an idealised entire granule. (D) The likely double helix structure taken up by neighbouring chains and giving rise to the extensive degree of crystallinity in the granule.

is one chain, referred to as the C chain, which is unique in carrying a reducing end-group. As shown in Figures 3.2B and 3.2C, the amylopectin molecules are believed to be radially orientated in the starch granule, with the terminal reducing group towards the centre, possibly close to the central spot sometimes visible by light microscopy, the *hilum*. The clusters of short chains are believed to be distributed on the B chain skeleton as shown in Figure 3.2B, leading to the formation of concentric domains of regularly orientated (crystalline) polysaccharide chains alternating with amorphous domains, shaded in Figure 3.2B, in which most of the $1 \rightarrow 6$ branch points are located. In the crystalline regions, pairs of neighbouring chains are believed to form short double helices with 6 glucose units per turn in each chain (Figure 3.2D).

When observed under the polarising microscope, starch granules show the Maltese cross pattern characteristic of birefringent materials. This confirms the presence of a high degree of molecular orientation. The crystallinity is confirmed by their X-ray diffraction patterns, which incidentally reveal that root starch granules are in general more crystalline than those from cereals. Another indication of the importance of the amylopectin fraction as the source of the crystallinity of starch granules is that waxy starches, which lack amylose, give X-ray diffraction patterns very similar to those of a normal starch. Surprisingly little is known of the arrangement of the amylose molecules within the starch granule. These are assumed to meander amongst the amylopectin molecules, orientated more or less radially. The granules of cereal starches contain approximately 1% of lipid material, mostly in the form of lysophospholipids (*see* Chapter 4). The hydrocarbon chains of the lipid molecules are believed to lie along the axis of helical amylose chains, forming a complex not dissimilar in structure from the iodine/amylose complex (*see below*).

Owing to the collective strength of the hydrogen bonds binding the chains together, undamaged starch granules are insoluble in cold water but as the temperature is raised to what is known as the initial gelatinisation temperature, water begins to be imbibed. The initial gelatinisation temperature is characteristic of individual starches but usually lies in the range 55–70 °C. As water is taken in the granules swell and there is a steady loss of birefringence. Studies using X-ray diffraction show that complete conversion to the amorphous state does not occur until temperatures around 100 °C are reached. As swelling proceeds and the swollen granules begin to impinge on each other, the viscosity of the suspension/solution rises dramatically. The amylose molecules are leached out of the swollen granules and contribute to the viscosity of what is best described as a paste. If heating is maintained together with stirring, the viscosity soon begins to fall again as the integrity of the granules is destroyed by the physical effect of the mixing. When the paste is allowed to cool, the viscosity rises again as new hydrogen-bonding relationships between amylopectin and amylose are re-established to give a more gel-like consistency. These changes will be familiar to anyone who has used starch to thicken gravy or to prepare desserts such as blancmange. They can be followed in the laboratory using

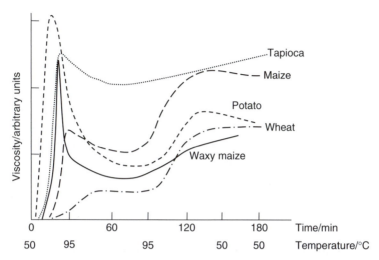

Figure 3.3   Brabender amylograph traces of the gelatinisation of a number of starches. The temperature is raised gradually from 50 to 95 °C over the first
half hour. It is then maintained at 95 °C for one hour before being allowed
to fall to 50 °C over the final half-hour period. (Reproduced by permission
from Dr D. Howling and Applied Science Publishers.)

viscometers such as the Brabender amylograph, which apply continuous
and controlled stirring to a starch suspension during a reproducible
temperature regime. Typical Brabender amylograph results for a number of different starches are shown in Figure 3.3. Although our understanding is increasing, we are still some way from being able to explain
all the distinctive features of a particular Brabender trace in terms of the
molecular characteristics of the starch.

Solutions of starch have a number of characteristic properties,
of which their reaction with iodine is the best known. Solutions of iodine
in aqueous potassium iodide stain the surface of starch granules
dark blue, but the reaction is most striking with the amylose component
of dissolved starch. In the presence of iodine the amylose chains
are stabilised in a helical configuration with six glucose units to
each turn of the helix and the iodine molecules complexed along its
axis. The blue colour of the complex is dependent on the length of the
chain involved: with around 100 units λ_{max} [ii] is at 700 nm, but with
shorter chains λ_{max} gradually falls, until with only 25 units it is down to
550 nm. In view of this it is not surprising that amylopectin (in the
absence of amylose) and glycogen both give a reddish brown colour with
iodine.

[ii] λ_{max} is the wavelength of maximum light absorption.

When starch solutions or pastes are allowed to stand for a few hours, they begin to show changes in their rheological properties. Dilute solutions lose viscosity, but concentrated pastes and gels become rubbery and exude water. Both types of change are due to a phenomenon termed retrogradation, which involves only the amylose molecules. Over a period of time these associate and effectively crystallise out. If one recalls that in starch solutions amylose tends to bind together the expanded amylopectin molecules within the granule structure, then the effects of retrogradation on rheological properties are clearly inevitable.

Special efforts have to be made in a number of food products to avoid the problems which retrogradation could cause. For example, starch is frequently used commercially to thicken the juice or gravy of pies. If after manufacture the pie is frozen, the amylose will undergo rapid retrogradation. On thawing, the starch paste will become completely liquid, revealing just how much or how little of the filling was actually meat or fruit, as well as making the pastry all soggy. One answer to this has been to use starch from the "waxy" varieties of maize,[iii] which consist only of amylopectin and therefore give pastes that will survive freezing and thawing.

The manufacturers of instant desserts must also be wary of retrogradation. These starch-based dry mixes are designed to be blended with milk to give a blancmange-like dish without the need for cooking. A slurry of raw starch in water is passed over steam-heated rollers, resulting in rapid gelatinisation, immediately followed by drying to a film which can be scraped off the roller and subsequently milled to a fine powder. The water is thus removed before the amylose molecules have had the chance to get organised. The resulting pregelatinised starch will disperse in water fairly easily, but no one would claim that this gives as good a texture as that of a proper blancmange.

In the manufacture of bread and other bakery products the behaviour of starch is obviously very important. When flour is milled it is inevitable that a significant proportion of the starch granules will be physically damaged, scratched, cracked or broken. When the dough is first mixed these damaged starch granules absorb some water. During bread making, the kneading and proving stages prior to the actual baking allow time for the α- and β-amylases naturally present in flour to break down a small proportion of the starch to maltose and other sugars. These are fermented by the yeast to give the CO_2 which leavens the dough. Other

---

[iii] The term "waxy" is derived from the appearance of the starchy endosperm of the maize kernel when it is broken in two.

baked goods such as cakes rely on the CO_2 from baking powder[iv] for the same purpose, together with air whipped into the mixture.

Once in the oven, the starch granules gelatinise and undergo varying degrees of disruption and dispersion. The α-amylase activity of flour persists until temperatures around $75\,^\circ C$ are reached, so that some enzymic fragmentation of starch granules also occurs, particularly in bread. Among the factors which affect the degree of starch breakdown are the availability of water and the presence of fat. Obviously in a shortbread mixture containing only of flour, butter and sugar the coating of fat over the starch granules will limit the access of the small amount of water present, and we find that the granular structure is still visible after baking. A roux is another example of fat being used to control the gelatinisation behaviour of starch.

After one or two days, bread, and cakes with a low fat content, start to go stale. Staling is often assumed to be a simple drying-out process, but of course it occurs just as readily in a closed container. What we are actually seeing is the retrogradation of the starch. The current view of bread staling is that it is a two-stage process. The first stage, marked by the transition from the soft "new-bread" state, takes place only a few hours after baking and is believed to involve the retrogradation of the unbranched amylose fragments. The second stage, staling proper, involves retrogradation of a proportion of the amylopectin fragments. The crystallinity of the retrograded starch gives the crumb of stale bread its extra whiteness, and the increased rigidity of the gelatinised starch causes the lack of "spring" in the crumb texture. Retrograded starch can be returned to solution only by heating, *i.e.* by regelatinisation; stale bread can be revived by moistening and then returning it to a hot oven for a few minutes.

The phenomenon of retrogradation is a reminder that all the starch in a foodstuff is not necessarily available to us, *i.e.* readily digested and absorbed from the small intestine. The starch content of a foodstuff falls into a number of categories:

- RDS: Rapidly digestible starch, dominant in starchy foods cooked by moist heat, *e.g.* bread or potatoes.
- SDS: Slowly digestible starch, like RDS completely digested in the small intestine, but more slowly. Includes physically inaccessible amorphous starch, raw starch granules, or retrograded starch in cooked foods.

[iv] Baking powder is a mixture of sodium hydrogen carbonate (sodium bicarbonate, $NaHCO_3$), plus a weakly acidic salt such as sodium dihydrogen phosphate or potassium hydrogen tartrate (cream of tartar), which gives off carbon dioxide in the moist, hot conditions of the oven.

- RS: Resistant starch, potentially resistant to digestion in the small intestine. Resistant starch is further subdivided as follows:
  - RS1: Resistant due to physical inaccessibility, as in partly milled cereals and some very dense processed starchy foods. Measured on the basis of enzyme digestibility before and after very thorough homogenisation or milling.
  - RS2: Uncooked granules, particularly resistant to enzyme action, *e.g.* raw potato and banana.
  - RS3: The most resistant starch, consisting of retrograded amylose formed during cooling of previously heated gelatinised starch

All types of resistant starch are considered to be digestible by the colonic microflora, at least to some degree. Although the overall proportion of resistant starch in many foodstuffs has been determined, little information is available about the proportions of the different subdivisions. However, the data in Table 3.2 give a fairly accurate impression, which is in reasonable agreement with nutritional observations, in spite of the limitations of laboratory simulations of natural chewing *etc.*

In partially milled cereal products which are eaten uncooked, such as muesli, some of the starch remains undigested simply because the amylases cannot gain access to it. The starch in raw potatoes and unripe bananas is in a crystalline form that is highly resistant to intestinal hydrolysis. Another type of resistant starch is formed when starchy materials are heated in the absence of abundant water, for example during the manufacture of breakfast cereals. This type also appears during the canning of beans or other legumes. There may be plenty of water in the can—but little penetrates into the beans. These prolonged high-temperature treatments damage the granule structure so that it becomes immune to the effect of digestive enzymes and is very reluctant to gelatinise. It has therefore been suggested that it is classified as an insoluble fibre component, even though it is not easy for analytical procedures to distinguish it from available starch.

There are indications that the bacterial fermentation of some classes of resistant starch, notably RS3, is an important source of butyric acid in the colon. Over 20% of dietary resistant starch may be fermented, whereas the corresponding figure for non-starch polysaccharides, NSP, dealt with later in this chapter, is less than 15%. Butyric acid is now recognised as having a role to play in reducing the risk of colon cancer and other disorders. The fact that over 10% of the total starch in canned beans, and 2% of that in bread, comes into the resistant category suggests that calculation of a food's nutritional value based on chemical composition needs in some cases to be viewed with caution. Further nutritional

Table 3.2 The classification of food starches on the basis of digestibility

A. Proportions of total resistant, *i.e.* RS1 + RS2 + RS3. (From: H. N. Englyst, S. M. Kingman and J. H. Cummings, *Euro. J. Clin. Nutr.*, 1992, **46**, p. S33.).

Food	Total starch % dry matter	Resistant starch % total starch	Food	Total starch % dry matter	Resistant starch % total starch
White bread	77	1.2	Potato, boiled, hot	74	6.8
Wholemeal bread	60	1.7	Potato, boiled, cold	75	13.3
Corn flakes	78	3.8	Spaghetti, freshly cooked	79	6.3
Porridge oats	65	3.1	Peas, cooked	20	25
Rye crisp bread	61	4.9	Haricot beans, cooked	45	40

B. Detailed subdivision of starch types in a limited range of foods. The expressions RDS, SDS, *etc.*, are explained in the text. (From: British Nutrition Foundation, *Complex Carbohydrates in Food: Report of the British Nutrition Foundation Task Force*, Chapman and Hall, London, 1990.)

Food	RDS	SDS	Starch type (%) RS1	RS2	RS3
White flour	38	59	–	3	trace
Spaghetti	55	36	8	–	1
Shortbread	56	43	–	–	trace
Canned chick peas	56	24	5	–	14
Canned red kidney beans	60	25	–	–	15
Wholemeal bread	90	8	–	–	2
White bread	94	4	–	–	2

implications of resistant starch are discussed under the heading Glycaemic Index in the Special Topics section at the end of this chapter.

PECTINS

Pectins are the major constituents of the primary cell walls of plant tissues and also the middle lamella. The middle lamella is the material which lies outside the cell walls between cells and provides the packing material which helps to bind the cells together. In these sites pectins comprise a substantial proportion of the structural material of soft tissues such as the parenchyma of soft fruit and fleshy roots. As the structure of pectins can only be explored after they have been separated from the rest of the plant tissue, we cannot be certain as to the exact nature of pectins as they occur in plant tissues, and the term protopectin is applied to the water-insoluble material in the plant which yields water-soluble pectin on treatment with weak acid. Most elementary accounts of pectins describe them as linear polymers of α-D-galacturonic acid (2.18), homogalacturonan,[v] in which a proportion of the carboxyl groups are esterified with methanol. The proportion of carboxyl groups which are esterified varies between plant species from around 60%, *e.g.* those from apple pulp and citrus peel, to around 10%, *e.g.* that from strawberries. The molecular weight of pectin molecules can best be described as uncertain but high, probably around 100 000. During recent years, many studies of pectins have concentrated on commercially significant pectin sources such as citrus peels and apple pomace (*i.e.* the pulpy residue after the juice has been pressed out) and a much more complex picture has emerged, especially with regard to the occurrence of neutral sugars and complex branching chain structures. Nevertheless, the homogalacturonan chains illustrated in Figure 3.4 are still regarded as the dominant structural feature, constituting some 65–80% of the total pectin fraction.

Although this simple structure was always recognised by carbohydrate chemists to be an over-simplification, it is only in recent years that food scientists have begun to appreciate that the behaviour of pectins in food systems such as jam and confectionery can be properly understood only if minor components and structural features are also taken into consideration.

The importance of pectin in food products such as jam and other fruit preserves is its ability to form gels. Gels consist largely of water, but are

---

[v] Terms such as galacturonan, galactans and arabinan describe polysaccharide chains consisting, respectively, of galacturonic acid, galactose and arabinose. The "an" suffix is derived from "anhydro", indicating that the links between the monosaccharide units are formed by loss of the elements of water.

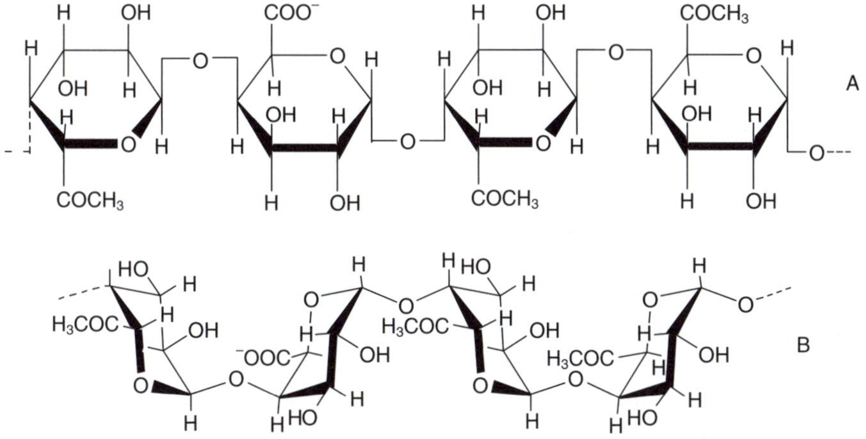

Figure 3.4 The polygalacturonan backbone of pectin. (A) shows four galacturonic acid units, three of which are methylated, drawn according to the Haworth convention. (B) shows the same structure with the galacturonate units drawn in their chair configuration to illustrate the zigzag shape of the chain.

stable and able to retain the shape given to them by a mould. We have already seen that sugar molecules, and by inference polysaccharides also, are effective at binding water. However, a stable gel also requires a three-dimensional network of polymer chains with water, plus solutes and suspended solid particles, held in its interstices. It has often been assumed that simple interactions such as single hydrogen bonds between neighbouring chains were all that was required to maintain the stability of the network. The fact that the lifetime of an isolated hydrogen bond in an aqueous environment is only a tiny fraction of a second was quite overlooked. It is obvious that attractive forces of this character provide an adequate explanation for the viscosity of the solutions of many polymers, such as the polysaccharide gums, but do not explain gel structure.

The concept of the junction zone was first introduced by Rees to explain the structure of these gels. His concept initially concerned gels formed by carageenans, a group of polysaccharides considered later in this chapter, but it can be applied to most gel-forming polymers, including the pectins. Although single hydrogen bonds are not sufficiently durable to provide a stable network, a multiplicity of hydrogen bonds or other weak forces acting cooperatively, is the basis for the stability of crystal structures. Rees showed that polymer chains in a gel interact in regions of ordered structure, involving considerable lengths of the chains. These regions, the junction zones, are essentially crystal-line in nature, and details of their structure have been revealed by X-ray crystallography.

The detailed structure of the junction zone is dependent on the nature of the polymer involved. In agar and carageenan gels the interacting chains form double helices which further interact to form "super-junctions". The details of pectin junction zones are less clear, but it is accepted that interacting homogalacturonan chains pack closely toge-ther in a crystal-like array. However, our understanding of gel structure requires more than merely the ability of the polymer chains to pack together. Other structures are required to disrupt the associations between chains, since without them all the chains would simply pack into a single junction and precipitate out of solution, and a gel would not be formed. Two types of disruptive structures are recognised. One of these is referred to as an *anomalous residue*.[vi] This comprises individual monosaccharide units with different linkages and/or other structural features which occur occasionally in the chain. The disruptive effect of a rhamnose residue inserted into a sequence of galacturonate residues is shown in Figure 3.5.

The second type of disruptive structure is due to branching in the polysaccharide chain. Even a single monosaccharide residue attached to an otherwise regular backbone would make the formation of a junction zone unlikely—and a long branch chain, like the ones already encountered in amylopectin, would make junction zone formation impossible. The picturesque terms, *smooth* and *hairy*, have been adopted to describe, respectively, uninterrupted regular regions of poly-saccharide chains and regions rich in branches. The potential effects of anomalous residues and branches on junction zone formation are illu-strated in Figure 3.6.

It is now possible to understand the contribution made by the hairy regions of the chain to the formation of gels. Without them there would be nothing to prevent interaction taking place along the entire length of the chains, resulting in precipitation. The polymer chains linked at the junction zones provide a stable three-dimensional network in which water, solutes and suspended particles are bound or entrapped by the branched or irregular polymer chains. The retrogradation of the amy-lose component of starch pastes (*see* page 56) is a perfect example of what happens when unbranched uninterrupted chains are able to interact.

[vi] In this context the terms residue and unit both refer to a monosaccharide molecule which is a component of a polysaccharide or oligosaccharide chain.

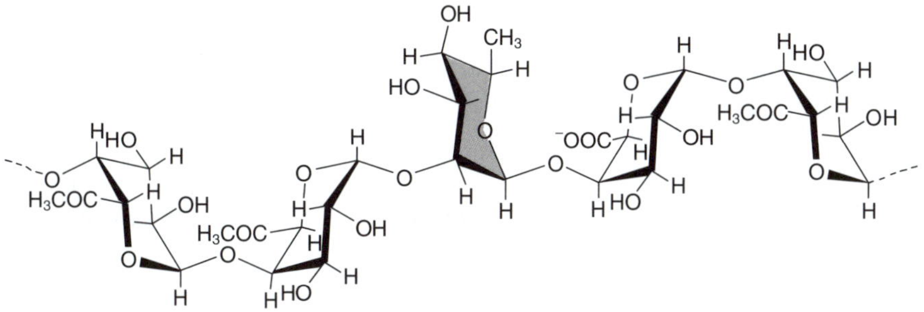

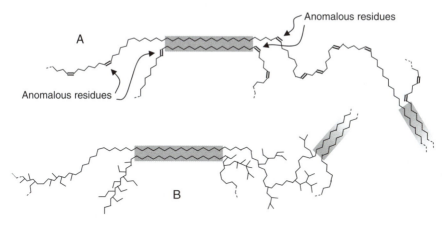

Figure 3.5 The insertion of an L-rhamnose residue (shaded) into a sequence of galacturonate residues:

[... →4)-GalA-α-(1→4)-GalA-α-(1→2)-Rha-α-(1→4)-GalA-α-(1→4)-GalA-α-(1→ ...]

can be seen to disrupt the regular zigzag pattern. As drawn here three of the galacturonate residues are esterified with methanol. The terms Rha and GalA are the conventional abbreviations for rhamnose and galacturonate, respectively.

Figure 3.6 The effect of (*A*) the insertion of anomalous residues, and (*B*) the attachment of side-chains in otherwise homogenous polymer chains on the formation of junction zones (shaded).

As many as 20% (but usually rather less) of the components of a pectin may be found to be neutral sugars, including L-rhamnose (2.20), D-glucose, L-arabinose (3.2), D-xylose (3.3) and D-galactose (3.4). L-Rhamnose residues are found in all pectins.

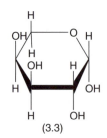

(3.2)   (3.3)   (3.4)

Particular neutral sugars tend to occur in particular situations such as in specialised chain structures. For example, rhamnose residues are most often found in rhamnogalacturonan chains, *i.e.* chains of alternating rhamnose and galacturonic acid residues (3.5).

[... →4)-α-GalA-(1 →4)-α-GalA-(1 →2)-α-Rha-(1 →4)-α-GalA-(1 →4)-α-GalA-(1 →....]

(3.5)

Arabinose units are found in highly branched chains which have an arabinan backbone carrying further short, and often themselves branched, arabinan chains linked to the 5 position of the backbone arabinose units (3.6).

[....→3)-α-Ara–(1→3)-α-Ara–(1→3)-α-Ara–(1→3)-α-Ara–(1→3)-α-Ara–(1→3)-α-Ara–(1→3)-α-Ara–

Ara–(1→3)-α-Ara–(1→3)-α-Ara–(1→5)α

Ara-(1→3)-α-Ara-(1→5)-α

α-(5←1)-α-Ara–(3←1)-α-Ara

(3.6)

More details of these and other pectin components are available in the Special Topics section at the end of the chapter.

It may have been inferred from the discussion so far that pectins form gels quite readily, but anyone who has been involved in jam making either at home or commercially will know how untrue this is. The presence of pectin, solubilised when the fruit is boiled, is not enough in itself. For successful jam making we must meet other conditions as well. Ideally the pectin should be highly esterified, with at least 70% of the carboxyl groups methylated, and present at a concentration of at least 1% w/v. The acid arising from the fruit, or added separately, must maintain the pH below 3.5 and ideally close to 3.0. Citric acid or malic acid, each of which occur naturally in many fruits (*see* Chapter 7), are the acidifying agents normally used. The total sugar content must be at least 50% w/v. While the relationships between these requirements and the procedures of jam making can be readily discerned, the reasons behind them are less obvious. In fact, pectin is a reluctant participant in gel formation. The successful formation of junction zones requires that the affinity of the pectin chains for each other must be maximised. Keeping the pH low has two valuable effects. Firstly, at low pH the concentration of hydrogen ions is high, which tends to repress the ionisation of the galacturonate carboxyl groups:

$$-COO^- + H^+ \leftrightarrow -COOH$$

This means that the tendency of the negatively charged COO^- groups to repel each other is eliminated. Furthermore, any charged groups such as COO^- tend to attract clusters of water molecules which would prevent the pectin chains getting close to each other. The methyl esterification of the carboxyl groups has the same beneficial effect in preventing ionisation and similarly encourages junction zone formation. The high concentration of sugar molecules also competing for water has the same beneficial effect on gel formation.

During the boiling phase of jam making a considerable proportion of the sucrose (both added and from the fruit) is hydrolysed to its parent monosaccharides, glucose and fructose (*see* Chapter 2). This increases the total number of bound water molecules, *i.e.* those not available for binding to the pectin chains, thereby enhancing the process of gel formation.

It is usual for commercial jam manufacture to require the addition of extra pectin to improve the gel strength. This pectin will normally have been extracted from apple pomace or citrus peel and treated to modify its properties to suit particular applications. So-called high-methoxy pectins are 50–75% esterified, and low-methoxy pectins are 25–50% esterified. The amount of pectin added varies with the proportion and type of fruit being used. For example, strawberries and raspberries, which are low in pectin, require higher levels of added pectin (approximately 0.1 to 0.4%, depending on the fruit content of the jam) than medium pectin fruits such as oranges (*i.e.* in marmalade) and apricots, or high pectin fruits such as apples or gooseberries, which require only 0.2% or none at all. Commercially pectins having a range of degrees of esterification (*i.e.* the proportion of galacturonate residues esterified with methanol) are available. Rapid-set pectins, around 70% esterified, are used in jam making. Slow-set pectins, with around 60% esterification, are used in confectionery. Low-methoxy pectins only form gels in the presence of calcium ions, either added as calcium salts or provided by the calcium phosphate component of milk (*see* Chapter 4). The junction zones formed closely resemble the so-called "egg-box" structure, considered in more detail when the gel forming capabilities of alginates are discussed later in this chapter. Amidated pectins, in which the galacturonate carboxyl group is converted to an amide ($-CONH_2$), are manufactured by treatment of high-methoxy pectins with ammonia. They are sometimes used in the formulation of milk-based dessert products.

The breakdown of pectins is an essential feature of fruit ripening and the enzymes involved occur widely in plant tissues, as well as being secreted by the fungi and bacteria which specialise in attacking and breaking down plant tissues. There are three important classes of enzymes (pectinases) which catalyse pectin breakdown:

- Pectin methyl esterases, which remove the methanol residues from methylated galacturonate units.
- Polygalacturonases, which hydrolyse the glycosidic links between unmethylated galacturonate units. There are two types of these enzymes. Exo-polygalacturonases release small fragments from the non-reducing end of the chain, whereas endo-polygalacturonases hydrolyse linkages within the chain and generate short fragments. Polygalacturonases only become active following the action of pectin methyl esterases, since they cannot work when the degree of methylation of the pectin molecule as a whole is higher than around 60–70%.
- Pectin lyases, which break the glycosidic links between methylated galacturonate units by an unusual β-elimination reaction in which a *trans* double bond is inserted between C-4 and C-5, rather than the usual hydrolysis (Figure 3.7).

As ripening takes place, these enzymes will break down the pectin surrounding the cells of the fruit tissue, causing it to soften and become more palatable. Of course, there are many other changes taking place at the same time, including the formation of sugars, pigments and flavour compounds.

Pectinase preparations containing these same enzymes, and others which attack the pectin side-chains, are manufactured commercially using cultures of bacteria and fungi isolated originally from rotting fruit and vegetables.

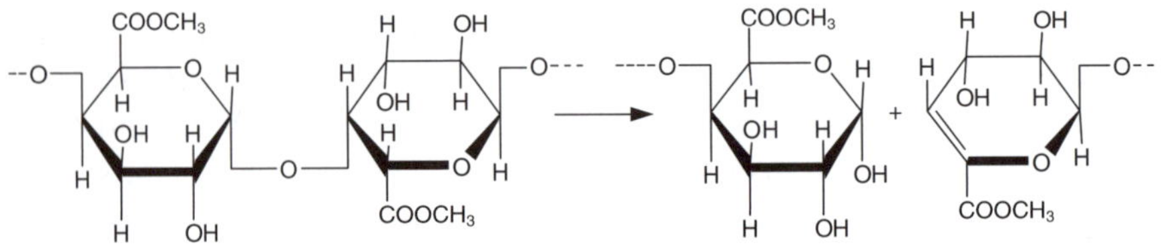

Figure 3.7 The β-elimination reaction catalysed by pectin lyase.

They are extensively used in wine and cider making and the fruit and vegetable juice industry to increase the degree of tissue damage, and therefore the juice yield, when the juice is extracted by pressing. They are also widely used to clarify fruit juices, cider and wines, since the large polygalacturonate molecules are very effective at binding protein particles and other cell debris, thereby producing an unwanted haze. Of course consumers do not expect all fruit juices to be clear, particularly tomato juice. Before tomatoes are fully ripe most of their pectinases remain within the cell membrane, kept away from the pectins of the middle lamella which holds the cells together. When the fruit is crushed the cell membranes are disrupted and the pectinases are able to reach their substrate. In the case of the tomato the ensuing breakdown is extremely rapid. Tomato juice is therefore made in a "hot break" process in which the tomatoes are rapidly heated either before or even during the crushing, in order to inactivate the enzymes.

Pectinases can be used for stripping off the albedo, or pith, of citrus segments before canning, but a curious variation of this process is to introduce the enzyme under the peel. The peel is punctured and the whole fruit is then placed in a sealed container from which the air is pumped out. As a result the air in the space under the peel is sucked out, so that when the fruit is then transferred to a bath of enzyme solution at atmospheric pressure the enzyme solution is sucked in. An hour or so later the residues of the peel can simply be washed off. Further enzyme treatment then separates the fruit into segments which, after another good wash, are ready for canning. With fruit such as peaches or apricots, pectinase treatments are replacing the traditional "lye peeling", in which the peel is dissolved away by immersion in very strong sodium hydroxide solution (lye).

SEAWEED POLYSACCHARIDES

As we have just seen, the pectins have a particular place in the structure of the tissues of higher plants. In the seaweeds their place is taken either by the alginates (in the brown algae, the *Pheophyceae*) or the agars and carageenans (in the red algae, the *Rhodophyceae*).

The alginates are linear polymers of two different monosaccharide units: β-D-mannuronic acid (M) (3.7a) and α-L-guluronic acid (G) (3.8a). When their structures are drawn using the Haworth convention (3.7b and 3.8b), these two seemingly rather different monosaccharides are seen in fact to differ only in their configuration at C-5, *i.e.* they are C-5 epimers of one other. This difference leads to the different pyranose ring conformations that are taken up, *C1* by mannuronic acid and *1C* by guluronic acid, the conformations which allow the bulky carboxyl groups to occupy equatorial positions. Both monosaccharide units occur together in the same chains, but linked in three different types of sequence:

$$-M-M-M-M-M- \quad -G-G-G-G-G-$$
$$-M-G-M-G-M-G-$$

in each case by 1,4-links.

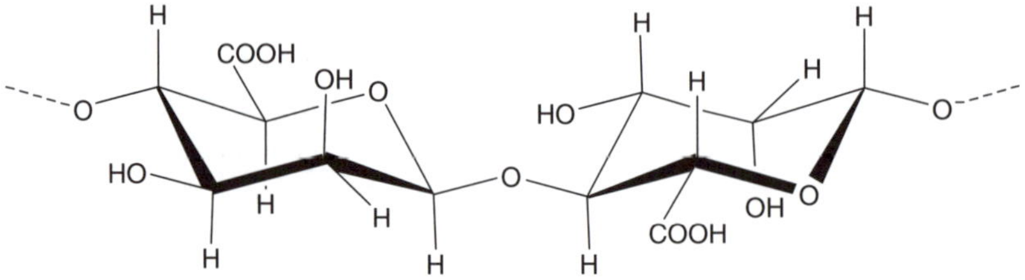

(3.7a)

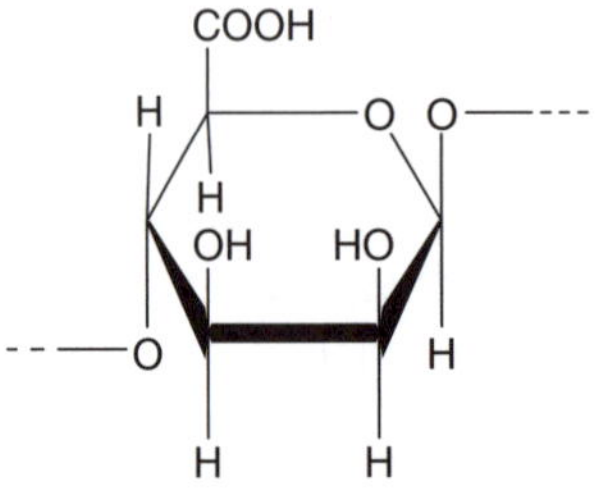

(3.7b)

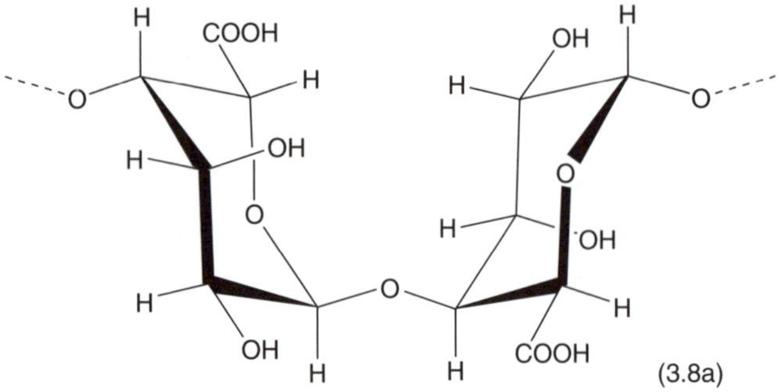

(3.8a)

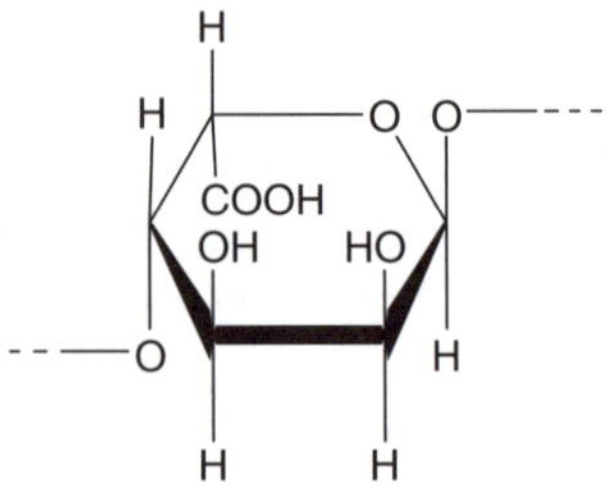

(3.8b)

Alginates from different species, and even from different parts of the same plant, differ in the ratios of the two components, as illustrated by

Table 3.3 The proportions of guluronate and mannuronate residues in different seaweed species and tissues (from K. I. Draget, in *Handbook of Hydrocolloids, see* Further Reading for details).

Species	Part	Guluronate : mannuronate ratio
Laminaria japonica	unspecified	35 : 65
L. digitata	unspecified	41 : 59
L. hyperborea	leaf (frond)	55 : 45
L. hyperborea	stem (stipe)	68 : 32
Macrocystis pyrifera	unspecified	39 : 61
Durvillea antarctica	unspecified	29 : 71

the data in Table 3.3. Alternating –G–M–G–M–G–M– sequences contribute approximately 10–25% of the total.

Alginic acid itself is insoluble, but its alkali metal salts are freely soluble in water. Gels readily form in the presence of Ca^{2+} ions. The junction zones only involve –G–G–G–G– sequences, in an egg-box structure in which a Ca^{2+} ion is complexed by four L-guluronate units, *see* Figure 3.8. As the proportion of guluronate residues rises, and the corresponding proportion of –G–G–G–G– sequences increases, the gel strength also rises.

Calcium alginate gels do not melt at a temperature below the boiling point of water, and this property gives them a number of applications in food products. If a fruit purée is mixed with sodium alginate and then treated with a calcium-containing solution, various forms of reconstituted fruit can be obtained. For example, if a cherry/alginate purée is added as large droplets to a solution of calcium ions the familiar synthetic cherries popular in the bakery trade can be made. A fairly convincing skin forms on the outside, where the alginate is exposed to the highest concentration of calcium ions. Diffusion of calcium ions through this skin gives lower concentrations inside, forming a softer jelly-like consistency. Rapid mixing of a fruit/alginate purée with a sparingly soluble calcium salt, such as calcium carbonate or sulfate, followed by appropriate moulding, gives the apple or apricot "pieces" sometimes sold as cheap pie fillings.

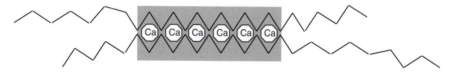

Figure 3.8 The junction zone of an alginate gel. The junction zone (shaded) consists of two strands of contiguous guluronate units terminated by sequences containing mannuronate residues.

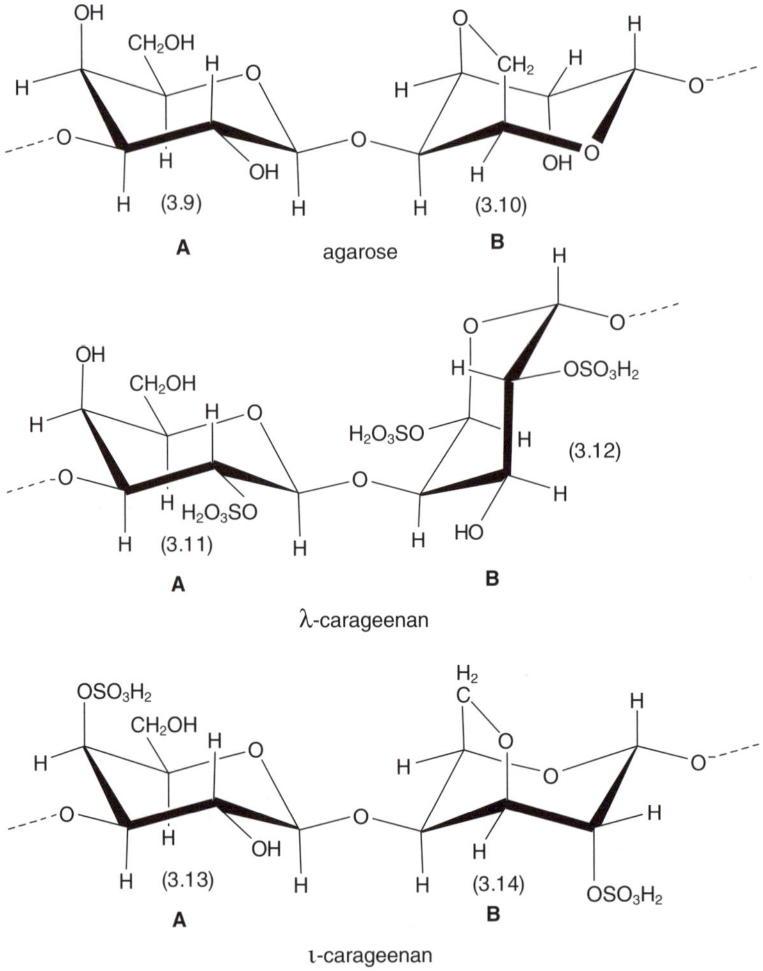

Figure 3.9   The repeating units of agar and carageenan. The role of the anhydro
                  bridge in maintaining the B residue in the unfavourable conformation (in
                  terms of the axial position of sulfate groups) is clear in these structural
                  formulae.

The agars and carageenans[vii] are a more complex group than the
alginates. Most are linear chains of galactose derivatives, with the following generalised structure:

$$\rightarrow 3)A\beta(1\rightarrow 4)B\alpha(1\rightarrow 3)A\beta(1\rightarrow 4)B\alpha(1\rightarrow 3)A\beta(1\rightarrow 4)B\alpha(1\rightarrow 3)A\beta(1\rightarrow 4)B\alpha(1\rightarrow$$

[vii] The red seaweed *Chondrus crispus* grows on the Irish coast, where it is known as Irish moss.
Carraigin is Gaelic for "rock moss".

Table 3.4 The monosaccharides of agar and carageenans. The structures of some of the monosaccharide units listed here are shown in Figure 3.9. There is a steady increase in the sulfate content as one goes down the list from agarose to μ-carageenan, matched by a steady decrease in gelling power; λ- and μ-carageenans do not gel at all.

		Position		
		A	B	
			Major	Minor
Agarose		I	IV	
κ-Carageenan		III	V	VI–VIII
ι-Carageenan		III	VI	VIII
λ-Carageenan		I, II	VIII	
μ-Carageenan		III	VII	V, VIII
I	β-D-Galactose			(3.9)
II	β-D-Galactose 2-sulfate			(3.11)
III	β-D-Galactose 4-sulfate			(3.13)
IV	3,6-Anhydro-α-L-galactose			(3.10)
V	3,6-Anhydro-α-D-galactose			
VI	3,6-Anhydro-α-D-galactose 2-sulfate			(3.14)
VII	α-D-Galactose 6-sulfate			
VIII	α-D-Galactose 2,6-disulfate			(3.12)

where A and B are residues from two different groups, as indicated in Table 3.4. Different monosaccharides predominate in particular carageenans, but the distinction between different carageenans in particular seaweeds is not always clear-cut. Commercial carageenan preparations sold for food use as gelling or thickening agents are blends designed to have properties appropriate to a particular application.

The junction zones which these polysaccharides form are quite unlike the simple egg-boxes we have met so far. Rees's group showed that gelation first involves the formation of double helices of galactan chains wound around each other. The formation of these helices is not sufficient in itself for the gel to form. This occurs only when the rod-like structures associate together to form superjunctions, as shown in Figure 3.10. Examination of carageenans by X-ray crystallography has shown that it is the intervention in the B-position of residues lacking the 3,6-anhydro bridge (VII and VIII) which disrupts the initial helix formation. As the sulfate content rises, the resulting negative charge on the helix requires the presence of neutralising cations such as K^+ (for κ-carageenan) or Ca^{2+} (for ι-carageenan) if the helices are to associate into superjunctions. Agar, with its very low sulfate content, gels well regardless of the cations present. The agar superjunctions are sufficiently large to

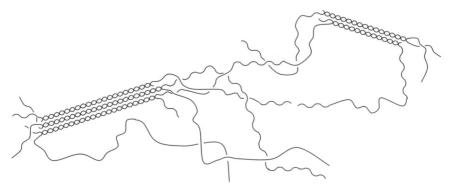

Figure 3.10   Carageenan or agar superjunctions.

refract light and are the source of the slight opalescence visible in even the purest agar gels.

Although agar has been used extensively in microbiology for a very long time, the development of food applications for this group of polysaccharides has been slow. They are beginning to be used both as gelling agents in milk-containing desserts and as stabilisers in a wide range of products. Concerns over their safety as food additives based on suggestions that carageenans could cause intestinal irritation have not been sustained, owing largely to the apparent freedom from this problem in Japan, where seaweeds form an important part of the diet. There are studies which suggest that carageenans degraded by high temperatures in acid conditions are potential carcinogens, and these are not permitted for food use.

CELLULOSE, HEMICELLULOSES AND FIBRE

Cellulose and hemicelluloses are major components of the fraction of plant foods, particularly cereals, commonly known as dietary fibre, or "roughage". The definition of these two terms is rather more problematical than for most food substances, as we are dealing with a group without a clear boundary either in terms of their chemistry or their contribution to human nutrition and their general association with the structure of plant cell walls. It would be useful to attempt to reach a proper understanding of these terms, and some of the others used in this area, before going on to describe potential members of the group in detail.

Besides cellulose and the hemicelluloses, all of the following substances could be described as fibre: lignin, gums, seaweed polysaccharides, pectins, resistant starch and inulin. What these have in common is that they are not broken down in the small intestines of mammals by digestive enzymes. However, most are at least to some degree broken down by the bacteria resident in the large intestine of omnivores such as humans. The colon bacteria go on to ferment the resultant sugars to provide energy for their growth and proliferation. The various fermentation processes give rise to a number of end-products. Besides the hydrogen, carbon dioxide and methane already mentioned on page 26, considerable quantities of short-chain fatty acids, notably ethanoic, propionic and butyric acids, are produced. Humans obtain little or no energy from them, so the term "unavailable carbohydrate" has often been applied, but this expression has its drawbacks. For example, stachyose (*see* page 25) is certainly an unavailable carbohydrate, but since it is a sugar it hardly qualifies. Conversely, lignin (*see* page 76) should really be included, although it is certainly not a carbohydrate.

The term fibre, like the older term roughage, implies insolubility, but the substances listed above range from the totally insoluble cellulose to the totally soluble gums. To the analyst, fibre was traditionally the insoluble residue remaining after a defatted plant foodstuff had been extracted, first with boiling dilute acid and then with boiling dilute alkali, a treatment which solubilises almost all plant substances apart from cellulose and lignin. Nowadays the extractions are carried out using enzymes. Amylases and proteases are used to break down starch and proteins, followed by extraction with 76–80% ethanol. This dissolves the amino acids and glucose derived from the proteins and starch, respectively, as well as removing other low molecular weight substances. The resulting insoluble residue is considered to be close in composition to the unabsorbed residue that passes from the small intestine into the colon during human digestion.

Food composition tables, such as those given in *McCance and Widdowson*, usually quote data obtained by one of two methods. In the slightly earlier one, the Southgate method, the undissolved residue from the enzymic digest is simply dried and weighed. In the more recent method developed by Englyst and Cummings (usually known simply as the Englyst method), the residue is subjected to acid digestion and the resulting monosaccharides are identified and quantified by GLC or HPLC), and related back to their parent polysaccharides. The Englyst method gives a result for what is referred to as the non-starch polysaccharide (NSP), and this method and terminology was officially adopted in Britain for nutritional data. Unfortunately, and with hindsight unsurprisingly, this terminology was never recognised by consumers as indicating what had been understood as dietary fibre. Since 1999 Britain has changed its position and has joined the rest of Europe in adopting the Association of Official Analytical Chemists

International (AOAC International) methodology as the preferred method. This is an updated and refined version of the Southgate method, corresponding to the now widely accepted definition of dietary fibre developed by the American Association of Cereal Chemists (AACC) in 2000:

> *"Dietary fiber is the edible parts of plants or analogous carbohydrates that are resistant to digestion and absorption in the human small intestine with complete or partial fermentation in the large intestine. Dietary fiber includes polysaccharides, oligosaccharides, lignin, and associated plant substances. Dietary fibers promote beneficial physiological effects including laxation, and/or blood cholesterol attenuation, and/or blood glucose attenuation."*

The notable feature of this definition is that, while it ignores the issue of laboratory analysis, it includes a statement on the physiological activity of dietary fibre in humans.

Table 3.5 summarises the relationships between these different approaches. In 2007 the Codex Alimentarius Commission, a committee operating under the aegis of the Food and Agriculture Organization (FAO) of the UN and the WHO, published draft guidelines for the application of this definition in food labelling. At the present time, however, unless specified otherwise, data on food labels currently refer only to NSP. Some data on the fibre contents of a variety of foodstuffs are presented in Table 3.6.

The broad distinction between soluble and insoluble fibre extends beyond its considerable nutritional importance. Insoluble fibre is the subject of the present section; one form of soluble fibre, the pectins, has already been considered but another, the gums, and the dietary role of soluble fibre generally, will be given further attention in the next section.

The insoluble fibre polysaccharides are subdivided into totally insoluble cellulose and the sparingly soluble hemicelluloses. Cellulose, said to be the most abundant organic chemical on Earth, is an essential component of all plant cell walls. It consists of linear molecules of at least 3000 $\beta 1 \rightarrow 4$ linked glucopyranose units. This linkage leads to a flat-ribbon arrangement maintained by intramolecular hydrogen bonding along the chain, as shown in Figure 3.11. Within plant tissues, cellulose molecules are aligned to form microfibrils several μm in length and about 10–20 nm in diameter. Although doubt still exists as to the exact arrangement of the molecules within the microfibrils, it is clear that the chains pack together in a highly ordered manner maintained by hydrogen bonding between the chains. It is the stability of this ordered

Table 3.5 The components of the different fibre classes. The descriptions "soluble fibre" and "insoluble fibre" refer to the earlier usage of these terms rather than suggesting that the fibre is or is not actually soluble. The soluble non-cellulosic polysaccharides include pectins, gums and seaweed polysaccharides.

	Lignin	*Cellulose (insoluble)*	*Hemicelluloses*	*Soluble noncellulosic polysaccharides*	*Resistant starch*
Southgate	+	+	+	+	+
Englyst (NSP)	–	+	+	+	–
Insoluble fibre	–	+	+	–	–
Soluble fibre	–	–	–	+	–
AACC	+	+	+	+	+

Table 3.6 The fibre components of different foods[a]

			g/100 g		
	Cellulose	*Other insoluble fibre[b]*	*Soluble fibre[c]*	*Lignin*	*Total*
Wheat flour:					
wholemeal	1.6	6.5	2.8	2.0	12.9
white	0.1	1.7	1.8	trace	3.6
Oatmeal	0.7	2.3	4.8	~2.0	9.8
Broccoli	1.0	0.5	1.5	trace	3.0
Cabbage	1.1	0.5	1.5	0.3	3.4
Peas	1.7	0.4	0.8	0.2	3.1
Potato	0.4	0.1	0.7	trace	1.2
Apple	0.7	0.4	0.5	-	1.6
Gooseberry	0.7	0.8	1.4	-	2.9

[a]Data quoted by D. A. T. Southgate in *Determination of Food Carbohydrates*, Elsevier Applied Science, London, 2nd edn, 1991.
[b]Largely hemicellulose.
[c]Including pectin.

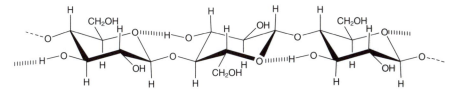

Figure 3.11 The flat-ribbon arrangement of the cellulose chain. Hydrogen bonds are shown between the ring oxygen and the C-4 hydroxyl group of the neighbouring glucose residue.

structure that gives cellulose both its insolubility in virtually all reagents and also the great strength of the microfibrils. Herbivorous animals, particularly ruminants, are able to utilise cellulose by means of the specialised micro-organisms which occupy the highly anaerobic sections of their alimentary tracts, such as the rumen. These secrete cellulolytic enzymes which release glucose; they then obtain energy by fermenting the glucose to short-chain fatty acids, such as butyric acid, which can be absorbed and utilised by the animal. Even in these specialised animals cellulose digestion is a slow process, as evidenced by the disproportionately large abdomens of herbivores compared with carnivores.

The hemicelluloses were originally assumed to be low molecular weight, and therefore more soluble, precursors of cellulose. Although this idea has long ago been discarded, we are left with a most unfortunate name to describe this group of structural polysaccharides. They are closely associated with the cellulose of plant cell walls, from which they can be extracted with alkaline solutions (*e.g.* 15% KOH). Three types of hemicellulose are recognised: firstly the xylans, secondly the mannans and glucomannans, and finally the galactans and arabinogalactans. Although the leaves and stems of vegetables provide a high proportion of the indigestible polysaccharide in our diet, very little is known about the hemicelluloses of such tissues. We know much more about the hemicelluloses of cereal grains, in which the xylans are the dominant group. The essential feature of these xylans is a linear or occasionally branched backbone of $\beta 1 \rightarrow 4$ linked xylopyranose residues with single L-arabinofuranose and D-glucopyranosyluronic acid residues attached to the chain, as indicated in Figure 3.12. There is every reason to suppose that the conformation of the xylan backbone resembles the flat ribbon of cellulose, but of course the side-chains will prevent tight packing and thereby allow some solubility.

Xylans are major constituents of the seed coats of cereal grains, which on milling constitute the bran. A typical wheat bran composition (on a dry basis) is: lignin[viii] 8%, cellulose 30%, hemicellulose 25%, starch 10%, sugars 5%, protein 15%, lipid 5%, with inorganic and other substances making up the remainder. Much less hemicellulose (about 5%) is found in the endosperm, which is the basis of white flour. Nevertheless, various effects of the hemicelluloses on the milling and baking characteristics of particular flours have been noted.

[viii] Lignin is the characteristic polymer of woody plant tissues. The structure is extremely complex and variable, based on the condensation of a variety of aromatic aldehydes such as vanillin (3.15), syringaldehyde (3.16) and *p*-hydroxybenzaldehyde (3.17).

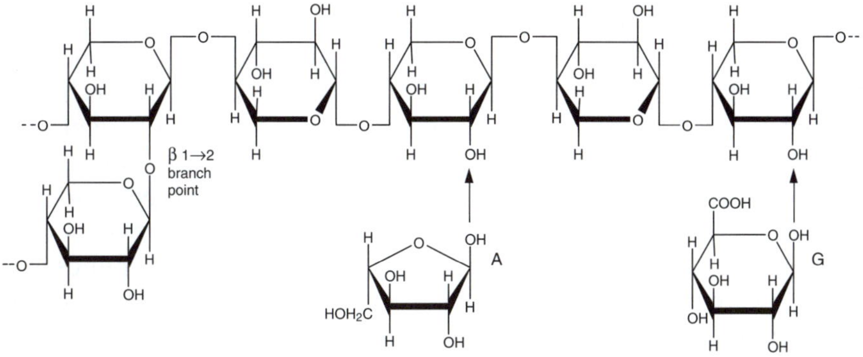

Figure 3.12 Typical structural features of a wheat xylan. The more soluble endo-sperm xylans are richer in L-arabinose, whereas bran xylans are less soluble, more highly branched, and richer in glucuronic acid. The arrows show the points of attachment of any L-arabinose (*A*) or glucuronic acid (*G*) residues.

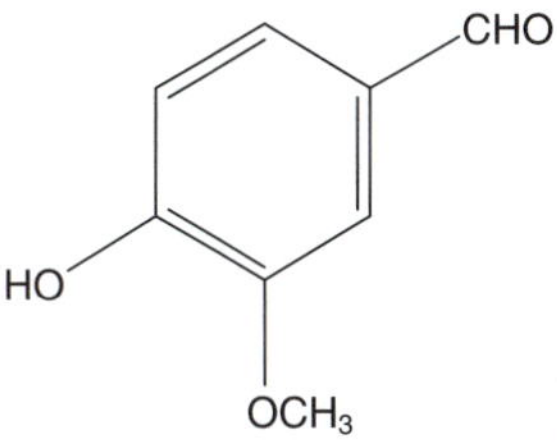

(3.15)

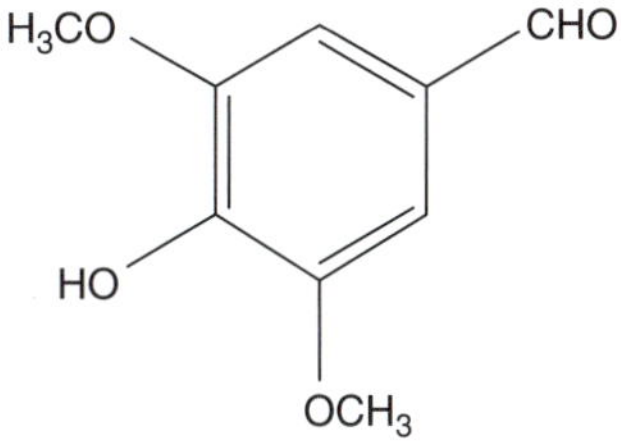

(3.16)

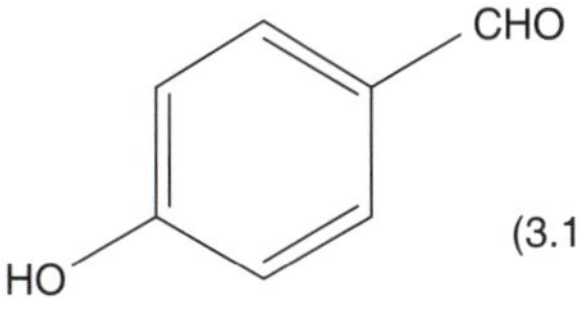

(3.17)

The nutritional benefits of high-fibre diets are strongly promoted in relation to many of the diseases which characterise modern urban man and his highly refined food. There are two quite distinct areas of human health where fibre is believed to exert a beneficial influence. The benefits of a high-fibre diet to the functions of the large intestine have been recognised for a long time. Significantly lower incidence of a number of bowel disorders, ranging from constipation and diverticulosis to cancer, is associated with high-fibre diets. These benefits appear to be derived from the greater bulk which fibre brings to the bowel contents. Besides improving the musculature of the bowel wall, this also reduces the time that potential carcinogens derived from our food spend in the bowel. Recent research suggests that a quite different mechanism may also be important. Some fibre components are broken down and fermented by the bacteria resident in the bowel to give rise to high concentrations of the short-chain fatty acids, ethanoic, propionic and butyric acids. Butyric acid has been shown to induce the death of bowel tumour cells in a process known as apoptosis. The possible role of fibre intake in the incidence of other types of cancer is a topic of ongoing, but inconclusive, interest. Interpretation of the results of epidemiological surveys is complicated by the fact that diets which are rich in fibre will almost always include large amounts of fruits and vegetables. It is obviously possible that the health benefits are due to other fruit and vegetable constituents rather than the fibre polysaccharides. Separating the two factors in epidemiological studies is extremely difficult.

It is becoming clear that the relationship between dietary fibre intake and the incidence of arterial disease is complex, and that not all forms of fibre are equally effective in lowering levels of different types of blood lipids. However, what does seem to be agreed is that it is soluble rather than insoluble fibre which is the more effective, and this aspect will therefore be included in the next section rather than here.

At the boundary between hemicelluloses and gums are the β-glucans, of which the β-glucan in oats is the best known. Oat glucan consists of long chains of glucose units (at least 5000) joined by both β1→3 and β1→4 glycosidic links As Figure 3.13 shows, the chain is dominated by cellulose-like β1→4 links, but interrupted every third or fourth glucose unit by a β1→3 linkage. The latter has the effect of disrupting the characteristic ordered cellulose-like structure and imparting solubility in water. Solutions of oat β-glucan are extremely viscous, as are rolled oats cooked to give porridge. This high viscosity, and the related property of water binding, has led to cereal glucans being investigated as a potential food

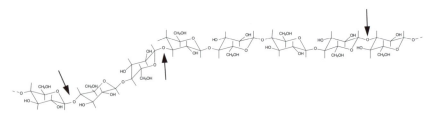

Figure 3.13   Oat β-glucan. The arrows indicate the β1→3 links, the remainder being β1→4.

ingredient, since it can stabilise oil-in-water emulsions (*see* Chapter 4) and, for example, reduce water loss from reduced-fat sausages (*see* Chapter 5). Other recent applications include its use in non-dairy lactose-free products such as non-dairy ice cream, yoghurt and cream.

There are strong indications that oat β-glucans are beneficial in relation to heart disease by lowering the level of cholesterol in the blood. The most likely mechanism is that their viscous character impedes the reabsorption of cholesterol and bile acids from the small intestine, thereby placing a drain on the body's cholesterol economy. The encapsulation of the porridge's starch by the oat cell walls and its consequently low Glycaemic Index (*see* pages 88–91) may also be a factor in its health benefits.

One should also be cautious of automatically regarding a high dietary fibre level as beneficial. Phytic acid, inositol hexaphosphate (3.18), is an important component of bran, which is a valued source of dietary fibre. Phytic acid is the phosphate reserve for germinating seeds, and its phosphate groups make it a potent complexer of divalent cations, implicating it in calcium- and zinc-deficiency diseases in those children who consume very little milk and large quantities of baked goods made from wholemeal flour.

$$H_2O_3PO \qquad OPO_3H_2$$

$$H_2O_3PO \qquad OPO_3H_2$$

$$OPO_3H_2$$

(3.18)

Inulin is another polysaccharide, and one which could be classified as a gum. It is the most important member of a group of plant poly-saccharides known as fructans, which are polymers of fructose. Although fructans occur widely in small amounts in leaves, inulin, whose structure is shown in Figure 3.14, is notable for its starch-like role in terms of plant physiology, in the tuberous roots of a number of food plants, especially chicory (*Cichorium intybus*) and Jerusalem artichokes (*Helianthus tuberosus*). These each contain around 18 g per 100 g fresh weight. Many other food plants, including asparagus, onions, garlic, bananas, and cereal flours such as wheat, contain just a few percent. The dominance of wheat consumption compared with chicory and Jerusalem artichokes naturally means that wheat is the more usual source of

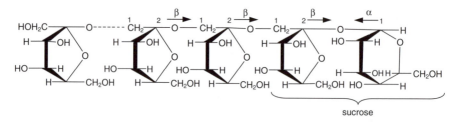

Figure 3.14 The β-(2 → 1) linked fructofuranose chain of inulin. A sucrose residue usually provides a non-reducing terminus for what would otherwise be the reducing end of the chain.

dietary inulin in most diets. Inulin as not a large polymer like starch or cellulose, and only a small proportion of the polymer chains contain more than 50 fructose units. A significant proportion of these are sufficiently short to qualify as oligosaccharides rather than polysaccharides.

Until recently inulin was little more than a phytochemical curiosity, but its consumption has now been shown to have a range of effects similar to those of other soluble fibre components. It resists digestion in the human stomach and small intestine but is readily hydrolysed and fermented by the bacteria of the colon. The fermentation generates considerable amounts of short-chain fatty acids such as ethanoic, propionic and butyric acids and promotes the growth of the bifidobacteria, important components of the colon flora. These bacteria are believed to impart considerable health benefits, but it remains to be seen whether these are sufficient to justify the manipulation of one's diet by consuming as much as 40 g of inulin daily from sources such as chicory.

GUMS

There are many substances in cereal products and vegetables which could properly be described as gums. However, the term is usually reserved for a group of fairly well-defined substances obtained as plant exudates or as "flours" made from certain non-cereal seeds. Two other gums secreted by bacteria, xanthan gum and gellan, have also become important food ingredients in recent years.

The distinctive feature of the gums is their great affinity for water and the high viscosity of their aqueous solutions. However, they do not form gels and their solutions always retain their plasticity, even at quite high concentrations. The reason for this becomes clear when the molecular structures of representative gums are examined. The key structural features of gum tragacanth (an exudate from the tree *Astralagus gummifer*) and guar gum (the storage polysaccharide in the endosperm of the

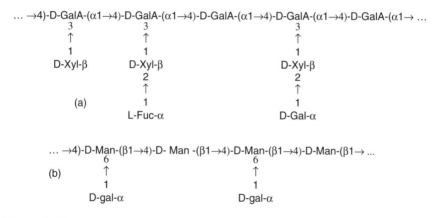

Figure 3.15 (*a*) Gum tragacanth. Shown here is the structure of its principal component, tragacanthic acid. The backbone of D-galacturonic acid residues has occasional insertions of L-rhamnose. Almost all the backbone residues carry one or other of the D-xylose-based side-chains shown. L-Fuc denotes L-fucose, which is 6-deoxy-L-galactose (3.19). (*b*) Guar gum. Alternate backbone residues carry single D-galactose residues. Another galactomannan, locust bean gum (from *Caratonia siligua*), differs from guar gum only in having a lower frequency of D-galactose side-chains.

seeds of a leguminous shrub, *Cyamopsis tetragonoloba*) are illustrated in Figure 3.15. The essential structural feature of all gums, clearly shown in both these examples, is extensive branching. This leaves no lengths of ordered backbone available to form the junction zones characteristic of polysaccharide gels. The branches are nevertheless capable of trapping large amounts of water, and interactions between them ensure that even quite dilute solutions will be viscous. If one examines the solution properties of polysaccharides in the sequence cellulose, amylose, agar, pectin, amylopectin and the gums, there is a clear parallel with the extent of branching and/or the diversity of the monosaccharide components.

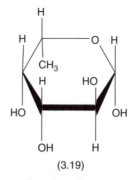

(3.19)

In the past, the usual application for gums in food products was as thickening agents, as a substitute for starch. Nowadays they are often used at quite low concentrations to assist in the stabilisation of emul-sions (*see* Chapter 4) and to smooth the texture of products such as ice-cream. However, the realisation that it is the soluble components which are responsible for the beneficial effects of a high-fibre diet on the incidence of arterial disease has in recent years focused attention on a nutritional role for gums. The favoured (but not universally accepted) explanation for the effect is that blood lipid concentrations are lowered due to a reduction in the reabsorption of bile acids (*see* Chapter 4) from the small intestine. Binding of the steroidal bile acids to soluble fibre components results in them being excreted with the faeces, creating a drain on the body's cholesterol metabolism.

Another valuable effect particularly associated with guar gum, but clearly involving other soluble fibre substances, concerns the behaviour of starch and sugars in the intestine. The presence of guar gum appears to retard the digestion and absorption of carbohydrates and thereby flattens the peak of glucose flow into the bloodstream which normally follows a carbohydrate-rich meal. The effect appears to be purely phy-sical, in that sugars and starch fragments get caught up in the three-dimensional network of fibre molecules, making them less accessible to digestive enzymes and slowing down their diffusion towards the absorptive surface of the small intestine. The application of this effect is believed to have considerable potential in the management of diabetes, particularly Type 2, the non-insulin-dependent form.

Xanthan gum is produced by the bacterium, *Xanthomonas campestris*. The gum encapsulates the cell and helps it to stick to the leaves of its favoured hosts, cabbages, as well as helping to protect it from it from dehydration and other environmental stresses. The gum is prepared commercially by growing the microorganism in large-scale fermentation processes. Steps are taken to ensure that the final product contains no viable cells of the bacterium. The enormous molecules of the polymer contain between 10 000 and 250 000 sugar units. As shown in Figure 3.16, the molecule has a backbone of glucose units joined by β-1,4 linkages, as in cellulose. However, every fifth glucose unit carries a trisaccharide side-chain which includes one, and sometimes two, carboxyl groups. These charged groups help to ensure the high affinity for water of xanthan gum.

The special properties of xanthan gum solutions result from the ease with which the polymer chains associate together, and also the ease with which these associations are reversibly broken. When a xanthan gum solution (typically in food products at around 0.1–0.3%) is both cool and stationary, *i.e.* not flowing, being stirred or shaken, the polymer chains take on a rigid zigzag shape (*see* Figure 3.12) of sufficient length for them to come together in junction zones (*see* page 61) and form a

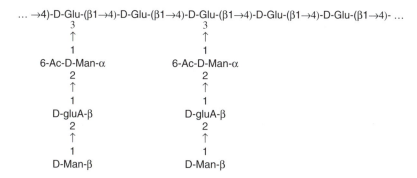

... →4)-D-Glu-(β1→4)-D-Glu-(β1→4)-D-Glu-(β1→4)-D-Glu-(β1→4)-D-Glu-(β1→4)- ...

```
        3                              3
        ↑                              ↑
        1                              1
  6-Ac-D-Man-α                   6-Ac-D-Man-α
        2                              2
        ↑                              ↑
        1                              1
    D-gluA-β                       D-gluA-β
        2                              2
        ↑                              ↑
        1                              1
    D-Man-β                        D-Man-β
```

Figure 3.16   Xanthan gum. The trisaccharide side-chains are carried on approximately every other glucose unit of the backbone. The mannose residues next to the backbone have acetyl groups (CH_3CO-) attached to the C-6 hydroxyl. Approximately half the terminal mannose residues have a pyruvate residue bridged across their C-4 and C-6 hydroxyl groups.

gel-like network. What makes these junction zones special is that the slightest hint of physical stress is sufficient to disrupt them. This stress, or shear, can result from any force imposed on the gel. This means that xanthan gum solutions are thixotropic, a property they share with non-drip paint. Such solutions are often described as being susceptible to "shear thinning".

The viscosity of a xanthan gum solution held in a bottle at rest can be sufficiently high to ensure that even quite large suspended particles do not settle out. However, as soon the bottle is inverted the solution instantly liquefies and readily flows out of the bottle. The most obvious applications of xanthan gum in food are in salad dressings containing suspended particles of herbs, *etc.*, and in ketchups. The concept of a tomato ketchup that flows easily from the bottle but does not dribble off one's chips has a certain appeal. Xanthan gum is also very effective in stabilising emulsions, and it is in this role that it is most frequently encountered in food products.

Gellan gum comes from the bacterium *Sphingomonas elodea*,[ix] manufactured by similar processes to those for xanthan gum. Although it has now been in use around the world for some years, it was cleared for food use in Europe only in 1994. Gellan gum (Figure 3.17) has an unbranched polymer chain consisting of a repeating tetrasaccharide unit (–glucose–glucuronic acid–glucose–rhamnose–), with a total length of around half a million monosaccharide units. Individual polymer chains form double helices around each other. In the native form, referred to as "high acyl" the first of the glucose units in each repeating

---

[ix] It used to be classified as a pseudomonad and is still widely referred to as *Pseudomonas elodea*.

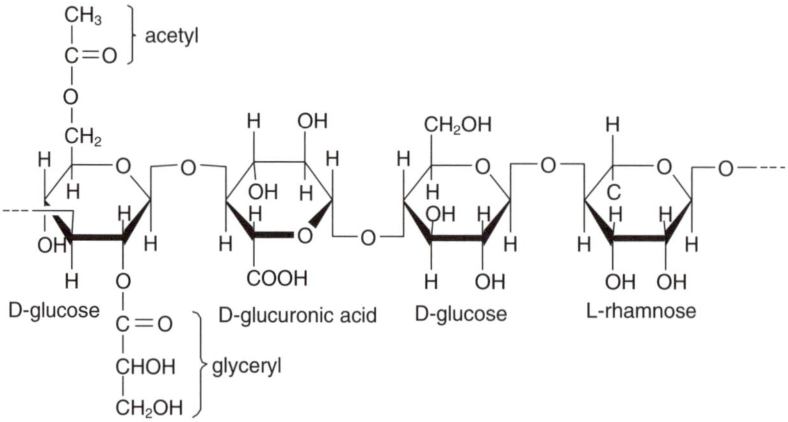

Figure 3.17   The repeating unit of gellan gum. In the native form, shown here, one of the glucose units in each repeating tetrasaccharide is esterified with glyceric acid and approximately half of these same glucose units are also esterified with ethanoic acid. As explained in the text, the gels formed by this "high acyl" form of gellan gum have different properties from those of the "low acyl" form, which have had the glyceryl and acetyl groups removed by treatment with alkali.

tetrasaccharide element is esterified with glyceric acid, and approximately half of these glucose units are also esterified with ethanoic acid. Treatment with alkali removes the glyceryl and acetyl groups to give the "low acyl" form. The characteristics of the two types differ in ways that can be exploited in food product formulations, although both will form gels at quite low concentrations ($<0.2\%$).

High acyl gellan gum solutions set quickly once the temperature falls below 70 °C, to give a gel which is soft and elastic. By contrast, low acyl gellan gels require the presence of cations, ideally divalent ions such as Ca^{2+} or Mg^{2+}, and do not set until the temperature drops into the 30–50 °C range. Their gels are harder and brittle or, by analogy with descriptions of pastry, short. The divalent cations associate with the carboxyl groups of the glucuronic acid residues, cross-linking the double helical sections to form the junction zones of the gel. Monovalent cations such as Na^+ give a similar result by simply neutralising acid carboxyl groups and allowing the chains to come together. The levels of sodium and calcium normally found in food systems are usually sufficient to promote gelation.

The low concentration of high acyl gellan gum required for it to gel is the basis of one of its most appealing characteristics. Flavours carried by the gel are released in the mouth much more quickly than from other types of gel. Since the gel is also rather brittle it tends to crumble in the mouth, giving the impression of melting. At much lower concentrations, around 0.03%, gellan gum does not form solid gels but instead produces

Table 3.7 The gums and non-starch gelling agents permitted for use in the EU. Locust bean gum, (from the seeds of the Mediterranean carob tree, *Ceratonia siliqua*), and tara gum (from those of the South American tara tree, *Cesalpinia spinosa*) are galactomannans structurally related to guar gum. Karaya gum, obtained commercially from the Indian tree, *Sterculia urens*, is a highly branched polysaccharide dominated by galacturonic acid. Gum arabic is a variable and complex mixture of polysaccharides and glycoproteins obtained commercially from *Acacia senegal* and *A. seyal*, trees found in sub-Saharan Africa. Konjac is a glucomannan obtained from the roots of perennial plant *Amorphophallus konjac*, grown in the Far East.

Gelling agent	E number
Agar	E406
Carrageenan	E407
Locust bean gum	E410
Guar gum	E412
Tragacanth	E413
Acacia gum (gum arabic)	E414
Xanthan gum	E415
Karaya gum	E416
Tara gum	E417
Gellan gum	E418
Konjac	E425

thixotropic fluids which are able to hold fruit particles in suspension. Both these effects are exploited in some of the creations of one of Europe's leading chefs, Heston Blumenthal.

This is a convenient place to summarise the full list of gums and gelling agents which are permitted for use as food additives in Europe. Table 3.7 lists all except the modified starches, which are shown later in Table 3.8.

SPECIAL TOPICS

1 Chemically Modified Starches

The use of waxy starches and pregelatinisation widens the range of behavioural characteristics available to food processors, but the most significant changes can obtained by chemical modification of the starch molecule. Three types of modification are used.

- **Depolymerisation**. Starch granules are treated with hydrochloric acid, followed by neutralisation. Only a very small proportion of the glycosidic links are broken, but the fragmentation of the amylopectin molecules is still sufficient for the granule structure to be

Table 3.8 The modified starches currently permitted for food use in Europe. The table is in two parts. On the left are the starch derivatives which are classified as additives and are therefore given E numbers, and on the right are derivatives whose pattern of use has led them to be classified as food ingredients, *i.e.* they make a nutritive contribution in their own right. The International Numbering System (INS) numbers for these are given. INS numbers are designated by The Codex Alimentarius, a subsidiary organisation of the World Health Organization (WHO) and the Food and Agriculture Organization of the United Nations (FAO). The system is not dissimilar from that adopted by the EU, but the possession of an INS number does not automatically imply approval. INS numbers frequently correspond to E numbers.

Permitted additives	E No.	Permitted ingredients	INS No.
Oxidised starch	E1404	Dextrin	1400
Monostarch phosphate	E1410	Acid treated starch	1401
Distarch phosphate	E1412	Alkaline treated starch	1402
Phosphorylated distarch phosphate	E1413	Bleached starch	1403
Acetylated distarch phosphate	E1414	Distarch glycerine	1430
Starch acetate	E1420	Hydroxypropyl distarch glycerine	1441
Acetylated distarch adipate	E1422		
Hydroxypropyl starch	E1440		
Hydroxypropyl distarch phosphate	E1442		
Starch sodium octenyl succinate	E1450		
Acetylated oxidised starch	E1451		

lost more easily on heating, and for gelatinisation to be much faster. Known as "thin-boiling" starches, they give stronger, clearer gels and are used in many types of jelly confectionery. Maltodextrins are produced by much more extensive starch hydrolysis, with DP values (degree of polymerisation, the average number of glucose units per molecule) within the range 5–100. The degree of hydrolysis is often expressed as the dextrose equivalent (DE, given by the expression):

$$DE = 100 \times \frac{Reducing\ sugar, expressed\ as\ glucose}{Total\ carbohydrate}$$

DE represents the percentage hydrolysis of the glycosidic linkages present. Pure glucose has a DE of 100, and pure maltose 50. Maltodextrin preparations with DE values around 5 (*i.e.* an average of around 20 glucose units per molecule) can be included in a wide range of products, including low-fat margarine, salad dressings and mayonnaise, where they provide some of the organoleptic properties (mouth feel) of fat.

- **Derivatisation**. Esterification of a small proportion (normally less than 1%) of glucose units with organic acids or phosphates gives "stabilised" starches. Depending on the particular groups attached, one can achieve gels with excellent freeze stability and resistance to retrogradation (acetyl or hydroxypropyl groups), high paste viscosity (phosphates) or emulsifying capabilities (1-octenylsuccinates).
- **Cross-linking**. Fewer than one cross-linkage per 1000 glucose units is sufficient to produce significant changes in starch properties. Cross-links between chains tend to strengthen the granule structure. This gives pastes which are much more resilient in the face of low pH, extended cooking or physical shear forces than those made from unmodified starch. Not surprisingly, cross-linked starches require much more cooking time to achieve gelatinisation. They are particularly valuable in canned products. Their slow gelatinisation allows the contents of the can to remain liquid during heating, so that heat transfer is not impaired by solidification but the final product has a very firm texture.

Commercial modified starch products frequently use both stabilisation and cross-linking in the same product, quite often applied to a waxy starch rather than the normal type. Some of the chemical reactions involved in derivatisation are shown in Figure 3.18. The extent of these reactions is restricted by food legislation but is always far too low to have any significant effect on the digestibility and therefore nutritive value. The diversity of applications of the different modified starches results in some being classified in current European legislation as additives, whereas others are regarded as ingredients. The key distinction is that those considered as additives are used in too small a proportion to make a significant contribution to a food product's nutritional properties. All the modified starches permitted are listed in Table 3.8.

2 Syrups from Starch

The conversion of starch into glucose and other sugars is a major industry. The old-fashioned process based on acid hydrolysis has been almost totally superseded by processes based on enzymic catalysed hydrolysis. The latter processes take advantage of the great specificity of enzymes for certain types of bond in particular substrates. Although enzymes, being proteins, are usually thought of as being inactivated at high temperatures, many enzymes used industrially are nowadays obtained from thermophilic bacteria which have thermostable enzymes.

In the usual two-phase process, starch, mostly from maize, and cold water are mixed to form a slurry at up to 40% solids content. A thermostable α-amylase from a thermophilic strain of bacillus (*Bacillus licheniformis*, *B. stearothermophilus* or *B. amyloliquefaciens*) is incorporated into the slurry, which is then brought rapidly up to a temperature of 105–110 °C and held within this range for a few minutes. The starch granules gelatinise, and in doing so the enzyme is able to gain access to the α1→4 glycosidic links of the amylose and amylopectin. As a result both polymers are fragmented to oligosaccharides (sometimes referred to as dextrins) averaging around 10 glucose units in size. Many of the fragments will include α1→6 branches derived from the branched structure of amylopectin, but these bonds are not attacked by the α-amylase. Unlike the impossibly stiff starch paste that would have resulted from simply gelatinising a 40% starch slurry, the solution of oligosaccharides is now manageably fluid.

Having gone through the initial liquefaction phase, the oligosaccharides are subjected to further enzymic hydrolysis in the "saccharification" phase of the process. The temperature is brought down to around 55–60 °C and preparations of the amyloglucosidase (otherwise known as glucoamylase) from the fungus, *Aspergillus niger*, are added. Amyloglucosidase is remarkable for its ability to catalyse the hydrolysis of both the α1→4 and α1→6 glycosidic links, and given time would break down the oligosaccharides almost entirely to glucose. However, the amyloglucosidase is often supplemented at this stage with the enzyme pullulanase, from *Bacillus acidopullulyticus*, which is specific for α1→6 links. Although the saccharification stage may take as long as 90 hours, the results are spectacular. After suitable clean-up procedures, yields of pure glucose of 95% of the theoretical result are obtained.

This need not be the end of the story. The high purity glucose syrup may be passed through a reactor in which the enzyme glucose isomerase has been immobilised on solid carrier particles. This enzyme is obtained from a number of different fungi in which its primary role is the interconversion of xylose and xylulose, *i.e.* interconversion of an aldose sugar and its ketose equivalent. This reaction is completely analogous to the interconversion of glucose and fructose and enables syrups with fructose : glucose ratios of 55 : 45 to be produced. Known as "high fructose corn syrups" these have sweetness comparable with invert sugar and are becoming increasingly popular, especially with soft drink manufacturers.

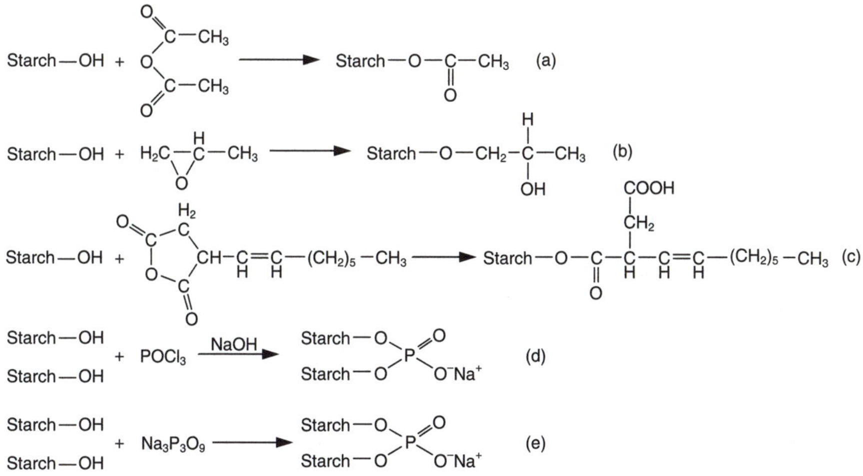

Figure 3.18 The chemical modification of starch. Reactions (*a*) to (*c*) are stabilising derivatisations, (*d*) to (*f*) are cross-linking reactions. The E numbers for some of the products are shown in brackets.

    (a) Acetylation with acetic anhydride (E1420).

    (b) Formation of a hydroxypropyl ether by reaction with propylene oxide (E1440).

    (c) Formation of the 2-(1-octenyl)succinyl ester (E1450).

    (d) Phosphate diester formation with phosphorus oxychloride (E1412).

    (e) Phosphate diester formation with sodium trimetaphosphate (E1412).

3   Glycaemic Index

Although the total dietary intake of carbohydrates is obviously a major factor in the development of obesity, and associated conditions such as diabetes and cardiovascular disease, it has become clear in recent years that the *rate* of influx of glucose into the bloodstream following carbohydrate consumption is also very important. Ordinarily the body responds to the rise in blood glucose concentration after a meal by releasing the hormone insulin from the pancreas into the bloodstream. This activates the transfer of glucose from the blood into the tissues, notably the muscles and liver, where it can be stored as the polymer glycogen, and adipose tissue, where it is stored as fat. Persistent consumption of foodstuffs which generate glucose in large quantities places the system under strain. One manifestation of this can be the onset of Type 2 diabetes, when the tissues become less responsive to insulin. The concept of the Glycaemic Index (GI) was introduced by Jenkins in 1994 to provide a quantitative measure of the rate of glucose influx into the bloodstream. The GI of a particular foodstuff is normally measured

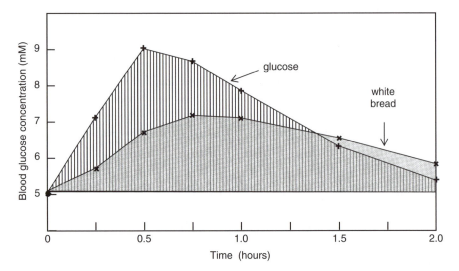

Figure 3.19   Determination of Glycaemic Index (GI). The shaded areas represent the increase in blood glucose concentration above the subject's baseline fasting level at the start of measurements, over a period of 2 hours. The area generated by a portion of a test foodstuff containing 50 g of carbohydrate is compared with that given by a standard 50 g portion of glucose. Areas are calculated from the sum of the areas of the series of trapeziums formed by the data points and the subject's baseline level. No attempt is made to smooth the curves before calculating the area. The inevitable natural variation between different test subjects means that large numbers of test curves have to be obtained and the results averaged, requiring the services of large numbers of volunteer subjects.

using healthy adult volunteers. Having fasted overnight, a food portion containing 50 g of carbohydrate or a 50 g portion of glucose (as a standard) is consumed and the blood glucose concentration measured at regular intervals over 2 hours. Figure 3.19 shows a typical result.

GI data for a selection of foodstuffs are given in Table 3.9. Also included in the table are the corresponding figures for the Glycaemic Load (GL). While GI data provide a useful characterisation of a foodstuff, they do not in themselves provide any information on the actual glycaemic effect of consuming the foodstuff in a realistic situation outside the laboratory. The concept of GL was introduced in 1997 to enable the probable elevation of blood glucose from a typical serving of a food to be estimated. The GL is obtained by multiplying the GI by the quantity of available carbohydrate in a standard serving or portion. The consumption of a diet with a relatively high GL figure has come to be associated with a higher probability of Type 2 diabetes (non-insulin dependent diabetes, NIDD) and coronary heart disease. Foods are often

Table 3.9 Glycaemic Index (GI) and Glycaemic Load (GL) for a range of foodstuffs, based on data assembled by Kaye Foster-Powell and colleagues (K. Foster-Powell, S. H. A. Holt and J. C. Brand-Miller, *Am. J. Clin. Nutr.*, 2002, **76**, p. 5). The values recorded in the table should not be treated as precise. A total of 135 different studies involving over 1300 human subjects contributed to the mean figures given. Different studies on the same foodstuff, each involving tests with considerable numbers of subjects, typically gave results varying by up to 5 to 10% from the mean values shown.

Food	GI	Portion size (g)	GL	Food	GI	Portion size (g)	GL
Cornflakes	81	30	21	Potato crisps	54	50	11
Potato, mashed	77	150	15	Banana	52	120	12
Wholemeal bread	71	30	9	Orange juice	50	250[a]	13
White bread	70	30	10	Green peas	48	80	3
Sucrose	68	10	7	Baked beans[c]	48	150	7
Rice, white	64	150	23	Spaghetti	44	180	21
Rye crispbread	64	25	11	Milk chocolate	43	50	12
Ice cream	61	50	8	Apples, raw	38	120	6
Cola	58	250[a]	15	Chickpeas	28	150	8
Porridge[b]	58	250	13	Lentils, red	26	150	5
Potatoes, boiled new	57	150	12	Peanuts raw	14	50	1
Rice, brown	55	150	18				
Honey	55	25	10				

[a] cm^3 rather than g.
[b] Made from rolled oats.
[c] Canned in tomato sauce.

labelled as high, medium or low GI. Although labelling legislation did not (in 2007) appear to have caught up with this issue, these three terms most often refer to GI values of 70 and above, 56 to 69, and below 55, respectively.

Close examination of these data reveals a number of unexpected results, notably for honey, sucrose, chocolate and ice-cream, which seem to be less glycaemic than might have been expected. In these cases a major factor is that the fructose component of the sucrose makes no direct or immediate contribution to blood glucose concentration. In the case of ice-cream and chocolate it is assumed that the fat content physically slows the rate of sugar absorption from the intestine. In more typical complete meals, in which the glycaemic carbo-hydrate components are less intimately associated with fat than in ice-cream or chocolate, it has proved possible to predict the glyca-emic effect from the GI values and the proportions of individual components.

4 Further Details of Pectin Structure

Earlier in this chapter the essential features of pectin molecules were considered, particularly in relation to pectin's contribution to the structure and texture of foods. Here we give a more detailed account of current views of the molecular structure of pectin.

The currently accepted models have a number of features in common. Firstly, the term pectin covers a number, usually three, of different broad types of polysaccharide structures, whose details can vary greatly between plant species and between different tissues within a single species. The difficulty of extracting intact (or supposedly intact) pectin molecules from plant tissues means that there is no absolute certainty about whether the different structures are always combined in the living plant into a single type of molecule, or whether they occur naturally as distinct molecules. At the present time the first of these alternatives seems to be preferred. The three major types are as follows:

- **Homogalacturonan (HG)**. This is the classical pectin structure, already shown in Figure 3.4, consisting of a linear backbone of α-1,4 linked galacturonic acid residues. HG is the site of the methylation crucial to the gel-forming capability of pectin. A feature of apple pectin which distinguishes it from citrus pectins is the presence of regions of the homogalacturonan backbone carrying numbers of D-xylose (3.3) residues, mostly as single residues linked α-1,3 to the galacturonate residues. Xylose-rich regions of the chain will still be classified as "hairy".
- **Rhamnogalacturonan I (RGI)**. This has a linear backbone based on repeating disaccharide units comprising galacturonic acid and L-rhamnose:

$$\ldots \rightarrow 4)\text{-}\alpha\text{-}\text{D-GalA}(1 \rightarrow 2)\ \alpha\text{-}\text{L-Rha-}(1 \rightarrow \ldots$$

A variety of different side-chains are attached to the rhamnose residues, probably in the 4- position. These can include galactans (*i.e.* chains of D-galactose (3.4) residues, only rarely branched, linked α-1,4) and arabinans (3.6), highly branched chains of L-arabinose residues (3.2).
- **Rhamnogalacturonan II (RGII)**. To say the least, this is a most unhelpful name, as RGII rarely contains more than a trace of rhamnose! RGII is based on a homogalacturonan backbone carrying relatively small but very elaborately branched side-chains. Besides galactose and xylose, these can include several otherwise

rare sugars such as L-fucose, 6-deoxy-L-galactose (3.19), D-apiose (3.20), L-aceric acid (3.21) and D-glucuronic acid (3.22).

(3.20) (3.21) (3.22)

The assumption that these different types of pectin chain are components of single large polymers still leaves open the issue of how they are connected together. The most generally accepted model, based initially on Pilnik's studies of apple pectin, is shown in Figure 3.20 (*A*). This has a central linear backbone mainly consisting of homogalacturonan, with smooth and hairy regions. Hairy regions could be characterised by the presence of xylose substituents or by concentrations of branches leading to the formation an RGII region. The galacturonan backbone is interrupted at intervals by a region of alternating rhamnose and galacturonate residues, forming an RGI region. An alternative to this arrangement, shown in Figure 3.20 (*B*), has been proposed by Vincken in 2003 (*see* G. T. Willats, *et al.*, in Recent Reviews at the end of this chapter). In this rhamnogalacturonan (RGI) forms the backbone of the pectin molecule, carrying a variety of different types of side-chain, including simple homogalacturonan (HG), and various RGII chains. The location of regions of homogacturonan extensively substituted with the xylose residues (xylogalacturonan) particularly associated with apple pectin remains uncertain. As shown in (*B*), these chains may be separately joined to the backbone, or they may simply represent regions of the homogalacturonan chain as shown in (*A*).

No attempt has been made in Figure 3.20 to illustrate the methylation of the galacturonate residues. A significant degree of acetylation, *i.e.* conversion of either the 2- or 3-hydroxyl group of galacturonic acid residues to an acetyl group:

$$-OH \rightarrow -COCH_3$$

is found in some pectins, notably that from sugar beet pulp. The residue after the sugar has been extracted from sugar beet is becoming another important commercial source of pectin.

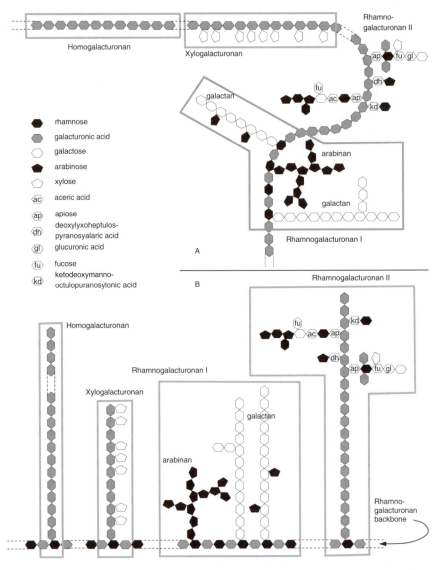

Figure 3.20 Pectin structures, as discussed in the text. (*A*), based on the work of
Pilnik and others since the 1980s, has the extended homogalacturonan
backbone carrying RGI and RGII side-chains. The bends shown in the
backbone of structure (*A*) are solely for artistic convenience. (*B*), based
on Vincken's 2003 model, has a backbone of rhamnogalacturonan carrying numerous homogalacturonan chains (only one is shown here),
together with various other branched chains, including those associated
with RGI and RGII.

FURTHER READING

Thickening and Gelling Agents for Food, ed. A. Imeson, Aspen, Maryland, USA, 2nd edn, 1997.

Handbook of Hydrocolloids, ed. G. O. Phillips and P. A. Williams, Woodhead Publishing, Cambridge, 2000 (in the USA: CRC Press, Boca Raton, Florida).

D. A. T. Southgate, *Dietary Fibre Analysis*, Royal Society of Chemistry, Cambridge, 1995.

S. Cho, J. W. Devries and L. Prosky, *Dietary Fibre Analysis and Applications*, Aspen, Maryland, USA, 1998.

V. S. R. Rao, P. K. Qasba, P. V. Balaji and R. Chandrasekaran, *Conformation of Carbohydrates*, Harwood Academic, Australia, 1998.

RECENT REVIEWS

R. F. Tester and X. Q. Karkalis, *Starch: Composition, Fine Structure and Architecture; J. Cereal Sci.*, 2004, **39,** p. 151.

M. G. Sajilata, *et al.*, *Resistant Starch: A Review*, in *Food Sci. Food Safety*, 2006, **5**, p. 1.

R. N. Tharanathan, *Starch: Value Addition by Modification*, in *Crit. Rev. Food Sci. Nutr.*, 2005, **45**, p. 371.

M. G. Sajilata and R. S.Singhal, *Specialty starches for snack foods; Carbohydr. Polym.*, 2005, **59**, p. 131.

F. Brouns, B. Kettliz and E. Arrigoni, *Resistant Starch and the "Butyrate Revolution"; Trends Food Sci. Technol.*, 2002, **13**, p. 251.

U. Lehmann and F. Robin, *Slowly Digestible Starch, its Structure and Health Implications: A Review; Trends Food Sci. Technol.*, 2007, **18**, p. 346.

J. A. Gray and J. N. Bemiller, *Bread Staling: Molecular Basis and Control; Comp. Rev. Food Sci. Safety*, 2003, **2**, p. 1.

W. G. T. Willats, J. P. Knox and J. D. Mikkelsen, *Pectin: New Insights into an Old Polymer are Starting to Gel; Trends Food Sci. Technol.*, 2006, **17**, p. 97.

M. Roberfroid and J. Slavin, *Nondigestible Oligosaccharides; Crit. Rev. Food Sci. Nutr.*, 2000, **40**, p. 461.

B. C. Tungland and D. Meyer, *Nondigestible Oligo- and Polysaccharides (Dietary Fiber): Their Physiology and Role in Human Health and Food; Comp. Rev. Food Sci. Safety*, 2002, **1**, p. 73.

K. N. Englyst and H. N. Englyst, *Carbohydrate Bioavailability; Br. J. Nutr.*, 2005, **94**, p. 1.

P. Wood, *Relationships between Solution Properties of Cereal β-Glucans and Physiological Effects: A Review; Trends Food Sci. Technol.*, 2002, **13**, p. 313.

A. Lazaridou and C. G. Biliaderis, *Molecular Aspects of Cereal β-Glucan Functionality: Physical Properties, Technological Applications and Physiological Effects; J. Cereal Sci.*, 2007, **46**, p. 101.

K. W. Waldron, M. L. Parker, and A. C. Smith, *Plant Cell Walls and Food Quality; Comp. Rev. Food Sci. Safety*, 2003, **2**, p. 101.

J. C. Brand-Miller, *Glycemic Load and Chronic Disease; Nutr. Rev.*, 2003, **61**, p. S49.

D. Verbeken, S. Dierckx and K. Dewettinck, *Exudate Gums: Occurrence, Production, and Applications; Appl. Microbiol. Biotechnol.*, 2003, **63**, pp.10–21.

L. Oatway, T. Vasanthan and J. H. Helm, *Phytic Acid; Food Rev. Internat.*, 2001, **17**, p. 419.

CHAPTER 4
Lipids

The term lipid refers to a group of substances which is even less clearly defined than the carbohydrates. Generally speaking it denotes a heterogeneous group of substances associated with living systems, which have the common property of insolubility in water but solubility in non-polar solvents such as the paraffins (hexane, *etc.*), diethyl ether and chloroform. Included in the group are the oils and fats[i] of our diet, together with the so-called phospholipids associated with cell membranes. While food materials such as oils and fats clearly consist of very little other than lipids, their presence in a variety of other foods is less obvious, as indicated by the data in Table 4.1. The lipids in all these foods are esters of long-chain fatty acids, but there are many other lipids which lack this structural feature. These include the steroids and terpenes, but, with the exception of cholesterol and its long-chain fatty acid esters, in cases where these substances do have significance in food they will be found under the headings of vitamins, pigments or flavour compounds.

FATTY ACIDS: STRUCTURE AND DISTRIBUTION

Monocarboxylic fatty acids (*i.e.* fatty acids with only one carboxyl group) are structural components common to most of the lipids of interest to food chemists, and since many of the properties of food lipids can be accounted for in terms of their component fatty acids they will be

---

[i]There is no formal distinction between oils and fats. The former are liquid at room temperature and the latter solid.

Food: The Chemistry of its Components, Fifth edition
By T.P. Coultate
© T.P. Coultate, 2009
Published by the Royal Society of Chemistry, www.rsc.org

Table 4.1 The total fat contents of a variety of foods. The figures are taken
from *McCance and Widdowson*. (*see* Chapter 1, Further Reading)
and in all cases refer to the edible portion. They should be regarded
as typical values for the particular food rather than absolute values
to which all samples of the food comply. The definition of fat used
here covers all lipids, but in fact only in the case of egg yolk and
products containing egg are polar lipids quantitatively significant.

Food	Total fat (%)	Food	Total fat (%)
Wholemeal flour	2.2	Low fat spread	40.5
White bread	1.9	Lard	99.0
Rich tea biscuits	16.6	Vegetable oils	99.9
Shortbread	26.1	Bacon, streaky, fried	42.2
Madeira cake	16.9	Pork sausage, grilled	24.6
Flaky pastry	40.6	Beef, sirloin, roast	21.1
Skimmed milk (cow's)	0.1	Beefburger, raw	20.5
Whole milk (cow's)	3.9	Lamb chop, grilled	29.0
Human milk	4.1	Chicken, roast	5.4
Soya milk	1.9	Turkey breast, roast	1.4
Clotted cream	63.5	Cod, raw	0.7
Cheese (Brie)	26.9	Cod, battered and fried	10.3
Cheese (Cheddar)	34.9	Mackerel, smoked	30.9
Dairy ice-cream	9.8	Salmon, smoked	4.5
Egg yolk	30.5	Taramasalata	46.4
Egg white	trace	Brazil nuts	68.2
Scotch egg	17.1	Peanuts, dry roasted	49.8
Butter	81.7	Plain chocolate	29.2
Margarine	81.6	Milk chocolate	30.3

considered in some detail. Almost without exception the fatty acids
which occur in foodstuffs contain an even number of carbon atoms in an
unbranched chain, *e.g.* lauric or dodecanoic acid (4.1). It is worthwhile
mastering both the trivial or common names and the systematic names
of the principal fatty acids.

$$CH_3-CH_2-CH_2-CH_2-CH_2-CH_2-CH_2-CH_2-CH_2-CH_2-CH_2-COOH \quad (4.1)$$

In addition to saturated fatty acids, of which lauric acid is an example,
unsaturated fatty acids having one, two, or sometimes as many as six
double bonds are common. The double bonds are almost invariably *cis*
(*see* Figure 4.1), and when the fatty acid has two or more separated
double bonds they are described as "methylene interrupted":

$$-CH_2-CH=CH-CH_2-CH=CH-CH_2-$$

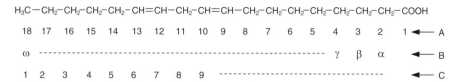

Figure 4.1 Numbering of the atoms in fatty acids, illustrated by linoleic acid. The most common system (*A*), and the one adopted to define the positions of double bonds in Table 4.2, simply numbers the carbon atoms starting from the carboxyl group (1) to the terminal methyl group, (18) in the example shown. Alternatively Greek letters may be used to label successive CH_2 groups, starting at the one adjoining the carboxyl group, as in (*B*). This system is rarely used beyond the γ- carbon, except that the terminal methyl can be designated ω to indicate that it is the final one regardless of the length of the chain. In some considerations of fatty acid metabolism, notably essential fatty acids, the carbon atoms are counted from the methyl terminus (*C*).

rather than conjugated. Conjugated carbon–carbon bonds are alternately single and double along the chain:

$$-CH_2-CH=CH-CH=CH-CH-CH_2-$$

There are a number of methods available for identifying the different carbon atoms in the chain and defining the positions of the double bonds. These are explained in Figure 4.1.

The structure of a fatty acid can be indicated using a convenient shorthand form giving the total number of carbon atoms followed by a colon and the number of double bonds; the positions of the double bonds is shown indicated after the symbol Δ. For example, linoleic acid, shown in Figure 4.1, would be written simply as 18:2Δ9,12. Since the double bond furthest from the carboxyl group is between the sixth and seventh carbon atoms counting from the ω-carbon, linoleic acid is sometimes referred to as an omega-6 fatty acid.

As shown in Figure 4.2, the double bonds in unsaturated fatty acids have two possible conformations, *cis* or *trans*. The double bonds of most naturally occurring fatty acids are in the *cis* conformation, which has significant implications for their physical properties, as will be explained later.

Table 4.2 gives the names and structures of a number of the fatty acids commonly encountered in food lipids.

The oils and fats are obviously the lipids of most interest to the food chemist. These consist largely of mixtures of triglycerides, *i.e.* esters of the trihydric[ii] alcohol glycerol (propane-1,2,3-triol) (4.2), and three fatty

---

[ii] Trihydric, having three hydroxyl groups.

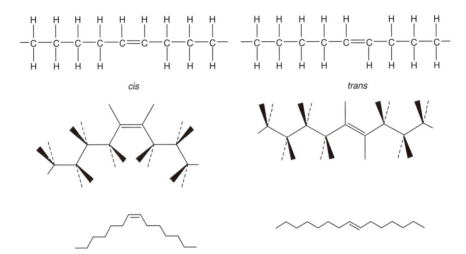

Figure 4.2   *Cis* and *trans* double bonds and their effects on the conformation of the
fatty acid chain. *Cis* and *trans* are Latin for "on this side" and "on the
opposite side", respectively. Unlike single bonds, double bonds cannot
rotate, so the four atoms involved in the double bond, two carbon atoms
and two hydrogen atoms, remain fixed in the same plane.

acid residues, not necessarily identical. The three hydroxyl groups are
usually identified by the numbers 1, 2 and 3, but sometimes the outer
hydroxyl groups, *i.e.* 1 and 3, are referred to as 1 and 1′, respectively, to
emphasise the point that under most circumstances they cannot be
distinguished from one another. Triglycerides are frequently referred to
as triacylglycerols (TAGs). Simple triglyceride molecules have three
identical fatty acid residues; "mixed" triglycerides are those containing
more than one species of fatty acid.

$$
\begin{array}{cc}
& H \\
& | \\
H-C-OH & 1 \\
& | \\
H-C-OH & 2 \\
& | \\
H-C-OH & 3 \text{ (or } 1') \\
& | \\
& H \qquad (4.2)
\end{array}
$$

A naturally occurring fat will be a mixture of quite a large number
of mixed and simple triglycerides. It is important to remember that
organisms achieve a desirable pattern of physical properties for the
lipids of, say, their cell membranes or adipose tissue by utilising a

Table 4.2 The fatty acids commonly found in foodstuffs. The meaning of the term *cis* and its opposite, *trans*, in the names of unsaturated fatty acids is explained in Figure 4.2. In most cases the common names are derived from the name (mostly Latin, *L*, but also Greek, *Gk*, or otherwise English) of a characteristic source.

Systematic name	Common name	Structure
Saturated acids		
n-Butanoic	Butyric (*L. butyrum* = butter)	4:0
n-Hexanoic	Caproic (*L. caper* = goat)	6:0
n-Octanoic	Caprylic (*L. caper* = goat)	8:0
n-Decanoic	Capric (*L. caper* = goat)	10:0
n-Dodecanoic	Lauric (*L. laurea* = laurel)	12:0
n-Tetradecanoic	Myristic (*L. myristica* = nutmeg)	14:0
n-Hexadecanoic	Palmitic (palm tree)	16:0
n-Octadecanoic	Stearic (*Gk. stear* = tallow)	18:0
n-Eicosanoic	Arachidic (*L. arachis* = peanut)	20:0
n-Docosanoic	Behenic (ben oil, *Moringa oliefera*)	22:0
Unsaturated acids		
cis-9-Hexadecenoic	Palmitoleic	16:1Δ9
cis-9-Octadecenoic	Oleic (*L. oleum* = oil)	18:1Δ9
cis, cis-9,12-Octadecadienoic	Linoleic (*L. linum* = flax, + oleic)	18:2Δ9,12
[a] all *cis*-9,12,15-Octadecatrienoic	α-Linolenic	18:3Δ9,12,15
all *cis*-6,9,12-Octadecatrienoic	γ-Linolenic	18:3Δ6,9,12
all *cis*-5,8,11,1-Eicosatetraenoic	Arachidonic (*L. arachis* = peanut)	20:4Δ15,8,11,14
all *cis*-5,8,11,14,17-Eicosapentaenoic	EPA	20:5Δ15,8,11,14,17
all *cis*-7,10,13,16,19-Docosapentaenoic	Clupanodonic (*L. clupea* = herring)	22:5Δ7,l0,13,16,19
all *cis*-7,10,13,16,19-Docosahexaenoic	DHA	22:6Δ4,7,l0,13,16,19

[a] "all" indicates that all the double bonds are *cis*.

mixture of different molecular species, rather than by utilising a single molecule, as is usually the case with proteins or carbohydrates.

The relative abundance and importance of the various fatty acids will be appreciated from Table 4.3, in which the molar proportions of the fatty acids in a number of fats and oils are presented.

The relationship between the fatty acid composition of a fat and its melting properties will be apparent from a comparison of the data in Table 4.3 with that in Table 4.4, which gives the melting points of the most important fatty acids. The lower melting temperature characteristic

Table 4.3 The fatty acid composition of fats and oils. The figures are typical values expressed in mol%. There is often considerable variation in composition between different samples of the same fat, due for example to variations in the animal's diet or the conditions during plant growth. Trace components representing less than 0.5% have been ignored.

Fatty acid	Beef fat	Lard	Goose fat	Cow's milk fat	Human milk fat	Herring oil	Maize (corn) oil	Cocoa butter[a]	Coconut oil	Palm oil	Olive oil
4:0				9							
6:0				5							
8:0				2					12		
10:0				4	1				8		
12:0				3	3				49		
14:0	4	2	1	10	5	7			16	1	
14:1	1			2	1						
15:-[b]	2			1	1						
16:0	28	26	22	23	26	13	14	29	7	48	11
16:1	5	4	3	2	5	9					1
17:-[b]	1			1	1						
18:0	20	15	6	12	7	1	2	35	2	6	3
18:1	34	44	57	23	37	12	34	32	5	34	79
18:2	3	9	9	2	11	2	48	3	1	11	5
18:3	2		2	1	1		1				1
20:0							1	1			
20:1						19					
20:5						8					
22:1						25					
22:5						2					
22:6						2					

[a]The principal component of chocolate.
[b]The sum of straight- and branched-chain, saturated and unsaturated fatty acids.

Table 4.4 Melting point (MP) of fatty acids.

Fatty acid	MP(°C)	Fatty acid	MP(°C)
Butyric	−7.9	Elaidic	43.7
Caproic	−3.4		
Caprylic	16.7	Oleic	10.5
Capric	31.6	Linoleic	−5.0
Lauric	44.2	α-Linolenic	−11.0
Myristic	54.1		
Palmitic	62.7	Arachidonic	−49.5
Stearic	69.6		
Arachidic	75.4		

of oils is associated with either a high content of unsaturated fatty acids, *e.g.* corn oil and olive oil, or a high proportion of short-chain fatty acids, *e.g.* milk fat and coconut oil.

The insertion of a *cis* double bond has a dramatic effect on the shape of the molecule, introducing a kink of about 42° into the otherwise straight hydrocarbon chain, as shown in Figure 4.2. The insertion of a *trans* double bond has very little effect on the conformation of the chain and has therefore very little effect on the melting temperature. For example, elaidic acid, the *trans* isomer of oleic acid, has similar physical properties to stearic acid. Although some *trans* fatty acids do occur in nature, their primary importance is as by-products of the hydrogenation process in margarine manufacture. They are considered in detail on pages 113–115.

The milk fats of ruminants are characterised by high proportion of short-chain fatty acids. These are derived from the anaerobic fermentation of carbohydrates such as cellulose by micro-organisms in the rumen. The branched-chain fatty acids which are also found in ruminant milk fats are also by-products of rumen metabolism. Most of these branched fatty acids belong to either the *iso* or *anteiso* series, depending on the position of the additional methyl group. For example the milk fat of both cows and goats contains around 0.5% of 15-methylhexadecanoic acid (4.3), an *iso* fatty acid, alongside a similar proportion of the *anteiso* fatty acid, 12-methyltetradecanoic acid (4.4).

$$H_2C \diagdown \atop H_2C \diagup CH-CH_2-CH_2--- \qquad \qquad H_3C-CH_2-\overset{\overset{\displaystyle H}{|}}{\underset{\underset{\displaystyle CH_3}{|}}{C}}-CH_2-CH_2---$$

(4.3)   (4.4)

Human milk fat, like that of other non-ruminant species such as the pig, is rich in linoleic acid. Although no direct benefits of this to human babies have yet been identified, it will be surprising if these are not present in view of the importance of linoleic acid in human nutrition (*see* later in this chapter).

Cod liver oil is most familiar to us as a source of vitamin D, but fish oils are particularly interesting for the diversity of the long-chain, highly unsaturated fatty acids they contain. The extremely cold environments in which fish such as cod and herring live may account for the nature of fish oil lipids.

There are a number of fatty acids with unusual structures which are a feature of particular groups or species of plants. For example, petro-selenic acid ($18:1\Delta11cis$) is found in the seed oils of celery, parsley and carrots. Ricinoleic acid (4.5) makes up about 90% of the fatty acids of castor oil.

$$CH_3(CH_2)_5CHCH_2CH{=}CH(CH_2)_7COOH$$
$$\underset{\displaystyle OH}{|} \qquad (4.5)$$

Fatty acids containing cyclopropene rings are mostly associated with bacteria, but traces of sterculic acid (4.6) and malvalic acid (4.7) are found in cottonseed oil. These are toxic to non-ruminants, and the residual oil in cottonseed meal is sufficient to produce an adverse effect in poultry fed exclusively on the meal. It must be assumed that the low levels man has consumed for many years in salad dressing and margarine manufactured from cottonseed oil have had no deleterious effects.

$$\overset{\displaystyle CH_2}{\overset{\displaystyle \diagup\diagdown}{CH_3(CH_2)_7C{=}C(CH_2)_7COOH}}$$

(4.6)

$$\overset{\displaystyle CH_2}{\overset{\displaystyle \diagup\diagdown}{CH_3(CH_2)_7C{=}C(CH_2)_6COOH}}$$

(4.7)

Erucic acid, *cis*-13-docosenoic acid ($22:1\Delta13$), is a characteristic component of the oils extracted from the seed of members of the genus *Brassica*, notably varieties of *B. napus* (in Europe) and *B. campestris* (in Canada and Pakistan), which are known as rape. Rapeseed meal is an important animal feedstuff, and rapeseed oil is important both for human consumption (*e.g.* as a raw material for margarine manu-facture) and for industrial purposes such as lubrication. Most rapeseed oil at one time contained some 20–25% erucic acid, to which was attributed a number of disorders of the lipid metabolism in rats fed with high levels of the oil. The problem of possible toxicity in humans has been largely eliminated by plant breeders who have developed strains of rape giving oil with acceptably low erucic acid levels (less than 5%), which is utilised in margarine manufacture. Oils rich in erucic acid remain important for many industrial, *i.e.* non-food, applications.

Bacteria, especially those that occupy the rumen of cattle, synthesise a considerable variety of fatty acids which do not fit the patterns associated with most of the higher animals and plants. Small amounts of some of these fatty acids eventually reach the depot and milk fats of these animals, and ultimately traces find their way into the human diet. Elaidic acid has already been mentioned, and vaccenic acid, *cis*-11-octadecenoic acid, is another.

Conjugated Linoleic Acids

Of all the naturally occurring fatty acids containing *trans* double bonds, the conjugated linoleic acids (CLAs) are the among the most interesting due to their apparent anti-cancer functionality.

The two double bonds in CLAs are not in their usual positions, but are conjugated (*see* page 98), starting at carbons 9, 10 or 11, and most commonly with at least one of the two bonds in the *trans* configuration. The most abundant CLAs are $18:2\Delta cis9$, *trans*11 (4.8) and $18:2\Delta$ *trans*10,*cis*12 (4.9). The CLA content of dairy products based on cow's milk is around 5 to $10\,\mathrm{mg\,g^{-1}}$ of total fat. Similar levels are found in the depot fat and milk of other ruminants but much lower levels are found in non-ruminants (monogastrics), including horses, pigs and humans. CLAs arise as by-products of the hydrogenation of polyunsaturated fatty acids by specialist rumen bacteria such as *Butyrovibrio fibrisolvens*. Vegetable oils and seafoods contain much lower levels of CLAs, particularly the most abundant *cis*9, *trans*11 type, which have been shown to be effective anti-mutagens[iii] and antioxidants (*see* page 123). In experiments on laboratory rodents, dietary levels containing around 1% CLAs have proved effective anti-tumour agents; this represents 5 to 10 times the dosage normally found in a human diet containing normal levels of CLA-rich fats. Details of the proposed mechanisms of these effects are outside the scope of this book. Research in this area is some way from the stage of clinical studies with humans, but this has not prevented CLAs (prepared in laboratories by a chemical isomerisation) being enthusiastically taken up as dietary supplements.

[iii] Using the well-known Ames test for mutagens and potential carcinogens.

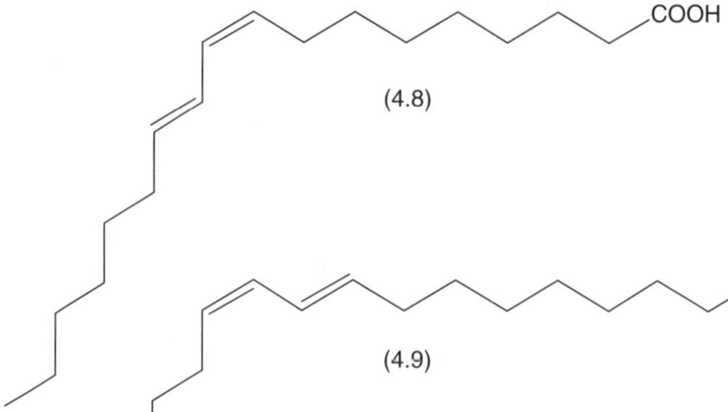

(4.8)

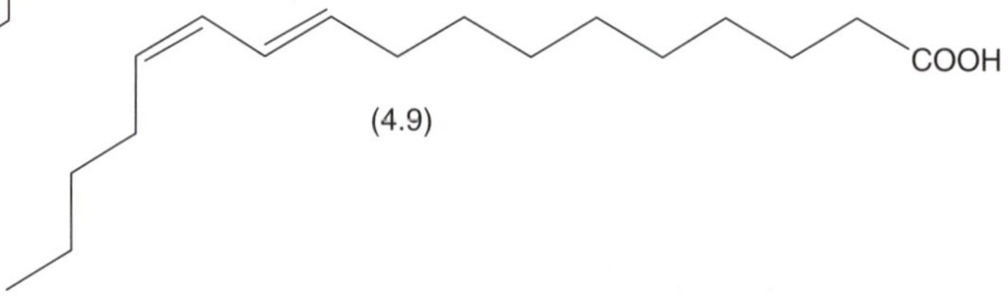

(4.9)

ESSENTIAL FATTY ACIDS

Fatty acids in the form of the triglycerides of dietary fats and oils provide a major proportion of our energy requirements. Unfortunately when present in excess they contribute to the unwelcome burden of superfluous adipose tissue that so many of us carry. In recent years we have begun to appreciate that certain dietary fatty acids have a more specific function in human nutrition. Rats fed on a totally fat-free diet show a wide range of acute symptoms affecting the skin, vascular system, reproductive organs and lipid metabolism. Although no corresponding disease state has ever been recorded in a human patient, similar types of skin disorder have been observed in children subjected to a fat-free diet. The symptoms in rats can be eliminated by feeding linoleic or arachidonic acids (which in consequence became known for a time as vitamin F), and it is generally accepted that 2–10 g of linoleic acid per day is sufficient to meet an adult human's requirements.

The identification of these two "essential fatty acids" in the 1930s preceded by some 25 years their identification as precursors of a group of animal hormones, the prostaglandins.[iv] Although normally regarded as the province of biochemists interested in metabolic pathways, an insight into the various interconversions of unsaturated fatty acids is useful for food chemists seeking to understand their nutritional status,

---

[iv] Hormones are defined as substances which affect cellular processes at sites in the body distant from the tissue that produces them. Prostaglandins are not therefore true hormones, in that they exert their effects locally and the effects are specific to that tissue, often in the mediation or facilitation of the action of other hormones.

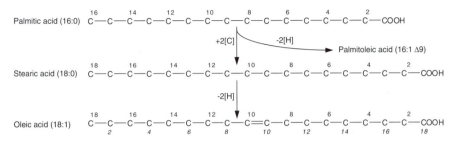

Figure 4.3 The biosynthesis of oleic acid from palmitic acid. For clarity the hydrogen atoms of the hydrocarbon chain have been omitted. In the illustration + [2C] indicates the addition of (CH_2CH_2) at the carboxyl end of the chain and -[2H] represents oxidation to form a double bond. A biochemistry textbook will provide details of these reactions, which also feature below in Figures 4.4 and 4.5. The numbering of the carbon atoms is the usual one, starting from the carboxyl end. However, when complex structural relationships of unsaturated fatty acids are being considered it is common to classify them by the distance of the last double bond from the terminal methyl group of the chain, the so-called ω-carbon. Oleic acid, as shown here, has this bond at the ω-9 position, making it a member of the ω-9 series. (Some authorities use the letter *n* in place of ω.)

aided by the grouping of unsaturated fatty acids into three series, known as ω-3, ω-6 and ω-9.

Animals, including humans, are equipped to synthesise saturated fatty acids up to 18 carbon atoms in chain length. However, desaturation, *i.e.* the insertion of double bonds, normally only takes place between C-9 and C-10, as shown in Figure 4.3, or much more rarely at other positions nearer to the carboxyl group. This means that animals cannot synthesise linoleic acid, but humans are able to convert it to arachidonic acid, Figure 4.4. Arachidonic acid is not found in plants. This is of little consequence to humans but of great significance to cats. In the course of evolution they have lost, but not missed, the ability to convert linoleic acid into arachidonic acid. Being carnivores, cats have always been able to obtain sufficient arachidonic acid from their diet. This only becomes a serious problem with cats whose owners attempt to convert them to vegetarians.[v] Fortunately most cats should be in a position to supplement such a misguided and unnatural diet with the arachidonic acid supplied by eating small mammals, birds, *etc.*

The normal isomer of linolenic acid, sometimes referred to as α-linolenic acid (18:3Δ9,12,15), is formed in plants by insertion of a third double bond into linoleic acid.

---

[v] . . . and by inference the domestic cat's larger cousins. The author is unaware of any attempts to discover whether lions or tigers might be persuaded to become vegetarians.

The fatty acids shown in Figure 4.4 are all members of the ω-6 series, since their first double bond lies between the sixth and seventh carbon atoms, counting from the terminal methyl group.

α-Linolenic acid is the starting point for the formation of the ω-3 series of fatty acids, Figure 4.5. This sequence occurs in unicellular algae, and one might be excused for questioning their relevance to food chemistry. However, fish consume algae, and the oils of fatty fish such as herring, mackerel, salmon and trout provide ω-3 fatty acids in human diets.

The prostaglandins are now recognised as being just one class of a group of hormones known as the eicosanoids. Other eicosanoids include the thromboxanes and leukotrienes. There are a great many different prostaglandins but their formation always follows the same pattern, typified by that for prostaglandin E_2, shown in Figure 4.6. Although the list of physiological activities in which prostaglandins are known to be involved continues to grow, they are best known for their role in inflammation and in the contraction of smooth muscle. Thromboxanes play a role in the aggregation of platelets, part of the process of blood clot formation.

Fatty Acids and Coronary Heart Disease

The involvement of dietary fatty acids in the occurrence of athero-sclerosis, particularly heart disease, is a complex issue. Nutritional guidelines focus on reducing the contribution which fats make to the total energy content of the diet and on lowering the proportion of saturated fatty acids in dietary fat. However, it is becoming increasingly clear that the apparently simple relationships (which formed the scientific basis of these guidelines) between dietary lipids, blood cholesterol levels and the risk of coronary heart disease, are not simple at all. A diverse range of other risk factors and refinements are being identified which help to explain the apparent inconsistencies of the so-called "lipid hypothesis" of coronary heart disease (CHD). While some of these risk factors have an obvious nutritional element others, such as ethnicity, employment, smoking and family history, do not.

Thus, when the fatty acids in the diet are considered it is now believed important to discriminate between saturated fatty acids of different chain lengths, as well as those which are mono- and polyunsaturated. Currently the most popular candidates for the true villains of the piece appear to be myristic, lauric and palmitic acids. Stearic acid is considered to be much less effective at raising plasma cholesterol levels. The difficulty is that fats relatively rich in stearic acid are usually well

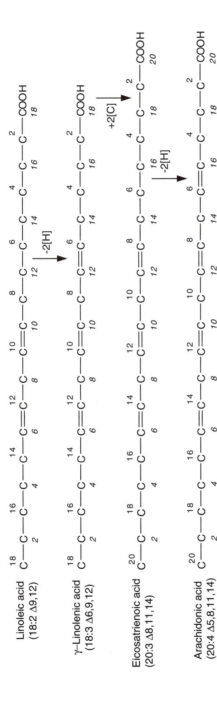

Figure 4.4 The biosynthesis of arachidonic acid from linoleic acid: the ω-6 series. The two intermediate fatty acids are found only in trace amounts in most species, except that γ-linolenic acid replaces the α-analogue in a handful of higher plants, notably the evening primrose and borage, and the *Phycomycete* fungi, the class which includes the familiar moulds, *Mucor* and *Rhizopus*. The theoretical justification for the quasi-pharmaceutical use of evening primrose oil is that our natural synthesis of arachidonic acid may need some encouragement and might benefit from an enhanced supply of its precursor, γ-linolenic acid.

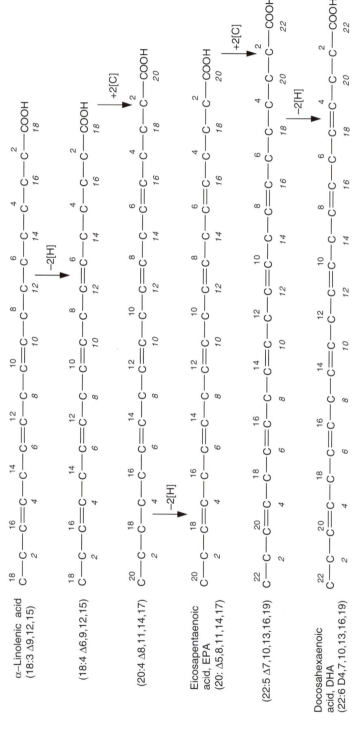

Figure 4.5 The formation of ω-3 series fatty acids in algae, the origin of those found in fatty fish.

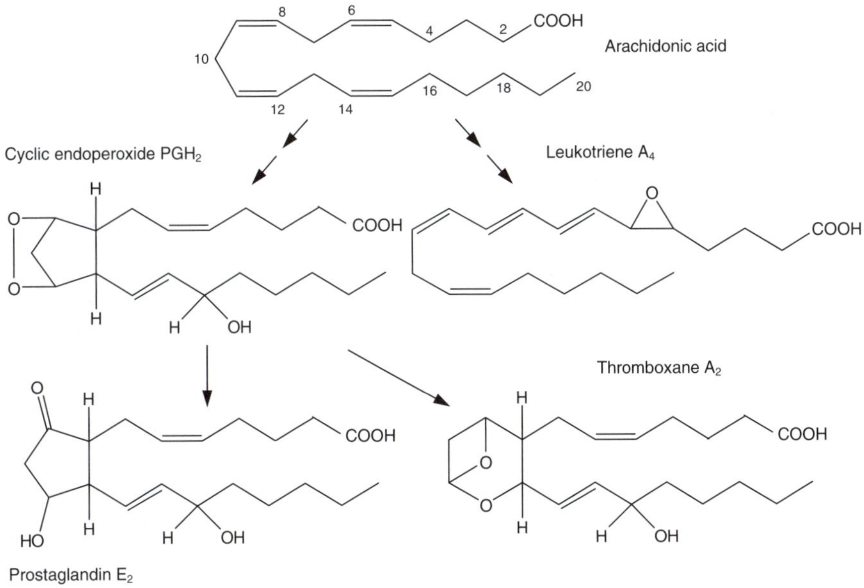

Figure 4.6 The biosynthesis of eicosanoids from arachidonic acid. The numbering of arachidonic acid carbon atoms will facilitate comparison with the unfolded structure in Figure 4.4. The F series prostaglandins differ in that both oxygen atoms attached to the ring are present as hydroxyl groups. This pathway also shows why the position of the double bonds in essential fatty acids is critical. Prostaglandin E_1 is derived from eicosatrienoic acid (C20:3Δ8,11,14) (Figure 4.4) and therefore lacks the C5–6 double bond. Similarly, prostaglandin E_3, derived from eicosapentaenoic acid (Figure 4.5), has an extra double bond at C17–18.

endowed with palmitic acid as well. The pattern of unsaturation is also important, as indicated by the remarkably low incidence of arterial disease among the Inuit people of the Arctic, in spite of a diet that appears to break all the usual nutritional rules. However their traditional diet, although extremely fatty, is very rich in polyunsaturated fatty acids of the ω-3 series. Eicosanoids from ω-3 fatty acids are generally less potent than those from the ω-6 series in promoting the formation of the blood clots involved in CHD. This observation, and subsequent studies of the relationship between fatty fish consumption and the incidence of heart disease, forms the basis of the recommendation that we should all eat more fatty fish.

The observation that the occurrence of CHD is linked to the peroxidation (*i.e.* oxidation involving the formation of peroxides) of blood lipids has drawn attention to the possible role of the "antioxidant vitamins", A, E, and C, in addition to other naturally occurring antioxidants, and hence fruit and green vegetable consumption. The function of antioxidants, substances which interfere in the oxidation of unsaturated fatty acids, is discussed in detail later in this Chapter. There are now indications that many of the differences in the incidence of CHD between population groups (*e.g.* between the north and south of the British Isles) which have defied explanation in terms of dietary fat consumption can be correlated with the intake of fruit and green vegetables.

The evidence that changing one's diet, *e.g.* by replacing butter with margarine, actually lowers the risk of CHD by a significant amount is also less clear-cut than the advertisements for margarine "rich in poly-unsaturates" might imply. The account of *trans* acids in the next section shows that the benefits of margarine may not be all that they appear to be. Similarly, the fatty acid compositions presented in Table 4.3 show that animal fats can be highly unsaturated (*e.g.* fish oils) and plant fats can be the reverse (*e.g.* coconut oil). The observation that the level of cholesterol in the diet has little or no influence on the level of cholesterol in the blood is also widely ignored by the advertisers of products derived from plant oils.

The brevity of this brief account of the complexity of the CHD issue confirms that it has now moved well beyond the confines of food chemistry.

REACTIONS OF UNSATURATED FATTY ACIDS

Unsaturated fatty acids take part in a number of chemical reactions important to the food scientist, the hydrocarbon chain of the saturated fatty acids being essentially chemically inert under the conditions encountered in food. These reactions do not involve the carboxyl group, and since most occur regardless of whether or not the fatty acid is esterified with glycerol, it is convenient to deal with them at this point rather than when the properties of the fats and oils are considered.

Halogens react readily with the double bonds of fatty acids, and the stoichiometry of the reaction has been utilised for many years as an indication of the proportion of unsaturated fatty acids in a fat. The iodine number of a fat, *i.e.* the number of grams of iodine which react with 100 g of fat, is still quoted in older literature as a measure of the overall degree of unsaturation. However, for most purposes it has been replaced by gas chromatography of the volatile methyl esters of the fatty acids obtained by saponification and methylation of the fat. This can provide complete data on the fatty acid composition of an unknown fat.

Hydrogenation, Margarine and *trans* Fatty Acids

In the presence of hydrogen and a suitable catalyst the double bonds of unsaturated fatty acids are readily hydrogenated, a reaction at the heart of modern margarine manufacture.[vi] The aim of hydrogenation is to convert a liquid vegetable or fish oil to a fat with a butter-like consistency, by reducing the degree of unsaturation in its component fatty acids. At the present time soybean oil and palm oil, which together account for approximately 60% of worldwide consumption of fats and oils, are the major raw materials used in the manufacture of margarine and other hydrogenated fats.

After preliminary purification to remove polar lipids and other substances which tend to poison the catalyst, the oils are exposed to hydrogen gas at high pressures and temperatures (2–10 atmospheres, 160–220 °C) in the presence of 0.01–0.2% of a finely divided nickel catalyst. Under these conditions the reduction of the double bonds is fairly selective. Those furthest from the ester link of the triglycerides and those belonging to the most highly unsaturated fatty acid residues are the most reactive. This selectivity fortunately results in trienoic acids being converted to dienoic acids and dienoic to enoic acids, rather than producing an accumulation of fully saturated acids. The margarine manufacturer is of course able to control the degree of hydrogenation to give the desired characteristics to his particular finished product.

The hydrogenation reactions are by no means as straightforward as one might imagine. (The first of the Special Topics at the end of the chapter has additional details.) Isomerisation of *cis* to *trans* occurs under the extreme conditions of the hydrogenation process, *trans* being the thermodynamically preferred configuration of the double bond. Double bonds also tend to migrate during the process, so for instance a fatty acid chain starting with a *cis* double bond at C-9 might end up with either a *cis* or *trans* double bond anywhere between C-4 and C-16.

The proportions of these isomeric fatty acids vary widely between different samples of margarines and other hydrogenated fats, but typical data are shown in Table 4.5. The total amount of *trans* fatty acids can range from 5% in lightly hydrogenated vegetable oils, to more than 40% in commercial frying fats. (Light hydrogenation of cheaper vegetable oils is often carried out to reduce their polyunsaturated fatty acid content and thereby their tendency to rancidity, as discussed below.

[vi] The first margarine was invented by a French chemist, Hippolyte Mège Mouriès in 1869 under the name *beurre économique*. It was a mixture of milk, chopped cow's udder and a low-melting fraction of beef fat. The later name, margarine, was derived from the Greek *margaritas*, meaning "pearl-like"!

Table 4.5 Isomers of C18 monoene and diene fatty acids in margarine. (Derived from data quoted by Craig-Schmidt in *Fatty Acids and their Health Implications*, ed. C. K. Chow, Dekker, New York, 1992.)

Isomer	Total fatty acids (%)	Isomer of C18:1 trans	Total fatty acids (%)
		$\Delta 7$	0.5
		$\Delta 8$	1.3
		$\Delta 9$	2.5
		$\Delta 10$	3.1
C18:1 *trans*	14.8	$\Delta 11$	2.8
		$\Delta 12$	2.1
		$\Delta 13$	1.4
		$\Delta 14$	0.7
		$\Delta 15$	0.4
C18:2 *trans, cis*	1.3		
C18:2 *cis, trans*	1.3		
C18:2 *trans, trans*	0.3		

Until the late 1990s the total consumption of *trans* fatty acids was quite high. The average intake figure for UK adults was around 5 g per day, corresponding to over 2% of the total energy content of the diet. Individual consumption of over 12 g per day was not uncommon. The contribution of different foods to *trans* acid intake differs widely from one individual to another. Spreading fats, used on bread, contribute about one third and shortening fats, used in pastry and biscuits, contribute a similar amount. Although the *trans* acid content of animal fats is low,[vii] dairy products and meat can provide around one-tenth of total consumption. A typical "fast food" meal comprising cheeseburger, French fries and a milk shake, with a pastry for dessert, could have supplied almost 4 g of *trans* acids.

In spite of their presence in our diet since the early years of this century, the metabolism and the possible effects on health of *trans* acids have not received much attention until recently. It is reasonable to conclude that any clearly defined toxic effects would have come to light by now. Humans are equipped with appropriate enzymes to metabolise at least some *trans* acids, but it remains to be shown how we can cope with some of the more exotic positional isomers. There are hints from epidemiological studies of a positive correlation between mortality from arterial disease and the consumption of hydrogenated fats. A possible

---

[vii] The fats of ruminants, including the intramuscular and milk fats, contain traces of these fatty acids, which, like the short-chain and branched fatty acids considered earlier in this chapter, are by-products of the hydrogenation reactions carried out by bacteria in the rumen.

explanation appears to be that the body reacts towards *trans* unsaturated fatty acids as though they were saturated. This would certainly apply in the case of their contribution to the physical properties of membrane lipids. However there is evidence that dietary *trans* fatty acids tend to raise levels of low density lipoprotein (LDL) cholesterol and reduce levels of high density lipoprotein (HDL) cholesterol. It is elevated levels of LDL cholesterol which are specifically associated with the incidence of atherosclerosis, and in this respect *trans* unsaturated fatty acids should be regarded as being worse for health than saturated fatty acids. It has thus become desirable to exclude *trans* fatty acids from the total of mono- and polyunsaturated fatty acids when analyses of margarines, *etc.*, are quoted on labels. Since 2006 legislation in the USA requires that the *trans* fatty acid content of food products containing more than 0.5 g per serving must be declared on the label. Unfortunately, there appear to be no plans at present for similar legislation in Europe. The measurement of the *trans* acid contents of food products is relatively easy. In addition to gas–liquid chromatography they may be detected by their characteristic infrared absorption bands (965–990 cm^{-1}).

The *trans* acid problem is likely to prove short-lived. Later data (1999) from the oils and fats industry show that very few hydrogenated fat products, all of which are shortenings (in food industry parlance "white fats"), have *trans* acid contents as high as 10%. Most, including a wide range of margarines and low fat spreads (yellow fats!), now contain 1% or less. The most recently published data (2007) suggest that in Britain at least the contribution of *trans* fatty acids to total energy consumption has fallen from its 1980s figures of over 2% to 1.5% or less. This has been achieved by means of an adaptation of the interesterification process, to be discussed later in this chapter once the properties of triglycerides have been considered in more detail.

Rancidity

Rancidity is a familiar indication of the deterioration of fats and oils, leading to unpleasant smells. There are two quite distinct mechanisms by which this can arise, lipolytic rancidity and oxidative rancidity.

Lipolytic rancidity can be a problem in dairy fats, particularly butter. It results when lipases secreted by the microbial flora catalyse the hydrolysis of the triglycerides in the fat, liberating short-chain fatty acids (*see* Table 4.3). Although low concentrations of these acids make an important contribution to the desirable flavour of butter, too much can produce flavours resembling over-ripe cheese. On the other hand these enzymic reactions do in fact have an essential role in cheese making. Lipases from the moulds which develop as the cheese matures liberate butyric acid, caproic acid, *etc.*, which constitute a major element of the flavour of cheese. Proteolytic enzymes (proteases) from the same moulds liberate peptides and free amino acids (*see* Chapter 5) from the milk proteins, which provide the other major component of cheese flavour. For many of us, Stilton, Camembert and Gorgonzola, and their innumerable country cousins, are what we understand by cheese—the yellow plastic slice in the cheeseburger is not. Nevertheless we have to accept that vast numbers of cheeseburgers do get sold every day, and for such a large market any reduction in the maturation period of cheese makes financial sense. So-called "accelerated cheese ripening" preparations, containing both lipases and proteases, are included nowadays and are used extensively in the manufacture of cheese for applications such as cheeseburgers.

Oxidative rancidity, which has no redeeming features, affects fats and oils and the fatty parts of meat and fish, and is the result of the autoxidation of unsaturated fatty acids. The course of the reaction has been studied in free fatty acids (as shown here) but there is no reason to suppose that the behaviour of fatty acids esterified in the form of triglycerides is significantly different.

This sequence of reactions is traditionally presented in three stages, initiation, propagation, and termination, shown in outline in Figure 4.7. The initiation reactions give rise to small numbers of highly reactive fatty acid molecules containing unpaired electrons, *free radicals*. These are shown as $R^\bullet$ in Figure 4.7a. The dot ($^\bullet$) shows the presence of an unpaired electron and R denotes the remainder of the molecule. Most free radicals are short-lived and highly reactive, as they seek a partner for their unpaired electron.

Initiation: $X^\bullet + RH \rightarrow R^\bullet + XH$ (a)

Propagation:
$$R^\bullet + O_2 \rightarrow ROO^\bullet \tag{b}$$
$$ROO^\bullet + RH \rightarrow ROOH + R^\bullet \tag{c}$$
$$2ROOH \rightarrow RO^\bullet + ROO^\bullet + H_2O \tag{d}$$

Termination: $R^\bullet, RO^\bullet, ROO^\bullet \rightarrow$ stable, non-propagating species (e)

Figure 4.7   An outline of the reactions involved in the autoxidation of unsaturated fatty acids. The various stages, (a) to (e), are discussed in the text.

In the propagation reactions, atmospheric oxygen reacts with $R^{\bullet}$ to generate peroxy radicals, $ROO^{\bullet}$ (Figure 4.7b). These are also highly reactive and react with other unsaturated fatty acids to produce hydroperoxides, ROOH, and regenerating the free radical, $R^{\bullet}$ (Figure 4.7c). Do not confuse the hydroperoxide group with a carboxyl group; in a hydroperoxide the hydroxyl group –OH is attached directly to the other oxygen atom, not to a carbonyl carbon as in a carboxyl group.

This free radical can then repeat the process, forming a chain reaction (Figure 4.7b), but the hydroperoxide is able to break down to give other free radicals (Figure 4.7d). These too can behave much as $ROO^{\bullet}$ did. The result is that increasing numbers of free radicals accumulate in the fat, and these absorb considerable quantities of oxygen from the air. Eventually the concentration of free radicals reaches the point when they react with one other to produce stable end-products (Figure 4.7e). These are known as termination reactions.

There used to be great uncertainty about the nature of the reactions that give rise to the first free-radical species $(X^{\bullet})$ which initiates the autoxidation sequence. It is now accepted that the initiation reaction involves a short-lived but highly reactive, high-energy form of oxygen known as singlet oxygen, $^1O_2$. Singlet oxygen molecules arise from the normal low-energy ground-state oxygen (triplet oxygen, $^3O_2$) in a number of reactions, but most relevant to food systems are those in which oxygen reacts with pigments such as chlorophyll, riboflavin (very important in the deterioration of fresh milk), and haem, in the presence of light. The second of the Special Topics at the end of this chapter provides more details of the nature of singlet and triplet oxygen. As shown in Figure 4.8, the singlet oxygen reacts with the double bond of an unsaturated fatty acid. In the reaction the double bond changes its configuration from *cis* to *trans*, and at the same time it moves along the chain.

Autoxidation proper begins as these first hydroperoxides decompose to hydroperoxy and alkoxy radicals, Figure 4.9. These are sufficiently

Figure 4.8 The reaction between a singlet oxygen and a double bond to form a hydroperoxide. The curly arrows show serious organic chemists how the electrons migrate during the course of the reaction, but they can be ignored by the rest of us.

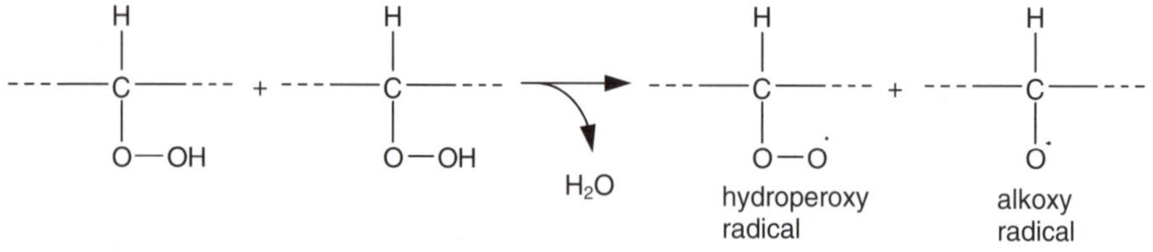

Figure 4.9 The breakdown of hydroperoxides.

reactive to abstract H•, *i.e.* a hydrogen atom, or more correctly a hydrogen radical, from particularly vulnerable sites in monoenoic (*e.g.* oleic) and polyenoic (*e.g.* linoleic) fatty acid residues in the fat.

The most vulnerable sites are the CH_2 groups adjacent to double bonds, known as α-methylene groups. A hydrogen atom of an α-methylene group is particularly labile, *i.e.* easily abstracted. Not surprisingly, the most vulnerable α-methylenes are those lying between neighbouring double bonds in a polyunsaturated fatty acid, *e.g.* number 11 in linoleic acid, as shown in Figure 4.10a and b. Once the hydrogen radical has been abstracted, the unpaired electron can migrate, leading to the formation of *trans* double bonds. Reaction with oxygen then leads (Figure 4.10c) to the formation of hydroperoxy radicals. These abstract hydrogen from other fatty acids, forming hydroperoxides and continuing the chain reaction. Oleic acid gives rise to 8-, 9-, 10-, and 11-hydroperoxides in roughly equal amounts. Linoleic acid, which reacts with oxygen 10 times faster than oleic acid, gives only 9- and 13-hydroperoxides, since the stability of the conjugated diene system favours attack by oxygen at the ends of the system rather than in the middle.

As hydroperoxides accumulate in the autoxidising fat considerable quantities of oxygen are absorbed from the atmosphere, but as the chain reaction proceeds the build up of their breakdown products becomes increasingly important. As shown in Figure 4.11, the alkoxy radicals give rise to aldehydes, ketones and alcohols, in addition to a continuing supply of free radicals which maintain the chain reaction.

Transition metal cations (inevitable trace contaminants in oils which have been processed, stored or utilised in metal vessels) are important catalysts of hydroperoxide breakdown:

$$R-OOH + M^+ \rightarrow R-O^\bullet + OH^- + M^{2+}$$
$$R-OOH + M^{2+} \rightarrow R-OO^\bullet + H^+ + M^+$$
$$\overline{2R-OOH \rightarrow R-O^\bullet + R-OO^\bullet + H_2O}$$

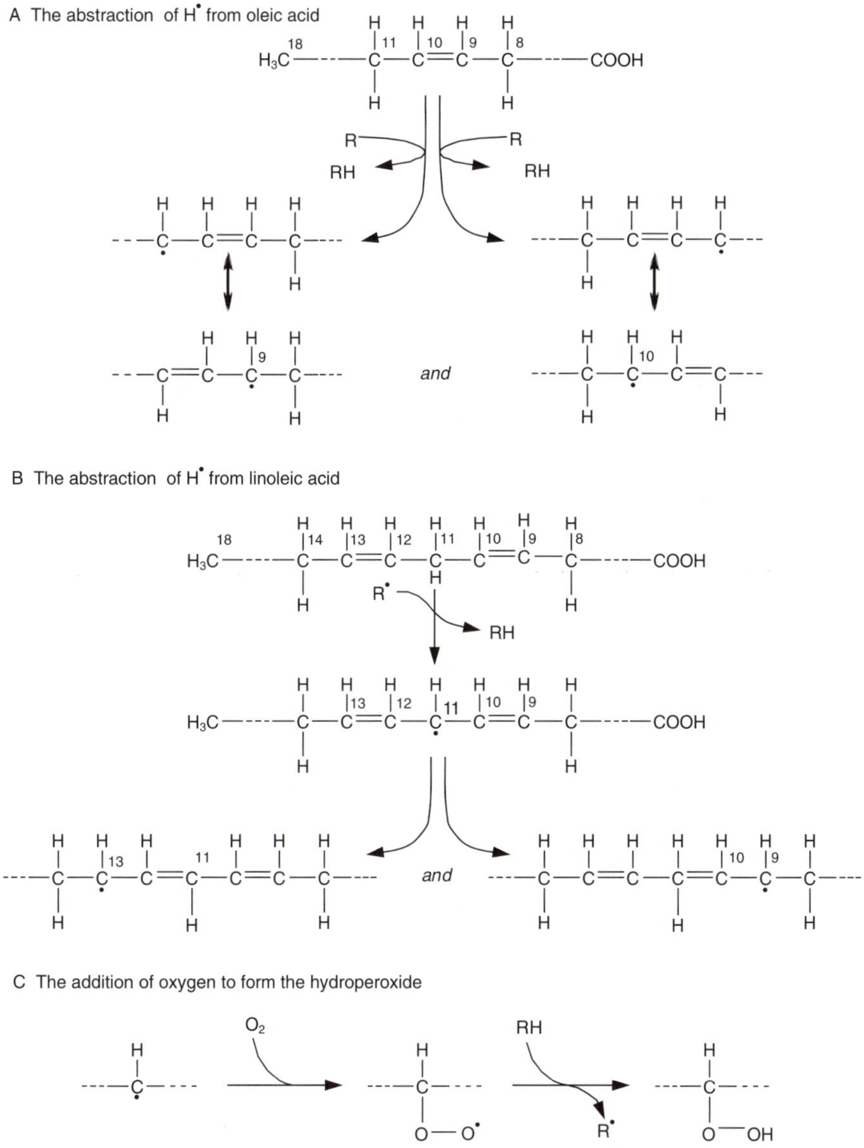

A The abstraction of H˙ from oleic acid

B The abstraction of H˙ from linoleic acid

C The addition of oxygen to form the hydroperoxide

Figure 4.10 The formation of hydroperoxides in the propagation phase of autoxidation.

Aldehydes arising from cleavage of the carbon chain on either side of the alkoxy radical are the source of the characteristic odour of rancid fat. For example, the 9-hydroperoxide from linoleic acid will give either 2-nonenal or 2,4-decadienal, as shown in Figure 4.12.

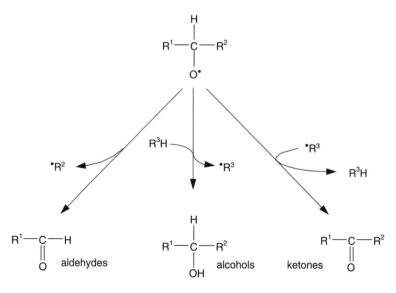

Figure 4.11 The formation of stable end-products from alkoxy radicals.

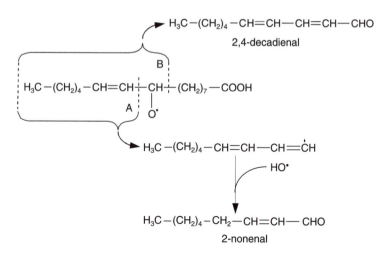

Figure 4.12 The formation of unsaturated aldehydes during autoxidation. Cleavage can occur on either side of the alkoxy group, at A or B.

One particular compound, malonaldehyde (4.10), results from cleavage at both ends of a diene system. It forms a pink colour with thiobarbituric acid, a reaction which forms the basis of a useful method for assessing the deterioration of fats. Other tests include measurements of the carbonyl content using 2,4-dinitrophenylhydrazine, the hydroperoxide content by reaction with iodine, and the characteristic absorbance of conjugated dienes and trienes at 230 and 270 nm, respectively.

$$HC \overset{\overset{\displaystyle H_2}{|}}{=} C - CH$$

(with the structure showing HC=O and CH=O groups)

(4.10)

Use of tests such as these, as well as monitoring the oxygen uptake, has shown that the course of oxidation of a fat is marked by an induction period of slow oxygen uptake while the level of initiating free radicals builds up, followed by a period of rapid oxidation. A pronounced rancid odour is not usually detectable until the rapid oxidation phase is well established, although some oils can develop off-flavours and odours even when little oxidation has occurred. Soya bean oil is particularly susceptible to this phenomenon, termed reversion, apparently due to its content of the highly labile linoleic and linolenic acids.[viii] Among the compounds which have been implicated in the "flavour" of reverted soya bean oil are *cis*-3-hexenal, diacetyl (butanedione), 2,3-pentanedione, and 2,4-pentadienal. Reduction in the linolenic acid content of seed oils is an important objective for plant breeders.

Fats and oils exposed to the atmosphere and to heating over a long period exhibit the final stage of the oxidation sequence, polymerisation. The highly unsaturated oils used in paint, *e.g.* linseed oil, show the phenomenon even more readily. The cross-linking reactions can be of various types, as shown in Figure 4.13.

Polymerisation reactions are particularly significant in frying oils, where prolonged use leads to high molecular weight compounds which cause foaming and increased viscosity. Discarded oils are sometimes found to contain as much as 25% of polymerised material. The chain reaction characteristic of autoxidation reactions provides the reason why a half empty bottle of oil approaching the end of its useful life should be discarded in its entirety rather than topped up with fresh oil.

The possibility that lipid oxidation products are toxic to humans remains unresolved. The most important source in the human diet is deep-fried foods, although fatty fish and their oils may be significant in some diets. In order to produce measurable levels of adverse effects,

---

[viii] In recent years plant breeders, using genetic engineering techniques, have developed soya bean varieties in which the proportions of linoleic acid have been reduced from 56% to 2% and linolenic acid from 8% to 2%.

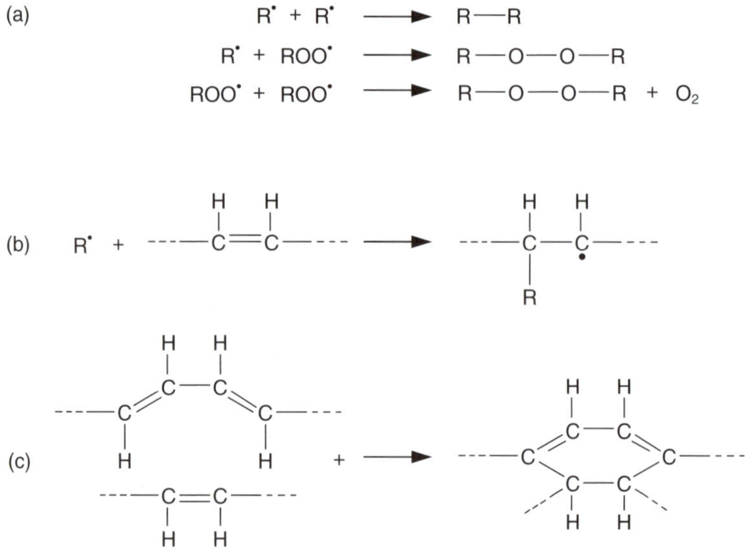

Figure 4.13 Cross-linking reactions in the formation of end-products. Free radicals may react directly together (*a*) or with other alkenic systems (*b*). Diels–Alder type reactions also occur, to give cyclic structures (*c*).

experiments with animals use much larger amounts (in relation to body weight and life span) than would be encountered by humans. Feeding these large amounts of highly oxidised fats induces a wide range of symptoms, most of them corresponding to vitamin E deficiency. It is assumed that the high intake of oxidised fats overwhelms the body's natural antioxidant systems (*see* page 111).

In recent years attention has been focused on the potentially harmful involvement of lipid oxidation products in arterial disease, particularly those derived from cholesterol. The mechanisms of these effects are complex and beyond the scope of this book. As is so often the case the link between experimental results using laboratory animals and the incidence of human disease is a tenuous one and is vulnerable to criticism from scientists and others who may have different priorities. Suggestions, not wholly endorsed by other workers, have been made that the elevated levels of cholesterol oxides in clarified butter (ghee) used in cooking may account for the high level of atherosclerosis among Britain's Asian community. Cholesterol oxides have also been recorded at 50 ppm in French fries from fast-food outlets.

It is commonly assumed that at least some of the oxidation products of fats may be carcinogenic, indeed some have been definitely shown to be mutagenic. The observation that many antioxidants have some anti-carcinogenic activity supports this view. However, there is little evidence at the present time to suggest that oxidised fats do cause cancer in humans. Perhaps the adverse effects of a diet over-rich in fats generally swamp these more subtle influences.

Antioxidants

Although the development of rancidity in bulk fats and oils can be retarded by careful processing procedures avoiding high temperatures, metal contamination, *etc.*, such measures are never wholly effective. Fatty foods such as biscuits and pastry are particularly susceptible to rancidity, as their structure necessarily exposes the maximum surface of the fat to the atmosphere. The shelf life of these types of foods (on the customer's shelf as well as the shopkeeper's) can be massively extended by the inclusion of antioxidants in many of the fats we buy, particularly in lard. Antioxidants are substances used as food additives to retard the autoxidation reactions. Some are wholly synthetic products of the chemical industry, but antioxidants of natural origin are becoming more significant as consumers and the food industry seek out what are perceived to be healthier alternatives. As mentioned earlier in this chapter, the involvement of lipid peroxidation reactions in arterial disease has focused attention on the role of antioxidants as dietary components in their own right, not merely as protectors of fats prior to consumption.

There are two well recognised mechanisms of antioxidant action. In the best known, particularly associated with the synthetic antioxidants such as BHA and BHT (*see* Figure 4.14), the antioxidant blocks the propagation phase of the autoxidation process (Figure 4.15).

Many other antioxidants, most famously β-carotene, lycopene and other carotenoids (*see* Chapter 6), operate by a singlet quenching mechanism. These molecules react with singlet oxygen, $^1O_2$, returning it to the relatively unreactive triplet state:

$$^1O_2 + \text{carotenoid} \rightarrow {}^3O_2 + \text{carotenoid}^*(\text{excited})$$

$$\text{carotenoid}^* \rightarrow \text{carotenoid} + \text{heat}$$

In biological systems a number of highly reactive free radical forms of oxygen are important, such as the superoxide anion, $O_2^{\bullet-}$ (an oxygen molecule carrying an extra unpaired electron, giving it a negative charge), and the hydroxyl radical, $^{\bullet}OH$ (a hydroxyl ion lacking an electron). These arise as by-products of enzyme and other reactions

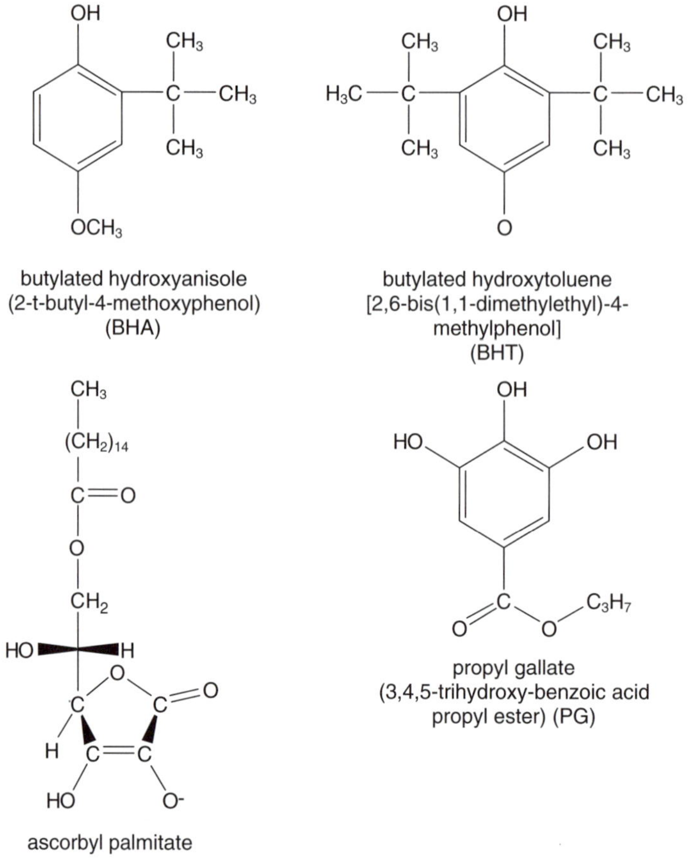

Figure 4.14 The chemical structure of the best known synthetic antioxidants used in food.

involving oxygen (*see* Figure 5.12) and it appears likely that singlet quenching antioxidants also react with them.

Antioxidants do not reduce the ultimate degree of rancidity; they merely extend the induction period in rough proportion to their concentration. Thus the inclusion of 0.1% BHA in the lard used to make pastry could be expected to add at least a month to the period, otherwise only a few days, before obvious signs of rancidity develop. The three entirely synthetic antioxidants shown in Figure 4.14 (BHA, BHT and PG) are normally added to fats at rather lower levels than this, normally up to about 200 ppm. There is growing concern about the safety of synthetic antioxidants: some laboratory studies with animals have hinted at carcinogenicity. Although this issue is far from resolved there is growing pressure for their withdrawal. However, it seems likely that they will not

(a)   AH + ROO$^\bullet$ $\longrightarrow$ ROOH + A$^\bullet$

(b)

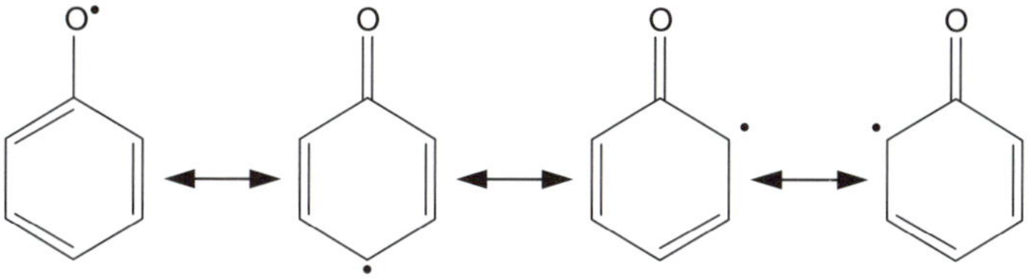

(c)   A$^\bullet$ + A$^\bullet$ $\longrightarrow$ A—A   A$^\bullet$ + ROO$^\bullet$ $\longrightarrow$ ROO—A

Figure 4.15   Free radical stabilisation by resonance. (a) A free radical intermediate in the lipid autoxidation reaction is stabilised when an antioxidant molecule, AH, donates a hydrogen atom; (b) the antioxidant free radical is stabilised by resonance and is consequently insufficiently reactive to propagate the autoxidation sequence; and (c) antioxidant radicals interreact to terminate the process.

be banned completely until (a) the carcinogenicity question is clarified by more research, and (b) more progress is made towards their replacement with more natural antioxidants such as the tocopherols and fatty acid derivatives of ascorbic acid.

Tocopherols occur in most plant tissues, as much as 0.1% in vegetable oils. Animals require tocopherols in their diet, as vitamin E, and these are therefore discussed more fully in Chapter 8. Two other vitamins, ascorbic acid and retinol, also have antioxidant activity, as discussed in Chapter 8. There are indications from both epidemiological studies and the results of experimental dietary supplementation that the "antioxidant vitamins" can have a collective beneficial effect on arterial disease. The explanation for this effect is that the progress of atherosclerosis involves the peroxidation of the lipid components of lipoproteins. It remains to be seen whether the dramatic differences in the incidence of heart disease between different populations will prove to be at least as much a function of vitamin intake as it is of fat consumption.

Some other sources of antioxidants are beginning to attract attention, notably tea and spices. In the Far East consumption of green (*i.e.* unfermented, *see* 240) tea has long been associated with good health. Catechin (*see* Figure 6.11), one of the principal polyphenolic compounds in green tea, has been shown in laboratory experiments to be a most effective free radical scavenger and inhibitor of lipid peroxidation. Theaflavins (6.20), also polyphenolics and the major red pigments

in black (*i.e.* fermented) tea, are also very effective free radical sca-
vengers. It remains to be seen whether these activities translate from the
laboratory to normal human nutrition. The same reservations must
apply to the enthusiasm currently being shown for the polyphenolic
anthocyanin pigments found in red and purple species of fruit. These
will be considered in detail in Chapter 6.

Spices have traditionally been used to add to the flavour of foods,
or to mask off-flavours, but modern research has shown that many
spices are rich in antioxidants that have the potential for use in curbing
rancidity. Some naturally occurring antioxidants are illustrated in
Figure 4.16. Their use poses a number of practical problems, including
the high costs of producing them in useful amounts, and in many
cases flavours that while desirable in some foods will be unwelcome
in others. Rosemary extracts are already finding applications in meat
products, where their antioxidant activity can be regarded as a beneficial
side-effect alongside their principal flavouring function.

The synthetic antioxidants currently permitted for use in Europe are
listed in Table 4.6.

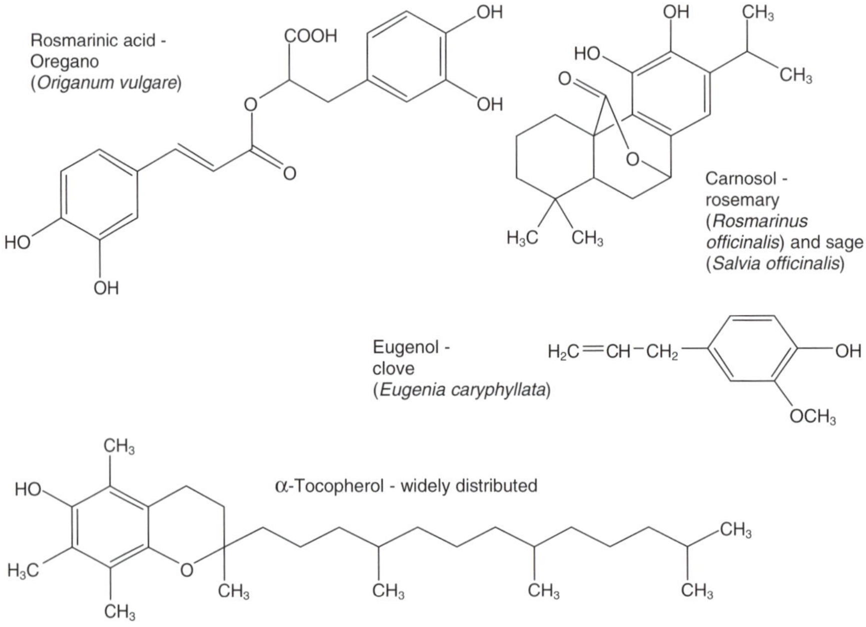

Figure 4.16   Some naturally occurring antioxidants.

Table 4.6 The synthetic antioxidants currently (2007) permitted for use in the EU.

Antioxidant	Number	Antioxidant	Number
Ascorbic acid	E300	Propyl gallate	E310
Sodium ascorbate	E301	Octyl gallate	E311
Calcium ascorbate	E302	Dodecyl gallate	E312
Fatty acid esters of ascorbic acid	E304	Erythorbic acid	E315
Tocopherols	E306	Sodium erythorbate	E316
α-tocopherol	E307	Butylated hydroxyanisole (BHA)	E320
γ-tocopherol	E308	Butylated hydroxytoluene (BHT)	E321
δ-tocopherol	E309		

TRIGLYCERIDES

Having considered the properties of the fats and oils from the point of view of the chemistry of their component fatty acids, we can now examine them in terms of their component triglycerides. The first descriptions of the glyceride structure of fats assumed that all their component triglycerides were simple compounds. Thus, a fat containing palmitic, stearic and oleic acids would be a mixture of the three triglyceride species, tripalmitin, tristearin and triolein. The first attempts to separate the component glycerides of fats employed the laborious process of fractional crystallisation from acetone (propanone) solution at low temperatures. This made it clear that much greater numbers of species of triglycerides were present than expected from this simple concept. Fats and oils became recognised as clearly defined mixtures of mixed, *e.g.* 1-palmityl, 2-linoleyl, 3-oleyl glycerol (4.11) and simple triglycerides.

(4.11)

If we reconsider our simple fat comprising only three fatty acids, there should be 27 possible triglycerides. However, as the 1- and 3-positions of the glycerol molecule are usually treated as indistinguishable, the total

Table 4.7   Detailed triglyceride composition of lard, expressed as percentages. Simple three-letter codes identify the triglycerides; thus PMS has palmitic acid at position 1, myristic acid at position 2 and stearic at position 3. O = oleic acid; L = linoleic acid. As with the fatty acid composition of a particular fat, there is considerable variation in triglyceride composition from sample to sample, although the broad features remain consistent. (Based on data quoted in *The Lipid Handbook*, ed. F. D. Gunstone, J. L. Harwood and F. B. Padley, Chapman and Hall, London, 1986.)

No double bonds:

PMS	SMS	PPP	PPS	SPS	PSS	SSS	Others
0.4	0.4	0.5	2.0	2.0	0.4	0.4	0.5

One double bond:

POP	POS	SOS	PMO	SMO	MPO	PPO	SPO	PSO	SSO	Others
0.6	1.9	1.5	0.4	0.7	0.8	7.9	12.8	0.9	1.6	0.6

Two double bonds:

POO	SOO	OMO	OPO	OSO	PPL	SPL	Others
5.2	6.1	1.6	18.4	1.2	1.8	2.1	1.5

Three double bonds:

OOO	SLO	OML	OPL	OSL	Others
11.7	0.6	0.6	7.2	1.2	0.5

Four or more double bonds:

OLL	OLO	LPL	Others
1.4	1.5	0.5	0.5

comes down to 18. This range of possibilities is illustrated in Table 4.7. It is important to note that isomeric pairs of triglycerides such as POP and PPO are clearly differentiated. Such an abundance of data for a single fat forces us to simplify triglyceride composition data when we are comparing different fats. The most striking feature of Figure 4.17 is that fats of fairly similar fatty acid composition, such as lard and cocoa butter (*see* Table 4.3), can have very different triglyceride compositions. In lard there is a definite tendency for unsaturated fatty acids to occupy the outer positions of the glycerol molecule, whereas in cocoa butter the converse is true. This is in fact a general difference between animal and plant fats, as is borne out by the further triglyceride compositions given in Figure 4.1.

If the fatty acid residues at the two outer positions are different, the four substituent groups attached to the central carbon atom of the glycerol moiety will be different. In such a case the triglyceride

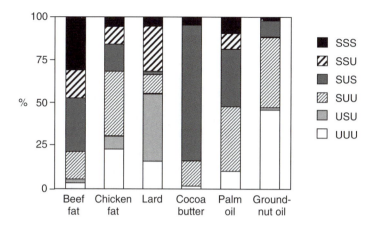

Figure 4.17 The simplified triglyceride compositions of some animal and plant fats. All saturated fatty acids are shown as S and unsaturated fatty acids as U.

molecule is asymmetric and optical activity is to be expected. It has been possible to show that a particular species of triglyceride in a natural fat occurs as a single enantiomer rather than as a racemic mixture. This indicates that the 1- and 3-positions are differentiated during triglyceride biosynthesis. However, in many fats there is in fact very little difference between the fatty acid compositions of these two positions, but in some, as implied by the data in Figure 4.18, there is a marked difference. Milk fats are exceptional in that their triglycerides fall into three broad classes. In the first, all three positions are occupied by long-chain fatty acids. The second class has long-chain acids at the 1- and 2-positions, but short-chain at position 3. The third class has medium-chain fatty acids at the 1- and 2-positions and medium or short-chain fatty acids at the 3-position.

The determination of the triglyceride composition of a fat is not easy. The need to distinguish between isomeric pairs of glycerides such as SUS and SSU, crucial to the difference between cocoa butter and lard, for example, makes special demands on chromatographic procedures. High performance liquid chromatography (HPLC), using silica gel impregnated with silver nitrate as the stationary phase, has been very successful as it can resolve these isomeric pairs. The silver ions are complexed by the double bonds of unsaturated triglycerides, whose mobility is thus reduced. The identity of a purified triglyceride can be established by means of pancreatic lipase. This enzyme specifically catalyses the hydrolysis of ester links in the 1- and 3-positions of a triglyceride, and in the laboratory, as in the small intestine, the products from a triglyceride molecule will be the two fatty acids from the

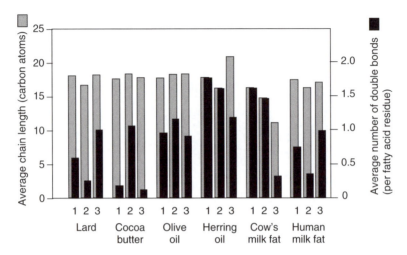

Figure 4.18   Distribution of fatty acids in a range of animal and plant fats. In order to simplify the presentation of a massive amount of data the characteristics of the fatty acids esterified at each of the three positions of the glycerides are expressed in terms of their average chain length and average number of double bonds per fatty acid residue.

outer positions and one 2-monoglyceride. The liberated fatty acids can be isolated from the reaction mixture and identified by gas chromatography.

Melting and Crystallisation

For the food chemist the melting and crystallisation characteristics of a fat are physical properties of prime importance. Although the melting points of pure triglycerides are a function of the chain lengths and the unsaturation of the component fatty acids, much as one might expect, the melting behaviour of fats is more complex. Since natural fats are mixtures, each component having its own melting point, a fat does not have a discrete melting point but a melting range. At temperatures below this range all the component triglycerides will be below their individual melting points and the fat will be completely solid. At the bottom of the range the lowest melting types, those of lowest molecular weight or most unsaturated, will liquefy. Some of the remaining solid triglycerides will probably dissolve in this liquid fraction. As the temperature is raised, the proportion of liquid to solid rises and the fat becomes increasingly plastic until, at the temperature arbitrarily defined as the "melting point", there is little or no solid fat left.

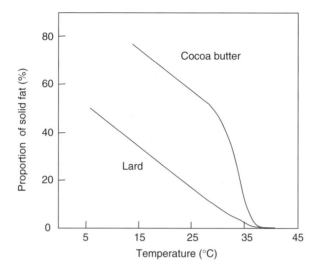

Figure 4.19   Typical melting curves for samples of cocoa butter and lard. The curves are based on data obtained by the use of the low-resolution NMR technique, the theory of which lies well beyond the scope of this book.

This melting behaviour is illustrated in Figure 4.19 for cocoa butter and lard. The key difference between the shapes of the melting curves for these two fats is the steepness of the cocoa butter curve in the 32–36 °C region. This almost unique property of cocoa butter is due to the fact that about 80% of its triglycerides are of the SUS type, namely POP ~15%, POSt ~40%, and StOSt ~25%.[ix] All of these have very similar melting points. By contrast, we have seen that lard contains a large number of different types, with no similarities between the major ones.

A second complication is that triglycerides are polymorphic, *i.e.* their molecules can pack into crystals in several different arrangements, each with its characteristic melting point and other properties. There are three basic types of crystal arrangement (polymorphic form), known as α, β′, and β, in order of increasing stability. A particular fat will sometimes show a small number of variants within the β′ and β classes, and some fats (*e.g.* cocoa butter) may demonstrate as many as six different polymorphic types, each with its own distinct melting point. The melting points of the principal forms of a number of simple triglycerides are given in Table 4.8. When a melted triglyceride is cooled rapidly it solidifies in the lowest melting, unstable, α-form. If this is slowly heated it will melt and then resolidify in the β′-form. Repetition of this

---

[ix] P = palmitic, O = oleic, St = stearic.

Table 4.8 Melting points (°C) of the polymorphic forms of simple triglycerides.

	α	β′	β
Tricaprin	−15	–	32
Trilaurin	14	43	44
Trimyristin	32	44	56
Tripalmitin	44	56	66
Tristearin	54	64	73
Triolein	−32	−13	4

procedure will bring about a transition to the final, stable, β-form. The β-form is also obtained by recrystallisation from solvent. Although each form, and many of their variations, has been fairly well characterised in terms of X-ray diffraction pattern and infrared spectrum, the actual arrangements of the triglyceride molecules in the crystals remain to be described, apart from a handful of purified triglycerides.

It is the occurrence of large numbers of different species of triglyceride in a natural fat which makes the situation so complicated, but as a rule particular fats are found to stabilise in either a β or β′ state. Important examples of β-stable fats and oils are cocoa butter, corn oil, groundnut oil, olive oil, safflower oil, and lard; β′-stable fats and oils include coconut oil, cottonseed oil, palm oil, palm kernel oil, rapeseed oil, beef fat (tallow), herring oil, whale oil and cow's milk fat. The third of the Special Topics, at the end of this chapter, provides further information about polymorphism.

The usefulness of an individual fat to a particular food application is crucially dependent on its melting and crystallisation characteristics. Fats to be spread on bread, or blended with flour for cakes or pastry, require the plasticity associated with a wide melting range. The term "shortening" is nowadays applied to all manufactured fats used in baking, *i.e.* the production of bread, cakes and pastry. The literal use of the term refers to the tendency of shortenings to reduce the cohesion of the wheat gluten strands in baked goods (*see* Chapter 5), and thereby "shorten" or soften them.

The different polymorphic types differ in the actual form of their crystals. The β′-stable fats crystallise in small needle-like crystals. At the correct mixing temperatures these fats consist of crystals embedded in a liquid fat matrix, giving a soft plastic consistency ideal for incorporating air bubbles and suspending the flour and sugar particles. The β-stable fats form large crystals, and these produce clumps which give a grainy texture. Although difficult to aerate they are valuable in pastry making.

If shortenings are blended from mixtures of β- and β'-stable fats, it is the latter which dictate the crystallisation pattern.

The fatty acid composition, and the distribution between the different positions of the triglyceride molecules, *i.e.* 1, 2 and 3, all help to determine the stable crystal form of a fat. Although there are exceptions, it is generally true that where most of the triglycerides have a broadly similar arrangement of fatty acids then it will readily form β crystals. However, fats with a more diverse mixture of triglyceride types, or those dominated by asymmetrical triglycerides (*i.e.* SSU or UUS), tend to form only β' crystals. Processes such as hydrogenation or interesterification (*see* Figure 4.17) provide the possibility of changing the physical characteristics of a fat by a number of different mechanisms. For example lard, which typically contains around 60% palmitic acid in the 2-position, normally crystallises in the β–form. Interesterification redistributes the palmitic acid around all three positions of the triglyceride, leading to a more diverse range of triglyceride types and a tendency to crystallise in the β'-form.

Cocoa Butter and Chocolate

Cocoa butter is another good example of the importance of melting properties to a foodstuff, in this case chocolate. The reason for the most notable feature of chocolate, its sharp melting point (it melts in your mouth but not in your hand!), has already been considered, but we also expect chocolate to have a very smooth texture and a glossy surface.

Cocoa butter can occur in six different polymorphic states, with melting points reported to range from 17.3 °C to 36.4 °C, as listed in Table 4.9. In fact the lowest melting states, γ and α, are only stable for longer than a few hours at temperatures below approximately −5 °C and +5 °C, respectively. The γ state has not been described in crystallographic terms. Only the fifth of these (a β-3 type, melting point 33.8 °C) has the properties required for chocolate, and the special skill of the chocolate maker lies in ensuring that the fat is in this particular state in the finished product. The liquid chocolate is first submitted to the picturesquely named "conching" procedure[x] in which the liquid chocolate is stirred continuously for several hours at 50–55 °C. This allows some undesirable volatile flavour compounds to be driven off. The next stage is tempering. The liquid chocolate is cooled to initiate crystallisation and reheated to just below the melting point of the desired

---

[x] The name comes from the conch shell shape of the vessel in which the procedure is carried out.

Table 4.9 The melting points of the different polymorphic phases of cocoa butter. Details of the crystal structure of the γ state have not been established. The two β' phases, III and IV, are nowadays regarded as the extreme members of a continuum of crystal forms. (From K. van Malssen, *et al.*, *J. Am. Oil Chem. Soc.*, 1999, **76**, p. 669.)

Phase	Crystal structure	MP (°C)
I	γ	∼ −5
II	α	∼ 5
III	β'	25.5
IV	β'	27.5
V	β	33.8
VI	β	36.3

polymorphic type in order to melt out any of the undesirable types. The chocolate is then stirred at this temperature for some time, to obtain a high proportion of the fat as very small crystals of the desired type by the time it finally solidifies in the mould, or when used to coat biscuits or confectionery.

Chocolate which has been incorrectly tempered, or has been subjected to repeated fluctuations in temperature, for example in a shop window, develops a bloom. This is a grey film which resembles mould growth, but is actually caused by the transition of some of the fat to a more stable polymorphic form which crystallises out on the surface. The migration of triglycerides from the nut centres of chocolates or the crumb of chocolate-coated biscuits can cause similar problems. Milk fat is an effective bloom inhibitor and small amounts are often included in plain chocolates, and accounts for about a quarter of the total fat in a typical milk chocolate. Cocoa butter is obviously a very difficult fat to handle successfully in the domestic kitchen and chocolate substitutes are marketed for home cake decorating. These contain fats such as hardened (*i.e.* partially hydrogenated) palm kernel oil. Although they have an inferior, somewhat greasy texture, these artificially modified fats are stable in the β' polymorphic state and therefore do not present the cook with the problem of tempering.

A fat such as hardened palm kernel oil is referred to as a "cocoa butter replacer" (CBR). Such a fat cannot be used in combination with cocoa butter, as their triglyceride composition and stable polymorphic states are different. A novel enzymic interesterification process (*see* below) for producing cocoa butter extenders (CBEs), which can be blended with cocoa butter but are rather cheaper, is now in commercial use. The inclusion of CBEs in chocolate can have other advantages, such as making it less prone to the formation of bloom. It is worth noting that since 2003 fats produced by enzymic interesterification have not been permitted for use in chocolate in the EU. Some countries such as Japan still permit them, but many other countries, including the USA, do not allow non-cocoa vegetable fats in chocolate at all.

Fractionation

Fats and oils have over the years found use in a wide range of food applications, in the kitchens of homes and restaurants and also in the industrial plants of food manufacturers. Although modern science has helped us to appreciate the reasons, the relationship between the physical properties of a fat or oil and its suitability for spreading on bread, frying chips or making cakes is probably just as well understood by cooks and chefs as it is by food technologists. Unfortunately there has always been a poor match between what is available in world or local markets and what consumers require. The "margarine" of Mège Mouriès was an early attempt redress the balance, but whether the primary aim was the betterment of the working classes or the supply of food to the French military is not clear. While hydrogenation has become the crucial process for converting abundant but relatively useless vegetable oils into plastic spreading fats, other processes are becoming increasingly important. Nutritional goals such as low levels of *trans* fatty acids have to be achieved, along with ever more subtle manipulations of physical properties. One fundamental approach has been to separate a fat or oil into different fractions of different melting properties, by virtue of differing proportions of more or less saturated triglycerides. A fraction thus obtained can then be used either as it is or combined with other fats or oils, by simple mixing or by interesterification, a process to be considered in the next section.

Cottonseed oil was one of the first oils to be subjected to a fractionation process. Cottonseed oil stored in large tanks out of doors over the winter months would become cloudy as saturated triglycerides crystallised out. Based on this experience, all that is required to make the oil more acceptable for domestic use as a salad oil or mayonnaise is to chill the oil to around 0 °C for a few hours and then remove the crystalline material by vacuum filtration. This technique, known as "winterization", is commonly carried out with olive oil. It is also sometimes applied to these and other oils after a hydrogenation stage, to obtain fractions with particularly desirable characteristics. The two fractions obtained by winterization are always referred to as "olein" and "stearin". The sought-after less saturated olein fraction is usually obtained in yields of 80–85%, but the more saturated stearin fraction is

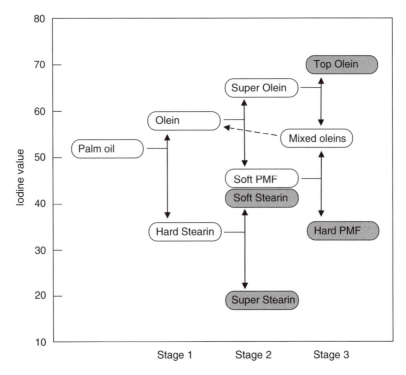

Figure 4.20   The fractionation of palm oil. Iodine value is an indication of the degree of unsaturation of the triglyceride fatty acids (*see* page 112). (Based on data quoted by V. Gibon in *Modifying Lipids for Use in Food*, *see* Further Reading.)

not discarded, but is incorporated into the feedstock for production of hydrogenated fats.

What are known as dry fractionation processes are a much more sophisticated version of winterization, applied to higher melting oils such as palm oil (MP 21–27 °C). The temperature of the oil is first raised to ensure that it is completely molten and then slowly cooled under continuous stirring (up to 24 hours may be needed) until the requisite proportion has crystallised. The crystals are separated from the liquid oil by vacuum filtration. Depending on the objective, further rounds of fractionation may be carried out on the oil or crystal fractions. Figure 4.20 shows the fractionation stages which may be applied to palm oil. The oleins obtained are useful components of frying and salad oils. The first stage should completely separate the trisaturated triglyceride (mostly PPP[xi]) content from the olein fraction, and yields oleins free

---

[xi] The abbreviations used for triglyceride types are explained in Table 4.7.

from cloud formation. The stearin fractions are used in margarine and shortening formulations, in which they replace the need for hydro-genated triglycerides. The most valuable final fraction is the hard palm mid-fraction (hard PMF). This largely consists of the monosaturated triglyceride, POP. It has almost identical melting properties to those of cocoa butter and is therefore valuable as a cocoa butter extender in chocolate manufacture.

Technologies such as fractionation and interesterification, which will be discussed in the next section, have undoubtedly helped the oil and fat industry towards the nutritional goal of reduced *trans* fatty acid content. Since soya bean and palm oils account for some 60% of worldwide fat and oil consumption, the influence of oilseed manufacture cannot be underestimated. There are increasing concerns about the environmental impact of massive plantings of oil palms in South East Asia and soya in South America.

Interesterification

The previous sections have shown that while we do have some capability for changing the properties of a fat by manipulating its fatty acid composition, quite often the properties we wish to alter depend more on the distribution of the fatty acids amongst the triglycerides than the overall fatty composition. Interesterification is a procedure by which the fatty acids within the triglycerides of a fat or oil can be rearranged. The ester links between the fatty acids and glycerol are broken and then reformed, but the fatty acids do not necessarily return either to their original position or even to their parent glycerol molecule. When the reaction is carried out in the presence of a chemical catalyst, the fatty acids end up randomly distributed among all the glycerol hydroxyl groups in the fat (Figure 4.21).

After "clean-up" (as with hydrogenation) and careful removal of all traces of water, the fat is exposed to the catalyst, usually sodium methoxide, at temperatures of 100–110 °C for some 30 minutes. At the end of the reaction any residual traces of catalyst are readily washed out of the fat with water. In addition to the manipulation of lard crystal-lisation referred to earlier, this type of interesterification has recently been brought to bear on the *trans* acid problem in hydrogenated fats (*see* page 113). Some of the vegetable oil converted into margarine or shortening is fully hydrogenated, so that it contains only saturated fatty acids such as stearic and palmitic and of course no residual unsaturated *trans* acids. It is then mixed in appropriate proportions with oil and subjected to interesterification. This is able to introduce just the right

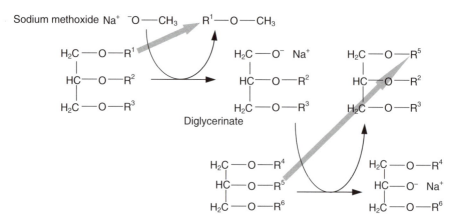

Figure 4.21   Triglyceride interesterification catalysed by sodium methoxide. The methoxide ion extracts a fatty acid from a triglyceride molecule to generate a diglycerinate. This then extracts a fatty acid from another triglyceride molecule, a process which can be repeated indefinitely. As fatty acids at all three positions on the triglyceride react, continuation of the sequence eventually leads to a completely random fatty acid distribution.

proportion of saturated acids into the oil to give an acceptable spreading fat without elimination of desirable polyunsaturates or the introduction of *trans* acids. It is possible to avoid hydrogenation completely if fractions of fats or oils rich in saturated fatty acids, such as the stearin fractions from palm oil, are used in place of fully hydrogenated fats in interesterification.

The specificity of enzyme-catalysed reactions offers interesting variations on chemically catalysed interesterification. Many of the lipases secreted by fungi (for the purpose of utilising fatty materials) share the specificity of pancreatic lipase (*see* page 129), in that they only hydrolyse triglycerides at the 1- and 3-positions. In the enzyme-catalysed interesterification reaction a fraction of palm oil rich in POP triglycerides is mixed with stearic acid. After diluting with hexane to ensure that both fat and fatty acid are in solution and that the resulting solution is not too viscous, the mixture is pumped slowly through a bed of polymer beads in a reactor at around 60 °C. The beads are coated with a fungal lipase, usually from *Mucor meihi*. The enzyme attacks the palmitic acid-rich 1- and 3-positions, but in the almost total absence of water the equilibrium of the reaction is such that the enzyme catalyses not only the removal of fatty acids from these positions but also the reverse reaction. The end result is the formation of a low-cost fat with almost identical proportions of POP, POS and SOS to the natural, and expensive, cocoa butter. As it also shares cocoa butter's polymorphic behaviour it can be used as a cocoa butter extender (the commercial term) to enhance the profitability of chocolate manufacture (the consumer's term).

Another interesting application of this technique is the preparation of fats with special nutritive characteristics. Human digestion of fats depends upon pancreatic lipase, with the fatty acids liberated from the 1- and 3-positions, together with 2-monoglycerides, being absorbed by the mucosa of the small intestine. It is observed that long-chain saturated fatty acids (*i.e.* 16 or more carbon atoms) are less well absorbed than their unsaturated counterparts, especially when they are in the outer positions of dietary triglycerides. This is because when the lipase liberates them they readily form insoluble salts with any calcium ions in the vicinity. The calcium salts of unsaturated fatty acids are much more soluble. As we have seen, animal fats, including human milk fat, tend to have their saturated fatty acids in the 2-position, but plant fats are the reverse. This reduces the digestibility of plant fats when used in formula feeds for infants. The solution has been to use enzymic interesterification with tripalmitin and oleic acid as feedstock, yielding a product with oleic acid in the 1- and 3-positions and palmitic acid in the 2-position. This is expected to find ready acceptance in infant formula feeds since it mimics human milk much more closely than does cow's milk. Cow's milk has more or less the correct distribution of fatty acids in its triglycerides, but compared with human milk it is low in linoleic acid, an essential fatty acid.

POLAR LIPIDS

The membranes of all living systems, not only the plasma membranes surrounding cells but also those which enclose or comprise organelles such as mitochondria, vacuoles or the endoplasmic reticulum, are composed of an essentially common structure of specialised proteins and lipids. A detailed account of the structure of membranes is outside the scope of this book, but the nature and properties of many of the lipid components are very important to food chemistry. Besides occurring in membranes, many polar lipids occur in small amounts in crude fats and oils and are also classically associated with egg yolk, in which they are major constituents.

The defining feature of polar lipids is that they are *amphiphilic*. This is to say that while their molecules share with the triglycerides (and waxes) an affinity for non-polar environments, *i.e.* they are *lipophilic*,[xii] their molecules also include structural elements which provide an affinity for

---

[xii] Alternatively, *hydrophobic*.

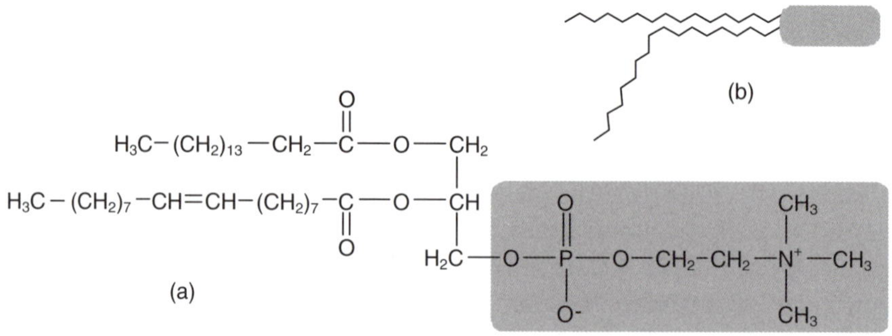

Figure 4.22   Phosphatidyl choline. (a) The molecule has two long chain fatty acids, typically oleic and palmitic acids, as shown here, esterified to the central glycerol. This also carries choline attached through a phosphate group. The hydrophilic polar area of the molecule is shaded. (b) The small diagram shows that the molecule as a whole has a large hydrophilic head, carrying a long hydrophobic tail.

aqueous polar environments, *i.e.* they are *hydrophilic*. These two contrasting elements are apparent in the structure of phosphatidyl choline (Figure 4.22), also known as lecithin. Commercial lecithin actually contains a mixture of glycerophospholipids separated from crude soya bean oil. Typically it contains around 15% phosphatidyl choline, 12% phosphatidyl ethanolamine and 10% phosphatidic acid, the remainder being mostly soya bean oil triglycerides.

Like phosphatidyl choline, most of the important naturally occurring polar lipids are glycerophospholipids, having two long-chain fatty acid residues esterified with a glycerol molecule which also carries a phosphate group. The phosphate group is in turn esterified with one of a number of organic bases, amino acids or alcohols, which add to the hydrophilic character of the molecule. The structures of a range of glycerophospholipids are shown in Figure 4.23. Although a great many other phospholipid types have been described, from the point of view of the food chemist their properties are not significantly different from those shown here.

One other polar lipid that should be mentioned is cholesterol (4.12). Although not immediately obvious from the usual presentation of its chemical structure, shown here, the hydrocarbon part of the cholesterol molecule has similar dimensions to the paired fatty acid chains of the glycerophospholipids, and the molecule is able to occupy a similar position to the other polar lipids in membranes, *etc.* Esters of cholesterol with fatty acids occur as components of the lipoproteins in the blood, but in membranes and in egg yolk cholesterol is unesterified.

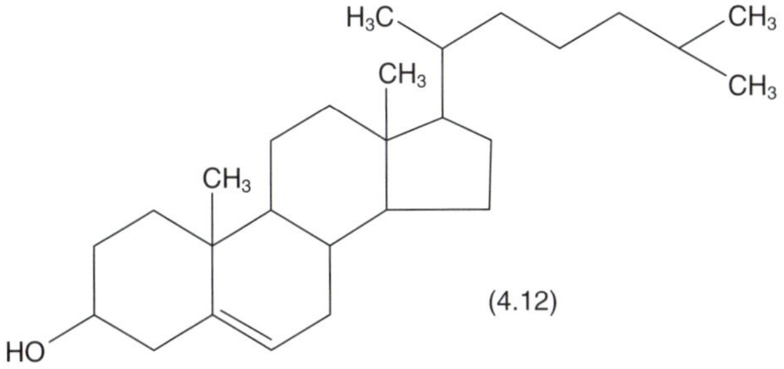

(4.12)

The traditional importance of polar lipids in food systems is due to their ability to stabilise emulsions. Emulsions are colloidal systems of two immiscible liquids, one dispersed in the other, the continuous phase. Foods can provide examples of both oil-in-water emulsions, *e.g.* milk and mayonnaise, and water-in-oil types, *e.g.* butter. In many cases the structure is complicated by the presence of suspended air, forming a foam, or solid particles.

Emulsions are prepared by vigorous mixing of the two immiscible liquids such that droplets of the disperse phase are formed. However, a simple emulsion usually breaks down rapidly as the dispersed droplets coalesce to form a layer, which either floats on the surface or settles to the bottom of the vessel. Emulsion stability is enhanced by the presence

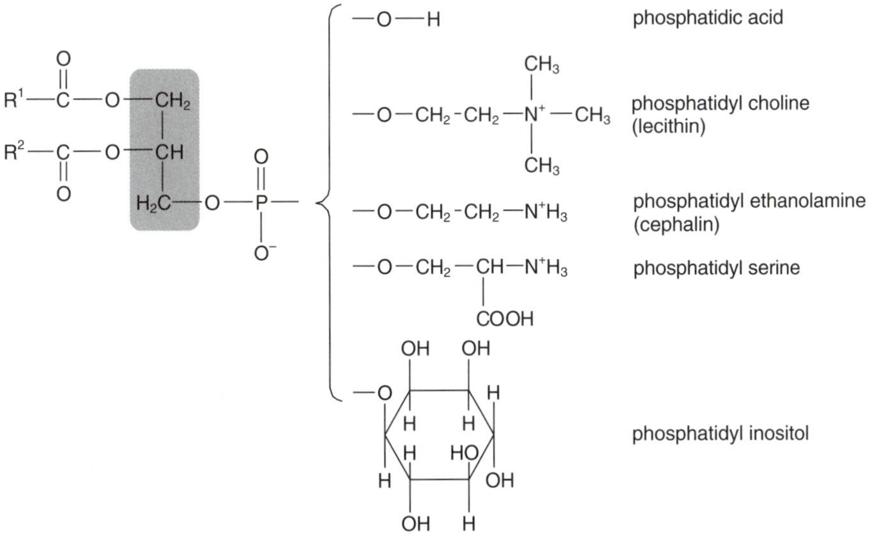

Figure 4.23 Glycerophospholipids. R^1-COO– and R^2COO– are fatty acids. The glycerol component is shaded.

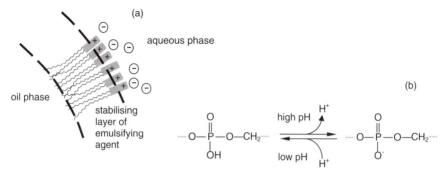

Figure 4.24   (*a*) Part of the surface of an oil droplet in an oil-in-water emulsion stabilised by a phospholipid such as lecithin. The phospholipid molecules
orientate themselves at the oil–water interface to allow their polar groups
to interact with the aqueous continuous phase. In the majority of food
situations the prevailing pH is low and the ionisable groups of the
phospholipid will carry a net positive charge, as shown in (*b*). This in turn
attracts negatively charged ions present in the aqueous phase. The
mutual repulsion of the layers of negative charges keeps the droplets
apart and prevents them from coalescing.

of substances such as phospholipids whose molecules contain both polar
and non-polar regions. Figure 4.24 shows how such molecules are able
to orientate themselves at the interface between the two phases and thus
inhibit the coalescence of the droplets.

Emulsifiers, and their non-food relatives the soaps and detergents, are
often described as *surfactants (surface active agents)*. This is because
one manifestation of their amphiphilic character is the ability to reduce
the surface tension of water (*see* Chapter 12) as they dissolve. Not all
surfactants have large lipophilic and hydrophilic groups. Polar organic
solvents which are miscible with both water and non-polar solvents,
such as ethanol (C_2H_5OH) or acetone (($CH_3)_2CO$) are each regarded as
surfactants in that they lower the surface tension of water. The hydrophilic groups in many phospholipids include both phosphate groups
(negatively charged at the pH normally found in food) and quaternary
nitrogen atoms (positively charged) and are therefore classified as
zwitterionic (*see* Chapter 5). Many other surfactants are acidic in character (*i.e.* they are anions) or have hydrophilic groups which cannot
ionise, the so-called nonionic surfactants. There are no cationic surfactants, in which the hydrophilic group is positively charged at normal pH
values, which are permitted as food additives.

To enable fat to be digested in the small intestine it is emulsified with
bile salts, derivatives of cholic acid secreted by the liver. Sodium
taurocholate (4.13) is a typical example. It may not be obvious how

taurocholate can operate as an emulsifier, since its formula appears to show no clearly marked polar region. However, a three-dimensional portrayal of its structure (4.13a) reveals that it is able to present a lipophilic face to a fat globule surface and a hydrophilic face to the surrounding water.

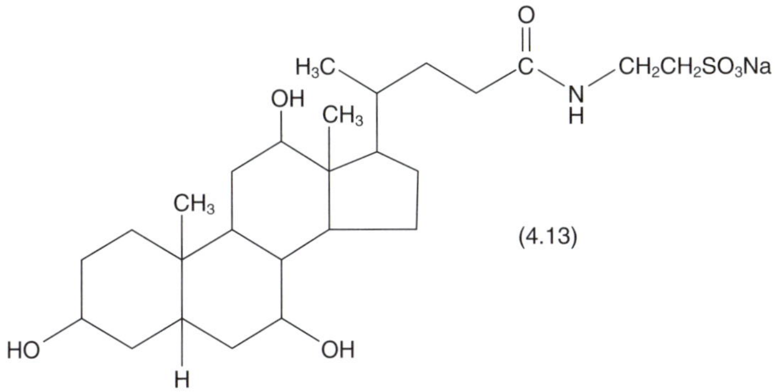

(4.13)

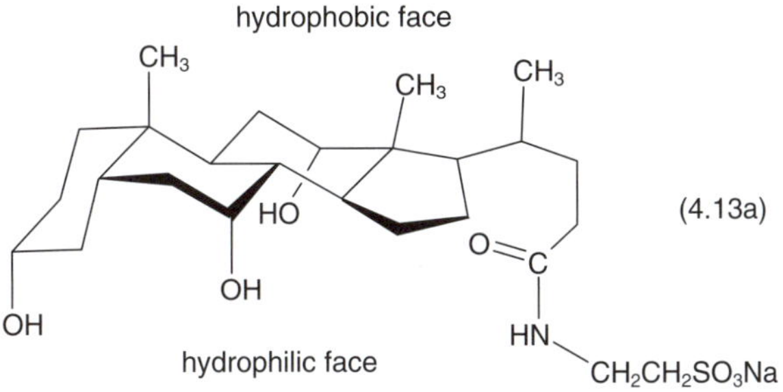

(4.13a)

The best known source of emulsifying properties in food preparation is egg yolk. Approximately 33% of the yolk of a hen's egg is lipid (protein amounts to a further 16%), of which about 67% is triglyceride, 28% is phospholipid, and the remainder is mostly cholesterol. Lecithins are the predominant phospholipids, with small amounts of cephalins, lysophosphatidyl cholines (*i.e.* lecithins with a single fatty residue, on the 1-position), and sphingomyelins (4.14).

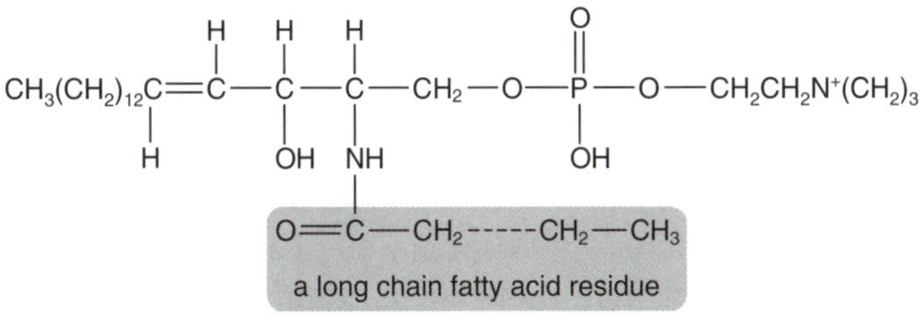

(4.14)

Yolk actually consists of a suspension of lipid/protein particles in a protein/water matrix. Specific associations of lipids and proteins such as these are known as lipoproteins and are employed in animal systems whenever lipid material has to be transported in an aqueous environment such as the blood. The best known applications of egg yolk as an emulsifier are in mayonnaise and in sauces such as hollandaise, but its role in cake making is just the same. The lecithin recovered from soya bean oil during its preliminary purification is increasingly valued as an alternative natural emulsifying agent, particularly for use in chocolate and other confectionery, and also where a specifically vegetarian product is demanded.

In 1965 Griffin introduced the HLB (hydrophile–lipophile balance) system, a scale of 1 to 20 for quantifying the emulsifying capability of surfactants. Although originally proposed for use with cosmetics, HLB values have become widely applied to food emulsifiers. The HLB for an emulsifier was originally defined by the equation:

$$HLB = 20[1 - (S/A)]$$

where S is the *saponification number* of the ester and A is the *acid number* of the resulting acid. These are, to say the least, outdated chemical parameters and it is common nowadays to regard S as the overall molecular weight of the emulsifier and A as the molecular weight of the hydrophilic part of the molecule. These definitions give no recognition to the differences in affinity for water between different hydrophilic groups, for example the two hydroxyl groups of a monoglyceride compared with the phosphate and quaternary nitrogen of lecithin. Many quoted HLB values have been derived by empirical methods based on ascribing values to particular chemical groups, or from laboratory-based experimental comparisons. The HLB values of a selection of commercially important emulsifiers are given in Table 4.10. Surfactants with low HLB values, those between 4 and 6, are the most lipophilic and best suited to forming water-in-oil emulsions, whereas for oil-in-water emulsions HLB values in the 8–18 range are required. High HLB values indicate a high degree of hydrophilicity and solubility in water.

Along with the uncertain origins of their values, the principal reason for the limited application of HLB values in food is that food systems are far too complicated for such a simplistic approach. In particular, almost all foods which contain emulsified fat also contain protein, and protein molecules are extremely effective emulsifying agents. The normal folding of the polypeptide chains of proteins (*see* Chapter 5) ensures that the outer surface of the protein molecule is dominated by hydrophilic

Table 4.10 HLB values of a range of permitted emulsifiers. The structures of the semi-synthetic emulsifiers asterisked are shown in Figure 4.25, later. The fact that the value for sodium stearoyl-2-lactylate is greater than 20 is a typical case of the more empirical approach to calculation of HLB values often adopted nowadays.

Emulsifier	E number	Type	HLB value
Oleic acid	470a	anionic	1.0
Glycerol dioleate	471	nonionic	1.8
Sorbitan tristearate (Span 65)*	492	nonionic	2.1
Glycerol monooleate	471	nonionic	3.4
Sorbitan monooleate (Span 80)	494	nonionic	4.3
Soya lecithin (crude)	322	zwitterionic	8.0
Diacetylated tartaric acid esters of monoglycerides*	472e	anionic	8.0
Sucrose monolaurate*	473	nonionic	15.0
Polyoxyethylene sorbitan monopalmitate* (Tween 40)*	434	nonionic	15.6
Sodium stearoyl-2-lactylate*	481	anionic	21.0

amino acid side-chains. In contrast, the interior of the molecule is dominated by hydrophobic amino acids. A major contributor to the maintenance of the correct folding of the protein is the drive to separate these hydrophobic side-chains from the aqueous environment surrounding the protein. When a protein molecule encounters an oil droplet it naturally unfolds to expose its hydrophobic elements to the oil surface and at the same time it presents a hydrophilic surface to the surrounding water—Figure 4.24 on a larger scale. Loops consisting of hydrophilic sections of the polypeptide chains will protrude some distance above the surface of the oil droplet. The adsorption of the protein on to the oil surface is effectively an irreversible denaturation.

The major role of food emulsifiers is usually assumed to be in foods which are most obviously emulsions, such as mayonnaise, in which an oil-in-water emulsion (actually olive oil and vinegar) is stabilised by egg yolk. However, food emulsions are much more widespread. For example, the fat in cake batters is dispersed by the egg yolk phospholipids. Soya lecithin is often incorporated into commercial cake recipes to reduce the amount of egg required, since this is much more expensive. The proteins from the meat are responsible for emulsifying the fat in sausages, especially the featureless pink objects revered in Britain. One reason why so much fat pours out of sausages and burgers when they are grilled is that the thermal denaturation of the meat proteins (in particular the so-called sarcoplasmic proteins, *see* Chapter 5) destroys

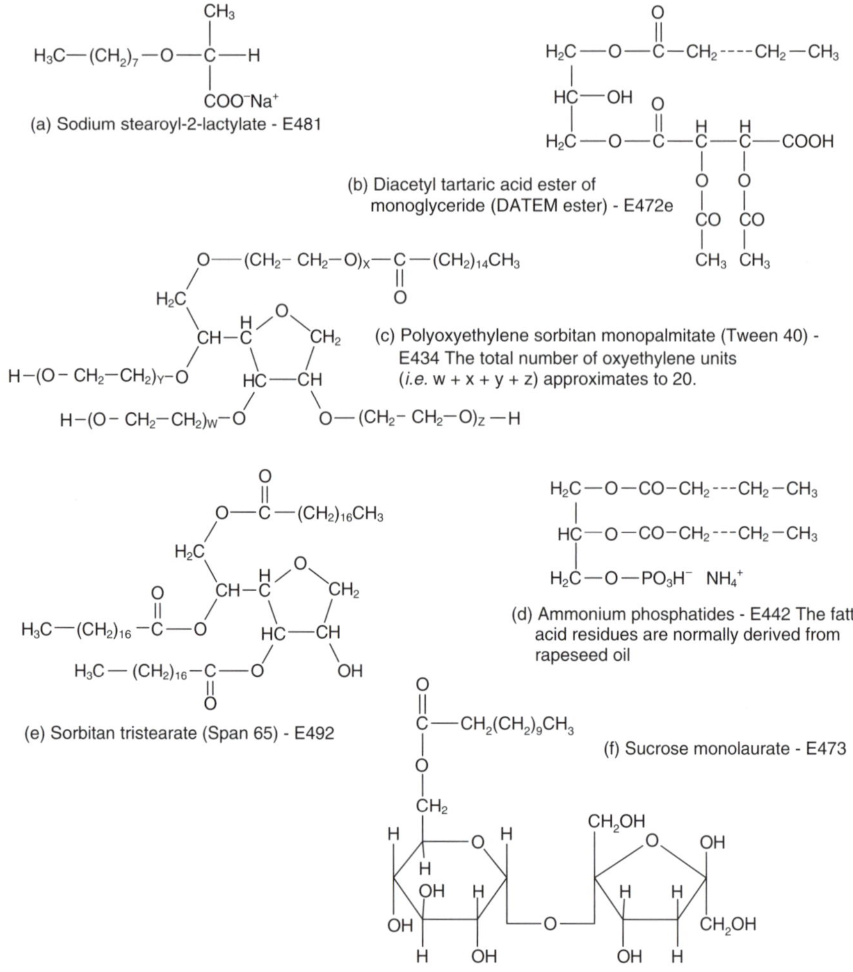

$H_3C-(CH_2)_7-O-\overset{\overset{\displaystyle CH_3}{|}}{\underset{\underset{\displaystyle COO^-Na^+}{|}}{C}}-H$

(a) Sodium stearoyl-2-lactylate - E481

$H_2C-O-\overset{O}{\overset{||}{C}}-CH_2----CH_2-CH_3$

$HC-OH$

$H_2C-O-\overset{O}{\overset{||}{C}}-\overset{\overset{H}{|}}{\underset{\underset{O}{|}}{C}}-\overset{\overset{H}{|}}{\underset{\underset{O}{|}}{C}}-COOH$

$\underset{CH_3}{\underset{|}{CO}}\ \underset{CH_3}{\underset{|}{CO}}$

(b) Diacetyl tartaric acid ester of monoglyceride (DATEM ester) - E472e

(c) Polyoxyethylene sorbitan monopalmitate (Tween 40) - E434 The total number of oxyethylene units (*i.e.* w + x + y + z) approximates to 20.

$O-(CH_2-CH_2-O)_x-\overset{O}{\overset{||}{C}}-(CH_2)_{14}CH_3$

$H-(O-CH_2-CH_2)_Y-O$

$H-(O-CH_2-CH_2)_W-O \qquad O-(CH_2-CH_2-O)_Z-H$

$O-\overset{O}{\overset{||}{C}}-(CH_2)_{16}CH_3$

$H_3C-(CH_2)_{16}-\overset{O}{\overset{||}{C}}-O$

$H_3C-(CH_2)_{16}-\overset{O}{\underset{||}{C}}-O \qquad OH$

(e) Sorbitan tristearate (Span 65) - E492

$H_2C-O-CO-CH_2---CH_2-CH_3$

$HC-O-CO-CH_2---CH_2-CH_3$

$H_2C-O-PO_3H^-\ NH_4^+$

(d) Ammonium phosphatides - E442 The fatty acid residues are normally derived from rapeseed oil

$\overset{O}{\overset{||}{C}}-CH_2(CH_2)_9CH_3$

O

CH_2

(f) Sucrose monolaurate - E473

Figure 4.25 The structures of a number of semi-synthetic emulsifiers permitted for use in food. As discussed in the text, many of these formulae should be regarded as representing a class of closely related substances, with variations in the fatty acid complement, *etc.* This is particularly the case in the sorbitan compounds (*c*) and (*e*), in which six-membered anhydride rings also occur and the locations and total numbers of fatty acid residues vary. "Tween" and "Span" are widely used quasi-commercial designations for the two types of sorbitan-based emulsifiers, regardless of the actual manufacturer.

their fat-binding capability. The term "emulsifying salts" refers to substances often included alongside added emulsifiers in meat products and processed cheese, such as the sodium or potassium salts of phosphates, citrates or tartrates. These promote emulsification by solubilising proteins, which are then able to act as the true emulsifying agent. Stabilisers, which also are not emulsifiers in their own right, are found in many food products which contain emulsions, such as ice cream and instant desserts. Most stabilisers are polysaccharides, such as karaya gum.

A small proportion of emulsifier, usually lecithin, is an essential ingredient in chocolate. Although chocolate contains very little water it does contain large quantities of sugar milled to very small ($<30\,\mu$m) particles. The sugar particles obviously have a very hydrophilic surface and would tend to clump together in liquid chocolate, making the liquid very viscous. The non-fatty particulate material from the bean, mostly polysaccharides and protein, would accentuate this. The emulsifier molecules bind to the surface of the particles and give them a lipophilic surface which keeps them apart, and keeps the viscosity of the liquid chocolate low. Melted chocolate needs to be free-flowing if it is to be used to coat biscuits or to flow into elaborate moulded shapes for Easter eggs.

Milk Fat, Cream and Butter

Dairy products provide examples of both oil-in-water and water-in-oil emulsions. The fat in milk occurs in very stable globules, mostly between 4 and 10 μm in diameter. There are $1.5–3.0 \times 10^{12}$ globules in one litre of milk. Globules of such small size should according to Stoke's Law take some 50 hours to float to the top of a pint bottle of milk and establish the familiar layer of cream, but we know that half an hour will in fact suffice. This implies that much larger particles are involved, with effective diameters up to 800 μm. This clustering of globules is clearly no ordinary coalescence, since evaporated milk which is sterilised after canning by holding it at temperatures above 100 °C for several minutes shows no tendency to "cream". The explanation for this curious behaviour lies in the nature of the "milk fat globule membrane" which surrounds each globule. This consists of specific lipoproteins and polar lipids arranged around the globule in a similar arrangement to that shown in Figure 4.24. The lipids in this membrane, including those associated with proteins to form lipoproteins, are typical of those found in the plasma membrane of animal cells, including cholesterol and its esters formed with long-chain fatty acids, phosphatidyl ethanolamine, choline, serine and inositol, and sphingomyelin. The membrane is most probably acquired as the milk fat globule is secreted through the plasma membrane of the mammary gland cells into the milk duct. Clustering is caused by cross-linking through one specific type of protein which occurs in the aqueous phase of the milk in very small amounts, mac-roglobulin. Heating above 100 °C for a few minutes (but not pasteur-isation) denatures this protein, and creaming is prevented.

Creaming is also prevented in homogenised milk, in which the fat globules are reduced to about 1 μm in diameter by passage through very small holes at high pressures, and velocities around $250 \, m \, s^{-1}$. The vast increase in total surface area of the fat globules, from around $75 \, m^2 \, dm^{-3}$ of milk to around $400 \, m^2 \, dm^{-3}$ results in extra proteins being adsorbed from the aqueous phase of the milk. These proteins inhibit coalescence of the globules but do not interact strongly with the macroglobulin, so that creaming no longer occurs.

Butter making results in the formation of a water-in-oil emulsion. Cream with a fat content of 30–35% is inoculated with a culture of bacteria and incubated for a few hours. The bacteria produce char-acteristic flavour compounds, such as diacetyl (butanedione). The cream is then mechanically agitated ("churned") sufficiently to disrupt the membranes of the fat globules and cause coalescence.

During mixing, a proportion of the aqueous phase known as butter-milk is trapped as small droplets. These are prevented from coalescing by the rigidity of the fat and by the layer of proteins and polar lipids which forms at the fat–water interface. Butter normally contains about 20% water, to which salt is added, nowadays as flavouring but originally to deter the growth of unwanted micro-organisms. Margarine has but-termilk and salt added to it during the final blending stages, but since it lacks its own emulsifying agents, soya lecithin and other emulsifiers also have to be added.

Synthetic Emulsifiers

Figure 4.25 shows a range of semi-synthetic emulsifiers and their structures. These are semi-synthetic in the sense that they are based on naturally occurring fats or oils but subjected to a variety of chemical processes. The processes used do not result in the formation of a pure product consisting of a single chemical species. What is given a single name is in fact a mixture. For example, polyglyceryl poly-ricinoleate (PGPR) is often used in place of lecithin in chocolate. As shown in outline in Figure 4.26, PGPR is synthesised from glycerol and ricinoleic acid (from castor bean oil) but the final product will contain individual molecules differing in both number of glycerol residues and in the number of ricinoleic acid residues.

In terms of bulk usage monoglycerides are by far the most important, as well as being the starting point for the industrial synthesis of many

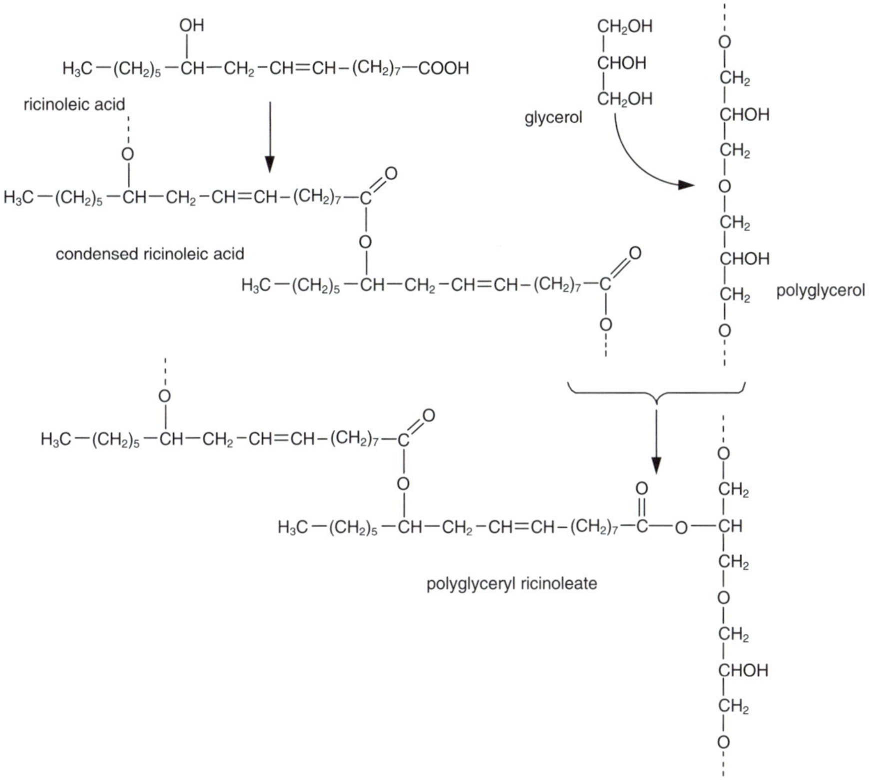

Figure 4.26 An outline of the synthesis of polyglyceryl polyricinoleate. The polyglycerol moiety normally contains two to four glycerol units, the polyricinoleate moiety about five fatty acid units. There may be several polyricinoleate chains attached to the polyglycerol moiety.

other types. They are manufactured by interesterification of triglycerides with excess glycerol in the presence of sodium hydroxide. The fatty acids become randomly distributed throughout the glycerol molecules present, resulting in a mixture of "monoglycerides" which also contains some glycerol and di- and triglycerides. If required, reasonably pure monoglycerides can be obtained from this mixture by fractional distillation under vacuum.

Baked goods, including bread, burger buns, cakes and biscuits, account for a high proportion of semi-synthetic emulsifier usage. Monoglycerides of saturated fatty acids are useful for slowing down the rate of staling. The straight hydrocarbon chains are able to occupy the central core of the amylose and amylopectin helices, slowing down retrogradation (*see* page 56). Diacetyl tartaric acid esters of monoglycerides (DATEMS) are included at levels around 0.5% of the weight of flour in many commercial bread recipes, where they not only delay staling but also increase the loaf volume and improve the crumb texture, apparently by binding to gluten proteins.

Phytosteroids

Cholesterol and bile salts were mentioned earlier in this section and provide the justification for including a brief consideration of a further group of steroids here, although the phytosteroids can hardly be classified as polar. Although their presence in plant tissues has been recognised for a long time their recent inclusion in some margarines and low fat spreads has brought them into prominence, since consumption of these can have a beneficial effect on serum cholesterol levels.

The three sterols shown in Figure 4.27 are abundant in plants, in which together with smaller amounts of the phytostanols they play the same role in cell membranes as cholesterol does in animal cells. Western diets normally deliver around 200–400 mg daily of these plant sterols

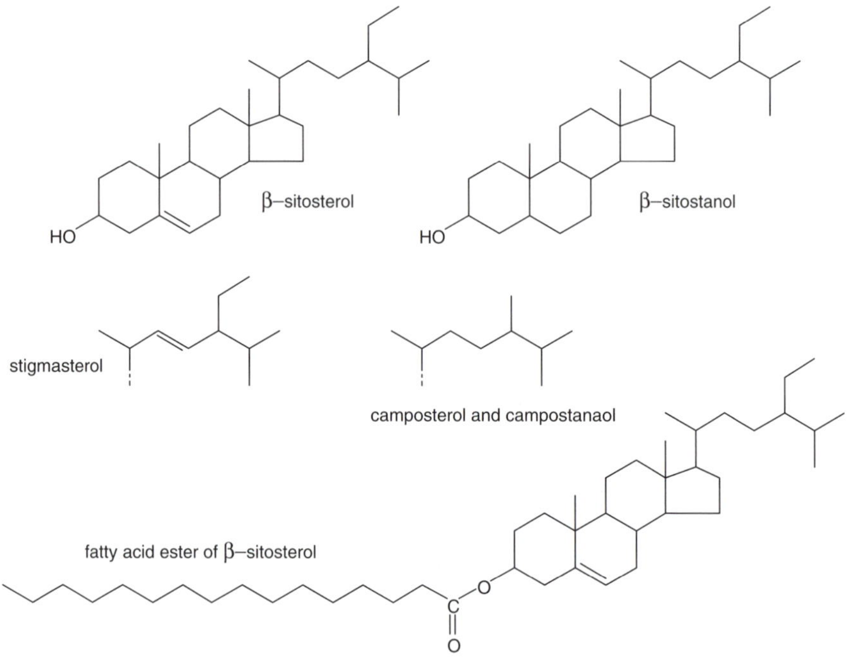

Figure 4.27   Structures of the common phytosterols and phytostanols. The phytostanols differ from the phytosterols in being fully saturated; otherwise the differences between them lie entirely within the side-chains. The fatty acid (normally stearic) esters are the form added to margarines.

(with vegetarians towards the top of this range), about the same as the amount of cholesterol, plus about one-tenth of this amount of phytostanols. Although one of them, β-sistosterol, has been used for the reduction of serum cholesterol levels in hypercholesterolaemic patients for many years, it is only recently that the possibility of their wider application as "nutraceutical" food additives has arisen. It was discovered that attaching a fatty acid chain greatly increased their solubility in oils and fats and made them effective at very much lower doses.

The usual sources of phytosterols are vegetable oils such as soya and corn, from which they are extracted along with lecithin and tocopherols during the refinement of the oil prior to hydrogenation. One important if unlikely source is tall oil. Tall oil is extracted from pine trees during the manufacture of wood pulp for paper making.

In studies on patients with elevated serum cholesterol, daily consumption of 2 g or more of phytosterols has been shown to reduce levels of serum cholesterol by 8–13%, particularly the low-density lipoprotein (LDL) form. It is an elevated level of LDL cholesterol rather than the overall cholesterol figure which is a major risk factor in heart disease. Bearing in mind that in Western Europe the typical daily consumption of spreading fats, butter, margarine, *etc.*, is around 20 g, levels of up to 20% of phytosterols (as esters) are being added to supplement margarines and low fat spreads. It has been calculated that consumption of 3 g per day of phytosterols could lower the risk of heart disease by between 15 and 40%, depending on other risk factors. The US Food and Drug Administration (FDA) has approved phytosterol addition at this level. It is not illegal to market them in Europe, but the EU legislative position is at present unclear.

Phytosterols work by inhibiting the uptake of cholesterol from the small intestine, but the exact mechanism remains unresolved. Only a small proportion of the phytosterol is absorbed, and even that is rapidly eliminated *via* the bile. There appear to be no significant safety issues associated with consumption at even the highest levels suggested here. However, the widespread adoption of margarines and low fat spreads supplemented by phytosterols faces a number of obstacles, in particular their cost. At the present time supplemented products (in Britain, at least) cost around five times as much as their unsupplemented counterparts! The question of their suitability for healthy young children does not seem to have been addressed as extensively as it has for hypocholesteraemic adults. In real families, offering two distinct piles of sandwiches may not be quite as straightforward as one would wish, and the development of yoghurt-type products as vehicles for phytosterols may be more practical.

SPECIAL TOPICS

1 Hydrogenation in Detail

The manufacture of margarines and shortenings is highly dependent on the process of hydrogenation for the conversion of liquid oils to plastic fats. The conditions used for the reaction (temperature, catalysis, *etc.*) were discussed earlier in this chapter but the following is a brief account of the chemical details. The process involves a number of reactions, illustrated in Figure 4.28:

I Hydrogen is absorbed on to the metal surface of the catalyst and in doing so is effectively converted from H_2 into individual hydrogen atoms bound to metal atoms. At the site of a double bond the fatty acid chain is bound both to the metal surface and to one hydrogen atom from the metal surface. This link may be

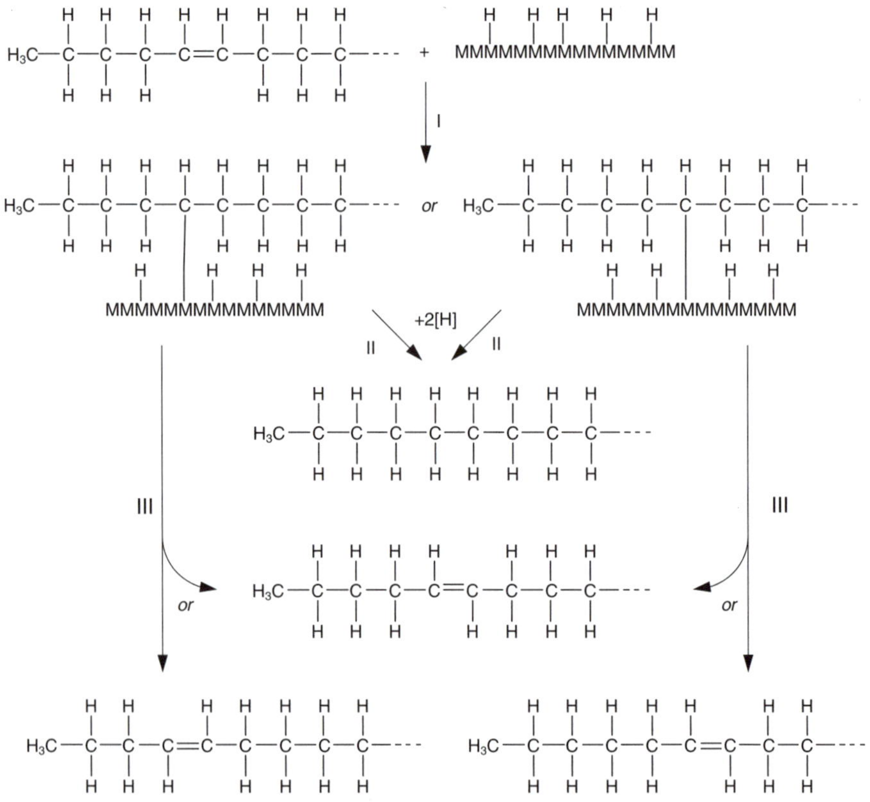

Figure 4.28 The hydrogenation reactions. (Reaction IV, the re-entry of isomerised by-products into the sequence, has been omitted.)

formed at either end of the double bond. From this point there are two possible outcomes:

II A second hydrogen atom reacts to fully saturate the double bond and the hydrogenated fatty acid chain dissociates from the catalyst, *or*

III The fatty acid chain dissociates from the catalyst without further reaction, and the double bond is restored. However there is no special reason why the hydrogen returned to the catalyst should be the original one donated by the catalyst, so the resulting double bond can be in the original position, or between a neighbouring pair of carbon atoms in the chain. The result is an apparent migration of the double bond, *e.g.* from 9,10 to 10,11 or 8,9. The newly formed bonds are most likely to be in the *trans* rather than the original *cis* configuration, since this is the thermodynamically favoured form.

IV A proportion of products of the migration will re-enter the sequence, some ending with *trans* double bonds where they originated, but others ending up even further down the chain.

The tendency of linoleic acid residues to be converted to oleic (desirable), compared with the conversion of oleic to stearic (undesirable), is expressed as the selectivity ratio (SR), the ratio of the rate constants for the two processes. The SR is affected by a number of operating parameters, including temperature, pressure (*i.e.* H_2 concentration), catalyst level and degree of agitation (stirring rate) in the reactor. Elevation of any of these parameters increases the rate of hydrogenation, but higher temperature and catalyst concentration each lead to high SR values and high proportions of *trans* acids, whereas high pressure and vigorous stirring have the opposite effect. If the catalyst surface is densely packed with hydrogen atoms hydrogenation is more likely and selectivity is low. On the other hand, if hydrogen atoms are sparse selectivity will be high but there will be a much greater likelihood of isomerisation, since there will not be a second conveniently placed hydrogen atom to complete the hydrogenation.

2 Singlet and Triplet Oxygen

These terms relate to the bonding between the two atoms in the oxygen molecule. The oxygen molecule, O_2, has a total of 12 valency electrons originating in the outer shells of the two oxygen atoms. In its normal low-energy ground state, eight of these are located in seven different molecular orbitals. In five of the orbitals the paired electrons spin in opposite directions, as indicated by the arrows in the first five of these boxes:

$$\boxed{\uparrow\downarrow}\quad\boxed{\uparrow\downarrow}\quad\boxed{\uparrow\downarrow}\quad\boxed{\uparrow\downarrow}\quad\boxed{\uparrow\downarrow}\quad\boxed{\uparrow}\quad\boxed{\uparrow}$$
$$(2s\sigma)^2\quad(2s\sigma^*)^2\quad(2p\sigma)^2\quad(2p\pi_x)^2\quad(2\,p\pi_y)^2\quad(2\,p\pi_x{}^*)^2\quad(2\,p\pi_y{}^*)^2$$

The asterisks indicate antibonding orbitals. The number of bonds between the atoms is obtained by subtracting the number of electrons in antibonding orbitals from those in bonding orbitals, and dividing the result by two.

The two highest molecular orbitals contain single electrons with parallel spin. This gives the molecule a *multiplicity* (S) of 3, from the equation:

$$S = 2s + 1$$

where s is the sum of the spins of all the electrons, counting $\boxed{\uparrow}$ as $+ 1/2$ and $\boxed{\downarrow}$ as $-1/2$. A multiplicity of 3 is known as the *triplet* state. The triplet state of oxygen molecules in their lowest energy condition, the ground state, is unusual. Most molecules have *singlet* ground states, in which all electron spins are paired. Triplet oxygen is denoted by the symbol, $^3O_2$. Besides its triplet ground state, oxygen molecules can occur in two *excited*, high-energy singlet states with their two highest orbitals filled in the following manner: $\boxed{\uparrow\downarrow}\boxed{}$ or $\boxed{\uparrow}\boxed{\downarrow}$, written as: $^1O_2$ and $^1O_2^*$, respectively. The others are as shown above. These excited states are extremely short-lived.

Only the first of these excited states, $^1O_2$, is of relevance to us. This has the two electrons which were unpaired in the singlet oxygen, in a $2p\pi$ orbital, leaving the corresponding antibonding $2p\pi^*$ orbital empty. The resulting molecule is an extremely reactive electrophile as it seeks to fill the empty orbital, for example by reaction with unsaturated fatty acids. The energy difference between this and triplet oxygen is $+ 24$ kcal per mole.

3   Triglyceride Crystals

The fatty acid chains in an individual triglyceride molecule are organised into either a "chair" or a "tuning fork" arrangement, as shown in Figure 4.29. Which of these is preferred depends on the symmetry of the triglyceride. The tuning fork arrangement is favoured in symmetrical triglycerides since it brings the two identical fatty acids in the 1- and 3-positions alongside each other. Similarly, the chair arrangement is

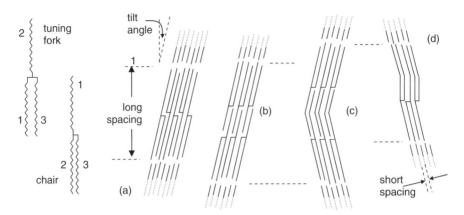

Figure 4.29 The organisation of triglyceride molecules in fat crystals. In the more stable crystal types shown here (β' and β, as discussed in the text) the triglyceride molecules stack alongside each other so that the terminal methyl groups form horizontal planes, the methyl terraces (shown as heavy dashed lines). The dimensions of the long and short spacings and the tilt angle, established by X-ray crystallography, characterise the different forms. The long spacing reflects the different overlap patterns, double chain length (DCL) in (a) and (d), and triple chain length (TCL) in (b) and (c).

favoured in asymmetrical triglycerides. Neighbouring triglyceride molecules in a fat crystal are stacked in layers, with the parallel hydrocarbon chains packed closely together and running more or less perpendicular to the parallel planes of the glycerol groups and the planes of terminal methyl groups. These planes are sometimes referred to as the "methyl terraces". Neighbouring triglyceride molecules overlap in different patterns to place the methyl terraces either two or three chain lengths apart. As with the choice between the tuning fork and chair arrangement, the overlap pattern is dictated by the need to accommodate fatty acids of different chain length and the distortions introduced by double bonds, as illustrated in Figure 4.29(b), (c) and (d).

The most important differences between the three crystal types are in melting point and crystal form. In the crystalline state the hydrocarbon chains are arranged in extended zig-zags with all the carbon atoms in the same vertical, or near vertical, plane. Figure 4.30(a), (b) and (c) show the chains viewed end-on, along their long axes.

In the unstable α-forms, Figure 4.30 (a) and (d), the chains are arranged in a hexagonal pattern but the planes of the hydrocarbon zigzag are oriented randomly. The crystal structures of the β' and β-forms are formally described as orthorhombic and triclinic, respectively.

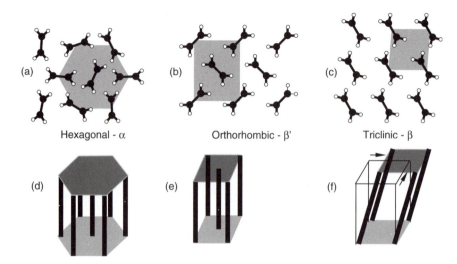

Figure 4.30   Crystal packing in triglycerides. In (a), (b) and (c) the zigzag hydrocarbon chains of the fatty acid residues are viewed end-on. The grey areas
emphasise the geometrical relationships of the chains. In hexagonal and
orthorhombic arrangements, (d) and (e), the chains are shown perpendicular to the planes formed by their CH_3 ends, the so-called methyl
terraces, while in the triclinic arrangement (f) the chains are tilted in both
vertical planes. It is generally assumed that in triglycerides the chains are
tilted in an orthorhombic arrangement.

The tilting of the chains in the triclinic form, shown in Figure 4.30 (*f*),
allows the most compact packing of the chains. Although not allowed
for in the formal definition of an orthorhombic crystal, these fatty acid
chains are also assumed to tilt.

Recent X-ray crystallographic studies of cocoa butter and its
constituent triglycerides provide an excellent illustration of how an
understanding of crystal structure can explain aspects of fat behaviour.
Figure 4.31 shows in simplified form the crystal packing of 1-palmityl-2-
oleyl-3-stearyl glycerol (POS), the dominant triglyceride of cocoa butter
in its two most stable polymorphic forms, β-V and β-VI. Essentially
similar patterns are also seen in the corresponding polymorphic forms of
pure cocoa butter in which POP and SOS, along with POS, make up the
bulk of the triglycerides. The first of these, β-V, is the form which gives
the physical characteristics desired by the chocolate maker. The second,
the β-VI form, of slightly higher melting point, is the one found in the
unwanted crystalline bloom formed on the surface of chocolate which
has been mishandled. The packing of the β-VI form is just sufficiently
favoured thermodynamically to provide the explanation for the transition between the two forms.

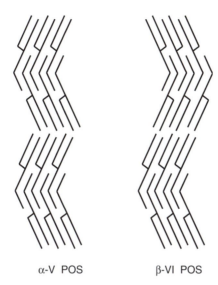

α-V POS   β-VI POS

Figure 4.31 The crystal packing of 1-palmityl-2-oleyl-3-stearyl glycerol (POS), redrawn from the data of J. B. van Mechelen, *et al.*, *Acta Crystallogr. B*, 2006, **62**, pp. 1121–1130 and 1131–1138, with the kind permission of the International Union of Crystallography, Chester, UK. The kinked oleic acid chains are packed between layers of straight saturated fatty acid chains, here illustrated as palmitic acid, but in cocoa butter also stearic acid. The difference between the two arrangements lies solely in the relationship between the successive layers.

FURTHER READING

Food Lipids: Chemistry, Nutrition, and Biotechnology, ed. C. C. Akoh and D. B. Min, Dekker, New York, 2nd edn, 2002.

Modifying Lipids for Use in Food, ed. F. D. Gunstone, Woodhead, Cambridge, 2006.

Physical Properties of Lipids, ed. A. G. Marangoni and S. S. Narine, Dekker, New York, 2002.

Advanced Dairy Chemistry: Lipids, ed. P. F. Fox and P. McSweeney, Springer Verlag, New York, 2006, vol. 2.

Egg Science and Technology, ed. W. J. Stadelman and O. J. Cotterill, Food Products Press, New York, 4th edn, 1995.

T. Sanders and P. Emery, *Molecular Basis of Human Nutrition*, Taylor and Francis, London, 2003.

C. Clarke, *The Science of Ice Cream*, Royal Society of Chemistry, Cambridge, 2004.

E. N. Frankel, *Antioxidants in Food and Biology: Facts and Fiction*, The Oily Press, Bridgewater, 2007.

S. T. Beckett, *The Science of Chocolate*, Royal Society of Chemistry, Cambridge, 2nd edn, 2008.

RECENT REVIEWS

H. Schenk and R. Peschar, *Understanding the Structure of Chocolate; Radiat. Phys. Chem.*, 2004, **71**, p. 829.

Wijendran and K. C. Hayes, *Dietary n-6 and n-3 Fatty Acid Balance and Cardiovascular Health; Annu. Rev. Nutr.*, 2004, **24**, P. 597.

M. H. Gordon, *The Development of Oxidative Rancidity in Foods,* in *Antioxidants in Food, Practical Applications*, ed. J. Pokorny, N. Yanishlieva and M. H. Gordon, Woodhead, Cambridge, 2001.

G. R. List, *Decreasing trans and Saturated Fatty Acid Content in Food Oils; Food Technol.*, 2004, **58**, p. 23.

R. E. Ostlund, *Phytosterols in Human Nutrition; Annu. Rev. Nutr.*, 2002, **22**, p. 533.

J. Salas-Salvadó, F. Márquez-Sandoval and M. Bulló, *Conjugated Linoleic Acid Intake in Humans: A Systematic Review Focusing on its Effect on Body Composition, Glucose, and Lipid Metabolism; Crit. Rev. Food Sci. Nutr.*, 2006, **46**, p. 479.

E. Choe and D. Min, *Chemistry and Reactions of Reactive Oxygen Species in Foods; Crit. Rev. Food Sci. Nutr.*, 2006, **46**, p. 1040.

J. B. German and C. J. Dillard, *Composition, Structure and Absorption of Milk Lipids: A Source of Energy, Fat-Soluble Nutrients and Bioactive Molecules; Crit. Rev. Food Sci. Nutr.*, 2006, **46**, p. 57.

CHAPTER 5
Proteins

The proteins are the third class of macrocomponents of living systems, and therefore of foodstuffs, that we are to consider.[i] Proteins are polymers with molecular weights ranging from around 10 000 to several million and are often described as having a highly complex structure. In actual fact there is a great deal about the structure of proteins that is quite straightforward. The monomeric units of which they are composed, the amino acids, are linked together by a single type of bond, the peptide bond, and the range of different amino acids is both limited in number and essentially common to all proteins. Furthermore, the polypeptide chain of proteins is never branched.

The special character of proteins lies in the subtlety and diversity of the variations, both in structure and function, which Nature works on this simple theme. The properties and functions of a particular type of protein depend entirely on the precise sequence of its amino acids, unique to that protein. Unlike the polysaccharides, there cannot be anything vague about the exact length of the chain. If even one amino acid in the sequence is out of place, then it is quite likely that the protein will lose its biological activity. It is the sequences of the amino acids in proteins which are defined by the sequences of bases in the DNA that make up our genes. The amino acid sequence of a protein, in this case the β–lactoglobulin of cow's milk, is shown in Figure 5.1. Although in some respects typical, this protein has been selected for illustration

---

[i] In spite of their fundamental importance to living systems the nucleic acids, RNA and DNA, are of almost no significance as components of our diet.

Food: The Chemistry of its Components, Fifth edition
By T.P. Coultate
© T.P. Coultate, 2009
Published by the Royal Society of Chemistry, www.rsc.org

NH₂ – Glu–GlN–Leu–Thr–Lys–Cys–Glu–Val–Phe–Arg–Glu–Leu–Lys–Asp–Leu–Lys–
     Gly–Tyr–Gly–Gly–Val–Ser–Leu–Pro–Glu–Trp–Val–Cys–Thr–Thr–Phe–His–Thr–
     Ser–Gly–Tyr–Asp–Thr–Glu–Ala–Ile–Val–Glu–Asn–AsN–GlN–Ser–Thr–Asp–Tyr–
     Gly–Leu–Phe–GlN–Ile–AsN–AsN–Lys–Ile–Trp–Cys–Lys–AsN–Asp–GlN–Asp–
     Pro–His–Ser–Ser–Asn–Ile–Cys–Asn–Ile–Ser–Cys–Asp–Lys–Phe–Leu–AsN–
     AsN–Asp–Leu–Thr–AsN–AsN–Ile–Met–Cys–Val–Lys–Lys–Ile–Leu–Asp–Lys–
     Val–Gly–Ile–AsN–Tyr–Trp–Leu–Ala–His–Lys–Ala–Leu–Cys–Ser–Glu–Lys–Leu–
     Asp–GlN–Trp–Leu–Cys–Glu–Lys–Leu–*COOH*

Figure 5.1   The amino acid sequence of bovine β-lactoglobulin, one of the principal
                whey proteins. The standard three-letter codes shown in Figure 5.2, below,
                are used. The start and finish of the sequence are known as the amino
                (NH₂) and carboxyl (COOH) ends, respectively. The use of these terms is
                explained in Figure 5.3.

solely because its unusually small size makes it easier to fit on the page.
Most proteins have polypeptide chains many times this length.
Sequences such as these may appear random, but are in fact very tightly
controlled and reproduced exactly in every molecule of the protein
synthesised by a particular organism. With possibly only a few changes
the sequence will normally be reproduced in related species, and often in
much larger taxonomic groups.

As we will see later in this chapter, the proportions of the various
amino acids in the proteins we consume are very important, but the total
quantity is also at least as important. Table 5.1 lists the protein content
of a wide variety of different foodstuffs.

AMINO ACIDS

All but two of the amino acids which occur in proteins have the same
general formula (5.1). The structural formulae shown in Figure 5.2 show
how the exceptions, proline and hydroxyproline do not quite fit this
pattern. The central carbon atom carrying the side-chain (R) which
characterises the particular amino acid, in addition to carrying the
carboxyl and amino groups, is known as the α-carbon.[ii] Except in the
case of glycine, where R = H, each of the four groups attached to the α-
carbon atom is different, *i.e.* the α-carbon is asymmetrically substituted.
This means that amino acids, like sugars (*see* page 10), are optically
active. All the amino acids found in proteins are members of the ʟ-series,
i.e. they have the optical configuration shown in Formula 5.2.

[ii] This use of the Greek alphabet was first introduced in the context of fatty acids, *see* Figure 4.1.

Table 5.1 The total protein contents of a range of foods and beverages. These figures are taken from McCance and Widdowson (see Chapter 1, Further Reading) and in each case apply to the edible portion. They should be regarded as typical for the particular food concerned rather than absolute figures to which all samples of the food will comply.

Food	Total protein (%)	Food	Total protein (%)
White bread	8.4	Tuna (canned)	27.5
Wholemeal bread	9.2	New potatoes	1.7
Rice	2.6	Canned baked beans	5.2
Pasta	3.6	Lentils (dried)	24.3
Cornflakes	7.9	Frozen peas (boiled)	6.0
Cow's milk (whole)	3.2	Bean sprouts (raw)	2.9
Human milk	1.3	Tofu (steamed)	8.1
Soya milk	2.9	Runner beans (boiled)	1.2
Cheese (Cheddar)	25.5	Cabbage (raw)	1.7
Cheese (Brie)	19.3	Mushroom (raw)	1.8
Cheese (Parmesan)	39.4	Sweet corn (canned)	2.9
Yoghurt (plain)	5.7	Eating apples (raw)	0.4
Ice-cream (dairy)	3.6	Bananas	1.2
Egg white	9.0	Raisins	2.1
Egg yolk	16.1	Almonds	21.1
Beef (lean, raw)	20.3	Peanuts (dry roasted)	25.5
Lamb (lean, raw)	20.8	Jam	0.6
Chicken (lean, raw)	20.5	Plain chocolate	4.7
Beefburgers (raw)	15.2	Milk chocolate	8.4
Pork sausages (raw)	10.6	Potato crisps	5.6
Bacon (back, grilled)	25.3	Beer (bitter)	0.3
Lamb's liver (fried)	22.9	Stout	0.3
Cod fillet (raw)	17.4	Lager	0.2

$$
\begin{array}{ccc}
& R & \\
& | & \\
H_2N-\!\!\!\!&C\!\!\!\!&-H \\
& | & \\
& COOH &
\end{array}
\qquad or \qquad
\begin{array}{c}
NH_3^+ \\
| \\
H \blacktriangleright\!\!-\!\!| \blacktriangleleft COO^- \\
| \\
R
\end{array}
$$

(5.1)   (5.2)

Formula 5.2 shows both the amino and carboxyl groups in their ionised condition, giving what is known as a zwitterion, the state which prevails at neutral pH values. D–Amino acids do in fact occur in nature, but not in proteins; they are found in bacteria as components of the cell wall, and in certain antibiotics. Free amino acids are not very important to food chemists, although they do contribute to the flavour of some foods, as indicated in Chapters 2 and 7.

The structures of the side-chains of the 20 amino acids found in most proteins are shown in Figure 5.2. Two other amino acids, hydroxylysine

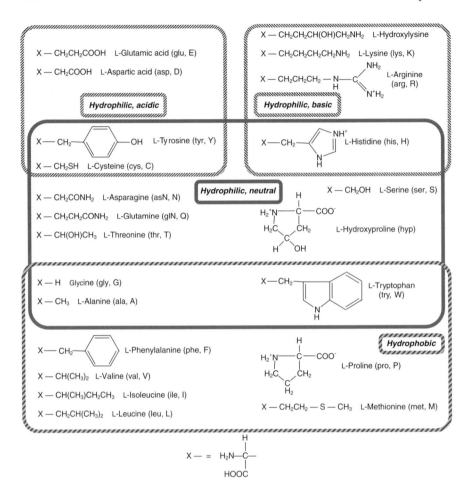

Figure 5.2   The structures of the amino acids commonly found in proteins. L-Proline and L-hydroxyproline are strictly not amino acids but imino acids, as their nitrogen atom is attached to two carbon atoms, the second of these being in effect the far end of the chain curled round and attached to form a ring. Apart from L-hydroxylysine and L-hydroxyproline, the names of all these amino acids have two abbreviated forms, shown here in brackets. The single letter codes are used when writing long amino acid sequences, as they are in entire proteins.

and hydroxyproline, are also included. These two occur in a number of the structural proteins of animals, notably collagen (*see* page 195). Various approaches are used to the classification of amino acids but the most useful is to consider them in terms of the properties of their side-chains rather than their chemical structures. The ionisation of some side-chains is obviously dependent on the pH of the immediate environment or the special conditions which prevail at the active site of an enzyme.

PROTEIN STRUCTURE

The amino acids of a protein are linked together by peptide bonds, formed by the loss of a molecule of water when the amino group of one amino acid reacts with the carboxyl of another. The peptide bond has the amide structure shown in Figure 5.3.

The actual reaction involved in protein synthesis in living systems is of course much more complicated than the simple arrow in Figure 5.3 might imply. This is not least because the enzymes concerned, part of the cell structures known as ribosomes, not only have to forge the links but also to ensure that the amino acids are combined in the correct sequence.

The electrons of the carbonyl group are delocalised and give the C–N bond considerable double-bond character. As a result, free rotation about the C–N bond is inhibited and, as shown in Figure 5.4, all six atoms lie fixed in a single "amide plane". Figure 5.4 shows how a length of polypeptide chain can be visualised as a series of amide planes linked at the α-carbon atoms of successive amino acid residues. The polypeptide chain in Figure 5.4 has been drawn with scant regard for the tetrahedral distribution of the bonds about the α-carbon atoms. In reality the

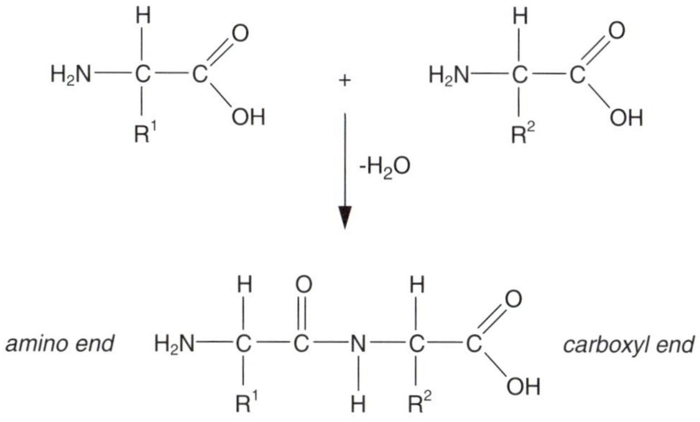

Figure 5.3 The formation of a peptide bond.

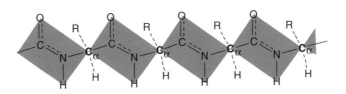

Figure 5.4 The "amide planes" of the polypeptide chain.

chain takes up more compact folding arrangements which can be defined by the angles of rotation about the C_α–N and C_α–C bonds. When a particular pair of such angles is repeated at successive α-carbon atoms the polypeptide chain takes on a visibly regular shape, usually a helix, but at the extreme will form a zigzag arrangement not unlike that in Figure 5.4.

The folding of the polypeptide chain in a protein is determined by its amino acid sequence in ways which molecular biologists are only just beginning to understand. Sophisticated computer programs are able to predict with considerable accuracy the most likely configuration that a given sequence of amino acids will take up. Within the molecule the unique patterns of folding of the peptide chain are maintained by bonds of various types. Covalent linkages, so-called sulfur bridges, are found linking cysteine residues (5.3).[iii]

$$
\begin{array}{ccccccc}
\vdots & & & & & & \vdots \\
| & & & & & & | \\
N & & & & & & N \\
| & & & & & & | \\
HC & \!\!-CH_2-\!\! & S & \!\!-\!\! & S & \!\!-CH_2-\!\! & CH \\
| & & & & & & | \\
O\!=\!C & & & & & & C\!=\!O \\
| & & & & & & | \\
\vdots & & & & & & \vdots
\end{array}
$$

(5.3)

Hydrogen bonding is especially important in maintaining ordered spatial relationships along the polypeptide chain. When these repeatedly join the carboxyl oxygen of one amino acid to the amino hydrogen of the amino acid next but three along the chain, the well-known α-helix results, as shown in Figure 5.5.

The side-chains of the amino acids in helical structures such as these point outwards (*i.e.* away from the axis of the helix) and some are able to form bonds between different regions of the polypeptide chain. These can stabilise the overall folding of the polypeptide chain in globular proteins. One example of such bonds is the cysteine–cysteine bridges already mentioned. Inspection of the amino acid structures in Figure 5.2 will reveal ample scope for hydrogen bonding between side-chains, particularly those involving the amide groups of glutamine and asparagine. Hydrophobic side-chains tend to be orientated in a manner which

---

[iii] The dimer of two molecules linked through their sulfhydryl groups is known as cystine.

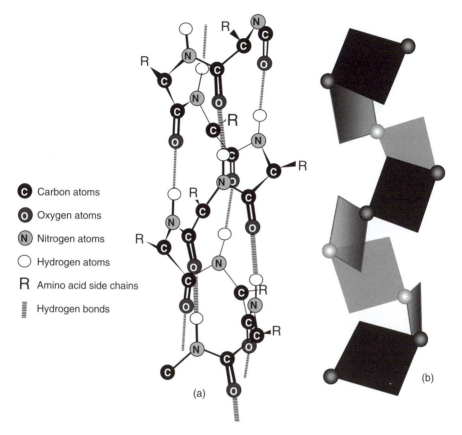

Carbon atoms

Oxygen atoms

Nitrogen atoms

Hydrogen atoms

R Amino acid side chains

Hydrogen bonds

(a)

(b)

Figure 5.5 The α-helix. Only the atoms of the backbone and those involved in hydrogen bonding are shown. There are 3.6 amino acid residues per turn of the helix. Other helical patterns occur with other patterns of hydrogen bonding, but these do not give such a compact structure without straining the hydrogen bonds away from the preferred linear arrangement of the four atoms, N–H···C=C.

minimises their interaction with the aqueous environment of the protein. Such residues are therefore usually directed towards the centre of the folded molecule, where they will be among other non-polar residues. The tendency of the amino acids to act in this way is a major factor in maintaining the correct folding of the polypeptide chain and is often referred to as hydrophobic bonding, unfortunately a far from ideal term (*see* Chapter 12).

Whatever biological role a particular protein may have, it is always dependent on the correct folding of the backbone to maintain the correct spatial relationships of its amino acid side-chains. Not surprisingly, extremes of pH and temperature disrupt the forces maintaining the correct folding and lead to "denaturation" of the protein. In a few cases highly purified proteins can be persuaded to return to their correct arrangement after denaturation, but in food situations this is highly unlikely to occur. It is much more probable that the unfolded proteins will form new interactions with their neighbours, leading to precipitation, solidification or gel formation. For example, white of egg is almost entirely composed of water ($\sim 88\%$) and protein ($\sim 12\%$). When egg white is heated, denaturation leads to the formation of a solid gel network in which the water is trapped. Similarly, when liver is cooked the proteins contained in the liver cells are denatured; if cooking is continued for too long an unpalatable hard texture results from the rigid network of polypeptide chains.

The denaturation of proteins in foodstuffs is not always undesirable, however. The steam or boiling water employed when vegetables are blanched before freezing will inactivate certain enzymes, particularly lipoxygenase (*see* page 302), whose activity can lead to the generation of off-flavours.

Proteins perform numerous roles in living systems. Enzymes, the catalysts upon which the chemical processes on which all life processes depend, are proteins. The carrier molecules, such as haemoglobin, which carry oxygen in the blood, are proteins. So also are the permeases, which control the transport of substances across cell membranes, often against concentration gradients. Another group of proteins are the immunoglobulins, which form the antibodies which provide an animal's defence against invading micro-organisms and certain toxins. These classes of proteins, of which enzymes are by far the most numerous, are all characterised by their physiological functions. In structural terms they are all globular, *i.e.* their polypeptide chains are folded up into a compact arrangement. Just as the exact amino acid sequence is characteristic of a particular protein, so also is their pattern of folding into the globular structure. Figure 5.6 is a diagrammatic representation of the protein myoglobin, which illustrates many of the structural features of more complex proteins which have larger molecules.

Like many enzymes and other carrier molecules myoglobin has a prosthetic group, *i.e.* a non-protein component, which is an essential participant in the catalytic or carrier function. In this case the prosthetic group is a porphyrin known as haem. The iron atom at the centre of the porphyrin ring can form a temporary coordinate link with the oxygen molecule, so that the myoglobin created can function as an oxygen carrier in muscle cells. It should be pointed out that the proportion of the polypeptide chain of myoglobin in the α-helix configuration is particularly high, about 75%; most proteins have no more than 10% as α-helix, with a significant proportion of other ordered structures. The sequence of amino

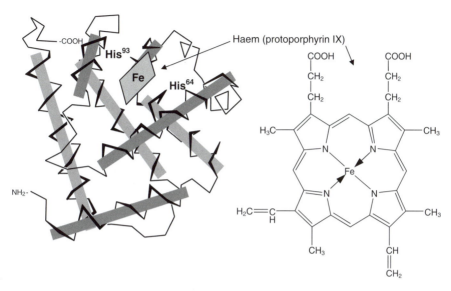

Figure 5.6 Myoglobin. Only the backbone $(-C_\alpha-N-C-C_\alpha-N-C-C_\alpha-)$ of the polypeptide chain is shown, with the angles between successive bonds showing the positions of the α-carbon atoms. The shaded rods mark the axes of regions of the polypeptide chain arranged in the α-helix configuration. The central shaded area shows the position of the haem prosthetic group, surrounded by amino acid side-chains forming a hydrophobic pocket. The two histidine residues linked to the haem iron atom are marked *His*.

acids in the polypeptide chain is referred to as the *primary structure* of the protein. The term *secondary structure* refers to regular locally-ordered arrangements of the chain, including the α-helix (Figure 5.5). The overall folding pattern (Figure 5.6) is referred to as the *tertiary structure*.

If a protein has angles of rotation about the C_α–N and C_α–C bonds repeated along the majority of its polypeptide chain, this will obviously rule out any sort of compact, globular folding arrangement. Instead we can expect extended, highly-ordered molecules. Proteins with such a configuration have structural roles in animal tissues, and one in particular, collagen, is a crucial factor in the texture of meat. We will return to meat and muscle proteins later in this chapter.

The third group of proteins have a nutritional function, either in the transmission of nutrients from mother to offspring (the casein in milk), or in the storage of nutrients to be utilised by an embryo (*e.g.* seed proteins of plants and the egg proteins in birds). In these cases the physical characteristics of the protein are of secondary importance to its overall chemical composition. For example gluten, the principal protein of wheat is rich in glutamine, so it has a higher nitrogen content than other proteins. Among their other distinctive characteristics, seed proteins and caseins share a tendency to form more or less ill-defined aggregates. Biologically, such behaviour ensures that large amounts of nutrient can be concentrated without the problems of osmotic pressure associated with high solute concentrations. This behaviour has hindered efforts to elucidate the structure of these proteins, and as we shall see, it is only recently that the structure of gluten has begun to be properly understood.

ESSENTIAL AMINO ACIDS AND PROTEIN QUALITY

The protein in our diet provides the amino acids from which the body synthesises its own proteins, the major constituent of our tissues. The proportions of the different amino acids supplied by a range of foodstuffs are presented in Table 5.2. The action of hydrolytic enzymes, firstly in the stomach and then in the small intestine, breaks down food proteins to their component amino acids. On absorption into the bloodstream they then become part of the body's amino acid pool. Breakdown of the body's own tissue proteins, an essential part of the process of the renewal of ageing or redundant cells, also contributes to this pool. The amino acid pool is drawn upon not only for protein synthesis but also to provide the raw materials for the synthesis of purines, pyrimidines, porphyrins and other substances.

Many amino acids can be synthesised by mammals provided adequate supplies of amino nitrogen and carbohydrate are available. On the other hand, other amino acids cannot be synthesised by mammals and must be supplied in the diet. The two groups are known respectively as *non-essential* and *essential amino acids*, and they are listed in Table 5.3. Two amino acids among those normally listed as non-essential, histidine and arginine, were originally regarded as essential in experiments carried out in the 1930s on laboratory rats. However, subsequent work showed that humans could synthesise these for themselves. The rate at which humans are able to synthesise histidine, although adequate for adults, is insufficient to meet the needs of rapidly growing children. Two other amino acids, cysteine and tyrosine, also have an intermediate status, and are sometimes referred to as *semi-essential*. Humans require supplies of the essential amino acid methionine in order to synthesise cysteine, plus similar supplies of phenylalanine for the synthesis of tyrosine. As implied in Figure 5.7, when the nutritional quality of food proteins is considered, histidine may be regarded as essential.

Ideally, the protein in the diet should provide amino acids in the same relative proportions as the body's requirements, but in practice such an ideal is never attained. Balancing the pattern of amino acid supply

Table 5.2 The amino acid composition of dietary proteins. The data (calculated from those in the First Supplement to McCance and Widdowson's The Composition of Foods, ed. A. A. Paul, D. A. T. Southgate and J. Russell, HMSO, London, 1980) show the proportions of each amino acid in the total protein content of the foodstuff.

	Fresh peas	Wheat flour	Chicken breast	Beef steak	Whole egg	Cow's milk	Human milk	Cod fillet
				g/100 g total protein				
Isoleucine	4.7	3.9	4.8	4.9	5.6	4.9	5.3	5.2
Leucine	7.5	7.0	7.8	7.6	8.3	9.1	9.9	8.3
Lysine	8.0	1.9	9.3	8.7	6.3	7.4	7.1	9.6
Methionine	1.0	1.6	2.5	2.6	3.2	2.6	v1.5	2.8
Cysteine	1.2	2.6	1.3	1.2	1.8	0.8	2.0	1.1
Phenylalanine	5.0	4.8	4.7	4.3	5.1	4.9	3.8	4.0
Tyrosine	3.0	2.6	3.6	3.7	4.0	4.1	3.0	3.4
Threonine	4.3	2.7	4.3	4.5	5.1	4.4	4.5	4.7
Tryptophan	1.0	1.1	1.1	1.2	1.8	1.3	2.3	1.1
Valine	5.0	4.4	5.0	5.1	7.6	6.6	6.8	5.6
Arginine	10.0	3.6	6.5	6.4	6.1	3.6	3.8	6.2
Histidine	2.4	2.1	3.1	3.5	2.4	2.7	2.5	2.8
Alanine	4.5	3.1	6.0	6.1	5.4	3.6	4.2	6.7
Aspartic acid[a]	11.9	4.4	9.4	9.1	10.7	7.7	9.1	10.2
Glutamic acid[b]	17.3	32.9	17.1	16.5	12.0	20.6	17.4	14.8
Glycine	4.3	3.2	5.1	5.6	3.0	2.0	2.5	4.6
Proline[c]	4.1	1.3	4.3	4.9	3.8	8.5	9.9	4.0
Serine	4.7	5.6	4.1	4.3	7.9	5.2	4.3	4.8

[a]Includes asparagine.
[b]Includes glutamine.
[c]Includes hydroxyproline. The older techniques used to obtain the data were not normally able to separate these pairs of amino acids.

Table 5.3 The status of amino acids regarded as essential and non-essential in humans.

Essential	Non-essential
Isoleucine	Alanine
Leucine	Arginine
Lysine	Asparagine
Methionine	Aspartic acid
Phenylalanine	Cysteine
Threonine	Glutamic acid
Tryptophan	Glutamine
Valine	Glycine
	Histidine
	Proline
	Serine
	Tyrosine

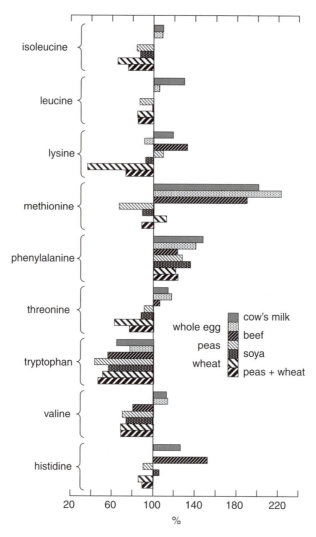

Figure 5.7   The essential amino acids in some important food proteins. The values shown for each amino acid indicate the proportion present (as a percentage) relative to that present in human milk. Thus the provision of methionine by animal proteins is seen to compare very favourably with this standard, whereas the provision of lysine by wheat protein is inadequate. The total amino acid composition of human milk and some other important foodstuffs was given in Table 5.2.

against the needs for protein synthesis is a major function of the liver. Excess supplies of particular amino acids are broken down and the carbon skeletons either oxidised to provide energy or converted to fat for energy storage. The nitrogen is either converted to urea for excretion *via* the kidneys or utilised in the synthesis of any non-essential amino acids which may be in short supply. Difficulties arise when it is supplies of one or more of the essential amino acids which are restricted. Absence of even one particular amino acid will effectively result in the cessation of all protein synthesis, since virtually all proteins contain at least one structural component of all the amino acids shown in Figure 5.2 (exceptions are hydroxyproline and hydroxylysine).

The diet of a typical European normally contains more than ample supplies of both total protein and the individual essential amino acids, but for many of the people of Africa and Asia the supply of protein in the diet is inadequate in both respects. Nutritionists have established that the proportions of the different amino acids required by the human infant correspond closely to the amino acid composition of human milk, and this is now accepted as the standard against which the nutritional value of the protein content of other foodstuffs is judged. Figure 5.7 shows the relative proportions (molar rather than by weight) of the essential amino acids of a number of key foodstuffs in comparison with human milk.

It is immediately apparent from these data that the universality of the roles of the different amino acid side-chains in the maintenance of protein structure leads to broadly similar amino acid compositions in foodstuffs of quite diverse origins. Animal protein foods (eggs, milk and meat) do not differ significantly from human milk in regard to any particular amino acid, but the plant protein foods do present some problems. The low proportion of lysine in wheat protein means that wheat is only about 50% as efficient as human milk as a source of protein. This means in turn that the amount of wheat protein which provides just adequate levels of lysine supplies wasteful levels of the other amino acids. Legumes such as soya and peas provide ample proportions of lysine but are deficient in methionine, so they also are inefficient as sources of protein. A diet containing a mixture of both cereal and legume proteins will obviously be much more efficient than either alone. The efficiency of an individual protein, foodstuff, or whole diet can be described numerically by calculation from the relative proportions of the amino acids present. The proportions of each essential amino acid are compared with those in human milk protein and the lowest of these (as a percentage) is the "chemical score". These values (a selection are shown in Table 5.4) provide a reasonable guide to the actual performance of proteins in human nutrition. Experiments which measure efficiency in terms of the proportion of the nitrogen that the body utilises from dietary protein indicate that we never actually attain the theoretical efficiency. The inevitable inefficiencies of our digestive system and the effects of cooking are usually blamed for this, and also for the wide variations in the published data,

Table 5.4　The nutritional efficiencies of proteins by "chemical score", determined experimentally. The wide range due to the potential influence of cooking means that a single experimental value for meat was not practicable.

Protein source	Chemical score (%)	Approximate experimental value (%)
Human milk	100	94
Whole egg	100	87
Cow's milk	95	81
Peanuts	65	47
Beef	57	–
Wheat	53	49

most of which have been determined in the context of the nutrition of farm animals rather than of humans. The protein content of a typical well balanced European diet gives a chemical score around 70%. Adults in the UK consume between 60 and 90 g of protein per day, so that any reasonably balanced diet (whether vegetarian or not) will provide at least 3 g of even the least abundant amino acids.

Although plant proteins are used less efficiently than animal proteins, one cannot ignore the much greater efficiency of agricultural processes in providing sources of plant protein.

The problems posed by the low levels of lysine and methionine in what would otherwise be ideal proteins are exacerbated by the tendency of these two amino acids to undergo reactions during the storage and processing of foods which destroy their nutritional activity. The Maillard reaction between the lysine amino groups and reducing sugars is the most important of these and has been considered in Chapter 2, but this is not the only route by which the lysine in a protein may be lost. During processing at high temperatures, especially under alkaline conditions, a number of amino acids undergo changes in their side-chains, resulting in nutritional loss. Cysteine and serine are both liable to be converted into dehydroalanine, which can form links with lysine side-chains that result in irreversible links being formed between different sections of the polypeptide chain. (Affluent readers may have observed that silver egg spoons are tarnished by the hydrogen sulfide resulting from the breakdown of cysteine in egg proteins.) Besides reducing the number of lysine residues available, formation of cross-links of this type between neighbouring polypeptide chains will tend to prevent the assimilation of much of the remainder of the protein molecule, since unfolding and access for the proteolytic enzymes of the intestine will be inhibited. The reactions involved in these processes are shown in Figure 5.8, which also indicates how arginine and glutamine may additionally be lost under similar conditions.

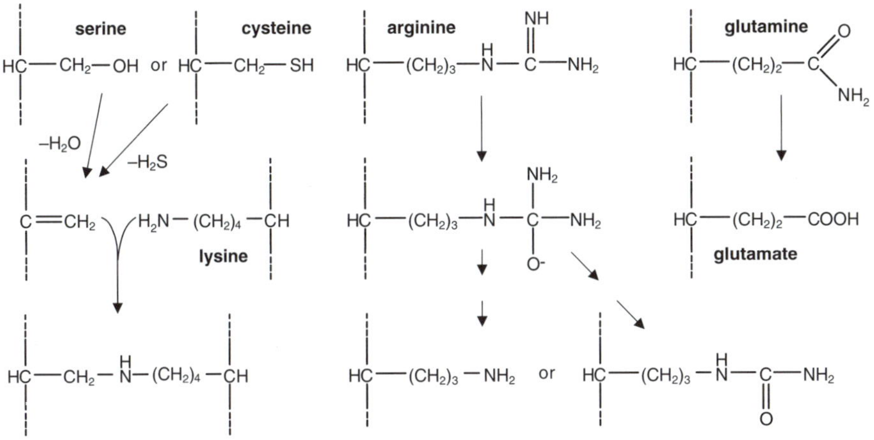

Figure 5.8 Pathways of amino acid breakdown which occur when proteins are subjected to heat at alkaline pH. Asparagine is converted to aspartic acid in a reaction similar to that shown for glutamine. The loss of water from serine is to some extent a reversible reaction, but the product of the reverse reaction is just as likely to be in the D-configuration, in which case its nutritional value is still lost.

ANALYSIS

Measuring the amount of protein in a food material is far from straightforward, and the method chosen tends to be strongly influenced by the subsequent application of the data obtained. For this reason the analysis of protein is given more prominence in this book than the analysis of other food components.

The nutritional quality of a food protein can really only be determined by feeding trials, but sufficient is now known about protein digestion and the effects of processing techniques to enable fairly accurate predictions to be made. These predictions require knowledge of both the total protein content of the food material and its amino acid composition.

The determination of the total protein content of a foodstuff or raw material is by no means as simple as might be imagined. To be applicable to proteins almost any chemical test will depend on the presence of a particular amino acid side-chain. Analytical results will therefore be subject to variations in the proportions of the "target" amino acid in the proteins under test and will generally need to be relative to some arbitrarily selected standard protein. An example of this approach is the Lowry method, more popular with biochemists than food chemists, as it is only applicable to fully dissolved proteins in solution at concentrations up to $300 \, \mu g \, cm^{-3}$. In alkaline solution in the presence of copper ions, tyrosine residues reduce phosphomolybdotungstate to a blue compound which can be estimated spectrophotometrically.

Another colorimetric–spectrophotometric procedure which is fairly uniform in its response to different proteins is the Biuret method. Like the Lowry method, this too is limited to proteins in solution. Under alkaline conditions, which ensure that the polypeptide chain is unfolded and fully accessible, Cu^{2+} ions are complexed by the nitrogen atoms of the polypeptide chain to give a characteristic violet colour (5.4). Potassium tartrate is included in the reaction mixture to ensure that excess Cu^{2+} ions are not precipitated as $Cu(OH)_2$. Although much less sensitive than the Lowry method, it gives a linear response up to around $10 \, mg \, cm^{-3}$ and has the merit of being very easily carried out.

(5.4)

By far the most common method for protein determination in food materials is the Kjeldahl procedure. This method gives a figure for total nitrogen and its use depends on the assumption that this can provide an accurate measure of total protein. In food materials the proportion of non-protein nitrogen is too small to be significant and neither nitrate nor nitrite ions are present. The proportion of nitrogen in most proteins is roughly 16% by weight, and a factor of 6.25 is used to convert nitrogen percentage to protein content. Variations in the amino acid composition of different foodstuffs naturally requires a slightly different factor to be used in accurate work on certain foodstuffs. For example, cereal proteins have an unusually high proportion of glutamine, resulting in a higher than average nitrogen content and consequently a lower conversion factor, 5.70. About one-third of the amino acid in gelatine is glycine, and this similarly raises the nitrogen content, a factor of 5.55 being needed. Meat requires the standard 6.25, but the highest factors, 6.38 and 6.68, are reserved for milk and egg, respectively. Almost without exception, the protein content of food products recorded on their labels has been derived from the total nitrogen content multiplied by an appropriate factor.

In the Kjeldahl procedure the entire food sample is digested in concentrated sulfuric acid at temperatures approaching 400 °C, with a heavy metal salt such as copper sulfate as catalyst. A quantity of sodium sulfate is also added to raise the boiling point of the sulfuric acid. Under these conditions organic material is oxidised, any organic nitrogen being retained in solution as ammonium ions. On completion of the digest an aliquot is withdrawn and transferred to an apparatus known as a Markham still. Here it is first made alkaline and then the liberated ammonia is steam distilled off into an aliquot of boric acid solution. The ammonia is estimated by titration. Modern laboratories use a semi-automated system for this procedure.

Near-infrared reflectance spectrophotometry is a valuable instrumental procedure, useful when large numbers of similar solid samples, for example of flour, are being routinely tested. Measurements are taken at 2100 nm (characteristic of the hydroxyl groups in starch), 2180 nm (characteristic of the amide groups in protein), and 1680 nm (a background reference value). Calculation of the protein content is based on subtraction of the reference value plus the starch shoulder from the amide peak. The test can be extended to give a measure of moisture content. Although the technique demands accurate control of particle size, it is very rapid and lends itself to automation. Samples taken from a lorry-load of wheat can be checked while waiting to be unloaded at the flour mill.

Over the years many colour reactions specific to particular amino acids have been developed. Unfortunately, few of them are quantitative and are unable to provide the comprehensive data required. For this reason chromatographic techniques are the obvious choice for determining the amino acid composition of a protein hydrolysate. Although well established systems are available for one- and two-dimensional paper chromatography, modern laboratories always prefer one of the techniques based on high-performance liquid chromatography (HPLC). In some methods the amino acids are first converted to a derivative which is more amenable to chromatographic separation and detection. Alternatively, the amino acids can be detected in the eluate by reaction with ninhydrin (triketohydrindene hydrate, 5.5), to give a blue colour. Proline and hydroxyproline similarly produce a yellow colour.

(5.5)

FOOD PROTEIN SYSTEMS

Milk

The milk from the domestic cow, *Bos taurus*, is an important protein source for man, and particularly for children. Outside Britain the milks of other ruminant species, notably sheep (*Ovis aries*) and goats (*Capra hircus*), also have considerable significance for human nutrition. Milk is an aqueous solution of proteins, lactose, vitamins and minerals, and it carries emulsified globules of fat. It also contains particles known as casein micelles, consisting of a number of related protein species collectively known simply as the casein proteins, together with phosphate, calcium and a trace of citrate. If the fat is removed from milk, the product is skimmed, or skim milk.[iv] If the casein micelles are also removed (they can be precipitated out at a pH around 4.6), the residual solution is known as serum (by analogy with blood) or, more commonly, whey. The whey obtained in cheese making has a slightly different composition, as some of the casein is solubilised and some of the lactose will have been converted into lactic acid by the action of bacteria.

The proportions of various casein and whey proteins vary quite considerably, as does the total protein content of milk, but the values shown in Table 5.5 are typical.

The Greek letters used in the names of the different proteins were originally assigned on the basis of the mobility of the protein during electrophoresis. The "s" in α_{S1}-casein refers to its sensitivity to precipitation by calcium ions. Although the prefixes α_{S1}, β, *etc.* define particular protein species, it is now recognised that there are a number of

---

[iv] Semi-skimmed milk, with approximately 2% fat, is not manufactured by removing half the fat. It is technically much easier to mix equal proportions of skimmed milk with whole milk.

Table 5.5 Typical proportions of major proteins in skimmed milk. The α_S-casein fraction consists of two proteins, α_{S1} and α_{S2}, present in the approximate ratio 3:1.

	Quantity (g/dm³)	Total protein (%)
Casein proteins (total)	~27	~80
α_S-Casein	10–16	31–45
β-Casein	8–12	24–34
κ-Casein	10–15	3–5
γ-Casein	4–6	1.0–1.5
Whey proteins (total)	~6	~20
α-Lactalbumin	1.0–1.5	3.3–5.0
β-Lactoglobulin	2–4	6.6–13.3
Immunoglobulins	0.7–0.8	2.3–2.7
Others	1–2	3–7

different versions of each that differ slightly in their amino acid sequence. For example, α_{S1}C–casein differs from α_{S1}B-casein (the most common variant) in having a glycine residue in place of a glutamic acid residue at position 192 in the polypeptide chain. One can actually identify the breed of cow from which a sample of milk has originated by examination of the proportions of these variants.

The casein micelles of milk are roughly spherical, with diameters up to 600 nm. However, approximately half of the total casein is found in micelles with diameters between 130 and 250 nm, with the remainder fairly equally divided between micelles above and below this range. The micelles are sufficiently large and abundant to be the source of the opalescence of skimmed milk. Milk contains about 10^{15} micelles per litre. A typical micelle contains 10^4–10^5 casein molecules (the molecular weights of α_S-, β-, and κ-casein are 2.35, 2.40 and 1.90×10^4, respectively). γ-Casein is believed to be an artefact of the laboratory procedures used to separate and purify casein molecules, and is a fragment of β-casein resulting from limited breakdown caused by proteolytic enzymes in milk.

For many years chemists have sought to describe the structure and organisation of the casein micelle in relation to both its natural nutritional role and the behaviour of milk proteins during processing operations such as cheese making. Milk has evolved as a vehicle for the nutrients required by a newborn mammal, and its value to humans for the manufacture of cheese and butter and for inclusion in tea and coffee is decidedly secondary. The organisation of much of milk's total protein, and all of its calcium and phosphate, into micelles overcomes two major problems. Firstly, in the absence of aggregation the quantity of protein present in milk would form highly viscous solutions which would pose problems for secretion and delivery. Secondly, the calcium and the phosphate, each essential for the growth of the newborn skeleton, would form large relatively insoluble crystals of calcium phosphate, again posing problems of secretion and also digestion. It is now recognised that the role of the casein proteins as a vehicle for calcium and phosphate has been a major determinant in the evolution of their structure.

For many years the most popular theory of casein micelle structure was based on the proposals of Slattery and Evard in 1973, and subsequent work by Schmidt. In this theory the micelle structure was composed of spherical aggregates of 25–30 casein molecules, the submicelles. Within the submicelles individual protein molecules were held together by hydrophobic interaction.[v] The phosphoserine side-chains (5.6) of the casein molecules provided the attachments for the calcium phosphate bridges which held the submicelles together. The submicelle theory has since been challenged by the results of various studies, both relating to the micelles themselves and also to the interactions between the casein, phosphate and calcium. The attention devoted to the submicelle theory in previous editions of this book is now transferred to what have now become known as the internal structure models. These were initially proposed by Carl Holt and subsequently developed by David Horne.

Studies which have told us a great deal about the three-dimensional structure of large numbers of different proteins have been much less successful when applied to the caseins. Although their amino acid sequences are well established, illustrated in simplified form in Figure 5.9, there has been no agreement about the secondary and tertiary structures (*see* page 167) of the caseins. Most authorities now agree that it is likely that they do not take up the compact tertiary structures associated with most proteins, instead tending to exist in irregular flexible conformations which link up to form extended three-dimensional networks. As shown in Figure 5.9, the polypeptide chains of all four caseins include extensive regions dominated by hydrophobic amino acids which tend to associate and form links between protein molecules. Hydrogen bonding between amino acid side-chains is also expected to strengthen these links.

---

[v] Hydrophobic interactions, sometimes referred to hydrophobic bonds, were described on page 165 in the context of the forces that maintain the folded of peptide chains, and are given further attention in Chapter 12.

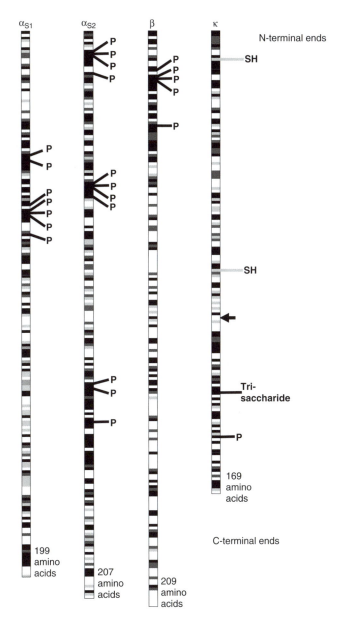

Figure 5.9 The polypeptide chains of bovine caseins. The characteristics of the amino acids are indicated by shading, black for the most hydrophilic and white for the most hydrophobic. Serine residues esterified with phosphate are identified by P, and the threonine (134) which carries κ-casein's tri-saccharide unit by Trisaccharide. The location of the cysteine residues of κ-casein is indicated by SH. The arrow marks the site between phenyla-lanine-105 and methionine-106, where rennin hydrolyses the peptide bond to liberate the glycopeptide during cheese making.

(5.6)   (5.7)

The phosphoserine residues (5.6) found in clusters in the hydrophilic regions of the polypeptide chains of α and β-caseins provide the second type of bonding which maintains micelle structure. Casein micelles contain a significant amount (about 6% on a dry basis), of colloidal calcium phosphate (CCP). The calcium phosphate occurs in nanoclusters a few nanometres (nm) in diameter, with a chemical composition approximating to $Ca_9(PO_4)_6$. The α- and β-casein molecules link to the nanoclusters through their own clusters of phosphoserine groups. The combination of these links and the association between the hydrophobic regions of the molecules allows the formation of a stable three-dimensional network. In contrast to the α and β-caseins, κ-casein has no more than a single phosphoserine. Instead, the extremely hydrophilic character of its *C*-terminal region is maintained by the presence of at least one trisaccharide unit, α-*N*-acetylneuraminyl-(2→6)-β-galactosyl-(1 → 6)-*N*-acetyl-galactosamine (5.7).

It is important to remember that in the formation of any branched or net-like polymer the monomeric units must each contain at least three functional, *i.e.* linking, groups, as illustrated in Figure 5.10. In the case of the casein micelle the α- and β-caseins both contain a number of regions capable of forming links with each other, either directly by hydrophobic bonding, or indirectly *via* the calcium phosphate nanoclusters. As the network grows it will be joined by κ-caseins. However these are only capable of integrating into the network *via* their *N*-terminal ends[vi] and their *C*-terminal ends will form a barrier to expansion in that

---

[vi] When separated from other caseins in the laboratory κ-casein forms dimers through its cysteine sulfhydryl groups, and it is quite likely that κ-casein also exists in this form in the casein micelle. However the dimers would still retain their chain-terminating function.

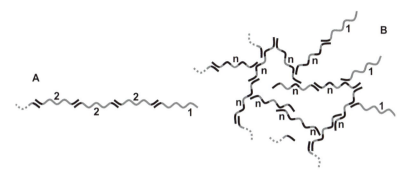

Figure 5.10   Polymer formation. In (A) monomers (grey) with two linking groups (shown
in black) form an unbranched chain which is terminated at one end when a
monomer with only one linking group joins the chain. In (B) monomers with
three or more linking groups (n) readily form a branched network. However,
when single linking monomers join the network, as here on the right-hand
side, further growth of the network on this side is prevented.

region. It is not difficult to see that eventually the growing micelle will
spontaneously acquire a surface dominated by κ-casein, and growth
will cease. Although the mathematics of the situation has yet to be formally addressed, even a superficial consideration of the process will
show that one can expect smaller micelles to contain a higher proportion
of κ-casein.

The model for the casein micelle that emerges from this is shown in
Figure 5.11. The core of the micelle is a three-dimensional network of α-
and β-casein chains, linked both by attraction between the extensive
hydrophobic regions of the polypeptide chains and by the binding of
their clusters of phosphoserine residues and the CCP nanoclusters. It has
been calculated that a typical micelle might contain some 800
nanoclusters. The surface of the micelle is dominated by the hydrophilic
C-terminal ends of the κ-casein.

The most abundant serum proteins, α-lactalbumin and β-lactoglobulin, have attracted considerable attention from protein chemists. They
have been interested in their physical properties and structure, but these
studies have revealed little about the reason for their presence in milk
beyond their contribution to the total protein content. The immunoglobulins in contrast have proved to be of great interest to food scientists. The role of one class of immunoglobulins, the macroglobulins, in
the creaming phenomenon has been discussed in Chapter 4, but the
biological role of immunoglobulins is much more important. Immunoglobulins are popularly known as antibodies, which are proteins
synthesised in various parts of the body in response to the invasion of

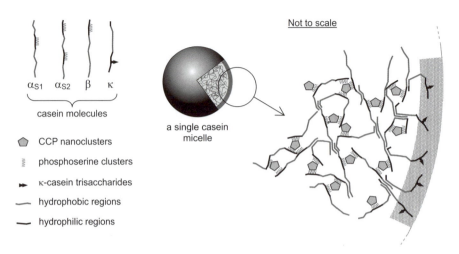

Figure 5.11   A diagrammatic impression of the structure of the casein micelle, based on the proposals of Horne and Holt. The surface of the micelle, shown by the shaded area in the enlarged cross-section, is dominated by the polar C-terminal regions of the κ-casein polypeptide chains carrying tri-saccharide units.

the tissues by foreign matter, particularly bacteria, viruses and toxins. Their reaction with these antigens is remarkably specific and facilitates their neutralisation or destruction by other elements of the body's defence mechanisms. It is now well established that exposure of a mother to many common pathogenic bacteria, especially those causing intestinal diseases such as diarrhoea which are very dangerous in the new-born, leads to the appearance of appropriate antibodies in the mother's milk. These undoubtedly play a part in protecting the newborn infant from many dangerous diseases. However closely the manu-facturers of infant milk products based on cow's milk are able to match the nutrient content of human milk, it is increasingly clear that they will not be able to match these unique advantages of human milk.

Cheese

The fundamental process involved in cheese making is the precipitation of casein to form a curd. In the case of yoghurt and some cottage cheeses, the precipitation is caused entirely by the effect of low pH. The growth of lactobacilli in the milk is followed by their fermentation of the lactose to L-lactic acid (5.8). The lactobacilli involved are sometimes those present naturally in the milk, but usually a starter culture of a strain with particularly desirable characteristics is added to the milk.

Complete precipitation of the casein does not occur at the pH of these types of fermented milk product. Association between the casein micelles gives yoghurt its characteristic gel-like texture.

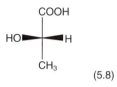

(5.8)

In the production of hard cheeses such as Cheddar, Stilton and Gruyère, microbial action is allowed to bring the pH down to around 5.5. At this point rennet is added to cause extensive precipitation and curd formation. Rennet is a preparation of the enzyme chymosin (known in the past as rennin) obtained from the lining of the abomasum (the fourth stomach) of calves. Chymosin specifically catalyses the hydrolysis of one particular peptide bond in κ-casein, as shown in Figure 5.12. The κ-casein is split into two fragments. One, *para*-κ-casein, remains as part of the micelle and includes the hydrophobic section of the molecules. The other, κ-casein glycopeptide, is lost into the whey. The macropeptide fragment carries the carbohydrate units. The loss of their carbohydrate coating means that the formation of strong cross-links between the micelles is no longer blocked and a curd rapidly develops.

The curd is held for several hours, during which time the acidity increases and the curd acquires the proper degree of firmness. Once this is achieved, the curd is chopped into small pieces to allow the whey to run out as the curd is stirred and finally pressed. Salt is added before

NH₂ -	D	D	Q	N	Q	D	Q	P	I	R	C	D	K	E	D	R	F	F	S	E	K	→
	I	A	K	Y	I	P	I	Q	Y	V	L	S	R	Y	P	S	Y	G	L	N	Y	→
-	Y	Q	Q	K	P	V	A	L	I	N	N	Q	F	L	P	Y	P	Y	Y	A	K	→
-	P	A	A	V	R	S	P	A	Q	I	L	Q	W	Q	V	L	S	E	T	V	P	→
-	A	K	S	C	Q	A	Q	P	T	T	M	A	R	H	P	H	P	H	L	S	F ‖→
-	M	A	I	P	P	K	K	N	Q	E	K	T	D	I	P	T	I	N	T	I	A	→
-	S	G	D	P	T	S	T	P	T	I	D	A	V	D	S	T	V	A	T	L	D	→
-	A	S	P	D	V	I	D	S	P	P	D	I	N	T	V	Q	V	T	S	T	A	→
-	V	-COOH																				

Figure 5.12   The amino acid sequence of κ-casein using the single letter code for amino acids. The bar marks the peptide bond between phenylalanine-105 and methionine-106 which is hydrolysed by the enzyme chymosin during cheese manufacture, liberating the fragment (shaded) known as the κ-casein glycopeptide into the whey. The boxed threonine residue (133) is the normal site of attachment for the trisaccharide. Trisaccharide units are also occasionally found attached at a number of nearby threonine and serine residues.

pressing to hasten the removal of the last of the whey and to depress the growth of unwanted micro-organisms. The final stage of cheese manufacture, ripening, requires low temperature storage for a considerable period. Rennet enzymes (notably pepsin) and other proteases from the micro-organisms cause a limited degree of protein breakdown, which is important for both the texture and flavour of the finished cheese (many small peptides have distinctive bitter or meaty tastes). Lipases from the milk liberate short-chain fatty acids from the milk fat triglycerides, which also contribute to the flavour.

In recent years the increasing yields of milk per cow have meant that fewer calves are born relative to the amount of milk converted to cheese. As a result, rennin supplies have not kept pace with demand. This has forced cheese makers to examine other enzymes as substitutes. Proteases from fungi and porcine pepsin have been tried, but these suffer from the disadvantage of causing excessive breakdown of other proteins.[vii] The modern solution brings genetic engineering to bear on the problem. The DNA coding for calf rennin has been successfully inserted into the chromosomes of the yeast *Kluveromyces lactis* and the bacterium *Escherichia coli*. Both micro-organisms can then be persuaded to produce large quantities of the enzyme, identical in every way to that produced by the calf.

The special relationship between rennin action and the structure of casein obviously did not evolve to enable us to make cheese, but the details of casein digestion in calves or in man have not yet been investigated in sufficient detail to provide an explanation for the origin of the relationship.

Egg

The yolk fraction of the eggs of the domestic hen, *Gallus domesticus*, was discussed briefly on page 143, and the present section will be devoted to the egg white, or albumen fraction. Egg white consists almost entirely of protein, 11–13%, and water. Many of the applications of eggs in cooking can be explained in terms of the properties of this protein, as we shall see.

In the raw egg the white is not homogeneous but occurs in four layers, two thick and two thin. The inner thick layer nearest the yolk is continuous with the chalazas, the stringy structures that suspend the yolk in the centre of the egg. The protein composition of the different layers is

---

[vii] Cheese manufactured using fungal rennet is often labelled as "vegetarian". Bearing in mind that cows have to give birth to a calf before they can produce milk and that only a small proportion of calves survive to enjoy a fulfilling career of their own in milk production, it seems rather eccentric to focus on the source of chymosin as the qualification for vegetarian status.

Table 5.6 The major proteins of hen's egg
white. Conalbumin is also known
as ovotransferrin.

	Approximate total protein (%)
Ovalbumin	54
Conalbumin	12
Ovomucoid	11
Ovomucin	1.5–1.3
Lysozyme	3.4
Ovoglobulins	8

identical, except that the more viscous layers have a higher proportion of
the protein, ovomucin. For all culinary purposes the different layers of
the white are thoroughly mixed together, and the white can therefore be
regarded as a simple aqueous solution of all but one of the globular
proteins listed in Table 5.6. The exception is the ovomucin, which
apparently forms fibres which make a major contribution to the viscosity of the white.

The predominant protein, ovalbumin, is a glycophosphoprotein, *i.e.* a
protein with both phosphate and carbohydrate groups attached. Up to
two phosphates are attached to serine residues (as in α- and β-caseins)
and an asparagine residue carries the carbohydrate unit. This consists of
two or more *N*-acetylglucosamine (3.1) units carrying a number of
mannose units. Ovalbumin is readily denatured, the process we observe
when egg white sets on heating. The polypeptide chain of ovalbumin
includes six cysteine residues. Two of these form a sulfhydryl bridge (5.3)
which helps to maintain the three-dimensional structure of the protein.
As the temperature is raised the polypeptide chain unfolds and new
sulfhydryl bridges become possible, not only within the molecule but
also linking neighbouring molecules together, forming the rigid gel of
cooked egg white.

Ovalbumin is also particularly susceptible to denaturation at air–
water and oil–water interfaces. It will be recalled that the polypeptide
chains of globular proteins tend to fold up, so that amino acids with
hydrophobic side-chains are located in the centre of the molecule away
from the surrounding water. If the water on one side of a protein
molecule is displaced by air or oil the advantage of keeping the hydrophobic amino acids tucked away in the centre of the molecule disappears
and the molecule unfolds, *i.e.* it becomes denatured. As before, the
denatured molecules readily interact together, but instead of forming a
solid gel they form a film around air bubbles or fat droplets, stabilising
respectively foams (as in meringue) or emulsions (as in cake batters).

Subsequent heating causes further denaturation which reinforces these films.

Beating egg white to a stable foam is a difficult process to control. If the beating is excessive too much of the ovalbumin becomes denatured, leaving insufficient undenatured protein to bind the water. Beating must stop as soon as peak stiffness is achieved, since further beating to incorporate more air will only weaken the foam.

Another glycoprotein, conalbumin, is a metal-binding protein analogous to lactoferrin in milk (*see* page 449). It behaves similarly to ovalbumin but has been implicated in one rather odd aspect of cookery technique. Glass bowls are preferred to plastic bowls for beating egg white, as the surface of even well cleaned plastic bowls is likely to be contaminated with traces of fat, which interferes with foam formation. However, copper bowls are claimed to be the very best. It has been suggested that the conalbumin binds with Cu^{2+} ions from the bowl and that the resulting copper–protein complex is more resistant to surface denaturation, thereby providing some protection against over-beating.

Ovomucoid, also a glycoprotein, gives the albumen its high viscosity. Although a certain level of viscosity is important for foam stability, if it is too high whipping becomes difficult. This is why it is important that eggs for meringue or cake making should not be used straight from the refrigerator, since their low temperature would cause too high a viscosity.

Foam stability is enhanced when the pH is close to the isoelectric points of the proteins. Many protein amino acid side-chains include ionisable groups, notably aspartic and glutamic acids, lysine, arginine and histidine. The proportion of a particular amino acid side-chain which is ionised is clearly a function of the hydrogen ion (H^+) concentration, *i.e.* the pH:

$$-COO^- + H^+ \leftrightarrow -COOH \qquad \text{aspartic and glutamic acids}$$

$$-NH_2 + H^+ \leftrightarrow -NH_3^+ \qquad \text{lysine}$$

$$=NH + H^+ \leftrightarrow =NH_2^+ \qquad \text{arginine}$$

$$\geqslant N + H^+ \leftrightarrow \geqslant NH^+ \qquad \text{histidine}$$

The isoelectric point of a protein is the pH at which the number of positively charged groups exactly equals the number of negatively charged groups, *i.e.* the protein molecule carries no net charge. Protein molecules do not repel each other so much when they are uncharged, and therefore interact more strongly. Cooks use cream of tartar (potassium hydrogen tartrate) or a trace of vinegar to achieve this effect.

Meat

When we consume the proteins in milk and eggs we are taking in proteins which have evolved purely as a food. When we consume meat, however, or if we are studying the processes involved in the conversion of muscle tissue into the slice of roast beef on our plate, we must remember that the biologist's view of its function, as well as the view of the animal itself, will be quite different from that of the food chemist or nutritionist. The structure of muscle tissue[viii] is exceedingly complex, and readers with biological or biochemical interests are strongly advised to supplement the present account with the details to be found in text books of cell or mammalian physiology, or biochemistry.

A typical joint of meat from the butcher's shop is cut from a number of muscles, each originally having had its own independent attachment to the skeleton, its own blood supply and its own nerves. Each muscle is surrounded by a layer of connective tissue consisting almost entirely of the protein, collagen. This layer contains and supports the contractile tissues of the muscle and at its extremities provides the connections to the skeleton. The contractile units of the muscle are the muscle fibres. These are exceptionally elongated cells, 10–100 µm in diameter, and can be as long as 30 cm. Examined in cross-section, the muscle fibres can be seen to be organised into bundles separated by connective tissue which is continuous with that surrounding the muscle and provides the connections to the skeleton. Blood vessels, adipose (*i.e.* fatty) tissue, and nerves are also embedded in the connective tissue of the muscle.

Muscle fibres share most of the features found in more typical animal cells. Surrounding the fibre is the cell membrane, known as the sarcolemma (the prefix *sarco* is derived from the Greek for *flesh*). Along the length of the cell, just inside the sarcolemma, are numerous nuclei. Muscle fibres have two elaborate intracellular membrane systems, the transverse tubules, or T-tubules, which are invaginations of the sarcolemma, and the sarcoplasmic reticulum, which is the counterpart of the endoplasmic reticulum of other cells. Both participate in the

---

[viii] Although the term meat usually encompasses fish and poultry as well as other mammalian tissues such as liver, kidney and intestine (the offals), this account is largely directed towards the properties of the skeletal muscles of mammals such as pigs, sheep and cattle.

transmission of the signal from the motor nerve endings on the surface of the fibre to the contractile elements of the fibre, the myofibrils. The T-tubules and sarcoplasmic reticulum wind around each myofibril, so that the contraction events in each myofibril are accurately synchronised. Innumerable mitochondria are also found lying between the myofibrils to ensure that the supply of adenosine triphosphate (ATP), the chemical fuel for the contraction process, is maintained.

The myofibrils themselves are composed of bundles of protein filaments arranged as shown in Figure 5.13a. The thin filaments are composed mostly of the protein *actin,* together with smaller amounts of *tropomyosin* and *troponin,* as shown in Figure 5.13b. The thick filaments are aggregates of the very large elongated protein *myosin,* molecular

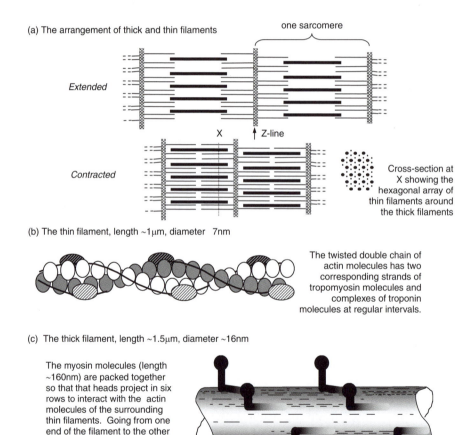

(a) The arrangement of thick and thin filaments

one sarcomere

Extended

X   ↑ Z-line

Contracted

Cross-section at X showing the hexagonal array of thin filaments around the thick filaments

(b) The thin filament, length ~1μm, diameter 7nm

The twisted double chain of actin molecules has two corresponding strands of tropomyosin molecules and complexes of troponin molecules at regular intervals.

(c) The thick filament, length ~1.5μm, diameter ~16nm

The myosin molecules (length ~160nm) are packed together so that that heads project in six rows to interact with the actin molecules of the surrounding thin filaments. Going from one end of the filament to the other the myosin molecules change direction, resulting in a central region lacking heads.

Figure 5.13   The myofilaments of skeletal muscle.

weight 5×10^5, arranged as shown in Figure 5.13c. The sequence of events in muscle contraction is as follows:

(i) The nerve impulse is received at the membrane surface of the fibre, the sarcolemma, transmitted throughout the cell by the T-tubules and provokes the release of Ca^{2+} ions from the reservoir of vesicles formed by the membranes of the sarcoplasmic reticulum.

(ii) The Ca^{2+} ions bind to the troponin of the thin filaments. This causes a change in the shape of the troponin molecules which move the adjoining tropomyosin protein complex. The change in the position of the tropomyosin exposes the active site of the actin molecules.

(iii) The activated actin molecules are now able to react repeatedly with the myosin of the thick filament and ATP, the energy source for contraction, as shown in Figure 5.14

(iv) The heads of the myosin molecules are at different angles to the thick filaments at different stages in the cycle, so that with each turn of the cycle, and with the hydrolysis of one ATP molecule, the myosin head engages with a different actin molecule, the two filaments move relative to each other, and contraction occurs.

(v) When the nerve impulse ceases, calcium is pumped back into the sarcoplasmic reticulum and the cycle comes to rest at stage (ii). The tropomyosin returns to its original position and the actin can no longer interact with a myosin head. The muscle is now relaxed and will be returned to its original extended state by the action of other muscles.

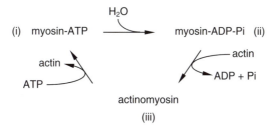

Figure 5.14 The myosin–actinomyosin cycle. When the muscle is in the process of contracting the cycle functions continuously, as shown here. When the concentration of Ca^{2+} ions drops the cycle comes to rest at stage (ii), as linkages between actin and myosin can no longer be formed. If the supply of ATP fails the cycle comes to rest at stage (iii).

The energy in the form of ATP, for contraction and for the active uptake of calcium by the sarcoplasmic reticulum, is derived from two sources. When moderate levels of muscle activity are demanded, pyruvic acid (2-oxopropanoic acid, 5.9), derived from carbohydrate or fatty acid breakdown, is oxidised in the mitochondria of the muscle fibres. The energy made available by the oxidation is utilised in the mitochondria for the phosphorylation of adenosine diphosphate to the triphosphate. Three molecules of ATP are formed for every $1/2O_2$ reduced to H_2O. The oxygen is carried from the blood to the mitochondria as an oxygen adduct of myoglobin, the source of the red colour of fresh meat.

$$
\begin{array}{c}
COOH \\
| \\
C = O \\
| \\
CH_3
\end{array}
$$

(5.9)

When short bursts of extreme muscle activity are demanded, for example when running for a bus, or when a bird like a pheasant suddenly takes to the air, the limited rate at which the lungs can supply oxygen to the muscles *via* the bloodstream is insufficient to give an adequate supply of ATP. At these times the muscle obtains its energy by the less efficient, but anaerobic and more rapid, conversion of glycogen or glucose to lactic acid by the metabolic process of glycolysis. Only two molecules of ATP per molecule of glucose are obtained in glycolysis, compared with 36 when the glucose is oxidised completely to CO_2. In the muscles of the living animal, anaerobic glycolysis cannot be maintained for many seconds before the accumulation of lactic acid becomes excessive and the resulting pain brings the exertion to a halt. As soon as demands on the muscles ease, the accumulated lactic acid is readily oxidised back to pyruvate; then either the pyruvate can be oxidised to CO_2 or some of it is reconverted to glucose in the liver.

On the death of the animal there is still a demand for ATP in the muscle from the pumping systems of the sarcoplasmic reticulum and other reactions, even though the muscles are no longer called upon to contract by signals from the nervous system. As there is obviously no further availability of oxygen from the blood stream, the supply of ATP is maintained by glycolysis for a time, until one of two possible circumstances brings it to a halt. If the animal was starved and otherwise stressed by bad treatment before slaughter, reserves of glycogen in the muscles will have been depleted and will rapidly be exhausted after slaughter, in which case glycolysis will quickly come to a halt. However, if the animal's muscles had adequate glycogen reserves glycolysis will still cease, but only after a few hours when the accumulation of lactic acid has reduced the pH to around 5.0–5.5, sufficient to inhibit the activity of the glycolytic enzymes. When the ATP level in the muscle has consequently fallen, calcium will no longer be pumped out of the sarcoplasm into the vesicles of the sarcoplasmic reticulum. This calcium allows the myosin to interact with the actin, but the lack of ATP will prevent the operation of the full cycle which would normally lead to contraction. The thick and thin filaments will therefore form a permanent link at stage (iii) and the muscle will go rigid—the state of *rigor mortis*. It is normally impossible to get meat cooked before it goes into rigor, and meat cooked during rigor is said to be exceedingly tough.

The answer to the problem of *rigor mortis* and, as we shall see, other aspects of meat quality, lies in butchers ensuring that the critical pH range of 5.0–5.5 is achieved. (Suffice to say, it is the application of sound pre- and post-slaughter practices rather than the possession of a pH meter which makes a good butcher.) The most obvious effect of low pH is to deter the growth of the putrefactive and pathogenic micro-organisms which spread from the hide, intestines, *etc.,* during the preparation of the carcass. During the conditioning (hanging) of the carcass the toughness due to *rigor mortis* disappears. The phrase "resolution of rigor" is often applied to the disappearance of toughness.

Certain specific changes in the myofibrillar proteins are now associated with the resolution of rigor. These involve the activity of a group of proteolytic enzymes (*i.e.* enzymes that hydrolyse peptide bonds) known as calcium-activated neutral proteinases, or calpains. These enzymes occur in the sarcoplasm of the muscle cell, but require levels of Ca^{2+} ions which only become available once the muscle's ATP supply is exhausted and the sarcoplasmic reticulum releases them. Several calpains are known to attack a group of minor proteins, sometimes referred to as "gap filaments", which are responsible for maintaining the organisation of the major myofibrillar proteins, actin and myosin. One of these, titin, appears to provide an elastic connection between the thick filament and the Z-line (*see* Figure 5.13). The breakdown of these proteins during conditioning creates planes of weakness across the muscle fibres which allow the muscle fibres to break when they are stretched—in effect, become tender.

Muscle tissue contains a great deal of water, between 55 and 80%. Some of this water is directly bound to the proteins of the muscle, both the myofilaments and the enzyme proteins of the sarcoplasm. The bulk of the water, however, is in the spaces between the filaments. Protein denaturation on cooking releases some of this water and gives meat its desirable moistness, and in the raw state a moist and shiny cut surface of a joint of meat is favoured by purchasers. Although the sarcoplasmic proteins will bind less water at pH values near to their isoelectric points, mostly around pH 5, the principal cause of moisture release (drip) from well-conditioned meat is that as the pH falls the gap between neighbouring filaments is reduced and some of the water is squeezed out. Although the butcher would obviously not wish to lose too much moisture from a joint before it is weighed and sold, the increased mobility of the muscle water has another beneficial effect besides giving the meat an attractive moist appearance.

The further effect is, as we shall see, on the colour of the meat. It has already been mentioned that the red colour of myoglobin varies with the age of the animal, from species to species, and from muscle to muscle. Some typical data for cattle and pigs are given in Table 5.7, based on results quoted by R. A. Lawrie and D. A. Ledward (2006), *see* Further Reading. As a general rule muscle used for intermittent bursts of activity, which are fuelled predominantly by anaerobic glycolysis, contains only low levels of myoglobin and few mitochondria, and will thus be pale in colour. The breast, *i.e.* the flight muscle, of poultry is an extreme example. In contrast, muscles used more or less continuously, such as those in the legs of poultry, will be fuelled by oxidative reactions and require high levels of myoglobin, and they are consequently much darker.

The state of the iron in myoglobin also affects its colour. The structure of myoglobin was shown in Figure 5.6, but we need to concentrate on the iron atom. As shown in Figure 5.15, the state of the iron atom differs in the three important forms of myoglobin. In its normal state, referred to simply as "myoglobin" and abbreviated to Mb, the iron is in its reduced (ferrous)

Table 5.7   The myoglobin content of different muscles.

Animal	*Age*	*Muscle*	*Myoglobin content (g/100 g fresh weight)*
Calf[a]	12 days	Longissimus dorsi[f]	0.07
Steer[b]	3 years	Longissimus dorsi	0.46
Pig[c]	5 months	Longissimus dorsi	0.030
Pig[d]	7 months	Longissimus dorsi	0.044
Pig[e]	7 months	Rectus femoris[g]	0.086

[a]Veal.
[b]Beef.
[c]Pork.
[d]Used for bacon.
[e]Used for pork sausage.
[f]The principal muscle of rib of beef or short back bacon.
[g]The principal muscle of leg of pork.

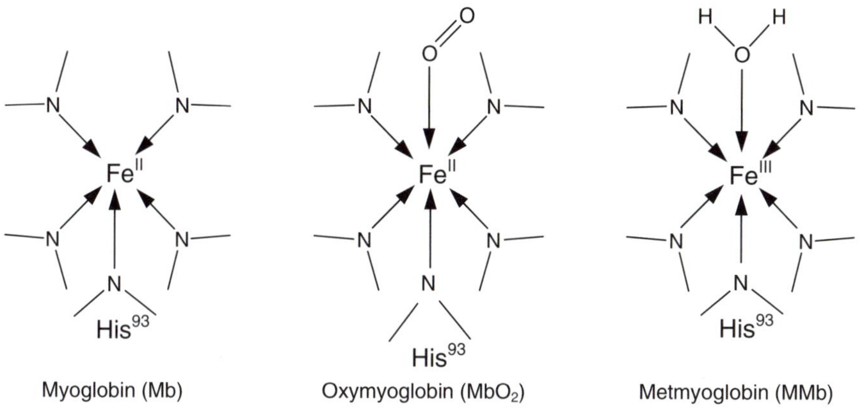

Myoglobin (Mb)   Oxymyoglobin (MbO$_2$)   Metmyoglobin (MMb)

Figure 5.15   The three states of iron in myoglobin. The four nitrogen atoms in a square around the iron atom are those of the porphyrin ring and the fifth nitrogen, marked His^{93} is part of the histidine imidazole ring, the 93rd amino acid counting from the NH$_2$ terminal. The sixth coordination position is occupied by oxygen or water. All six atoms attached to the iron are linked by coordinate bonds, in which the lone pair electrons of the nitrogen or the oxygen atoms (in H$_2$O or O$_2$) occupy the empty d-orbitals of the iron.

state, Fe^{2+} or Fe^{II}. When it binds an oxygen molecule in the living animal as part of the system for transferring oxygen from the haemoglobin of the bloodstream to the muscle cell mitochondria, the resulting compound is known as oxymyoglobin (MbO$_2$). In spite of the presence of the oxygen, the iron is not oxidised, and an oxygen molecule can be released again if the local oxygen concentration is low. Oxymyoglobin is bright red in colour (its spectrum has a double peak, λ_{max} 544 and 581 nm) whereas myoglobin is a dark purplish red (λ_{max} 558 nm).[ix]

In the living animal, the simple interchange:

$$Mb + O_2 \leftrightarrow MbO_2$$

is all that matters, since the myoglobin only encounters the oxygen in high concentrations when it collects oxygen from the blood, or in very low concentrations in the muscle cells to which the oxygen is delivered. However, in meat the essentially irreversible formation of metmyoglobin takes place at intermediate oxygen concentrations, a process discussed in detail in the first of the Special Topics at the end of this chapter.

[ix] The difference between the colours of arterial and venous blood is based on a similar relationship between the corresponding two forms of haemoglobin.

The interior of the muscle tissue of a carcass will obviously be devoid of oxygen, and the freshly cut surface of a piece of raw meat is, as one would expect, the corresponding shade of dark reddish-purple. Soon diffusion of oxygen into the surface begins to convert myoglobin (Mb) to oxymyoglobin (MbO_2) and the meat acquires an attractive bright red appearance as a layer of MbO_2 develops. The importance of an adequate *postmortem* fall in pH is that it is the free moisture in the meat through which the oxygen diffuses, and if the pH remains high there is much less free moisture available. The respiratory systems of the muscle continue to consume any available oxygen until the meat is actually cooked, so the MbO_2 layer never becomes more than a few mm deep. The relationship between oxygen concentration and the proportions of the different forms of myoglobin is illustrated in Figure 5.16

At the intermediate oxygen concentrations existing at the interface between the MbO_2 and Mb layers, the iron of the myoglobin is oxidised by the oxygen it binds to and converts it to the ferric state (Fe^{3+}, or Fe^{III}). The resulting metmyoglobin (MMb) is a most unattractive shade of brown (λ_{max} 507 nm). Meat kept too long at room temperature therefore develops a brown layer just below the surface, and as time goes on this widens until the surface of the meat looks brown rather than red, a sure sign that the meat is no longer fresh. Another consequence of the oxidation of Fe^{II} to Fe^{III} is that during the reaction a superoxide free radical is generated. Although very short-lived, this free radical may well have time to initiate the chain reactions of fatty acid autoxidation,

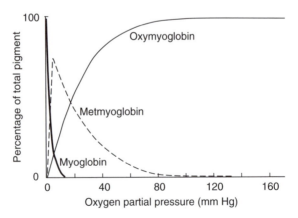

Figure 5.16 The effect of oxygen concentration on the proportions of the different forms of myoglobin. In air at atmospheric pressure the concentration of oxygen is at a partial pressure of 170 mm Hg. (From J. C. Forrest, et al., *Principles of Meat Science*, W. H. Freeman, San Fransico, 1975.)

described in Chapter 4, leading to rancidity of the fatty parts of the meat. There is therefore a clear association between the appearance of the raw meat and its flavour when cooked.

Although *postmortem* changes in the myofibrillar structures are an important factor in the texture of meat, the consumer is much more aware of differences in texture that stem from the maturity of the animal and the part of the carcass the meat came from. These variations arise in the connective tissue rather than the contractile tissues of the muscle. The connective tissue of the muscle is composed almost entirely of fibres of the protein collagen. Collagen fibres consist of cross-linked, longitudinally arranged tropocollagen molecules, each of which is 280–300 nm in length. The tropocollagen molecule itself consists of three very similar polypeptide chains. The conformation of each chain is an extended left-handed helix (remember that the classic α-helix shown in Figure 5.5 is a compact right-handed helix), and the three chains are then wound together in a right-handed helix, Figure 5.17a. The use of left- and right-handed helices at different levels of the organisation of the molecule mirrors that in rope and textile threads, in which it makes unravelling much less likely.

The amino acid composition and sequence of the tropocollagen chain are striking. Roughly one-third of all the amino acid residues are glycine; proline plus hydroxyproline account for 20–25%. The brief sequence given in Figure 5.17b shows that in fact glycine occurs at every third residue along the chain. Structural studies indicate that the glycine residues always occur at points in the helical structure where the three chains approach closely. Only the minimal side-chain of glycine, a solitary hydrogen atom, can be accommodated in the small space available. The two imino acids are important, since their particular form of the peptide bond gives the ideal bond angles for the tropocollagen helix. The structure is maintained by hydrogen bonding between the three chains and also by hydrogen bonds involving the hydroxyproline hydroxyl group. The hydroxylation of proline to hydroxyproline occurs after the protein has been synthesised. The observation that ascorbic acid, vitamin C, is the reducing agent in the hydroxylation reaction can be readily correlated with the clinical signs of the deficiency of this vitamin discussed in Chapter 8. However strong an individual tropocollagen molecule is, the strength of the fibre as a whole will depend on lateral cross-links to ensure that neighbouring molecules do not slide over each other under tension. A number of types of cross-links have been identified, of which two are shown in Figure 5.18. Cross-links are most abundant in connective tissues in which the greatest strength is required, such as the Achilles tendon.

(a)   (b)
                                                              |
                                                             Ser
                                                              |
                                                             Gly
                                                              |
                                                             Pro
                                                              |
                                                             Arg
                                                              |
                                                             Gly
                                          0.87 nm, |
                                          containing   Leu
                                          approximately |
                                          3.3 amino acid   Hyp
                                          residues |
                                                             Gly
                                                              |
                                                             Pro
                                                              |
                                                             Hyp
                                                              |
                                                             Gly
                                                              |
                                                             Ala
                                                              |
                                                             Hyp
                                                              |

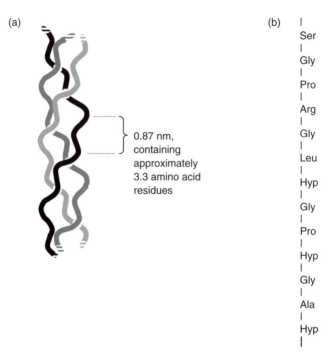

Figure 5.17 The molecular structure of collagen. (a) A section of a tropocollagen molecule showing only the – C – N – C – C – N – C – backbones of the three polypeptide chains. Each chain forms an extended left-handed helix. The three polypeptide chains twist around each other in a right-handed helix. (b) A section of the amino acid sequence of a tropocollagen polypeptide chain.

It is a common observation that meat from young animals is more tender than that from older ones. It has now been realised that this is due to an increase in the number of cross-links as the animal ages, rather than an increase in the proportion of total connective tissue. There are of course variations in the amount of collagen between different muscles in the same animal, and these are very important to meat texture. The most tender cuts, and the most expensive, are those with the lowest proportion of connective tissue. For example in beef, the connective tissue content of the *Psoas major* muscle (in fillet steak), is one-third of that in the *Triceps brachii* muscle (stewing beef).

When meat is heated the hydrogen bonds that maintain collagen structure are weakened. Very often the fibres shorten as the polypeptide chains adopt a more compact helical structure. If heating is prolonged, as in a casserole, not only the hydrogen bonds but also some of the weaker covalent cross-links will be broken. The result will be

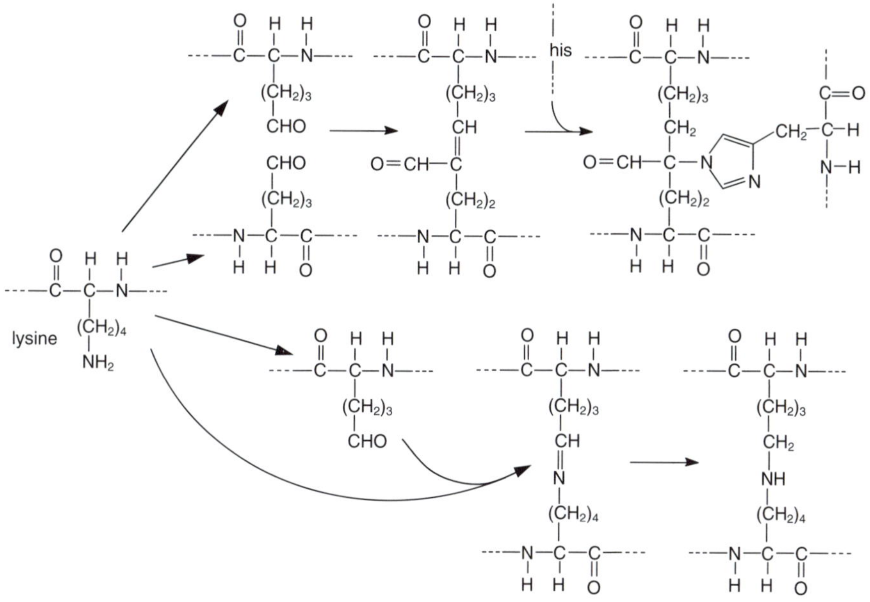

Figure 5.18 The formation of tropocollagen cross-links involving lysine side-chains. The initial conversion of lysine to an aldehyde derivative is catalysed by the enzyme lysine oxidase, but all the subsequent reactions are spontaneous. The two linkages shown here are normally found between neighbouring tropocollagen molecules. The attachment of the histidine side-chains provides a three-way linkage. Other types of link are present, supplementing the hydrogen bonds between chains within the same tropocollagen molecule.

solubilisation of the collagen, some of which will be leached out and causes gelatinisation of the gravy when it cools and the hydrogen bonds are re-established.

Muscle tissue accounts for only a modest proportion of an animal's total collagen; most is located in the skin (or hide) and the bones. Although these tissues are of little importance as food, they are the usual source of collagen for conversion into an important food ingredient, gelatin. Gelatin is widely used as a gelling agent, particularly in confectionery, desserts (the jelly of children's parties in Britain, and "jello" in the USA), and in meat products.

When gelatin is produced from bones they are first crushed and then steeped in dilute hydrochloric acid for up to two weeks to solubilise and remove the calcium phosphate. The residue can then be processed along similar lines to cattle hides. This involves an initial treatment with lime (calcium hydroxide, $Ca(OH)_2$) for several weeks. The alkaline conditions break down the cross-links between the collagen chains. When the lime is eventually washed out, other proteins and fatty residues are also washed away. After neutralisation, the residue is then extracted with hot water at progressively higher temperatures. The gelatin, as it now is, dissolves and after filtration the solution can be concentrated and finally dehydrated.

Rather like the polysaccharide gel formers described in Chapter 3, gelatin is insoluble in cold water but dissolves readily in hot water. When the solution is cooled the polymer molecules revert from their dis-organised "random coil" state to a helical arrangement similar to that in the original connective tissue. In this state segments of the polymer chains associate together, forming junction zones based on hydrogen bonding, as occurs in polysaccharide gels.

Bread

Cereals have been an important source of human food for thousands of years. Archaeologists have been able to relate the development of human civilisations, particularly those in the Middle East, to the evo-lution of what is now regarded as the modern species of bread wheat, *Triticum aestivum,* from its primitive ancestors. The other major cereal crops grown around the world today are as follows:

rye	*Secale cereale*
barley	*Hordeum vulgare*
oats	*Avena sativa*
sorghum	*Sorghum bicolor*
maize (corn)	*Zea mays*
rice	*Oryza sativa*

Compared with these, it is only the flour made from wheat that can be made into loaves with a leavened, open, crumb structure. Wheat bread has been eaten in the temperate zones of the world for thousands of years. Some aspects of the ancestry of *T. aestivum* are given more consideration in the second of the Special Topics at the end of the chapter.

The grains (*i.e.* the seeds) of the cereals have a great deal in common. As shown in Figure 5.19 they consist of three major structures: (i) the embryo or germ of the new plant, (ii) the endosperm, which is the store of nutrients for the germinating plant, and (iii) the protective layers of the seed coat, which are regarded by the miller as bran. The endosperm is about 80% of the bulk of grain. White flour is almost pure endosperm, and since we are primarily interested in bread, what follows refers to the endosperm and the endosperm constituents of wheat.

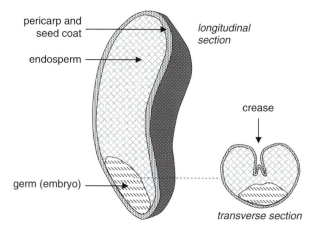

Figure 5.19 Sections through a wheat grain.

The endosperm consists of tightly packed, thin-walled cells of variable size and shape. The cell walls are the origin of the small proportions of cellulose and hemicelluloses in white flour. The cells are packed with starch granules lying in a matrix of protein. The protein, some 7–15% of the flour, is of two types. One of these (about 15% of the total) consists of the residues of the typical cytoplasmic proteins, mostly enzymes, which are soluble in water or dilute salt solutions. The remaining 85% are the storage proteins of the seed, insoluble in ordinary aqueous media and responsible for dough formation. These dough-forming proteins are collectively referred to as gluten. The gluten can be readily extracted from flour by adding water and mixing it to form a dough, leaving the dough to stand for half an hour or so, then finally kneading the dough under a stream of cold water, which washes out all the soluble material and the starch granules. The resulting tough viscoelastic and sticky material is about one-third protein and two-thirds water.

Flours from different wheat varieties vary in protein content. In general, flours that are good for bread making (*i.e.* which give a good loaf volume) are produced from the spring-sown wheat varieties grown in North America. These varieties tend to have higher protein content (12–14%). Good bread-making wheats are usually the type described by the miller as "hard", *i.e.* the endosperm is brittle and disintegrates readily on milling. The baker's description of a good bread-making flour as "strong" refers to the characteristics of the dough: it is more elastic and more resistant to stretching than dough made from a "weak" flour.

Weak flours are essential for making biscuits and short-crust pastry. These flours are usually obtained from "soft" winter wheats (actually sown in the autumn!) grown in Europe.[x] Their protein content is usually less than 10%. In spite of the general correlation between protein content and milling/baking properties, there are enough exceptions to demonstrate that it is the properties of different wheat proteins rather than their abundance which is the major factor. The so-called "patent" or "household" flour that is sold for domestic use is blended from different types to give a product with intermediate properties which provides a reasonable compromise for home baking.

The gluten proteins can be fractionated on the basis of their solubilities. The most soluble, the gliadins, can be extracted using 70% ethanol. These constitute between a third and a half of the total gluten. The remainder are the glutenins, which are extremely difficult to dissolve fully. A solution consisting of $0.1 \, mol \, dm^{-3}$ ethanoic acid, $3 \, mol \, dm^{-3}$ urea, and $0.01 \, mol \, dm^{-3}$ cetyltrimethylammonium bromide (cetrimide, a surfactant) will dissolve all but 5% of the glutenins. What is known of the molecular structure of the gluten proteins goes a long way towards explaining their remarkable dough-forming and solubility characteristics. Using elaborate electrophoretic techniques, it has been shown that a single variety of wheat may contain more than 40 different gliadin proteins.

The glutenins are much more difficult to characterise. They form very large aggregates (molecular weights above 1 million, mostly around 2 million) of individual proteins known as glutenin subunits. Two different classes of these are recognised, high molecular weight (HMW) and low molecular weight (LMW) subunits. The distinctive elastic properties of dough are largely attributable to the molecular features of the HMW subunits. The glutenin "macropolymers" can be broken down into these subunits by treatment with reagents such as mercaptoethanol, which break the disulfide bridges between cysteine sulfhydryl groups:

$$\ldots - protein - S - S{-}protein - \ldots + 2HSCH_2CH_2OH \rightarrow$$
$$\ldots - protein{-}S - SCH_2CH_2OH + HOCH_2CH_2S - S - protein - \ldots$$

---

[x] Nowadays British bread-making grists, *i.e.* the blends of grain being milled for a particular batch of flour, contain significant quantities of European-grown wheat varieties which do give strong flour. These varieties have been developed as a result of the increasing tendency for the EU countries to import less of their wheat from outside Europe. Another technique for utilising European wheat is the manufacture of so-called "vital gluten". This is manufactured by first mixing the flour with water and forming a dough. The starch (and soluble proteins *etc.*) is then washed out of the dough by prolonged kneading in a stream of cold water, and the remaining gluten is dried and milled. As long as high temperatures are avoided at the drying stage the gluten retains its dough forming properties, hence the adjective "vital". The manufacture of Quorn™, the mycoprotein food ingredient used as a meat substitute, was originally devised as an outlet for the considerable quantities of starch which were an inevitable by-product of gluten manufacture.

Table 5.8 Essential features of the gluten proteins.

		Total gluten protein (%)	Molecular weight (×1000)	Distinctive features of amino acid composition
Gliadins	Sulfur-rich (α, β and γ)	51–64	32–42	glutamine 32–42%, proline 15–24%, cysteine, ~2% phenylalanine 7–9%
	Sulfur-poor (ω)	7–13	55–79	glutamine 42–53%, proline 20–31%, cysteine 0%
Glutenin subunits	Low MW	19–25	36–44	glutamine 33 – 39%
	High MW	7–13	90–124	glycine 13 – 18% cysteine 0.5 – 1%

Some features of the gliadins and glutenin subunits are shown in Table 5.8.

The amino acid composition of these proteins explains their lack of solubility in water. Compared with more typical proteins they contain a low proportion of charged amino acids (*e.g.* in gliadins, glutamic acid is around 2% and the three basic amino acids, histidine, lysine and arginine, in total around 3%). The unusually high proportions of glutamine and proline indicated in Table 5.8 are readily explained. These two amino acids contain an above average proportion of nitrogen, and are also particularly easily utilised by the germinating seed as sources of both nitrogen and carbon. By comparison, the metabolic pathways for the breakdown of the other nitrogen-rich amino acids, arginine, lysine and histidine, are much more complex. Although glutamine may be regarded as a hydrophilic amino acid, its abundance leads to the formation of large numbers of hydrogen bonds between polypeptide chains (5.10), many with an intermediate water molecule (5.10a).

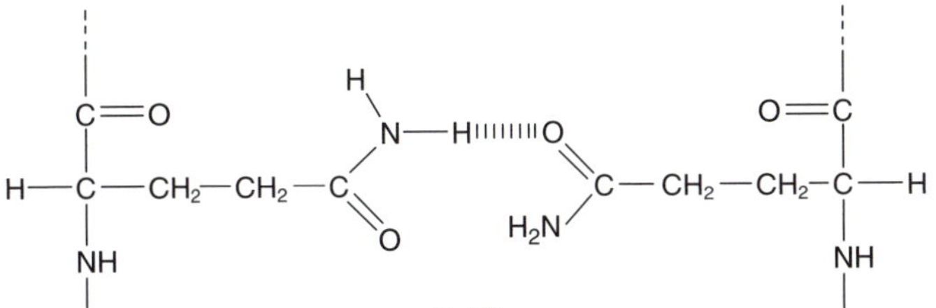

(5.10)

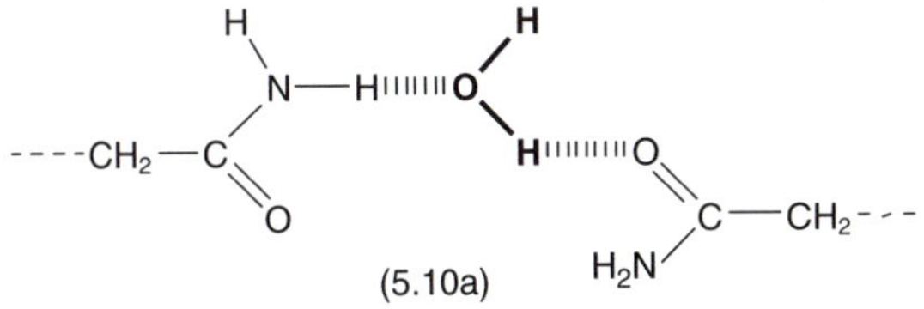

(5.10a)

Complete amino acid sequences have now been obtained for large numbers of gluten proteins and reveal a number of fascinating features. The first is the extensive homology between the amino acid sequences of different proteins within the same class. For example, the first 22 amino acids (counting as usual from the *N*-terminal end) of α-gliadins A, 2, 10, 11 and 12, and γ-gliadin 1, are identical except at one single position in β-gliadin 5. Similar homologies have been found within the ω-gliadin class and between many HMW glutenin subunits. These homologies indicate that in spite of the extensive changes in the wheat brought about by natural evolution and plant breeders over the centuries, their common genetic ancestry is still evident.

Another feature of all classes of gluten proteins is that their amino acid sequences are highly repetitive, as illustrated by the example of α-gliadin A. The sequence of its first 95 amino acids is based on a 12-amino acid sequence (–PQPQPFPPQQPY–), repeated eight times.[xi] The repetition may not be immediately obvious, since in each of the repeats there are often a small number of changes from the "consensus" sequence. These changes may arise by the insertion of an extra amino acid, a deletion from the sequence, or a substitution of one amino acid for another. Continuous sequences of 10–20 glutamine residues also occur in many gluten proteins. The third of the Special Topics at the end of this chapter provides further insights into the amino acid sequences of gluten proteins.

Comparison of these sequences with those in proteins whose secondary and tertiary structures are well established has suggested that gluten proteins are unusually rich in an arrangement known to protein chemists as a "β-bend". β-Bends are normally found in proteins in which the polypeptide chain undergoes a sharp change in direction (such as at the corners of the α-helical sections of myoglobin, Figure 5.6). The hydrogen bonds linking successive turns in an α-helix (*see* Figure 5.5) instead cut across the corner. Scanning tunnelling electron microscopy of a HMW gluten subunit, and physical measurements using the sophisticated techniques of Fourier transform infrared spectroscopy and far-UV circular dichroism,[xii] have confirmed this suggestion. The only other protein known to possess an abundance of β-bends is the mammalian connective tissue protein, elastin. Ligaments and artery walls are very rich in elastin, and it is now believed that they owe their elasticity to this protein thanks to its large numbers of β-bends

---

[xi] The single letter codes used for the amino acids are as follows: P or pro = proline, Q or glN = glutamine, F or phe = phenylalanine, Y or tyr = tyrosine.
[xii] The theory of these techniques is beyond the scope of this book.

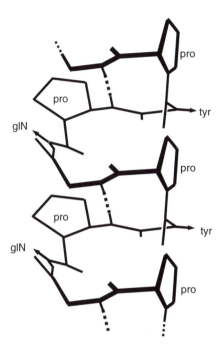

Figure 5.20 The gluten β-spiral. The conformation of the polypeptide chain illustrated is proposed for the repeating pentapeptide units (–pro–tyr–pro–glN–glN–) of ω-gliadin, but the β-spiral regions of other gluten protein are unlikely to be very different. The –N–C–C–N–C– backbone of the polypeptide chain is shown with proline rings, including the N–H----O=C hydrogen bonds and (arrowed) the positions of the tyrosine and glutamine side-chains. The spiral is stabilised by hydrogen bonding across the β-spiral rather along it, as in the α-helix. This allows some degree of stretch by rotation of the bonds around the glutamine residues. Interaction between the π-electron clouds of the stacked tyrosine side-chains gives further stabilisation.

arranged in sequence to form a β-spiral. There is therefore plenty of support for the idea that the viscoelasticity of a dough can be accounted for by the amino acid sequences of its proteins. The β-spiral structure which has been proposed for one class of gliadin proteins is shown in Figure 5.20.

The current view of the structure of glutenin which emerges from these studies is of a backbone of HMW subunits, linked nose-to-tail by disulfide bridges and carrying short, possibly branched, chains of LMW subunits.

Extensive comparison of the bread-making performance of a wheat cultivar with its complement of gliadins and glutenin subunits has shown that HMW glutenin subunits are a major factor. Close genetic linkage (*i.e.* genes within the same chromosome lying very close to each other) means that subunits occur in pairs. For example, subunits 5 + 10 impart much better baking performance than 4 + 12. It is possible that the best subunit types are those possessing multiple cysteine sulfhydryl groups for the formation of branched gluten macropolymer chains, producing strong doughs and high loaf volumes.

When water is added to flour, dough can be formed as the gluten proteins become hydrated. The viscoelastic properties of the dough depend upon the glutenin fraction, which is able to form an extended three-dimensional network. Links between the glutenin chains depend on various types of bonding. Hydrogen bonding between the abundant glutamine amide groups is probably the most important of these, but ionic bonding and hydrophobic interactions also play a role. The gliadin molecules are assumed to have a modifying influence on the viscoelastic properties of the dough. It is now well established that the relative proportions of high and low molecular weight proteins, the glutenins and gliadins, are a major factor in the bread-making character of a particular flour. In general, a high proportion of glutenins results in doughs which require more mixing, are stronger, and give loaves of greater volume, and hence a more open, lighter crumb. For biscuits (except crackers) and many types of cakes and pastry, a good bread-making flour is quite unsuitable. Made with strong flour, some types of biscuit would be hard rather then crisp and some tend to shrink erratically during baking. However these are problems for the commercial baker rather than the domestic cook.

Traditionally, the bread-making properties of a flour have been found to be improved by prolonged storage. Autoxidation of the polyunsaturated fatty acids of flour lipids (*see* Chapter 4) results in the formation of hydroperoxides, which are powerful oxidising agents. One consequence is the bleaching of the carotenoids in the flour, giving the bread a more attractive, whiter crumb. However, the most important of the beneficial effects of ageing are on loaf volume and crumb texture. During the first twelve months the flour is stored there is a steady increase in loaf volume, and the crumb becomes finer and softer. On the other hand, longer storage than this results in worsening baking properties. For many years it has been common practice to simulate the ageing process by the use of oxidising agents to treat the flour, either at the mill or as additives in the bakery. These have included chlorine dioxide (applied as a gaseous treatment by the miller), benzoyl peroxide [di(benzenecarbonyl) peroxide, 5.11], ammonium or potassium persulfate (5.12), and potassium bromate or iodate. However, none of these are now permitted within the EU.[xiii]

$$K^+ \text{ or } NH_4^+ \qquad O-S-O-O-S-O^- \text{ or } NH_4^+$$

(5.11)   (5.12)

The elimination of these oxidising processes has been made possible by the introduction of ascorbic acid (*see* Chapter 8) as an improving agent. This has come about largely by adoption of the Chorleywood Bread Process in commercial breadmaking, discussed in detail at the end of this chapter. Its acceptance has undoubtedly been enhanced by the status of ascorbic acid as vitamin C, giving it a whiter-than-white image with consumers, even though none of its vitamin activity survives the oven. The consequent use of unbleached flour has led to the appearance of bread with a distinctly yellow tinge to its crumb, but this is nowadays equated with naturalness rather than lack of purity as once would have been the case.

A satisfactory dough is one which can accommodate a large quantity of gas and retain it as the protein "sets" during baking. The achievement of such a dough requires more than simple mixing of the ingredients, and mechanical work must also be applied. In traditional bread making the kneading of the dough provides some of this work, but the remainder comes from the expanding bubbles of carbon dioxide evolved during fermentation of the yeast. At the molecular level these processes, collectively referred to as dough development, are not well understood and the following account is unlikely to represent the final word on the subject.

It is generally believed that during development the giant glutenin molecules are stretched into linear chains which interact to form elastic sheets around the gas bubbles. The HMW glutenin subunits are involved in a number of different reactions as formation and development of

---

[xiii] Prior to being banned in 2000, chlorine was used in the manufacture of commercial "high ratio" flours used for certain types of cakes, notably American-style muffins. These were made using very high proportions of water and sugar, and the chlorine treatment damaged the protein membrane around the starch granules, allowing them to imbibe and retain the extra water. A similar effect is now achieved by subjecting the flour to an intense dry heat treatment. The oxidising agent azodicarbonamide ($H_2NCON=NCONH_2$) is no longer permitted in Britain but it remains in use in the USA, for example in the flour for burger buns.

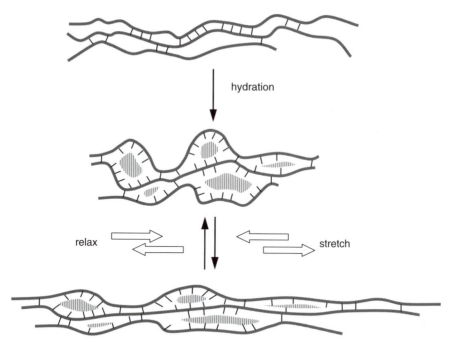

Figure 5.21   The formation of loops in glutenin subunits when flour is mixed with
water, and their contribution to the elasticity of the dough. There is
evidence that the extended sections of the polypeptide chains change
their conformation from a β-spiral to a β-pleated sheet.

dough proceeds. Initially before the addition of water, the central
domains of the protein molecules are bound to one another by the
hydrogen bonding referred to earlier. As the proteins hydrate and water
molecules become attached increasing numbers of these inter-chain
bonds will be broken. The result, shown in Figure 5.21, is that loops are
formed and these are believed to account for much of the dough's
elasticity. The β-spiral structure is assumed to be the other source of
elasticity in dough, and may in fact be more important in determining
the texture of the crumb after baking.

During the early stages of mixing the dough the polypeptide chains of
both gliadins and glutenins will tend to become aligned alongside each
other. This gives much more opportunity for hydrogen bond formation
and the resistance of the dough to mixing shows a sharp increase. Other
reactions involve the sulfhydryl groups of the proteins. Under
mechanical stress, exchange reactions between neighbouring sulfhydryl
groups allow the glutenin subunits to adopt a more extended arrangement. Exchange reactions (Figure 5.22) require the involvement of a low

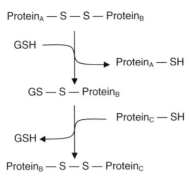

Figure 5.22 Disulfide exchange reactions in dough. The principal cysteine residues involved in reactions with GSH are those in the low molecular weight subunits. These subunits are suspected of forming links between the extended glutenin macropolymers.

molecular weight sulfhydryl compound. Glutathione (γ–glutamylcysteinylglycine, GSH, 5.13), which is found in flour in sufficient quantities (10–$15\,mg\,kg^{-1}$), has been identified as playing this role. It occurs naturally in flour in three forms: the free form (GSH), an oxidised dimer (GSSG), and bound to a protein molecule (PSSG).

$$
\begin{array}{c}
NH_2 \qquad\qquad\qquad H \quad H \\
\mid \qquad\qquad\qquad\qquad \mid \quad \mid \\
HOOC\!-\!C\!-\!CH_2\!-\!CH_2\!-\!C\!-\!N\!-\!C\!-\!CH_2\!-\!SH \\
\mid \qquad\qquad\qquad\quad \parallel \qquad\quad \mid \\
H \qquad\qquad\qquad\quad O \qquad\quad C\!=\!O \\
\qquad\qquad\qquad\qquad\qquad\qquad \mid \\
\qquad\qquad\qquad\qquad\qquad\qquad NH_2 \\
\qquad\qquad\qquad\qquad\qquad\qquad \mid \\
\qquad\qquad\qquad\qquad\qquad\qquad CH_2 \\
\qquad\qquad\qquad\qquad\qquad\qquad \mid \\
\text{(5.13)} \qquad\qquad\qquad\qquad COOH
\end{array}
$$

Traditional baking processes require up to 3 hours of fermentation before the dough is ready for the oven, and a great deal of research has taken place to find ways of accelerating the process. After a number of attempts both in Britain and the USA, it was at the laboratories of the Flour Milling and Baking Research Association (at Chorleywood, Hertfordshire, UK) that the first completely successful process was developed. Known internationally as the Chorleywood Bread Process (CBP), it depends upon the use of ascorbic acid (AA) as an improver, coupled with high-speed mixing of the dough. When the flour, water and other ingredients are mixed, the ascorbic acid is oxidised to dehydroascorbic acid (DHAA, *see* Chapter 8) by ascorbic acid oxidase, an enzyme present naturally in flour:

$$AA + \tfrac{1}{2}O_2 \rightarrow DHAA + H_2O$$

The dehydroascorbic acid, an oxidising agent, owes its effectiveness to the presence in flour of a second enzyme, which catalyses the conversion of glutathione to its oxidised dimeric form:

$$DHAA + 2GSH \rightarrow AA + GSSG$$

This is of course inactive in the disulfide exchange reactions during dough development. The elimination of long fermentation periods has made the CBP very popular with bakers, even though it requires special mixers capable of 400 rpm and careful control of the energy imparted to the dough, 11 watt hours (4×10^4 J) per kg of dough being required. The resulting bread is not easily distinguishable from that produced by traditional bulk fermentation methods. However, it has often been observed that the flavour of CBP bread is inferior, partly because the yeast has had less time to contribute ethanol and other metabolic by-products to the dough. Another factor is the tendency of bakers to under-bake their bread, which means that less of the Maillard reaction takes place as the crust forms. For bakers this has two advantages—the absence of a crust makes the loaf easier to slice, but even more importantly it means that more water is left behind in the loaf, allowing less flour to be used to produce a loaf of the same weight.

SPECIAL TOPICS

1 Myoglobin and Free Radicals

Figure 5.15 showed the basic structure of the three forms of myoglobin found in meat, myoglobin itself, oxymyoglobin with the iron atom present as the Fe^{2+} ion, and metmyoglobin with the iron in the oxidised Fe^{3+} condition. The actual mechanism of the reactions involved in the formation of the oxymyoglobin and metmyoglobin is somewhat more complex than implied in Figure 5.15.

The iron in the haem part of the molecule depends upon the surrounding protein to moderate its reaction with oxygen. In isolation, the Fe^{II} in haem would simply be oxidised irreversibly by molecular oxygen (O_2). In order to carry out its biological role of binding oxygen reversibly, however, this must be prevented. It can be achieved by occupation of the fifth coordination position of the iron atom by the nitrogen atom of the imidazole ring of histidine-93, together with the hydrophobic environment of the "haem pocket" of the protein. The oxygen adduct is believed to be further stabilised by formation of a hydrogen bond with the imidazole ring of histidine-64 (5.14).

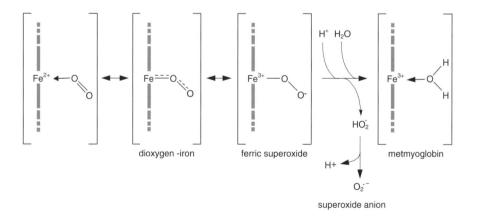

There is general consensus that the actual structure of the oxygen adduct represents an equilibrium between the three structures on the left of the sequence shown in Figure 5.23. It is probable that only a very small proportion of the myoglobin exists in the covalent form, on the for left in Figure 5.23, at any one time. In the equilibrium between the dioxygen and ionic forms, increasing temperature favours the formation of the ionic ferric superoxide.

In common with other free radicals the superoxide anion is extremely unstable, but in its short life it is still capable wreaking havoc. With the exception of the bacteria which are "obligate anaerobes", *i.e.* those which cannot survive in the presence of oxygen, all other organisms have

Figure 5.23 The conversion of oxymyoglobin to metmyoglobin and formation of the superoxide anion. The grey bars represent the porphyrin nucleus seen edge-on.

a pair of enzymes, superoxide dismutase (SODM, pronounced as though associated with Gomorrah) and catalase, whose joint role is to destroy this dangerous by-product of oxygen metabolism:

$$4O_2^{\cdot -} \xrightarrow[\text{SODM}]{2O_2} 2O_2^{2-} \xrightarrow[\text{catalase}]{2H^+} 2OH^- + O_2$$

(peroxide anion)

2   Wheat Genes and Chromosomes

The numbers of different but closely related gliadins and glutenin sub-units, found not only in the single species, *Triticum aestivum*, but even in a single cultivar (variety) of wheat, are a consequence of the special genetic character of the modern bread wheat plant. The nuclei of the vast majority of animal and plant cells (excluding of course the germ cells required for sexual reproduction) contain one double set of each chromosome, *i.e.* their cells are *diploid*. The cell nuclei of *T. aestivum* are exceptional in having three copies of each set (referred to as genomes AA, BB and DD) of six paired chromosomes, and we therefore refer to bread wheat as *hexaploid*. This is the result of nuclear fusion rather than conventional cross-breeding between wheat ancestors. The ancestors of bread wheat are believed to have included relatives of a number of modern species, for example:

T. tauchii	goat-faced grasses	diploid	DD
T. speltoides		diploid	BB
T. monococcum	einkorn wheat	diploid	AA
T. turgidum	emmer wheat	tetraploid	AABB

Modern durum wheat (*T. durum*), used in pasta, is a *tetraploid* relative of emmer wheat. Archaeologists believe that that a hexaploid ancestor of modern bread wheat had arisen in the eastern Mediterranean area by 7500 BC. *T. aestivum* had a much larger grain and higher protein content than its ancestors, and the impact of this event on early agriculture and on the history of mankind was considerable. Perhaps through the effects of human intervention during its cultivation, the species shed its other characteristics of a wild grass and it became easier to reap and thresh. At the same time, however, it became dependent on human cultivation for its seed distribution and its ability to compete with weeds.

The genome and chromosome locations of the genes for most gliadins and HMW glutenin subunits are shown in Table 5.9.

Table 5.9 The loci of the gliadin and glutenin genes.

	Chromosome		Genome	
		AA	BB	DD
HMW glutenin subunits	1	1 and 2*	6, 7, 8, 9, 13, 14, 15, 16, 17, 19, 21 and 22	2, 3, 4, 5, 10, 11 and 12
Gliadins	1	ω	most γ	a few β
Gliadins	6	α	most β	a few γ

3 Further Details of Gluten Proteins

The different classes of gluten proteins have different consensus sequences. For example, the consensus sequence P–Q–Q–P–F–P–Q–Q– is found in ω–gliadins, and several others, including G–Q–Q–, P–G–Q– G–Q–Q– and G–Y–Y–P–T–S–L–Q–Q– are found in HMW glutenin subunits. In most departures from the consensus involving a change of amino acid (as opposed to an insertion or deletion,) the replacement amino acid has similar physical properties and changes in the overall structure of the protein are minimal. In the case of α-gliadin$_A$, two short sections of which are shown in Figure 5.24, the full DNA sequence has been established and we know that of the seven changes in these two sections (Figure 5.24), five (at positions 8, 14, 56, 58 and 66) are the result of single base changes in the codon for the amino acid.

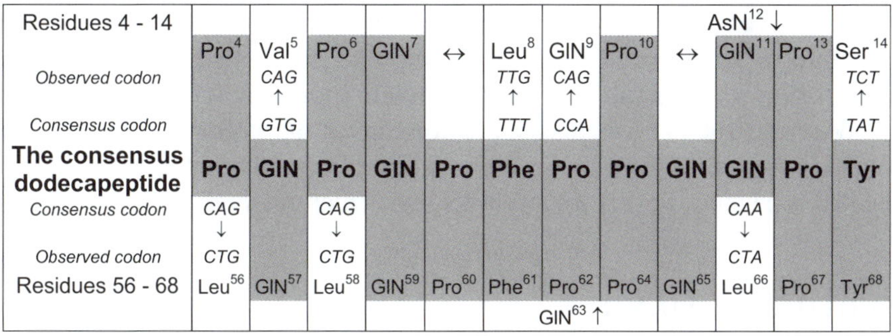

Figure 5.24 Masked repetitions in the amino acid sequence of α-gliadin$_A$. Two typical sections of the sequence are compared with the postulated consensus sequence. Shading shows where the residues are identical. There are two insertions, asparagine at position 12 and glutamine at position 63, and two deletions in these sections, between positions 7 and 8, and 62 and 63. The DNA codons for the positions involved in mutations are italicised.

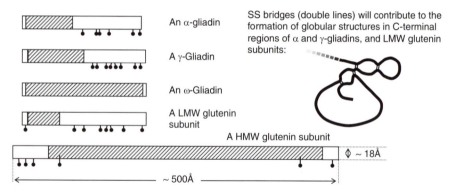

Domains I and III are unshaded, Domain II is shaded, cysteine residues are shown.

Figure 5.25 The domain structure of gluten proteins. The C-terminal regions of α- and γ-gliadins are believed to contain intramolecular SS bridges between some of their cysteine residues, which results in these molecules being more globular in shape. The central domains are dominated by repetitive sequences such as that shown in Figure 5.24, and these tend to result in the polypeptide chain taking on the β-spiral configuration. In the HMW subunits the central domain appears to consist entirely of a single β-spiral, giving these molecules a dumbbell shape, with a central rod-like spiral region terminated by globular regions consisting largely of α-helix and containing most or all of the cysteine residues. However, as discussed on page 206, the central region is flexible and is involved in the formation of loops as the gluten becomes hydrated.

The polypeptide chains of ordinary globular proteins are often organised into *domains*. These are usually regions which fold up together to form compact units in which some of the biological activity is localised. Although very little is known for certain about the three-dimensional structure of gluten proteins, it is common to divide their sequences into three (or more) domains. The central domain (normally II) is dominated by the β-spiral structure, whereas in the domains at the extremities (normally I and III) the polypeptide chain is more globular in character, with a great deal of α-helix and other less well-defined structures Figure 5.25.

FURTHER READING

Advanced Dairy Chemistry, ed. P. F. Fox and P. L. H. McSweeney, Kluwer Academic, New York, 2003.

Egg Science and Technology, ed. W. J. Stadelman and O. J. Cotterill, Food Products Press, New York, 4th edn, 1995.

R. A. Lawrie and D. A. Ledward, *Lawrie's Meat Science*, Woodhead, Cambridge, 7th edn, 2006.

The Chemistry of Muscle Based Foods, ed. D. A. Ledward, D. E. Johnstone and M. C. Knight, Royal Society of Chemistry, Cambridge, 1992.

Wheat Structure, Biochemistry and Functionality, ed. J. D. Schofield, Royal Society of Chemistry, Cambridge, 2000.

S. P. Cauvain and L. S. Young, *The Chorleywood Bread Process*, Woodhead, Cambridge, 2006.

L. G. Phillips, D. M. Whitehead and J. Kinsella, *Structure, Function and Properties of Food Proteins*, Academic Press, San Diego, 1994.

RECENT REVIEWS

D. S. Horne, *Casein Structure, Self-assembly and Gelation; Curr. Opin. Colloid Interface Sci.*, 2002, **7**, p. 456.

E. Smyth, R. A. Clegg and C. Holt, *A Biological Perspective on the Structure and Function of Caseins and Casein Micelles*; Int. J. Dairy Technol., 2004, **57**, p. 121.

D. S. Horne, *Casein Micelle Structure: Models and Muddles; Curr. Opin. Colloid Interface Sci.*, 2006, **11**, p. 148.

D. A. Ledward, *Gelatin,* in *Handbook of Hydrocolloids*, ed. G. O. Phillips and P. A. Williams, Woodhead, Cambridge, 2000.

G. G. Giddings, *The Basis of Colour in Muscle Foods*; *Crit. Rev. Food Sci. Nutr.,* 1977, **9**, p. 81. (Hardly recent, but still the definitive account of the topic. A short version of this long article was published at the same time in *J. Food Sci.*, 1977, **42**, p. 288.)

N. J. King and R. Whyte, *Does It Look Cooked? A Review of Factors that Influence Cooked Meat Color*; J. Food Sci., 2006, **71**, p. 31.

J. McCorriston, *Wheat*, in *The Cambridge World History of Food*, ed. K. F. Kiple and K. C. Ornelas, Cambridge University Press, Cambridge, 2000.

P. S. Belton, *On the Elasticity of Wheat Gluten; J. Cereal Sci.*, 1999, **29**, p. 103.

P. R. Shewry, N. G. Halford, P. S. Belton and A. S. Tatham, *The Structure and Properties of Gluten: An Elastic Protein from Wheat Grain; Philos. Trans. R. Soc. London, Ser. B*, 2002, **357**, p. 133.

F. M. Anium, *et al., Wheat Gluten: High Molecular Weight Subunits: Structure, Genetics, and Relation to Dough Elasticity; J. Food Sci.*, 2007, **72,** p. 56.

CHAPTER 6
Colours

Colour is very important in our appreciation of food. Parents who may well deplore their children's selection of the most lurid items at the sweetshop will later be found in the greengrocer's applying identical criteria to their own selection of fruit. Since we can no longer expect to sample the taste of food before we buy it, appearance is really the only guide we have to quality other than our past experience of the particular market stall, shop or manufacturer.

Ever since food preservation and processing began to move from the domestic kitchen to the factory there has been a desire to maintain the colours of processed and preserved foods as close as possible to the colours of the original raw materials. Some foodstuffs acquire their recognisable colour as an integral part of their processing, for example the brown of bread crust and other baked products, considered in Chapter 2, and the pink of cured meats, mentioned in Chapter 5. In this chapter we will largely be concerned with the most colourful foodstuffs, fruit and vegetables, and the efforts made by chemists to mimic their colours. In recent years consumers have shown increasing concern over the use of chemically synthesised dyestuffs as food colorants, and some attention is also paid to the increasing use of naturally occurring pigments such as annatto and cochineal.

CHLOROPHYLLS

The chlorophylls are green pigments of leafy vegetables. They also give the green colour to the skin of apples and other fruit, particularly when

Food: The Chemistry of its Components, Fifth edition
By T.P. Coultate
© T.P. Coultate, 2009
Published by the Royal Society of Chemistry, www.rsc.org

unripe. Chlorophylls are in fact the functional pigments of photo-synthesis in green plants. They occur alongside a range of carotenoid pigments in the membranes of the chloroplasts, the organelles which carry out photosynthesis in plant cells. The pigments of the chloroplast are intimately associated with other lipophilic components of the membranes such as phospholipids, as well as the membrane proteins. Algae and photosynthetic bacteria also contain a number of different types of chlorophyll, but the higher plants that concern us contain only chlorophyll a and b, in an approximate ratio of 3 to 1.

The structures of chlorophylls a and b are shown in Figure 6.1. They are essentially porphyrins, similar to those in haem pigments such as myoglobin, apart from the substituents, the presence of magnesium rather than iron, and the formation of a fifth ring by the linkage of position 6 to the γ-methine bridge. By analogy with myoglobin we can deduce that the porphyrin ring will associate readily with the hydrophobic regions of chloroplast membrane proteins. The extended phytol

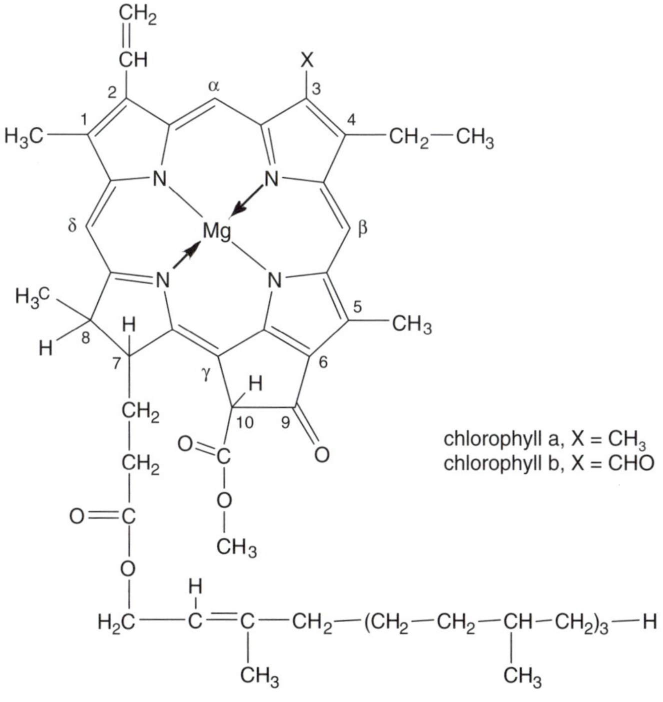

Figure 6.1 Chlorophyll. The double bonds of the porphyrin ring system are shown arbitrarily, and in fact the π-electrons in such systems are completely delocalised. The C_{20} isoprenoid alcohol esterified to the propionyl substituent at position 7 is known as phytol.

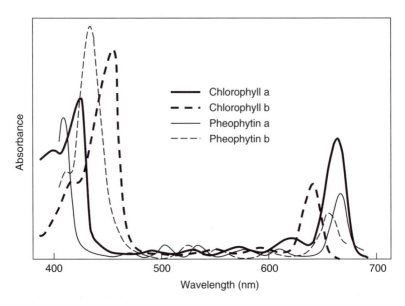

Figure 6.2 Absorption spectra of chlorophylls and pheophytins.

side-chain facilitates close association with carotenoids and the membrane lipids.

The absorption spectra of chlorophylls a and b are shown in Figure 6.2. The absorption maxima and extinction coefficients are not easily defined, since they vary with the organic solvent used and the presence of traces of moisture. These variations make the accurate determination of chlorophyll levels in leafy tissues by spectrophotometry rather difficult. Another difficulty is that in solution, especially if methanol is present, there is a tendency for oxidation or isomerisation to occur, with replacement of the hydrogen at position 10 with OH or OCH_3. It is the stability of chlorophyll in vegetable tissues, or its lack of it, which is of particular interest to food chemists. Chlorophyll is lost naturally from leaves at the end of their active life in the plant. Their breakdown accompanies a general breakdown of the chloroplast membranes, but the carotenoids are more stable, and this gives autumn leaves—and vegetables which are no longer fresh—a residual yellowish colour. A number of different reactions have been implicated in the breakdown of chlorophyll, including cleavage and opening of the porphyrin ring and removal of the phytol residue by the enzyme chlorophyllase, but there is little hard evidence to support these hypotheses.

The enzyme chlorophyllase is denatured by blanching. When green vegetables are blanched prior to freezing, during ordinary cooking or in canning, there is evidence that the phytol side-chain is lost, giving the corresponding chlorophyllides a and b. The most significant event, however, is the loss of the magnesium ion. This occurs most readily under acid conditions, the Mg^{2+} ion being replaced by protons (hydrogen ions), to give pheophytins a and b. Pheophytins are a dirty brown colour (*see* Figure 6.2) and are familiar as the dominant pigments in vegetables such as cabbage which have been overcooked. The acidity of the contents of plant cell vacuoles makes it difficult to avoid pheophytin formation, for example during the vigorous heating involved in canning peas. One approach is to keep the cooking water slightly alkaline by the addition of a small quantity of sodium bicarbonate. This is successful in retaining the colour, but the alkaline conditions have an unhappy effect on texture and flavour, and the vitamin C content also suffers.

In the canning of peas some chlorophyll loss is inevitable and has to be replaced with artificial colour. Cupric sulfate was occasionally—and dangerously—used for this purpose in the early years of the 20th century, but this has been universally replaced by organic dyes such as mixtures of tartrazine and Green S. In the 18th and 19th centuries vegetables were preserved by pickling them in vinegar. The resulting bright green colour was caused by Cu^{2+} ions leached from the copper pans (*see* Chapter 11). Since around 1860 legislation has ensured that pickles are copper-free. As far as chutney-type mixed pickles are concerned, we have grown to expect their brown colour and nowadays caramel is added to enhance the effect of the pheophytins already present.

Growing doubts about the safety of synthetic dyestuffs, especially tartrazine, are encouraging food chemists to examine the possible use of chlorophyll derivatives as alternatives. Chlorophyll itself is difficult to use. Not only is it unstable, its insolubility in water makes it difficult to incorporate except in fatty foods. However, a derivative known as sodium copper chlorophyllin is now being used. This is a sodium salt of chlorophyllin (*i.e.* chlorophyll from which the phytol side-chain has been removed) and has had the magnesium replaced by copper. Various plant materials are used as sources of chlorophyll by manufacturers of food pigments, including grasses, alfalfa (*Medicago sativa*), spinach (*Spinacea oleracea*) and stinging nettles (*Urtica dioica*). The product has an acceptable blue-green colour, is moderately soluble in water and, most importantly, it is able to survive the heating involved in canning. The amount of copper it contains is far too small to represent a toxicity hazard. Both chlorophyll and copper chlorophyllin are permitted food additives in Europe, E140 and E141, respectively.

CAROTENOIDS

Carotenoid pigments are responsible for most of the yellow and orange colours of fruits and vegetables. They are found in the chloroplasts of green plant tissues alongside chlorophyll, and in the chromoplasts of other tissues such as flower petals. In chloroplasts carotenoids are associated with the membrane proteins of the photosynthetic machinery. In the chromoplasts of carrots and tomatoes they occur as crystals, whereas in some other orange-coloured tissues such as paprika and mangoes the carotenoids are dissolved in lipid droplets. Chemically, carotenoids are classified as terpenoids, derived in Nature from the metabolic intermediate, mevalonic acid (6.1). This is the natural precursor of the basic terpenoid structural element, the isoprene unit (6.2). Terpenoids having one, two, three or four isoprene units (hemi-, mono-, sesqui-, and diterpenoids, respectively) are well known, but to food chemists the steroids, which are triterpenoids (*i.e.* containing 30 carbon atoms) and the carotenoids, the only known tetraterpenoid compounds, are the most important.

$$HOH_2C-CH_2-\underset{\underset{OH}{|}}{\overset{\overset{CH_3}{|}}{C}}-CH_2-COOH \qquad\qquad \underset{C}{\overset{C-C=C-C}{|}}$$

(6.1)   (6.2)

The carotenoids are divided into two main groups, the carotenes, which are hydrocarbons, and the xanthophylls, which contain oxygen. The structures of the carotenoids most important in foodstuffs are shown in Figures 6.3 and 6.4. The simplest carotene is lycopene, and the numbering system illustrates clearly how the molecule is constructed from two diterpenoid subunits linked "nose to nose". Other carotenes are formed by cyclisation at one or both the ends of the chain. One of two possible ring structures results, the α-ionone (6.3) or the β-ionone (6.4).

The acyclic end-group occurring in lycopene and γ-carotene is sometimes referred to as the ψ-ionone structure.

(6.3)   (6.4)

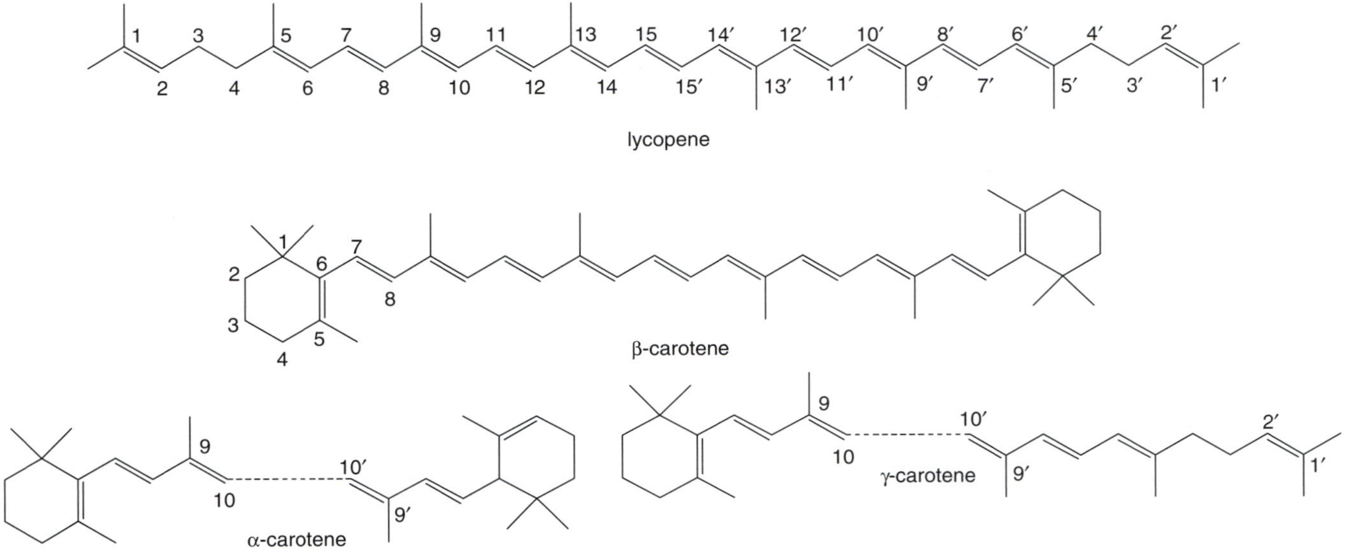

Figure 6.3 Carotene structures. In the conventional numbering system used here, the backbone carbon atoms are numbered from 1 to 15 and 1′ to 15′. The terminal and branch methyl groups along the chain are numbered 16 and 16′, *etc.* The central region common to all carotenoids (11 to 11′) has been omitted from the formulae of α- and γ-carotene, and in Figure 6.4.

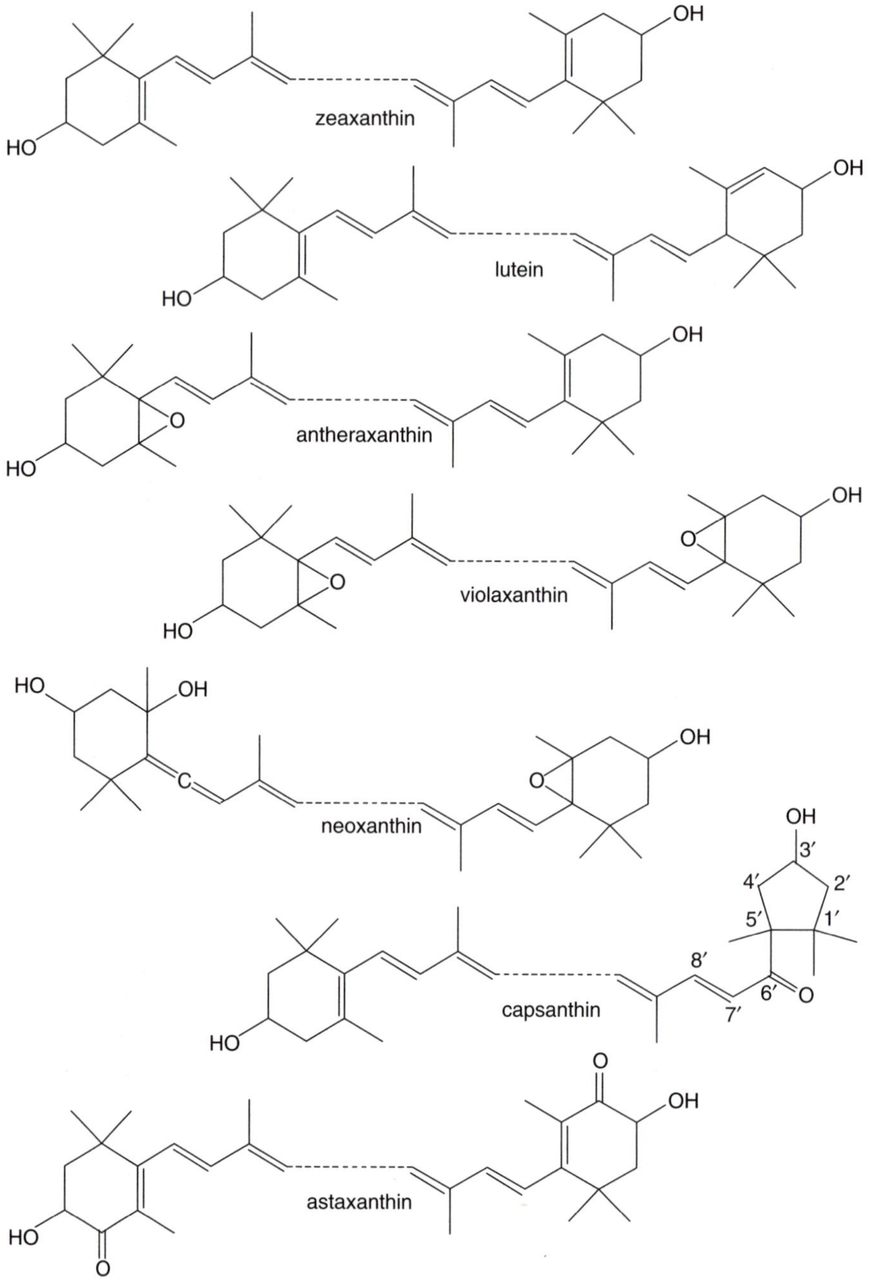

Figure 6.4 Xanthophyll structures.

The xanthophylls arise initially by hydroxylation of carotenes and most plant tissues contain traces of cryptoxanthins, the monohydroxyl precursors of the dihydroxyl xanthophylls, such as zeaxanthin and lutein, shown in Figure 6.4. Subsequent oxidation leads to the formation of epoxides such as antheraxanthin. Neoxanthin is an example, rare in Nature, of an allene, in which one carbon atom carries two double bonds.

The apocarotenoids are a small group of xanthophylls in which fragments have been removed from one or both ends of the chain. Three examples are shown in Figure 6.5.

With the exception of the two acidic apocarotenoids, which form water-soluble salts under alkaline conditions, carotenoids are freely soluble only in non-polar organic solvents. Their absorption spectra are generally quite similar, and the nature of the solvent causes almost as much variation in λ_{max} and extinction coefficient as occurs between individual carotenoids. The absorption spectra shown in Figure 6.6 bear out the observation that xanthophylls are the dominant pigment in yellow tissues, whereas carotenes give orange colours, or in the case of carotene a bright red.

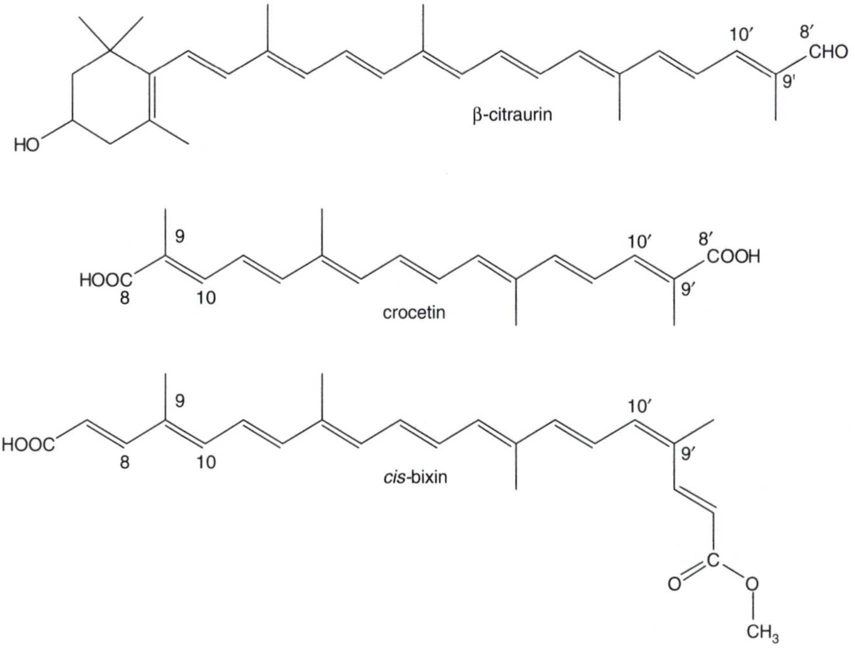

Figure 6.5 Apocarotenoid structures.

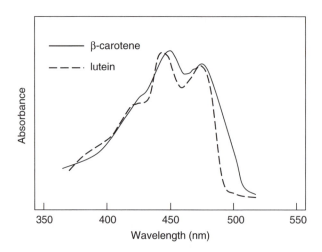

Figure 6.6 Absorption spectra of carotenoids.

The distribution of the carotenoids between the different plant groups and types of food materials shows no obvious pattern. Among the leafy green vegetables the carotenoid content follows the general pattern of all higher plant chloroplasts, with β-carotene dominant and the xanthophylls lutein, violaxanthin and neoxanthin all prominent. Zeaxanthin, α-carotene, cryptoxanthin (3-hydroxy-β-carotene), and antheraxanthin also occur in small amounts. The quantity of β-carotene in leaf tissue is usually between 200 and 700 µg g^{-1} (dry weight).

Among fruits, a wider range of carotenoids is found. Only rarely (*e.g.* mango and persimmon) are β-carotene and its immediate xanthophyll derivatives, cryptoxanthin and zeaxanthin, predominant. In the tomato, lycopene is the major carotenoid. Orange juice contains varying proportions of cryptoxanthin, lutein, antheraxanthin and violaxanthin, together with traces of their carotene precursors. The apocarotenoid β-citraurin has also been reported. During the processing of juice the acidic conditions promote some spontaneous conversion of the 5,6- and 5′,6′-epoxide groups to 5,8- and 5′,8′-furanoid oxides (6.5). In the presence of oxygen, particularly in dried foods such as dehydrated diced carrots, oxidation and bleaching occur rapidly. Hydroperoxides resulting from lipid breakdown are very effective in bleaching carotenoids. The final breakdown of β-carotene leaves a residue of β-ionone (6.6), a volatile compound which gives sun-dried (and partially bleached) hay its characteristic odour.

(6.5)

(6.6)

Some carotenoids are restricted to a few or even a single plant species; the capsanthin of red peppers is a good example. The classic occurrence of carotenoids is, of course, in the carrot. Here β-carotene predominates, together with a proportion of α-carotene. The carotenoids in the usual varieties of carrot comprise only 5–10% xanthophylls, located mostly in the yellow core. The total carotene content of carrots is $60–120\,\mu g\,g^{-1}$ fresh weight, but some varieties contain more than $300\,\mu g\,g^{-1}$.

The nutritional functions of dietary carotenoids are beginning to receive greater attention from food scientists, and the contribution of different foodstuffs to the supply of particular carotenoids in the human diet needs to be recognised. Although leafy green vegetables also contain significant amounts of β-carotene, carrots remain its most important source in the temperate regions of the northern hemisphere such as Europe. In tropical and subtropical regions, however, mangoes are the major source of β-carotene. Leafy green vegetables are the predominant source of lutein in most diets, and tomatoes provide the bulk of the lycopene.

Carotenoids are becoming increasingly popular as food colorants, and many of them are now officially sanctioned (*see* Table 6.1). The two apocarotenoids, crocetin and bixin, have been used in a limited range of foods for many years. Crocetin is the major pigment of the spice saffron, in which it occurs in the form of a glycoside, each of its carboxyl groups being esterified by the disaccharide gentiobiose. Saffron is extremely expensive, as it requires some 150 000 stigmas from the flowers of the autumn-flowering crocus, *Crocus sativus*, to produce 1 kg of the dried spice. *Cis*-bixin (*see* Figure 6.5) is the major constituent of the colouring matter annatto, which is the resinous coating on the seeds of *Bixa orellana*, a large shrub grown in a number of tropical countries. Norbixin is a water-soluble version obtained by alkaline hydrolysis, which gives the dicarboxylic acid. The high temperature of the hydrolysis results in considerable isomerisation of the *cis* double bond, and the end-product contains both *cis*- and *trans*-isomers. Annatto is probably the most frequently used natural food colour in the UK, and is the traditional colouring used in red cheeses such as Red Leicester. It is specifically forbidden for use in spices such as curry powder, in case the red colour might give a false impression of quality. Palm oil contains high levels of β-carotene (13–120 mg per 100 g), and is also used to colour margarine.

Three carotenoids have been synthesised commercially, β-carotene, β-apo-8′-carotenal (6.7) and canthaxanthin (6.8). Their commercial preparations are increasingly used in a wide range of products, including margarine, cheese, ice cream, and some baked goods such as cakes and biscuits. Canthaxanthin is only permitted for direct use in one product, Strasbourg sausage. It is however allowed as an additive in certain feedstuffs for animals and fish intended ultimately for human consumption, such as farmed salmon and chickens, in which it improves the colour of the flesh and the egg yolk, respectively.

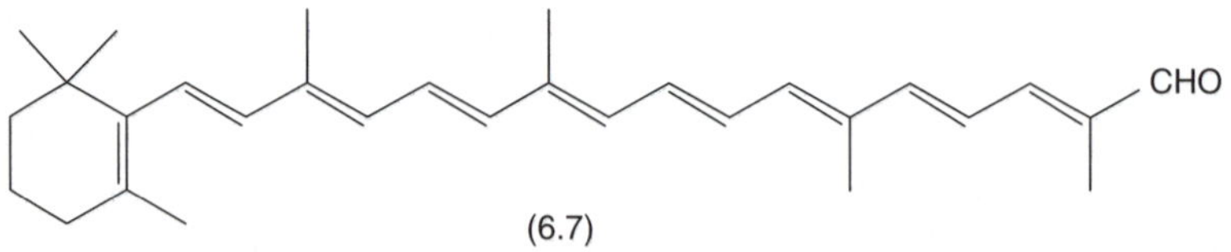

(6.7)

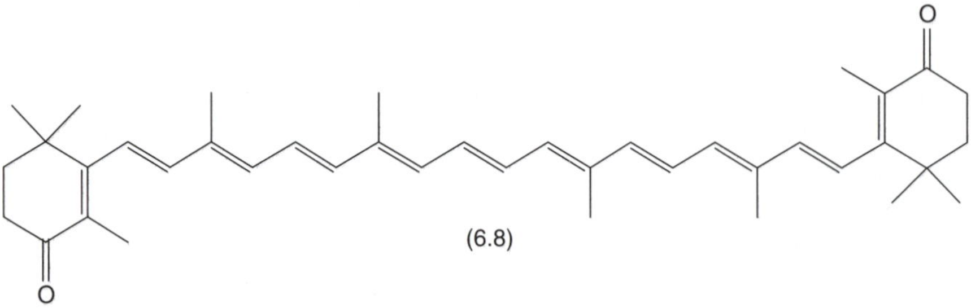

(6.8)

Although carotenoids are only synthesised by plants and micro-organisms, they are also found in some animal tissues. Egg yolk owes its colour to the two xanthophylls, lutein and zeaxanthin, with only a small proportion of β-carotene. These carotenoids, and the smaller amounts that give the depot fat of animals its yellowish hue, are derived from the vegetable material in the diet. The dark greenish-purple pigment of the lobster carapace (and the shells of other shellfish) is a complex of protein

Table 6.1 Carotenoids commonly used as food colorants. Extracts containing crocin or crocetin are treated as spices rather than colorants and are not classified as food additives in the EU.

Pigment	E no.	Commercial source
Mixed carotenes, β-carotene	E160a	Palm oil, carrots (*Dunaliella sp.*) and algae
Cis-bixin, cis-norbixin	E160b	Annatto
Capsanthin	E160c	Red peppers (*Capsicum annum*)
Lycopene	E160d	Tomato
β-Apo-8'-carotenal	E160e	chemical synthesis
Ethyl ester of β-apo-8'-carotenal	E160f	chemical synthesis
Lutein	E161b	Marigold (*Tagetes erecta*), alfalfa
Canthaxanthin	E161g	chemical synthesis
Crocin, crocetin		Saffron (*Crocus sativus*)

combined with astaxanthin. When lobster is boiled, the protein is denatured and the colour reverts to the reddish shade more typical of a carotenoid. Astaxanthin is also the natural source of the pink colour of salmon flesh. Algae at the bottom of the food chain leading to these marine animals will have been responsible for the initial formation of the tetraterpenoid raw material for astaxanthin synthesis.

Carotenoids are generally quite stable in their natural environment, but when food is heated or when they are extracted into solution in oils or organic solvents they are much more labile. On heating in the absence of air there is a tendency for some of the *trans* double bonds of carotenes to isomerise to *cis*, forming what are sometimes termed "neocarotenes". Although carotenes containing one or more *cis* double bonds have been detected in trace amounts in photosynthetic plant tissues, their formation is more often the result of high temperature food processing. As the number of *cis* double bonds increases there is a loss of colour intensity. The difference in colour between canned and fresh pineapple provides a demonstration of this effect. The potential effects of isomerisation on the biological activity of nutritionally significant carotenes is however of far greater importance than the colour of pineapple chunks. The provitamin A activity of 13-*cis*-β-carotene and 9-*cis*-β-carotene is about half that of the normal 100% *trans*-isomer.

The isomerisation of lycopene may also turn out to be nutritionally significant. Evidence is beginning to accumulate suggesting that lycopene can be useful in reducing the progress of prostate cancer. It may also be a factor in the benefits of the tomato-rich "Mediterranean diet" in reducing the incidence of cardiovascular disease. It is generally accepted that the heat processing of tomatoes, *e.g.* in the preparation of pasta sauces, enhances the bioavailability of lycopene. The effects of *cis–trans* isomerisation on absorption and subsequent biological activity remains unresolved, however.

In spite of the possibilities raised in the preceding paragraph, no specific physiological role for carotenoids has been formally identified in humans. The role of carotenoids containing at least one β-ring (collectively referred to as provitamins A) as the precursors of vitamin A is well established. Vitamin A activity is possessed by substances which are the precursors of retinal, the non-protein component of the visual pigment of the retina. The enzyme-catalysed formation of retinal from carotenoid precursors is illustrated in Figure 6.7. Subject to the availability of sufficient vitamin A or its precursors in the diet, the level of retinol in the tissues is physiologically regulated and high dietary levels (*i.e.* at the upper end of the normal nutritional range) do not raise the concentration of retinol in the tissues or the bloodstream. Vitamin A is discussed further in Chapter 8.

Two common xanthophyll carotenoids, lutein and zeaxanthin, are the major source of the yellow colour of the cells of the *macula lutea* of the eye. This is a small area at the centre of the retina where the photosensitive cells (the rods and cones) are most tightly packed. The yellow

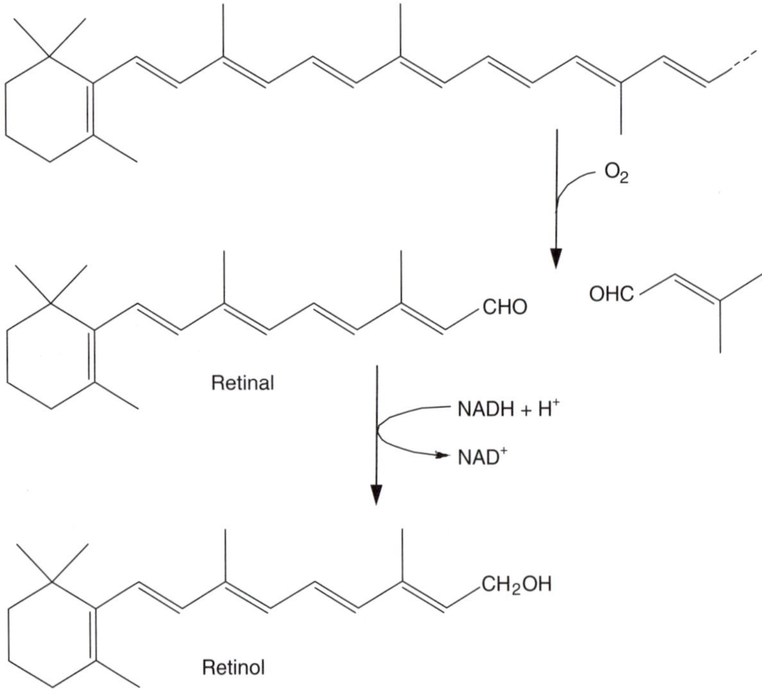

Figure 6.7   Formation of retinol and retinal.

pigments are believed to act as filters to protect the photosensitive cells from damage by ultraviolet light. Age-related macular degeneration (AMD) is a major cause of sight loss in the elderly, and has been attributed to inadequate levels of dietary xanthophylls. There is growing support for the possibility that dietary supplementation with lutein and zeaxanthin may offer a preventive or even a therapeutic role in dealing with AMD.

Epidemiological studies, in which the occurrence of a particular disease within a population is studied in relation to the proportions of particular dietary components, have suggested that the consumption of fruit and vegetables is associated with the reduced incidence of many types of cancer. This effect is most marked in the case of cancers of the gastro-intestinal tract (such as those of the mouth, oesophagus and stomach) and the respiratory system (*e.g.* larynx and lung). Although attention has been focused on β-carotene it has been recognised that other plant constituents such as other carotenoids, ascorbic acid, flavonoids and fibre may all be involved. Detailed analysis has associated dietary α- and β-carotene and lutein with reduced risk of lung cancer. Unfortunately the obvious follow-up to these observations, intervention trials in which β-carotene was given to smokers as a dietary supplement, produced the unexpected result of β-carotene supplements actually increasing the incidence of lung cancer. It therefore seems likely that other components of plant materials are involved in addition to particular carotenoids, and we would do better to concentrate on the quality of the diet as a whole rather than on the level of individual components. Whether the quasi-pharmaceutical "supplements" industry will embrace this concept remains to be seen.

The possibility that β-carotene and other carotenoids, along with vitamins C and E, might have an antioxidant function and therefore a bearing on the incidence of coronary heart disease, is a separate issue which was briefly considered in Chapter 4.

ANTHOCYANINS

The pink, red, mauve, violet and blue colours of flowers, fruit and a few vegetables are caused by the presence of anthocyanins. In common with other polyphenolic substances, anthocyanins occur in nature as the glycosides, the aglycones being known as anthocyanidins. They are located in the vacuoles of plant cells, kept apart from the enzymes and other delicate structures in the cytoplasm. The anthocyanidins are flavonoids,[i] *i.e.* substances based on a flavan nucleus (6.9).

---

[i] Often spelt *flavanoids*.

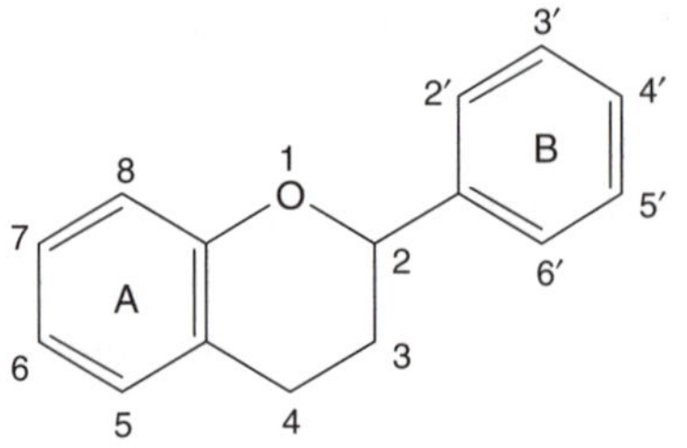

(6.9)

Only six different anthocyanidins occur in nature, but the diversity of the patterns of glycosylation (the attachment of glycosyl groups, *i.e.* sugars) and acylation (attachment of acyl groups in caffeic acid, *etc.*, below) means that there are innumerable different anthocyanins to be found in nature. A single plant species will also contain different anthocyanins. The structures of the six anthocyanidins are shown in Figure 6.8. The complete structure of one of the anthocyanins found in grapes, malvidin (3-coumarylglucoside) shown here (6.10) gives some impression of the complexity of these pigments.

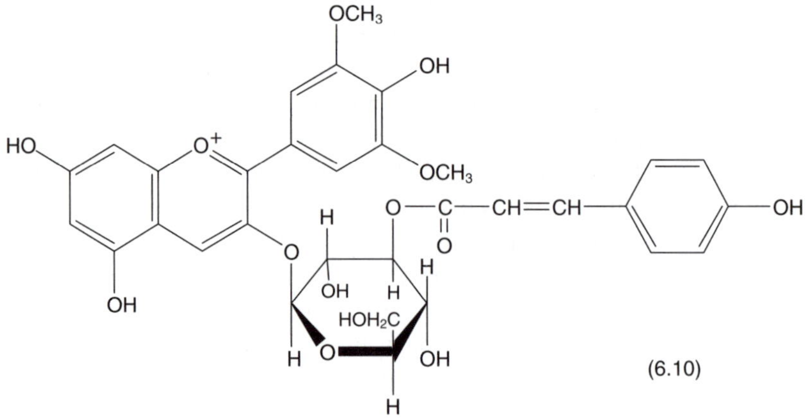

(6.10)

The form of the anthocyanidins shown in Figure 6.8 is termed the flavylium cation but, as we shall see, other forms occur depending on the pH. Anthocyanins always have the sugar residue at position 3 and glucose often occurs additionally at position 5, and more rarely at positions 7, 3′ or 4′. Apart from glucose, the monosaccharides most often found are galactose, rhamnose and arabinose. Some other unusual disaccharides also occur, such as rutinose [α-L-rhamnosyl-(1 → 6)-D-glucose] and sophorose [β-D-glucosyl-(1 → 2)-glucose]. It is not uncommon for the 6-position of the sugar residue located at position 3 to be esterified with phenolic compounds such as caffeic (6.11), *p*-coumaric (6.12) or

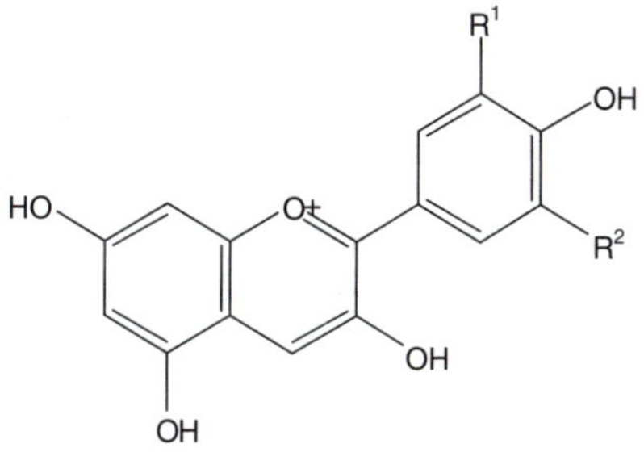

	R¹	R²	λ_{max}
pelargonidin	– H	– H	503
cyanidin	– OH	– H	517
peonidin	– OCH₃	– H	517
delphinidin	– OH	– OH	526
petunidin	– OCH₃	– OH	526
malvinidin	– OCH₃	– OCH₃	529

Figure 6.8 The structures of the anthocyanidins and the absorption maxima (λ_{max}) of the corresponding 3-glucoside anthocyanins at pH 3.

Table 6.2 Anthocyanins found in fruit.

Fruit	Principal anthocyanins
Blackcurrant (*Ribes nigrum*)	3-glucosides and 3-rutinosides of delphinidin and cyanidin
Blackberry (*Ribes fructicosus*)	3-glucosides and 3-rutinosides of cyanidin
Raspberry (*Rubus ideaus*)	3-glucosides, 3-sophorosides and 3-rutinosides of pelargonidin and cyanidin
Elerberry (*Sambucus nigra*)	3,5-diglucosides and 3-sambubiosides of cyanidin (sambubiose is β-D-xylosyl-(1 → 2)-β-D-D-glucose)
Strawberry (*Fragaria* spp.)	3-glucosides of pelargonidin and cyanidin, 3-arabinoside of pelargonidin
European grape (*Vitis vinifera*)	3-glucosides, 3-acetylglucosides and 3-coumarylglucosides of malvidin, peonidin, delphinidin, petunidin and cyanidin

ferulic (6.13) acids. The principal anthocyanins occurring in a range of fruits are listed in Table 6.2. Cyanidin is by far the most widespread anthocyanidin, and it is clear that there is no link between the taxonomic classification of a plant and the identity of its anthocyanins.

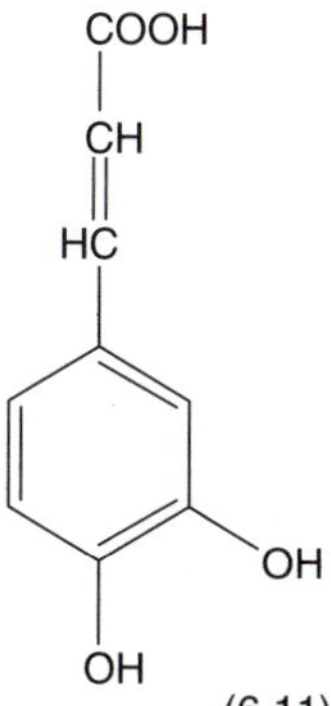

(6.11) 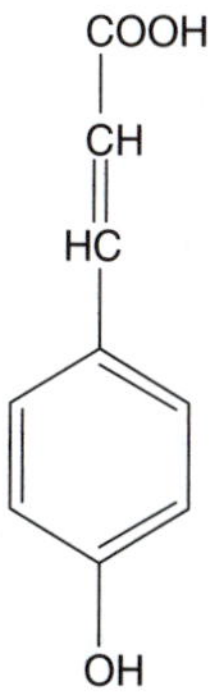   (6.12) 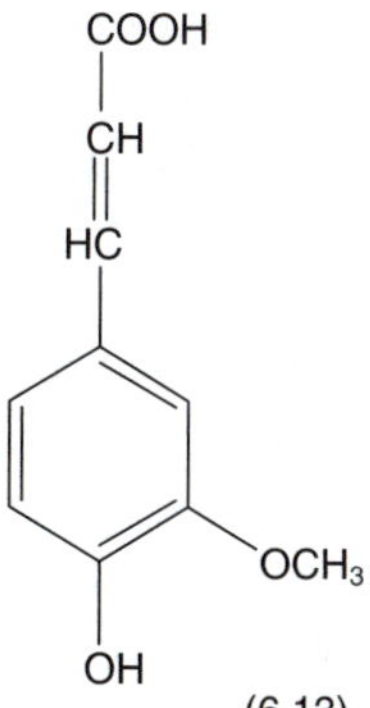   (6.13)

The black grape anthocyanins are particularly interesting. Of the total of six anthocyanidins only pelargonidin is not found in grapes, and there is much more variety in the pattern of glycosylation and acetylation than in most plants. The classic European grape species is *Vitis vinifera* which contains only 3-monoglucosides, whereas in non-European countries and the USA species such as *V. riparia* and *V. rupestris* and their hybrids with *V. vinifera* are grown. All these species, but not *V. vinifera*, contain both 3-monoglucosides and 3,5-diglucosides. The diglucosides can readily be detected by a chemical test and can also be separated from the monoglucosides by chromatography. Detection of diglucosides in a red wine is therefore conclusive evidence of its non-European origin.

Much less work has been done on the anthocyanins present in vegetables. Cyanidin is the anthocyanidin found in red cabbage, pelargonidin occurs in radishes and red-seeded varieties of beans, and delphinidin occurs in aubergines.

As seen in Figure 6.8, increasing substitution in the B-ring (identified in Formula 6.9) moves the absorbance maximum of anthocyanidins towards the red end of the spectrum, but these changes are not sufficient to account for the great range of colours that anthocyanins impart to flower petals and fruit. The pH and the presence of other substances have a much greater influence on their colour than the nature of the ring substituents. Anthocyanins form hydrogen-bonded complexes with the colourless flavonoids also common in plant tissues. Figure 6.8 shows the basic anthocyanidin structure in the flavylium cation form which predominates at low pH; at pH 1 this is the only significant form present. As the pH increases a proton is lost, a water molecule is acquired and the carbinol pseudo-base is formed (*see* Figure 6.9). The pK_a value (*i.e.* the pH value at which the ionisation is half complete) for this transition is around 2.6.

The red-coloured forms of the anthocyanins are the flavylium cations, and the carbinol pseudobases are colourless, so there is a gradual loss of colour intensity with rising pH, but strong blue colours are obtained at high pH values (Figure 6.10).

Work with purified anthocyanins has clarified the pattern of other ionisations, including those at higher pH. Above pH 3 the small residual proportion of the anthocyanins in the form of a flavylium cation begin to lose protons, forming first the non-ionised quinoidal base, which is weakly purple in colour, and then the ionised quinoidal base, which is deep blue. At the same time the carbinol pseudobase gives rise to a small proportion of the colourless chalcone forms. After extended periods at elevated pH quite high proportions of chalcones are present.

Although these equilibria have been described in some detail for purified anthocyanins, it must be remembered that in fruit juices and

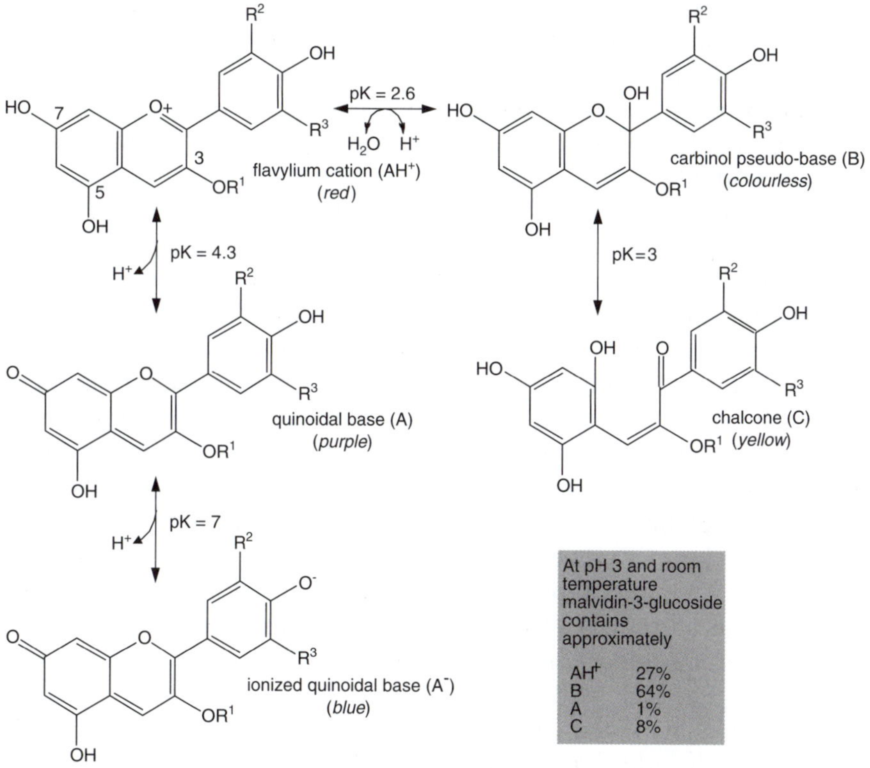

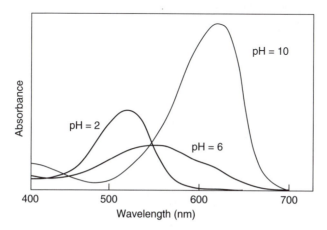

Figure 6.9 Transitions in the anthocyanin structure. The presence of sugar residues at 5 or 7 will obviously influence the formation of quinoidal forms.

Figure 6.10 The effect of pH on the absorption spectrum of an anthocyanin. (Redrawn from R. Brouillard, in *Anthocyanins as Food Colours*, ed. P. Markakis, Academic Press, New York, 1982.)

other natural plant materials the situation is neither as straightforward nor as well understood. Anthocyanins readily interact with other colourless flavonoids which abound in plant tissues. These associations, which depend on hydrogen bonding between hydroxyl groups, enhance the intensity of the colour, modify λ_{max} values, and tend to stabilise the blue quinoidal form.

Anthocyanins are also able to complex metal cations, but it is not clear how important these are to the colour of fruit. Occasional unusual colour developments in canned fruit are undoubtedly due to interaction between the anthocyanins of the fruit and the metal, iron or tin, of the can when the internal lacquer has failed. The unexpected appearance of pink colouration in canned pears has been attributed to the formation of iron or tin complexes with leucoanthocyanins (*see* Figure 6.18), which are normally colourless. Another important reaction of anthocyanins occurs with sulfur dioxide. Sulfur dioxide, usually in the form of sulfite (SO_3^{2-}) or metabisulfite ($S_2O_5^{2-}$) (*see* Chapter 9), is routinely used as an antimicrobial preservative in wine and fruit juices. At high concentrations (1–1.5%) these cause irreversible bleaching[ii] of anthocyanins, but at lower concentrations (500–2000 ppm) they react with the flavylium cation to form a colourless addition compound, the chroman-4-sulfonic acid (6.14). The lack of colour in the addition compound is explained similarly to that of the carbinol pseudo-base (*see* Figure 6.9). Acidification, or the addition of excess acetaldehyde (ethanal), will remove the sulfur dioxide. The reaction with sulfites provides the basis for an insufficiently well-known technique for dealing with red wine spills on white tablecloths—adding a few drops of white wine, which always contains sulfur dioxide, to the stain. If this requires another bottle to be opened, then so be it!

(6.14)

[ii] Maraschino cherries, used for garnishing cocktails, are bleached with sulfur dioxide and then dyed with synthetic colours.

The anthocyanins in red wine undergo a number of reactions with some of the colourless flavonoids also originally present in the grape skins and seeds, such as catechins (better known as a component of tea, *see* Figure 6.12). The formation of a link between the 4-position of the anthocyanin and the 8-position of the catechin leads to the formation of procyanidins, more strongly coloured compounds which are less affected by a change in pH or by sulfur dioxide. By the end of the fermentation period as much as a quarter of the anthocyanins may have formed oligomers with other flavonoids. The brown colour characteristic of older red wines is the result of extensive polymerisation of the anthocyanins and other flavonoids. This is accompanied by a mellowing of the flavour, as it is these "tannins" which give young red wine its characteristic astringency.

In most food processing operations the anthocyanins are quite stable, especially when the low pH of the fruit is maintained. Occasionally, however, the ascorbic acid naturally present can cause problems. In the presence of iron or copper ions and oxygen, the oxidation of ascorbic acid to dehydroascorbic acid is accompanied by the formation of hydrogen peroxide. This oxidises the anthocyanins to colourless malvones (6.15), a reaction implicated in the loss of colour by canned strawberries.

(6.15)

Anthocyanins are becoming increasingly popular as food additives, replacing synthetic dyes (discussed later in this Chapter) in confectionery, soft drinks and similar products. They are of little value in dairy products as their colour at less acid pH is either weak or an unappealing blue, or even purplish green if other yellow pigments are present. Good sources of anthocyanins are not very common. One which is being exploited is the residue, mostly the skins, from pressing black grapes for wine. Red cabbage extracts are being marketed, and elderberries are also likely to become useful anthocyanin sources.

BETALAINES

The characteristic red-purple pigments of beetroot,[iii] *Beta vulgaris*, used to be described as "nitrogenous anthocyanins". However, it is now established that these pigments, and those of the other members of the eight families usually grouped together as the order *Centrospermae*, are quite distinct from the anthocyanins and never occur with anthocyanins in any member of these eight families. The betalaines are divided into two groups, the betacyanins, which are purplish red in colour (λ_{max} 534–555 nm), and the less common yellow betaxanthins ($\lambda_{max} \sim 480$ nm). The betalaines of *B. vulgaris* are the only ones of food significance, and this account will therefore be largely restricted to these. Their structures are shown in Figure 6.11.

Firstly, we have the betacyanins, which account for about 90% of beetroot betalaines and occur at around 120 mg per 100 g of fresh beetroot. They are all either the glycosides or the free aglycones of betanidine or its C-15 isomer. In beetroot, *iso* forms account for only 5% of the total betacyanins. About 95% of the betanidine and iso-betanidine carry a glucose residue at C-5. A very small proportion of these glucose residues are esterified with sulfate, to give the *pre*-betanins. Secondly, in beetroot we have the betaxanthins, represented by the vulgaxanthins, which are characterised by the lack of an aromatic ring system attached to N-1 and by the absence of sugar residues. A great diversity of betaxanthins are found in the Centrospermae, but only the two known as vulgaxanthins I and II are found in beetroot, in roughly equal proportions.

Although of similar colour to the anthocyanins, betacyanins are different in that their colour is hardly affected by pH within the range normally encountered in foodstuffs. The reason for this is clear from the structure. Changes in the pattern of ionisation of an anthocyanin can be seen from Figure 6.9 to cause considerable changes in the arrangement of the double bonds. However, the state of ionisation of the carboxyl groups in betacyanin has no such effect. Below pH 3.5 the absorption maximum of betanin solutions is 535 nm, between pH 3.5 and 7.0 it is 538 nm, and at pH 9.0 it rises to 544 nm. Betacyanins are fairly stable under food processing conditions although heating in the presence of air at neutral pH causes breakdown to brown compounds. Concentrated beetroot extracts (containing around 0.3% pigment, the remainder being almost entirely sugar) are becoming popular as colorants for dairy and dessert products. Unfortunately their instability to heat at neutral pH means that they are unsuitable for colouring cakes, *etc.*

[iii] Also known as "red beet".

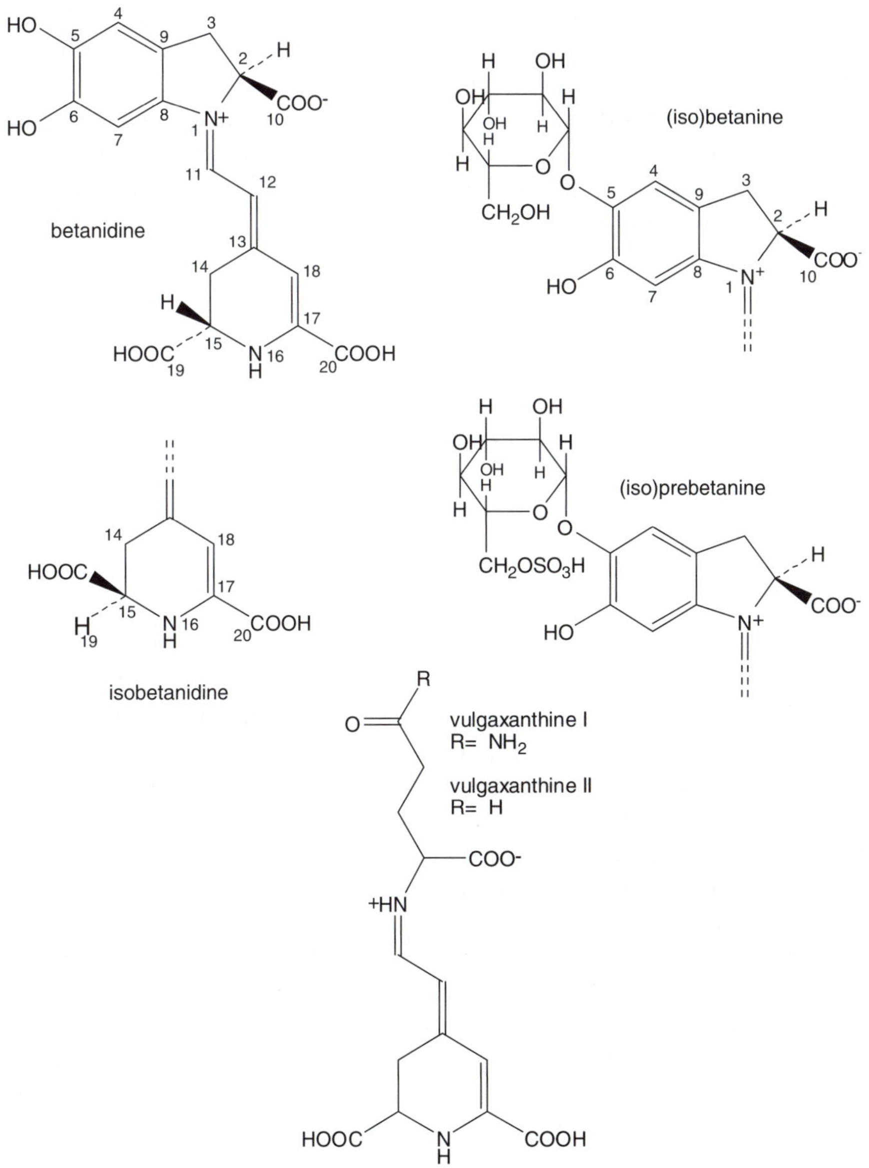

Figure 6.11 Betalaine and vulgaxanthin structures. The significance of *iso* and *pre* is explained in the text.

The processing of beetroot for use as a food colour poses some interesting problems. The betalaines are natural substrates for the enzyme phenolase (*see* page 237) which causes rapid browning of the pigment in the presence of oxygen, but early heating is sufficient to prevent this. Beetroot is naturally high in nitrates (*see* Table 10.3) and sugars, but both of these can be eliminated by allowing yeast to grow in the crude beet extract and then removing the yeast cells by filtration or centrifugation. The characteristic earthy odour of beetroot, which might limit the usefulness of its derivatives in delicately flavoured dessert products, is caused by geosmin (6.16). Fortunately this is distilled off at the time the beetroot juice is concentrated.

(6.16)

MELANINS

Although we do not necessarily regard them as desirable pigments in our fruits and vegetables, this is a convenient point to consider the brown melanin derivatives formed when plant tissues are damaged. Everyone is familiar with the way in which pale-coloured fruit and vegetables such as apples, bananas and potatoes quickly turn brown if air is allowed access to a cut surface. The browning occurs when polyphenolic substances which are usually contained within the vacuole of the plant cells are oxidised by the action of the enzyme phenolase, which occurs in the cytoplasm of plant cells. Tissue damage caused by slicing or peeling, fungal attack or bruising will bring enzyme and substrates together. The enzyme phenolase (originally known as polyphenol oxidase) occurs in most plant tissues. A related enzyme, tyrosinase, is found in mammalian skin. Phenolase is an unusual, but not unique, enzyme in that it catalyses two quite different types of reaction. The first of these, its cresolase activity, results in the oxidation of a monophenol (6.17) to an *ortho*-diphenol (6.18). The second, catecholase activity, oxidises the *o*-diphenol to an *o*-quinone (6.19).

(6.17)   (6.18)   (6.19)

o-Quinones are highly reactive and readily polymerise following their spontaneous conversion to hydroxyquinones (6.20).

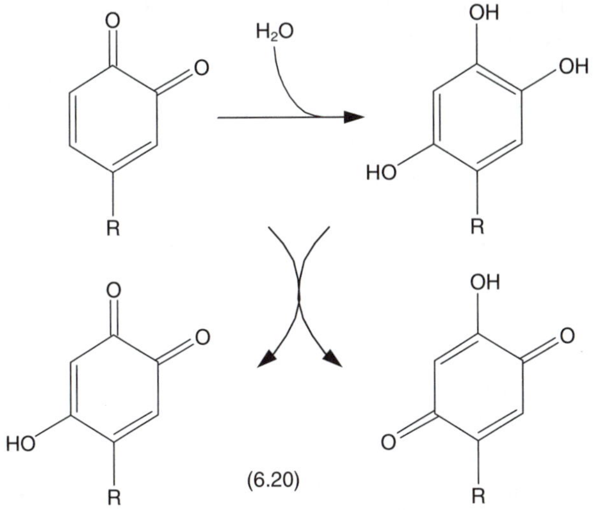

(6.20)

In animal tissues the only substrate is the amino acid tyrosine. The reaction in this case leads to the formation of the melanin pigments in skin and hair. The brown pigments which develop on the cut surface of fruit and vegetables may well be regarded as unsightly to consumers, but they do not detract from their flavour or nutritional value. In the plant the products of phenolase action, particularly the quinones, are very important as antifungal agents. The *o*-quinones interact with the tyrosine hydroxyl groups on the surface of fungal enzymes in much the same way as the reactions (6.20) shown above. This can cause the inhibition of these enzymes and reduces the ability of fungi to penetrate tissues which have suffered minor physical damage. The relative resistance to fungal diseases of different varieties of onions and apples has been correlated with their phenolase activity and their quinone content.

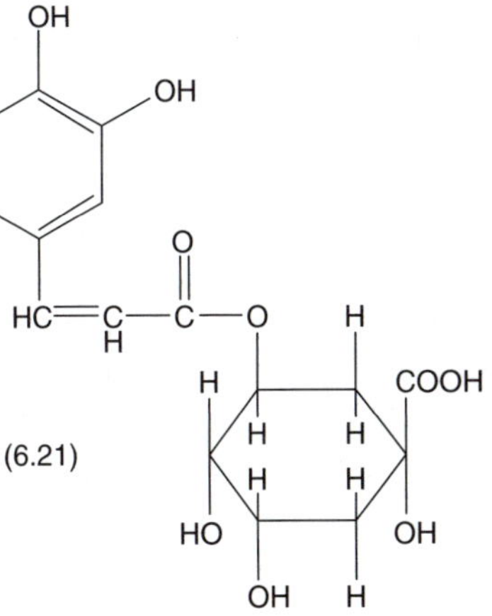

(6.21)

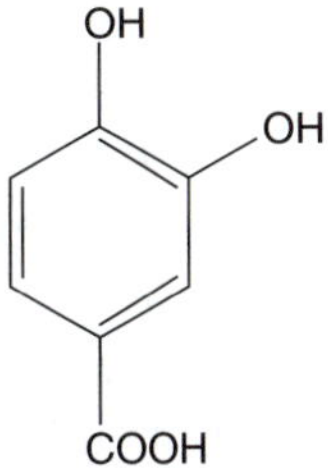

(6.22)

The most important substrate for phenolase in apples, pears and potatoes is chlorogenic acid (6.21), and in onions protocatechuic acid (6.22), but very little is known about the structures of the polymerised quinones derived from them.

During fruit and vegetable processing, both in the home and in the factory, we try to prevent phenolase action. Making sure that the fruit and vegetables are blanched as soon as possible after (or preferably before) any tissue-damaging operations will reduce phenolase action to a minimum. Reducing contact with air by immersion in water is also common practice. Some of the most effective specific inhibitors of phenolase activity are chelating agents, which act by depriving the enzyme of the copper ion which is a component of its active site. However, the substances used in analytical laboratories, such as diethylthiocarbamate, cyanides, hydrogen sulfide or 8-hydroxyquinoline, are hardly appropriate for food use.

Though less specific and less potent, the organic acids, citric (7.12) and malic acid (7.13), have advantages. Not only are they present in fruit juices, their acidity helps to keep the pH well below the optimum for phenolase action (approximately pH 7). Immediately after being peeled, and especially if strong sodium hydroxide solution (lye) has been used to assist this, fruit such as peaches are often immersed in neutralising baths containing dilute acids, frequently supplemented with ascorbic acid or sulfites.

Ascorbic acid is not only valued as a vitamin but also as a chelating agent and antioxidant. It interacts directly with quinones to prevent browning and, if sufficient is present, it can "mop up" the oxygen in closed containers such as cans:

$$o-\text{diphenol} + \tfrac{1}{2}O_2 \rightarrow o-\text{quinone} + H_2O$$

$$o-\text{quinone} + \text{ascorbate} \rightarrow o-\text{diphenol} + \text{dehydroascorbate}$$

Although it seems likely that manufacturers of dehydrated potato powders include ascorbic acid in their products to ensure their whiteness rather than for the benefits accruing from vitamin C, the motive does not belittle their nutritional value, particularly for the elderly, who often rely on such convenience foods in place of fresh fruit and vegetables.

Sulfur dioxide also has a specific, if not fully explained, inhibitory action against phenolase. A sulfite concentration sufficient to maintain a free SO_2 concentration of 10 ppm will completely inhibit phenolase. Being a reducing agent, sulfite has the additional benefit of preserving the ascorbic acid level, but it is not without its disadvantages, as it tends to bleach anthocyanins and to hasten can corrosion.

Tea

The enzymic oxidation of polyphenolic substances is a highly desirable feature of one particular foodstuff—tea. The tea we drink is a hot water infusion of the leafy shoots of the tea plant, *Camellia sinensis*, which are processed in various ways depending on the type of tea. That drunk most often in Europe and North America is known as black tea, to distinguish it from green tea, also known as China tea. The leaves of the tea plant contain enormous quantities of polyphenolic materials, over 30% of their dry weight. Caffeine (6.23), the stimulant found also in coffee and cola-type drinks, amounts to some 3–4% of the dry weight of the leaf. The freshly plucked leaves, normally just the bud plus the first and second leaves (the "flush"), are allowed to lose about one-fifth of their water content after they are macerated, a process known as "withering". This physical damage to the leaves allows the phenolase in the cytoplasm of the leaf cells to make contact with the polyphenolic substances contained in the cell vacuoles. The macerated leaves are then left at ambient temperature for a few hours to ferment, during which a high proportion of the polyphenolic substances are oxidised, and polymerisation occurs. The leaves are then "fired", *i.e.* dried at temperatures up to 75 °C, which inactivates the phenolase and brings the fermentation reaction to a halt. This gives the product we recognise as tea. Green tea, on the other hand, is fired before the leaves are macerated and the phenolase is inactivated before significant fermentation can occur. An intermediate type, Oolong tea, is fermented for only about 30 minutes before firing.

(6.23)

Almost all of the polyphenolic substances in tea are flavonoids, either catechin or its derivatives, as shown in Figure 6.12. Of these, *epi*-gallocatechin gallate is by far the most abundant, some 9–13% of the dry weight of the unfermented leaves.

The initial effect of phenolase action is in each case to convert the diphenols to the corresponding *o*-quinones (6.24 and 6.25).

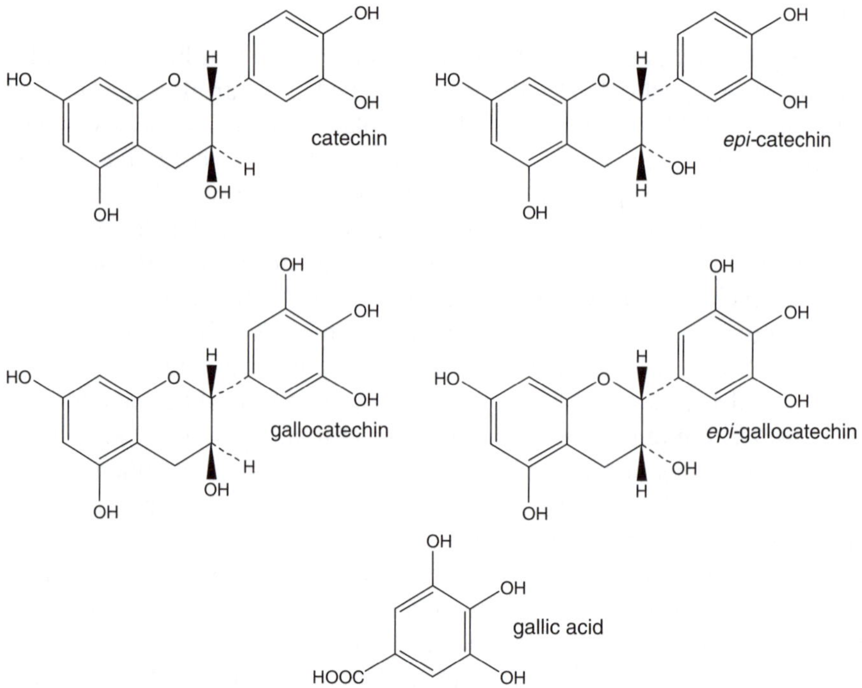

Figure 6.12   Gallic acid and the principal flavonoids of the unfermented tea flush. The prefix *epi-* refers to the optical configuration at the 3-position of the flavonoid ring system. *Gallo-* indicates three hydroxyl groups attached to adjacent carbon atoms in the aromatic ring. Both *epi*-gallocatechin and *epi*-catechin also occur as gallates, in which case gallic acid is esterified at the 3-position.

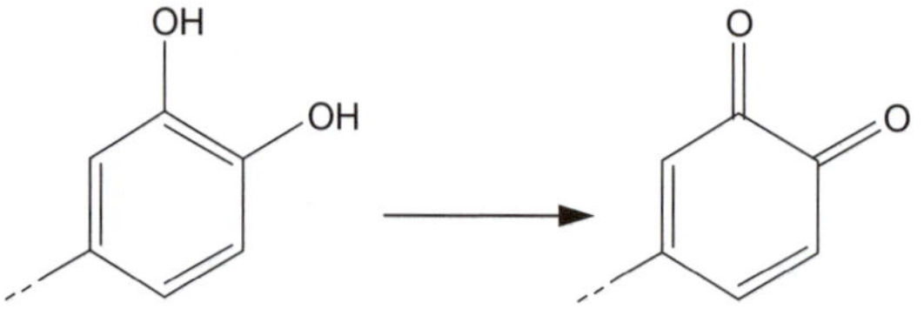

$$(6.24)$$

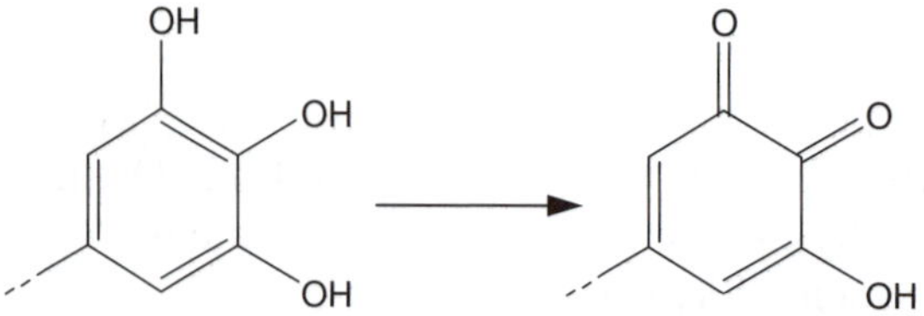

$$(6.25)$$

The quinones are highly reactive and readily form a wide variety of dimers during the fermentation period. Most of these interact further to produce polymeric thearubigins, but one class of dimers, the theaflavins (6.26), comprise about 4% of the soluble solids of black tea. Being bright red in colour, they make a major contribution to the "brightness" of the tea colour.

(6.26)
R^1 and R^2 may be either
H or gallate

The brown thearubigins are the principal pigments of black tea infusions. They contribute not only the bulk of the colour but also the astringency, acidity and body of the tea. Thearubigins extracted with hot water contain a wide range of molecular weights. Although the average corresponds to oligomers of four or five flavonoid units, molecules comprising as many as 100 flavonoid units may be encountered in tea infusions. Degradation studies have shown that all the principal flavanols shown in Figure 6.12 are represented in these oligomers. Although there is no certainty as to the nature of the links involved in the polymerisation, a proanthocyanidin structure is most likely, shown in Figure 6.13. The proanthocyanidins themselves are dimers, and occur in small amounts in many fruit juices, where they contribute to the astringency.

The proportions of the different classes of polyphenolics present in brewed tea vary considerably with the variety of tea, the proportion of leaves to water, and the brewing temperature and time. The data in Table 6.3 are reasonably representative. Data on the thearubigin content of brewed tea are sparse, but the indications are that the most common varieties of black tea, such as Ceylon and Assam, give brews containing 20–35 mg thearubigins per g of dry leaf.

In recent years there has been growing interest in the potential health benefits of drinking tea. Tea undoubtedly provides a valuable source of polyphenolic compounds, and thus antioxidants, in the diets of many communities around the world. Considerable benefits in relation to both cardiovascular disease and cancer have been attributed to tea, especially

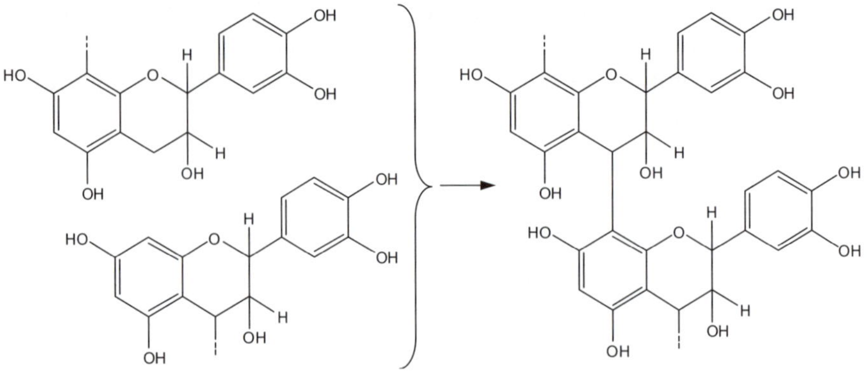

Figure 6.13   The polymerisation of tea flavonoids, forming a proanthocyanidin structure.

Table 6.3   Typical flavonoid content in tea extracted with boiling water for 5 minutes, expressed as $mg\,g^{-1}$ dry tea leaf (based on data presented by M. Friedman, *et al., J. Food Sci.*, 2005, **70**, C550). The data for the four classes of catechins (asterisked) represent the sum of the catechins and their respective *epi*-isomers.

| | Tea variety | | | | |
Flavonoid	Assam	Ceylon	Darjeeling	Oolong	Green
Catechins*	2.9	0.6	4.2	8.2	1.0
Gallocatechins*	1.2	trace	4.5	5.7	trace
Catechin gallates*	2.8	5.6	21.5	4.5	3.9
Gallocatechin gallates*	2.8	8.0	71.3	26.7	13.3
Theaflavin	0.6	0.8	0.7	0	0
Theaflavin gallates	1.1	1.3	0.7	0	0
Theaflavin digallates	3.0	3.1	3.8	0	0

green tea. Unfortunately epidemiological data do not offer as much support for this as we might like. In 2003 Higdon and Frei published a comprehensive review (*see* page 264 for details) and concluded:

> "*Overall epidemiological studies do not provide conclusive evidence for a protective effect of tea consumption on the risk of cardiovascular diseases, although several studies have demonstrated significant risk reduction in consumers of black and green tea.*"

Readers will find additional information in the first of the Special Topics at the end of this chapter.

TURMERIC AND COCHINEAL

Much of the consideration of colour so far in this chapter has been devoted to the colour of foods contributed by substances present in them naturally. The application of some of these natural colours as colorants for other foods has already been mentioned. This section is concerned with two important food colorants which do not fit into any of the major pigment types.

Turmeric, the powdered dried root of the turmeric plant, *Curcuma longa*, has been used for centuries both as a spice and as a food colour. In addition, it has found use in the Far East as a textile dye. Nowadays it is best known for its role in oriental cookery as an essential component of curry powder. As well as its colour, it also contributes a characteristic earthy, slightly bitter taste. Turmeric contains the yellow pigment curcumin (6.27), and is a traditional ingredient in the English pickle known as piccalilli.

(6.27)

Solvent extraction of the powdered spice yields the oil-soluble colouring matter curcumin. This is a bright yellow pigment with a slightly greenish cast. As synthetic dyestuffs have increasingly come to be regarded with suspicion as food additives, curcumin has proved invaluable for colouring dessert products, ice-cream, pickles and other manufactured foods.

In recent years there have been a large number of reports of the biological activity of curcumin and other minor constituents of turmeric. The spice has always been recognised in India as having medicinal properties and among these, attention is now being focused on its possible role in counteracting the progress of Alzheimer's disease.

Plants do not have a monopoly in providing natural colouring materials, however. Cochineal is a group of red pigments from various insects similar to aphids, most notably *Dactylopius coccus* (previously known as *Coccus cacti*), which belongs to the superfamily *Coccidoidea*. Cochineal, in the form of the dried and powdered female insects (about 100 000 are required per kg of the pigment), has been an object of commerce since ancient times. The insects are harvested from cactus plants of the prickly pear, or *Opuntia* family, in various parts of the world, including Central and South America and the Canaries. Apart from its use as a food colour, it has been used as a dyestuff for textiles and leather, and also as a heart stimulant. Although varieties of cochineal have been "cultivated" in many parts of the world, the bulk of production now comes from Peru.

The principal component of cochineal is carminic acid (6.28), an anthroquinone glycoside with an unusual link between the glucose and the aglycone.[iv] Cochineal itself has a purple colour, but textile dyers discovered that if used with tin salts it gave a strong scarlet dye. The deep-red pigment used as a food colour, variously referred to as carmine, carmines of cochineal or simply as cochineal, is obtained from the crude cochineal powder by first extracting the carminic acid with hot water. The extract is then treated with aluminium salts to precipitate the brilliantly coloured complex, "carmine lake". This can be precipitated out with ethanol to give a powder which can readily produce a suspension in water. In spite of its high cost, cochineal has retained popularity thanks to the public caution against the use of synthetic dyestuffs in food.

(6.28)

ARTIFICIAL FOOD COLORANTS

The use of artificial, non-natural colours in food has a long but far from glorious history. Wine has always been a target for the unscrupulous, and in the 18th and 19th centuries burnt sugar and a range of rather unpleasant vegetable extracts were commonly added to give young red wines the appearance of mature clarets. However, for sheer lack of scruple the confectionery trade had no equal. In 1857 a survey of adulterants to be found in food revealed that sweets were commonly coloured,

---

[iv] The term aglycone refers to the non-sugar component of any natural glycoside, see page 19.

for example, using lead chromate, mercuric sulfide, lead oxide (red lead), or copper arsenite! Legislation, and the availability of the newly developed aniline dyestuffs, rapidly eliminated the use of metallic compounds, together with some of the more dubious vegetable extracts.

The new synthetic dyestuffs had many advantages over natural colours, however. They tended to be much brighter, more stable and cheaper, and a wide range of shades was available. It did not take long before some of these dyes also turned out to have some toxic properties, though the effect was mainly on those engaged in making them rather than on consumers.[v] Since that time there has been a steady increase in the range of dyestuffs available, but at the same time we have become increasingly aware of their potential toxicity. Thus in 1957 there were 32 synthetic dyestuffs permitted for use in food in Britain. By 1973, 19 of these had been removed from the list and three added. In 1979, 11 of those remaining were granted E numbers, together with several others not permitted in the rest of the European Community. This situation has now been simplified, and Table 6.4 lists the E numbers of all the synthetic colours permitted in the EU.

The structural formulae and chemical names of a representative selection of permitted synthetic colours are shown in Figure 6.14. One complication with chemical details of dyestuffs is that sometimes the commercial name covers a mixture rather than a single substance. An extreme example is Brown FK, which has six components. Their elaborate ring structures ensure that they have rather intimidating chemical names. To make matters worse, both the common and chemical names vary between authorities and between Europe and the USA. For most dyestuffs, not just those used as food colours, the internationally accepted Colour Index (CI) numbers provide unequivocal identification. The reference numbers used in the USA (the equivalent of European E numbers) are also shown in Table 6.3. The common names of many dyestuffs often include suffixes derived from their intended application. The "FK" of Brown FK, for example, is an abbreviation of *for kippers*, since this dye has been traditionally used to enhance their colour.

All of the above colours are freely soluble in water, a property generally conferred by their sulfonic acid groups. The chromophore of the azo dyes is the azo group:

$$(-N{=}N-)$$

---

[v] It appears that the laxative effects of phenolphthalein, the popular laboratory pH indicator, were first discovered in Hungary in 1902, when wine to which it had been added to enhance its colour had this unanticipated effect on its consumers.

Table 6.4   The synthetic food colours permitted in Europe. The US numbering
system is only applied to synthetic dyestuffs. F, D and C indicate
that use is permitted in food, drugs and cosmetics, respectively, with
the prefix Ext meaning "External use only". In spite of the author's
efforts to ensure their accuracy, the data in this table should be
regarded as illustrative rather than authoritative. As discussed in
the text, restrictions are imposed on the types of foods in which
some of these colours may be used. Red 2G (E128) was removed
from the list of food colours permitted in Europe in August 2007.

Name	Chemical class	EU No.	US No.	CI No.
Yellow:				
Tartrazine	azo	E102	FD&C Yellow No 5	19140
Quinoline Yellow	quinoline	E104	D&C Yellow No 10	47005
Sunset Yellow FCF	azo	E110	FD&C Yellow No 6	15985
Red:				
Carmoisine	azo	E122	Ext D&C Red No 10	14720
Amaranth	azo	E123		16255
Ponceau 4R	azo	E124	FD&C Red No 2	16185
Erythrosine	xanthene	E127	FD&C Red No 3	45430
Allura Red AC	azo	E129	FD&C Red No 40	16035
Blue:				
Patent Blue V	triarylmethane	E131		42051
Indigo Carmine	indigoid	E132	FD&C Blue No 2	73105
Brilliant Blue FCF	triarylmethane	E133	FD&C Blue No 1	42090
Green:				
(Food) Green S	triarylmethane	E142		44090
Brown and black:				
Black PN	azo	E151		28440
Brown FK	azo	E154		
Chocolate Brown HT	azo	E155		20285

conjugated with aromatic systems on either side to give colours in the
yellow, orange, red and brown range. Three linked aromatic systems
provide the chromophore of the triarylmethane dyes, which unlike azo
dyes are anionic; they are characteristically bright greens, violets or
blues. Xanthene dyes are also anionic and have a similar range of hues
to the triarylmethane dyes; erythrosine, the only example permitted
in food, has a brilliant red shade. Resonance hybrids are important
in the colours of all these dyes. Indigo Carmine (Indigotine) in particular has many possible zwitterionic arrangements other that shown in
Figure 6.14. The absorption spectra of a number of the dyes are shown
in Figure 6.15.

The application of the synthetic colours in food processing demands
attention to other factors apart from the hue and intensity required. For
example, in sugar confectionery and baked goods high cooking

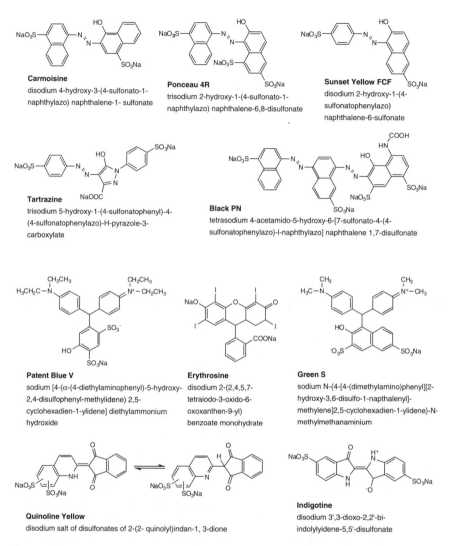

Figure 6.14 The structure and chemical names of a selection of synthetic food colours. The names are those currently used in official EU documentation.

temperatures may pose problems. In soft drinks, one may have to contend with light fastness (Food Green S, erythrosine, and Indigo Carmine are particularly vulnerable), and the presence of sulfur dioxide, ascorbic acid or low pH or may similarly cause difficulty.

Probably the widest use of food colours is in nominally fruit-containing products, including desserts, confectionery and ice-cream. Canned peas are coloured with a mixture of Food Green S and tartrazine. The colours

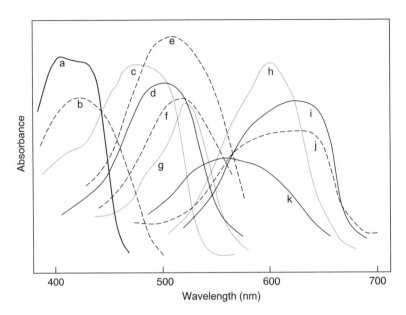

Figure 6.15 The absorption spectra of the synthetic food colours (at pH 4.2 in a
citrate buffer). a- Quinoline Yellow; b- Tartrazine; c- Sunset Yellow
FCF; d- Ponceau 4R; e- Carmoisine; f- Amaranth; g- Erythrosine;
h- Indigo Carmine; i- Patent Blue V; j- Food Green S; k- Black PN.

of chocolate products are obtained either with Chocolate Brown HT, or
with mixtures of Sunset Yellow and amaranth or Food Green S and
tartrazine. Canned meats can be coloured with erythrosine, but products
such as sausages or meat pastes which are likely to be exposed to light
were at one time coloured with Red 2G, but this was banned in Europe in
2007. Some products, particularly some types of confectionery, contain
very little free water in which the dyestuff can be dissolved. In these cases
insoluble lakes are used in which the colour has been adsorbed on a
hydrated alumina substrate. The powdered lake is milled down to a
particle size around 1 µm to avoid speckling. The intensity of colour
obtained using a given amount of a lake colour increases as the particle
size is reduced.

Current regulations in Britain forbid the addition of any colouring
matter to raw or unprocessed foods such as meat, fish and poultry, fruit
and vegetables, tea and coffee (including "instant" products), or wine,
milk or honey. The use of added colours in foods for babies and infants
is also prohibited.

In general, synthetic food colours such as the azo dyes are added
at a rate of 20–100 mg kg^{-1}. Rather higher levels of caramels (1000–
5000 mg kg^{-1}) are required to achieve a useful degree of colouring. It has

been calculated that in the UK there is a total annual consumption of some 9500 tons of food colours, of which all but 500 tons are caramels, which happily corresponds to a daily consumption rate for the various colours of less than 10%, and usually less than 1%, of the "Acceptable Daily Intake" figures now being established by toxicologists.

INORGANIC FOOD COLORANTS

There is one small group of food colours that do not fit into the classification used in this chapter. Listed in Table 6.5 are the inorganic colouring matters occasionally found in food products.

RESTRICTIONS ON THE USE OF COLOURS IN FOODSTUFFS

While so-called natural colouring matters have obvious attractions to consumers with concerns about the safety of the synthetic food colours, it should be noted that in the EU there are a considerable number of foods in which the only colours permitted are the "natural" type. These are listed in Table 6.6. In this context it is usual to include the caramels with the natural colours.

There are also restrictions on the range of foods in which some synthetic colours may be used, listed in Table 6.7.

The ever-increasing costs of safety testing of new food additives are making the appearance of new food colours extremely unlikely. It must be remembered that it is not only the food colour itself that must be shown not to cause death or disease. Any impurities that arise during its manufacture, and breakdown products which might be formed during food processing, cooking or digestion all have to be considered.

Over the last few years much concern has been expressed in the media, and by sections of the medical profession, about the safety of food

Table 6.5 Inorganic food colours.

Name	EC No.	
Carbon black	E153	*Obtained by limited combustion of vegetable material*
Titanium dioxide	E171	
Iron oxides and hydroxides	E172	
Aluminium	E173	*Only permitted for the surface decoration of sugar confectionery*
Silver	E174	
Gold	E175	

Table 6.6 Foods restricted in the EC to the use of natural colours. Note that neither blue nor green colours may be added to blue cheeses such as Stilton, Roquefort or Gorgonzola.

Foodstuffs	Permitted colours
Beer, cider, vinegar and spirits	Caramels (E150a–d)
Butter	Carotenes (E160a)
Margarine, *etc.*	Carotenes (E160a), annatto (E160b), curcumin (E100)
Cheese	As margarine, plus paprika (E160c) and the chlorophylls (E140, E141) for Sage Derby, cochineal (E120) for red marbled cheeses
Pickled vegetables	Chlorophylls, caramels, carotenes, betanin (E162), anthocyanins (E163)
Extruded, puffed and/or fruit-flavoured breakfast cereals	Caramels, carotenoids (E160a-c), betanin (E162), anthocyanins (E163), cochineal, betanin
Many types of sausages and pâtés	Carotenoids (E160a,c), curcumin, cochineal, caramels, betanin

Table 6.7 Synthetic colours restricted in the EU to particular foods.

Colour	Foods
Amaranth (E123)	Aperitif wines (vermouth), fish roe
Erythrosine (E127)	Cocktail and candied cherries
Brown FK (E154)	Kippers

additives, in particular synthetic food colouring matters. The stringency of the toxicity testing procedures now applied in Europe leads to the conclusion that the synthetic food colours permitted are at least as safe as any of the "natural" components of our diet. (The author is only too well aware of the diversity of interpretations that can be placed on this adjective when food is being discussed.) Food testing procedures are often looking primarily for evidence of carcinogenicity, and one question that conventional testing does not fully address is that of physical intolerance and allergy. Among the food colours, tartrazine is widely regarded as the most suspect, and there is some evidence that between 0.01 and 0.1% of the population may show intolerance towards this dye. Intolerance usually manifests itself as one or more of the symptoms of allergic reactions such as eczema and asthma (*see* page 408), and is most often encountered in those who already show a reaction to certain other groups of chemicals, notably the salicylates (including aspirin) and benzoates.

The possibility of food colours being involved in disorders such as hyperactivity in children has been given a great deal of attention by the popular press, but until recently has tended to be dismissed by most medical authorities. However, studies carried out at the University of Southampton have provided new evidence for the involvement of at least some food colours in the incidence of hyperactivity (correctly termed Attention Deficit Hyperactivity Disorder, ADHD) in susceptible children. The double-blind study used two different mixtures of food colours plus the preservative sodium benzoate in a fruit drink vehicle, along with a placebo. The trial was so designed as not to allow the participants to identify which particular additive(s) might be responsible for the effects observed. The colours included in the study were Sunset Yellow, tartrazine, carmoisine, Allura Red (all azo dyes) and Quinoline Yellow, a choice designed to reflect likely consumption by children. There were marked differences in the responses of individual children, some showing no behavioural effect at all but others showing definite adverse effects. However, the average effects in both groups of children were significant and judged to be worthy of further study. It remains to be seen whether these results will lead to legislation on the use of these additives, or merely advice to concerned parents.

THE MOLECULAR BASIS OF COLOUR

Many readers will be happy to accept that certain substances are coloured, without wishing to understand what it is about the molecular structure of such substances that imparts this particular property. Such readers can ignore the next few pages with a clear conscience. Others, however, will want to know a little more, possibly in order to understand why the colours of some food materials are what they are, and how they are subject to change.

Materials or objects appear coloured because they absorb some of the incident light that shines on or through them. This means that light of certain wavelengths is absorbed more than others. The light which subsequently reaches our eyes, having been reflected or transmitted, is therefore richer in light from some parts of the visible spectrum than from others. To appreciate the molecular basis of colour we need to understand firstly how molecules absorb light energy, and secondly what determines the wavelengths, or the colour, of the light they absorb.

Figure 6.16 shows the relationship between the wavelength of light and its colour. Although for most purposes physicists and chemists regard light as having the properties of waves, it can also be thought of as a stream of energy containing particles. These particles are called

Colour	Wavelength (nm)	Complementary colour
Violet	400	
		Yellow
Blue	450	
		Orange
Green	500	
		Red
Yellow	550	
		Violet
Orange	600	
		Blue
Red	650	
		Green
	700	

Figure 6.16 The visible spectrum. Substances that absorb light of the wavelength in the left-hand column have the colour shown in the right-hand column. For example, the yellow carotenoid pigments strongly absorb light in the purple/blue region (420–480 nm; see Figure 6.6). We describe the response of the human eye to the mixture of all the other visible wavelengths that the carotenoids do not absorb (480–800 nm) as yellow or orange.

"quanta". The amount of energy possessed by a quantum is related to the wavelength of the light. The shorter the wavelength, *i.e.* the nearer the violet end of the visible spectrum, the more energetic the quantum.

The energy in the form of light which a coloured molecule absorbs cannot simply be soaked up and vaguely converted into heat. Each single quantum must be utilised individually to provide exactly the right amount of energy to facilitate a particular change (or transition) in the molecule. The quantum will only be absorbed if its energy content matches, is neither too large nor too small, the energy demand of a transition that the molecule can undergo. The transitions that match the range of energies associated with visible light quanta involve the electrons that form the bonds between the atoms of the molecules of the absorbing substance. Of the many theoretically possible types of transition only those associated with the π-electrons of conjugated double bonds are common amongst the coloured compounds in food.

A single bond between two carbon atoms consists of one pair of shared electrons which form one σ-bond. The four bonding electrons comprising a double bond occur as one σ-bond and one π–bond. In the normal or ground state of σ- or π-bonds the two electrons are regarded as spinning in opposite directions; shown as (↑↓). However, if an appropriate quantum of energy is absorbed by the molecule one of the bonds may undergo a transition to an excited higher energy state in which both electrons are spinning in the same direction ($\uparrow\uparrow$). Such transitions are known as σ to σ^* and π to π^*. The σ to σ^* transitions require quanta of much higher energy levels than those of visible or ultraviolet light, and are therefore irrelevant to the question of colour.

However, π to π^* transitions require less energetic quanta and are relevant. Once an electron has undergone transitions like these it normally returns immediately to its original state, the energy it has absorbed being released as heat. In some molecules the return of the electron to its original state can lead to the release of a quantum with sufficient energy to appear as light, although obviously at a longer wavelength than that originally absorbed. This is the phenomenon of fluorescence.

The wavelength at which a substance absorbs most radiation, the position of the highest peak on the absorption spectrum, is referred to as the λ_{max} value.

When the bonds linking a chain of carbon atoms are alternately single and double:

$$-C = C - C = C - C = C-$$

the double bonds are described as *conjugated*. All the coloured organic substances described in this chapter have an abundance of conjugated double bonds. The π-electrons of conjugated systems interact together and in doing so particular pairs of π-electrons lose their association with a particular carbon–carbon bond. In this state they are said to be *delocalised* and require much less energy to undergo a π to π^* transition. In fact, the more double bonds are conjugated together in a molecule the less energy is required for the π to π^* transition and therefore the radiant energy absorbed by the process is of increasing wavelength. This is clearly illustrated by the sequence of carotenoid structures shown in Figure 6.17. Only when a carotenoid has its full complement of conjugated double bonds, as in lycopene, is visible light absorbed. Of course the exact λ_{max} values we observe, and the precise details of the absorption spectrum, are affected by other structural elements in the molecule such as the involvement of the double bonds in rings and also by interactions with external agencies such as solvents.

The need for at least seven conjugated double bonds for a substance to absorb visible light is also nicely demonstrated by the anthocyanins (*see* Figure 6.8). The flavylium cation (red) has a total of eight, but when a rise in pH converts it into the carbinol pseudo-base the conjugation in the middle of the system is lost. With at most four double bonds in any one conjugated system the molecule is colourless. The bleaching action

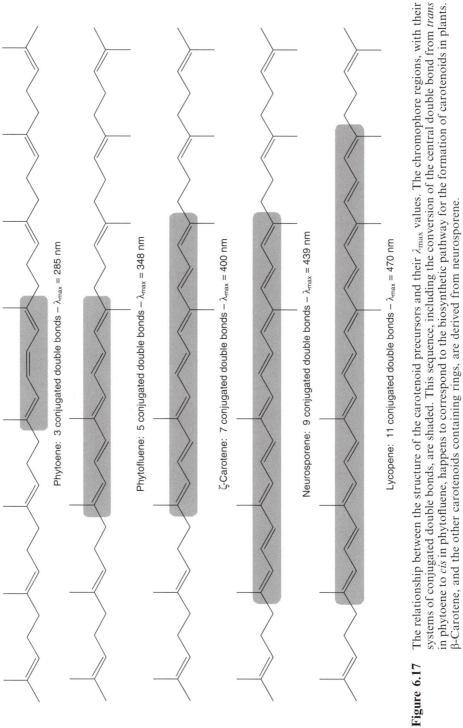

Figure 6.17 The relationship between the structure of the carotenoid precursors and their λ_{max} values. The chromophore regions, with their systems of conjugated double bonds, are shaded. This sequence, including the conversion of the central double bond from *trans* in phytoene to *cis* in phytofluene, happens to correspond to the biosynthetic pathway for the formation of carotenoids in plants. β-Carotene, and the other carotenoids containing rings, are derived from neurosporene.

Phytoene: 3 conjugated double bonds – λ_{max} = 285 nm

Phytofluene: 5 conjugated double bonds – λ_{max} = 348 nm

ζ-Carotene: 7 conjugated double bonds – λ_{max} = 400 nm

Neurosporene: 9 conjugated double bonds – λ_{max} = 439 nm

Lycopene: 11 conjugated double bonds – λ_{max} = 470 nm of sulfur dioxide (6.14) has a similar basis. In contrast, the ionisation of the carboxyl groups in betanidin with change in pH causes no change in its eight conjugated double bonds and there is no colour shift (*see* Figure 6.11). The term "chromophore" is used to denote the region of a molecule where delocalised electrons are responsible for light absorption. For example, the porphyrin ring system is the chromophore in chlorophyll.

Some food pigments contain metal atoms, for instance the chlorophylls (*see* page 215), myoglobin and its derivatives (*see* page 167), and vitamin B_{12} (*see* page 327). The structures of all of these have in common that they contain elaborate systems of fused rings providing large numbers of conjugated double bonds, and hence delocalised electrons. However, the electrons from the metal atom also become involved, with a considerable effect on the colour. For example, the loss of the magnesium from green chlorophyll gives us brown pheophytin. The colour of myoglobin in meat depends on whether the iron is in the ferric or ferrous state and which other substances are also co-ordinately bonded to the iron atom (*see* page 193).

SPECIAL TOPICS

1 Flavonoids, Tannins and Health

The anthocyanins in fruit and the catechins in tea are important members of the flavonoid group of plant components. They are all based on the flavan nucleus (6.9) but differ in state of oxidation and the substituents in the central oxygen-containing ring. Tannins are polymerised flavonoids. Flavonoids are difficult to categorise in terms of their function in food (and therefore their placement in this book) since in different situations they contribute colour, flavour or "mouth feel".[vi] Significant degrees of antioxidant and anti-tumour activity and the converse, carcinogenicity, have also been attributed to flavonoids and it is their effect on human health which has brought them to wider attention.

As shown in Figure 6.18, flavonoids can be grouped into classes which differ in the structure of the central ring. The A ring almost invariably carries hydroxyl groups at positions 5 and 7, and these hydroxyls frequently carry sugar residues. Similarly, the B ring carries one, two or three hydroxyl groups. When double bonds in the central ring provide a conjugated link between the aromatic systems of the A and B rings,

---

[vi] What scientists refer to as "mouth feel" is popularly described as "body" when applied to drinks.

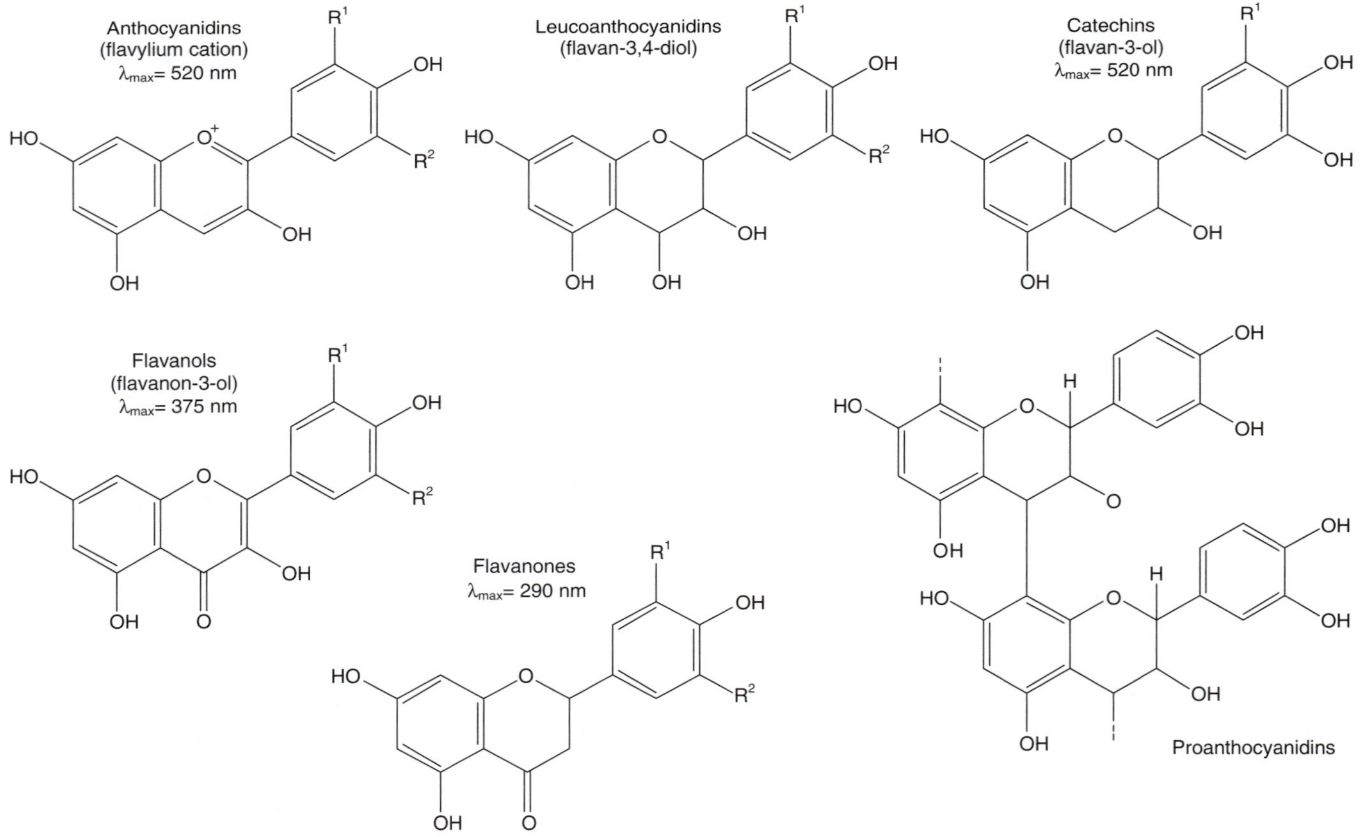

Figure 6.18 The basic structure of the major groups of flavonoids. Where available, typical values for the absorption maxima are shown. Except in the anthocyanins, where further details are shown in Figure 6.8, R^1 and R^2 are either H or OH. The proanthocyanidin illustrated contains two flavin-3-ol units, but proanthocyanidins commonly include flavin-3,4-diol and anthocyanidin units.

flavonoids absorb visible light, as indicated by the approximate absorption maxima (λ_{max}) shown in Figure 6.8.

The definition of tannins is imprecise and usually refers to poly-phenolic substances, *i.e.* substances containing aromatic rings carrying multiple hydroxyl groups, which can form links with proteins. These links usually result in the protein being rendered insoluble, as in the tanning of leather. Most often the polyphenolics concerned are flavo-noids, especially where two or more flavonoids are condensed together *via* their 4 and 8 positions. These oligomeric flavonoids are also known as "condensed tannins", or more precisely, proanthocyanidins. Most frequently proanthocyanidins are found to contain flavin-3,4-diols (leucoanthocyanidins) but flavin-3-ols (catechins), and anthocyanidins, occur commonly. The oligomers may contain as many as ten flavonoid units and have molecular weights in the range 500–3000.

In recent years the possible beneficial effect that flavonoids, including tannins, might contribute towards the health of consumers has received considerable attention, focusing particularly on red wine and tea. A complication is that both adverse and beneficial effects have been reported, and epidemiological studies tend to give less encouraging results than laboratory or clinical experiments.

Also on the negative side, the habit of chewing betel nuts, which have a very high condensed tannin content, has been implicated in the high incidence of mouth and oesophageal cancer in the Far East. The risk of oesophageal cancer in the Far East has also been linked to tea drinking, but in this case it is now accepted that the culprit is not the tea itself but cell damage caused by routinely swallowing the tea very hot ($> 55\,^{\circ}\text{C}$). Different studies of stomach cancer have shown contradictory results, both increased and decreased risks being linked to tea consumption. Various flavonoids have been shown to possess anti-tumour activity in laboratory experiments, but in the absence of unequivocal epidemiolo-gical data useful conclusions cannot be drawn.

More positive effects of dietary flavonoids are on arterial disease. The initial interest in their possible benefits arose from the observation that the French and Italians eat diets containing a lot of fat but suffer far less heart disease than expected. Reduced mortality rate from coronary heart disease (CHD) has been clearly linked to the intake of wine, a good source of flavonoids (1–$3\,\text{g}\,\text{dm}^{-3}$ in red, $0.2\,\text{g}\,\text{dm}^{-3}$ in white). However, total mortality rate does not show the same association, presumably due to the adverse effect of other factors such as high alcohol intake. The CHD rate is also low in the Far East, where consumption of green tea is high.

Two potential mechanisms have been implicated in these beneficial effects. Firstly flavonoids, including the quercetin (6.29) found in wine, tea, onions and apples, and the catechins in tea, are recognised as effective antioxidants and many laboratory experiments and some clinical studies indicate that they can be effective inhibitors of the oxidation of low-density lipoproteins (LDL), an important adverse factor in the development of atherosclerosis. Unfortunately there is a considerable gulf between demonstrating antioxidant activity *in vitro* in laboratory experiments and showing the same substances to be capable of reaching the relevant human tissues—and exhibiting antioxidant activity once they arrive. It is unfortunate that many of those who advocate the consumption of expensive and/or exotic fruit on the basis of their apparent antioxidant content are either unaware of or untroubled by this gulf. The second potential mechanism is the inhibition of platelet aggregation by wine flavonoids, another important stage in CHD. Quercetin is particularly effective, but the catechins are not. Resveratrol (*trans*-3,5,4′-trihydroxystilbene, 6.30), a non-flavonoid present in wine ($\sim 4\,\mathrm{mg\,dm^{-3}}$ in red, $\sim 0.7\,\mathrm{mg\,dm^{-3}}$ in white), has similar activity and has attracted plenty of attention recently.

2 Colour Measurement

The measurement of colour is a major issue in many fields of food science, and strictly speaking is a subject for physicists rather than chemists. However, in the real world it is almost invariably the analytical chemists who are required to cope with this aspect of food science. The aspect of colour being measured depends on the circumstances. Very often we need to be able to express the hue of a food product, ingredient or raw material in quantitative terms. To describe a tomato as red may be adequate for some purposes but expressing the concise difference in colour between a tomato and a strawberry to someone who has never seen either fruit would place impossible demands on even our greatest poets. In addition to the hue, we may find that the intensity of the colour is expressed using terms such as "lightness" or "darkness", for example in matters of fruit juice quality.

Chemists confronted by this problem would normally reach for the spectrophotometer to obtain an absorption spectrum, like those shown in Figures 6.2 and 6.6. This approach fails because food materials are often solids, or more or less opaque liquids. This difficulty can be overcome by adapting the instrument to record light reflected by the object rather than that transmitted through it, giving a reflectance spectrum. However, this does not solve the real problem with absorption or reflectance spectra, which is that they present *too much* information.

The human eye uses the cone cells of the retina to perceive the colour of an object. There are three different types of cone cells, each responding best to light of different wavelengths, 445, 535, and 570 nm, respectively. There is considerable overlap between the response of the different cone types. This means that light of a given wavelength, or mixture of wavelengths, will stimulate at least two and possibly all three types of cone cells. Our perception of the colour of light is based on the ratio between the three responses, combined with the contribution of the rod cells. The rods are some 10 000 times more sensitive than the cones and have peak sensitivity at 505 nm.

It should therefore be possible to describe any colour perceived by the human eye simply in terms of such ratios. In fact, the wavelengths used in colour definition using scientific instruments are more evenly distributed across the visible spectrum and correspond to the primary colours, red, green and blue.

It is well known that an appropriate mixture of red, green and blue lights will appear white, and it might be assumed that any colour could be similarly achieved. However, this turns out to be not quite true. In 1931 the CIE (Commission International de l'Eclairage) introduced the concept of *theoretical* primary colours, known respectively as X, Y and Z. Although no real lamps could produce them, they were able, *theoretically,* to match any shade. Their spectra are shown in Figure 6.19. The red, green and blue curves are labelled X, Y and Z, respectively. An important characteristic of the Y curve is that it matches the response of the eye to the total amount of light energy, in other words it represents the response of the rod cells of the retina.

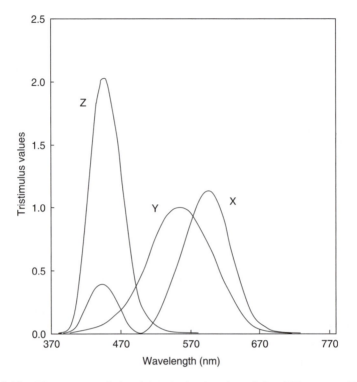

Figure 6.19   The spectra of the theoretical primaries of the CIE system. The subsidiary peak on the X curve results from the mathematical ingenuity used to ensure that all shades can theoretically be matched, as explained by Billmeyer and Saltzmann (*see* Further Reading).

The amounts of the three theoretical primaries required to produce a given colour can be calculated from its spectrum and the curves in Figure 6.18, and are known as its *tristimulus* values. The ratios between these are given by:

$$x = X/(X+Y+Z),\ y = Y/(X+Y+Z)\ \text{and}$$
$$z = Z/(X+Y+Z)$$

Since $x+y+z=1$, any colour can actually be described by just two of them, for example, by x and y.

Figure 6.20 shows the famous horseshoe-shaped CIE chromaticity diagram, in which x is plotted against y. On this plot the quantities x and y for a colour are called its chromaticity coordinates. The theoretical red primary is located at $x=1$, $y=0$, blue at $x=0$, $y=0$, and green at $x=0$, $y=1$. The outer curved line shows the positions of monochromatic

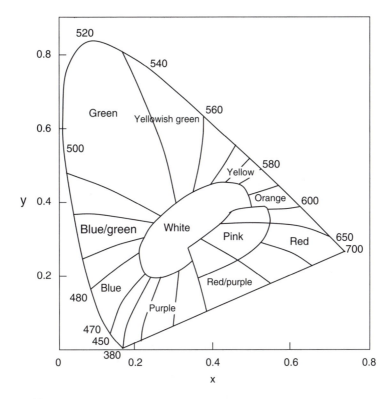

Figure 6.20 The CIE chromaticity diagram. The colour names shown on the diagram describe the colour perceived by a standard observer looking at a coloured surface illuminated by a standard illuminant (light source) having the theoretical spectral characteristics of ordinary daylight. A rather non-picturesque approach is usually adopted towards the naming of intermediate colours, largely dependent on the suffix "ish". All shades of grey are merely more or less bright white.

colours at different positions in the spectrum from 380 to 770 nm. The straight line closing the horseshoe represents the various purplish shades resulting from mixtures of red and blue. Towards the centre of the horseshoe the colours approach white.

The CIE system is linked closely to theoretical concepts of colour perception but has a number of practical drawbacks. One of these is the status of black. To handle this properly the diagram would have to be three-dimensional, with values for the tristimulus value, Y, expressing the total amount of light, plotted perpendicular to the plane of the paper. Another problem is that our perception of degrees of difference between colours is not uniform over the whole diagram. Both problems are dealt with better by the Hunter and Munsell systems.

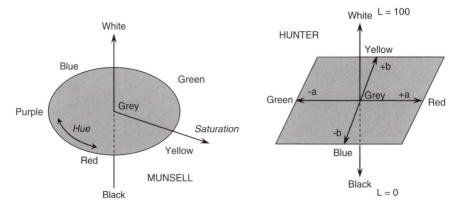

Figure 6.21   The Munsell and Hunter colour spaces. The mathematical relationships between these and the CIE system are well established, and the parameters of a colour determined in one system can be mathematically related to those in the others.

The Hunter *L, a, b* system is based on the concept of a colour space, with the colour defined by three coordinates. The vertical coordinate (*L*) runs from *L*=0 (black), *via* grey to *L*=100 (white). The horizontal coordinate (*a*) runs from −*a* (green) to +*a* (red). The other horizontal coordinate (*b*) runs from −*b* (blue) to +*b* (yellow). Purple and orange can be visualised as being located at +*a* −*b* and +*a* +*b*, respectively, as shown in Figure 6.21.

The Munsell system, also shown in Figure 6.21, has a similar structure but it describes colour in terms of the three attributes, hue, value and chroma. Five basic hues are distributed around the circumference of a circle. The vertical axis of the circle defines the value, or *lightness*, from black below the circle, through grey at the centre, to white above the circle. The distance from the centre to the circumference is given a figure for the chroma, or *saturation*. It is most important to remember that it is extremely unlikely that the colours of two real objects would differ in only one of these parameters; in most cases all three would differ. Users of modern computers with colour monitors and printers will have encountered some of the practical issues associated with these colour systems.

Armed with the theory, one can then approach the actual task of recording the colour of a foodstuff. One simple approach is by visual matching against a standard colour sample. Although apparently simple, this approach still requires a standardised light source. The outdoors sky, incandescent (tungsten) bulbs, and fluorescent tubes may all nominally give "daylight illumination" but they differ considerably in their distribution of energy across the spectrum. Two samples may appear identical under a fluorescent lamp but quite different in a different type of illumination. This phenomenon is known as *metamerism*. In addition, visual assessment is affected by the colour sensitivity of the observer, including his age, and whether he has "normal" colour vision. A further factor is the difficulty in describing small visual colour differences.

The instrumental alternatives are colorimeters and spectrophotometers. A colorimeter uses filters to generate three light beams whose spectra match as closely as possible the CIE curves. Photocells measuring the quantity of light reflected (or transmitted) by the sample will then directly give the tristimulus values. The inevitable inclusion of the microcomputer in modern instruments facilitates instant conversion to Hunter or Munsell values. More sophisticated spectrophotometers record the absorbance or transmittance of light at intervals, or even continuously, throughout the visible spectrum and then compute the tristimulus values.

FURTHER READING

Flavonoids: Chemistry, Biochemistry and Applications, ed. Ø. M. Andersen and K. R. Markham, CRC Press, Boca Raton, Florida, 2006.

S. Carmen, *Food Colorants: Chemical and Functional Properties*, CRC Press, Boca Raton, Florida, 2007.

Food Colours, ed. V. Emerton, Blackwell, London, 2007.

F. Delgado-Vargus and O. Paredes-Lopez, *Natural Colorants for Food and Nutraceutical Uses*, CRC Press, Boca Raton, Florida, 2002.

Anthocyanins as Food Colors, ed. P. Markakis, Academic Press, New York, 1982.

F. W. Billmeyer and M. Saltzmann, *Principles of Color Technology*, Wiley, New York, 2nd edn, 1981.

RECENT REVIEWS

A. Schieber and R. Carle, *Occurrence of Carotenoid cis-Isomers in Food: Technological, Analytical and Nutritional Implications; Trends Food Sci. Technol.*, 2005, **16**, p. 416.

M. M. Calvo, *Lutein: A Valuable Ingredient of Fruit and Vegetables; Crit. Rev. Food Sci. Nutr.*, 2005, **45**, p. 671.

F. C. Stintzing and R. Carle, *Functional Properties of Anthocyanins and Betalains in Plants, Food, and in Human Nutrition; Trends Food Sci. Technol.*, 2004, **15**, p. 19.

F. J. Francis, *Colorants,* in *Food Chemical Safety*, ed. D. H. Watson, Woodhead, Cambridge, 2002.

H. Wang, G. J. Provan and K. Helliwell, *Tea Flavonoids: Their Functions, Utilisation and Analysis; Trends Food Sci. Technol.*, 2000, **11**, p. 152.

J. V. Higdon and B. Frei, *Tea Catechins and Polyphenols: Health Effects, Metabolism, and Antioxidant Functions; Crit. Rev. Food Sci. Nutr.*, 2003, **43**, p. 89.

B. Halliwell, J. Rafter and A. Jenner, *Health Promotion by Flavonoids, Tocopherols, Tocotrienols, and Other Phenols: Direct or Indirect Effects? Antioxidant or Not? Am. J. Clin. Nutr.*, 2005, **81**, p. 268S.

B. Joe, M. Vijaykumar and B. R. Lokesh, *Biological Properties of Curcumin: Cellular and Molecular Mechanisms of Action; Crit. Rev. Food Sci. Nutr.*, 2004, **44**, p. 97.

CHAPTER 7
Flavours

However much nutritionists and "health food" enthusiasts may wish otherwise, it is the flavour and appearance of food rather than its vitamins and fibre that win compliments at the dinner table. This attitude on the part of the diners is not as short-sighted as it might at first appear. The sense organs concerned in the detection of taste and odour, collectively referred to as flavour, have evolved to perform an essential function—to establish what is and what is not suitable as food. As a general rule palatability can be equated with nutritional value, and most of the exceptions are related to excess. For example, to the genuinely hungry sweetness is a clear indicator of the presence of energy-yielding sugars, whereas to the mother of a well-fed European child sweetness implies dental caries and obesity. Conversely, their bitterness and astringency will prevent us from eating many poisonous plants, but these flavours are valued elements in the flavour of beer and tea.

We usually regard taste as a property of liquids or solids, and gases in solution, which can be detected in the mouth, not only by receptor cells in the taste buds of the tongue but elsewhere in the oral cavity. Odour (or aroma, or smell) is similarly regarded as a property of volatile substances detected by the receptor cells of the olfactory systems of the nose. Very few flavours, and even fewer complete foodstuffs, allow a clear distinction to be made between odour and taste.

For the food chemist the main interest is to identify the substances responsible for particular elements of flavour. One obvious motive for such work is that it enhances the possibility of natural flavours

Food: The Chemistry of its Components, Fifth edition
By T.P. Coultate
© T.P. Coultate, 2009
Published by the Royal Society of Chemistry, www.rsc.org

being simulated in processed foods. Related to this is identification of the features of chemical structure which lead to a particular response from our sense organs. This line of enquiry was hinted at many years ago by the Greek philosopher Theophrastus (*c*. 372–287 BC) who declared:

τὸν δὲ πικρὸν ἐκ μικρῶν καὶ περιφερῶν τὴν περιφέρειαν εἰληχότα

καὶ καμπὰς ἔχουαν. διὸ καὶ γλίσχρον καὶ κολλῶδη. ἁλμυρὸν δὲ

τὸν ἐκ μεγάλων καὶ οὐ περιφερῶν, ἀλλ᾿ ἐπ᾿ ἐνίων μὲν σκαληνῷ ...[i]

The research chemist faces a number of special difficulties in investigating flavour. The first of these is that there is no physical or chemical probe available for detection of the specific substances of interest. There is no flavour equivalent of the spectrophotometer, which quantifies the light-absorbing properties of the substances we perceive as colour. Here the success of an analytical procedure may depend on the skill of a trained taste panel or the nose of an experimenter. Another complication is that a food flavour is rarely due to a single substance. It is commonly found that when the mixture of substances involved in a particular flavour has been elucidated and the components reassembled in the correct proportions, the flavour is not the same as the original. This is usually because certain substances present in exceedingly small proportions, without obvious flavour of their own and overlooked by the analyst, can have a great influence on the overall flavour. These difficulties are not likely to be underestimated if it is realised just how sensitive the nose can be. For example, vanillin (7.1), the essential element of vanilla flavour, can be detected by most individuals at a level of 0.1 ppm.

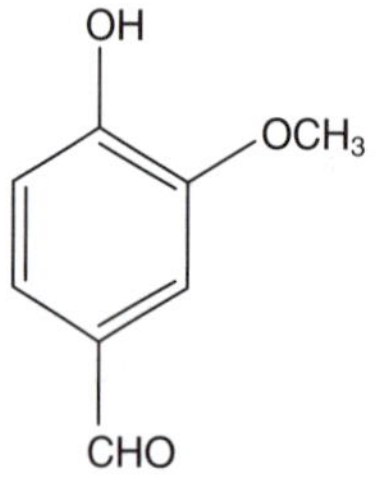

(7.1) 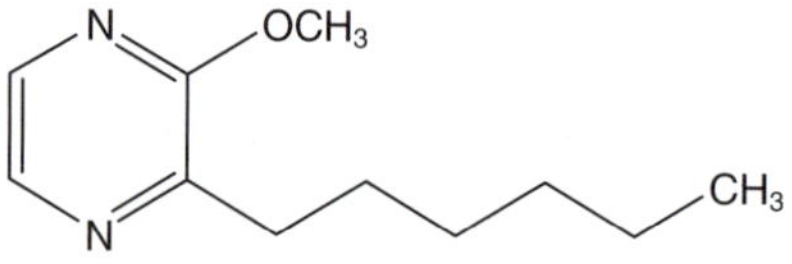   (7.2)

---

[i] Bitter taste is therefore caused by small, smooth, rounded atoms, whose circumference is actually sinuous; therefore it is both sticky and viscous. Salt taste is caused by large rounded atoms, but in some cases jagged ones ...”

There are many other substances to which we are far more sensitive than this. For example, the odour threshold in air (*i.e.* the lowest concentration at which most humans can detect a substance) of 2-methoxy-3-hexylpyrazine (7.2) is reported to be 1×10^{-6} ppm, which is 1 part in 1 000 000 000 000! It is interesting that as the length of the pyrazine side-chain shortens we become progressively less sensitive to the odour. The corresponding figures for the 2-methoxy-3-ethyl-, 2-methoxy-3-methyl, and 2-methoxypyrazines are 4×10^{-4}, 4×10^{-3}, and 7×10^{-1} ppm, respectively. The earthy smell of geosmin (6.16), mentioned in the previous chapter in connection with beetroot, is reputed to be detected in air at 0.5×10^{-6} ppm.

Gas chromatography (GC) has proved to be the most valuable analytical technique available to the flavour chemist. The most sophisticated modern instruments are capable of detecting as little as 10^{-14} g of a component in a mixture, and 10^{-10} g is within the capability of many commercially available gas chromatographs. The peaks on a gas chromatogram can tentatively be identified by comparison with the elution characteristics of known substances. Firm identification requires the isolation of the substances emerging from the chromatography column (for example by condensation at very low temperature) and their identity confirmed by infrared spectroscopy, nuclear magnetic resonance or mass spectrometry. Instruments are now available in which the link between GC and mass spectrometry is handled automatically and the substance immediately identified by comparison against a library of known spectra. Unfortunately, the vast amount of information that sophisticated GC makes available does not automatically make the food chemist a great deal wiser. For example, in a study of Scotch whisky 313 different volatile compounds were identified, including 32 alcohols and 22 esters, but this information would be of limited interest either to manufacturers or to drinkers.[ii]

Since it is the flavour more than anything else which enables us to distinguish one foodstuff from another, it would be surprising if the total number of different flavours was much smaller than the total number of different foodstuffs. Moreover, the flavour compounds are much harder to classify according to their chemistry than by their contribution to the chemistry of particular foods. However, we can make a reasonable distinction between tastes and odours, and this forms the basis of this chapter. As we shall see, the scale of the subdivisions tends to reflect the intensity of research and the volume of knowledge available, rather than

---

[ii] Not least because the finest Scotch whiskies, especially unblended single malts, may spend as long as twenty or more years maturing between distillation and bottling.

the relative importance of one flavour compared to another, which is an impossible judgement.

TASTE

The flavour sensations detected in the mouth, and particularly on the tongue, are described as tastes. A number of different sensations can be identified. Firstly, there is the classic quartet of saltiness, sweetness, bitterness and sourness mentioned in school biology textbooks and often associated with different parts of the tongue. Then we have three other tastes, astringency, pungency (*i.e.* hotness), and meatiness (also known as *umami*). The concept of seven basic tastes is actually nothing new—the Greek philosopher Aristotle has been credited with this number. However, his list omits meatiness and includes "sandiness", but it is not clear what he meant by this. The substances involved in all these sensations have a number of characteristics in common which distinguish them from those associated with odours. Taste substances are usually polar, water-soluble and non-volatile. Besides their necessary volatility, odour substances are generally far less polar and elicit a much broader range of flavour sensations. Each of these will be considered in turn.

Sweetness

Of the tastes detected by the taste buds of the tongue, sweetness has received by far the most attention from research workers. We usually think of sweetness as being the special characteristic of sugars, and this is certainly a major reason for their incorporation into so many foods. What is frequently overlooked is that most sugars are much less sweet than sucrose, and many are not sweet at all. Furthermore, our pursuit of low-calorie diet foods has increased our awareness of the fact that sweetness is in no way unique to sugars: saccharin has become a household word. The measurement of sweetness is peculiarly problematic. There are no laboratory instruments to perform the task, and no absolute or even arbitrary units of sweetness except perhaps, "One lump or two?" in a teacup. Instead we have to rely on the human tongue and the hope that if we average the findings of a large number of tongues we can arrive at useful data. The data we obtain will still not be absolute, but is expressed relative to some arbitrary standard, usually sucrose. Table 7.1 gives relative sweetness data for a number of sugars and other substances. It can be a revealing exercise for students to devise for themselves procedures for establishing the relative sweetness of, say, glucose relative to sucrose. The variations in the results from individual

Table 7.1 The relative sweetness of sugars and sugar alcohols. As discussed in the text, these values are subjective rather than absolute.

Sugar	*Relative sweetness*		Sugar alcohol	*Relative sweetness*	
	By weight	*By molarity*		*By weight*	*By molarity*
Sucrose	1.00	1.00	Xylitol	1.00	0.44
D-Fructose	1.52	0.80	Maltitol	0.90	0.90
D-Glucose	0.76	0.40	Sorbitol	0.60	0.32
D-Galactose	0.50	0.26	Isomalt	0.55	0.55
Lactose	0.33	0.33	Mannitol	0.50	0.27
Maltose	0.33	0.33	Lactitol	0.37	0.37
Trehalose	0.25	0.25			
D-Ribose	0.15	0.06			

tongues can then be considered and a "class average" obtained to compare with the data in Table 7.1. Any practical work in this field should be carried out in the workrooms of the home economists, not chemistry laboratories!

A vast amount of work has been devoted to identifying the molecular features of sweet-tasting substances which may be responsible for the sensation of sweetness. One approach is simply to survey as many sweet-tasting substances as possible to identify a common structural element. A second approach has been to prepare large numbers of derivatives of sweet-tasting substances in which potentially important groups have been either blocked or modified. The second approach led Shallenberger's group in 1967 to propose a general structure for what they termed the "saporous unit". They suggested the AH,B system shown in Figure 7.1 where A and B represent electronegative atoms (usually oxygen) and AH denotes hydrogen-bonding capability.

It has been shown that for sweetness the distance between the electronegative atom B and the hydrogen atom H on A must be close to 3 Å (*i.e.* 0.3 nm). Not only can it be assumed that the corresponding groups on the surface of a receptor protein (X and YH) of the taste bud epithelial cells are a similar distance apart, but also that if AH and B were any closer they would form an intermolecular hydrogen bond of their own rather than with the receptor protein. A likely candidate for the role of receptor protein (*i.e.* one having the ability to bind substances in proportion to their sweetness) has been isolated from the epithelium of the taste buds of a cow. It is assumed that when a sweet-tasting molecule binds to the receptor protein it undergoes a small change in shape. (Biochemists will recognise this as similar to the changes in regulatory enzymes when ligands bind at allosteric sites.) Various theories have been proposed to describe subsequent events, but the most popular view is that the initial

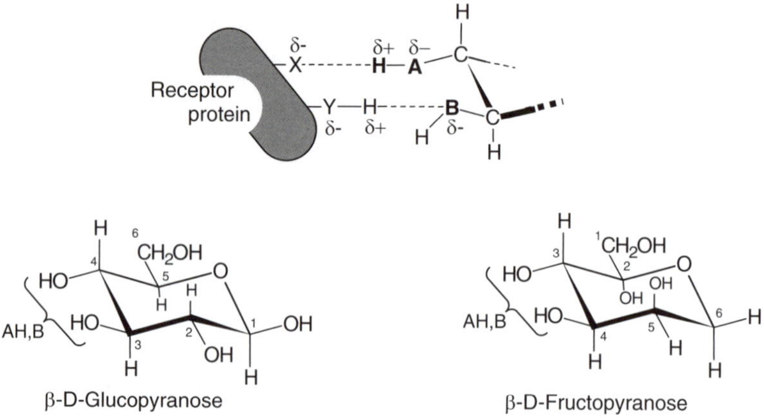

Figure 7.1   Shallenberger's "saporous unit". By analogy with the chromophores of coloured substances (page 255) the term "glycophore" is now commonly used as an alternative to saporous unit. The recognised AH,B elements of glucose and fructose are also shown. Although not immediately obvious, this portrayal of β-D-fructopyranose is entirely compatible with that on page 29 (2.28); it is shown here from an unusual viewpoint.

change in the receptor protein leads to a cascade of changes in neighbouring proteins in the cell membrane, much as one person fidgeting about in a packed train compartment can lead passengers some distance away to adjust their position. Ultimately these effects lead to the cell membrane briefly opening up to the passage of ions such as Ca^{2+} or Na^+ and a nerve impulse being dispatched to the brain.

Studies of numerous sugar derivatives have established that in aldohexoses such as glucose the 3,4-α-glycol structure is the principal AH,B system. The conformation of the sugar ring is of crucial importance to the level of sweetness. For the α-glycol AH,B system to have the correct dimensions, the two hydroxyl groups must be in the skewed (or *gauche*) conformation. If the α-glycol is in the eclipsed conformation, the two hydroxyl groups are sufficiently close for the formation of an intramolecular hydrogen bond, which excludes the possibility of interaction with a receptor protein. In the *anti* conformation the distance between the hydroxyl groups is too great for the correct interaction with the receptor protein. Inspection of the conformation of C1 β-D-glucopyranose (*see* page 28) shows that all the hydroxyl groups attached directly to the ring (at C-1, C-2, C-3 and C-4) are potential saporous units. In contrast, 1C β-D-glucopyranose has all its α-glycol structures in the *anti* conformation, making the distance between the hydroxyl groups too large for binding to the receptor. If D-glucose might be persuaded to take up the 1C conformation we would predict that in this form it would not be sweet.

The presence of an AH,B system of the correct dimensions is clearly not the only factor. For example, the axial C-4 hydroxyl of D-galactose is able to form an intramolecular hydrogen bond with the ring oxygen, which explains why D-galactose is less sweet than D-glucose. The furanose form of β–D-fructose is believed to lack sweetness, which is explained by the ease with which a hydrogen bond can form across the ring between the hydroxyl groups on C-2 and C-6. The observation that the disaccharide, trehalose, is not twice as sweet as D-glucose tells us that although a sugar may have more than one saporous unit, only one of these at a time can actually bind with the receptor. It should not then be surprising that high molecular weight polymers of glucose, such as starch, are completely without taste. The relative sweetness of a range of sugars is shown in Table 7.1.

There are clearly vast numbers of substances containing an AH,B system of the correct dimensions but which are actually tasteless, or even bitter. These include the L-amino acids, in which it is the ionised carboxyl and amino groups which could constitute a well-defined AH,B system. The fact that the corresponding D-amino acids are sweet led to the concept of a third binding site, and with it a three-dimensional receptor site. At the third site, known as γ, the binding depends on the hydrophobic character of both the γ site of the receptor protein and the corresponding position on the saporous unit. Figure 7.2 shows the most probable dimensions of the receptor and how this principle applies to leucine, in which only the D isomer is able to bind.

Sugars are not of course the only substances with a sweet taste, and we are now familiar with saccharine, aspartame, *etc.*, as "non-sugar sweeteners". These have the advantage of a sweet taste without the calorific disadvantage of sugars. In addition a small number of sweet-tasting proteins, notably thaumatin and monellin, have also been identified. The relative sweetness of some of the important non-sugar sweeteners are shown in Figure 7.3. These values are inevitably rather variable, partly because they have to be measured subjectively by human subjects, but also since the sweetness of a substance in solution is not linearly related to its concentration. Thus, if one prepares solutions of two different sweeteners of apparently equal sweetness, diluting both solutions by the same factor will probably not produce two equally sweet solutions.

The AH,B and γ concept has been extended to include many of the non-sugar sweeteners listed in Figure 7.3. The differences in relative sweetness are ascribed to both closeness of fit in geometrical terms and also the strength of hydrogen bonds and hydrophobic interactions at the γ site. Identifying the AH,B and γ sites within a molecule is far from straightforward. For example many authorities still record the AH and B sites of D-fructopyranose to be the C-2 and C-1 hydroxyl groups,

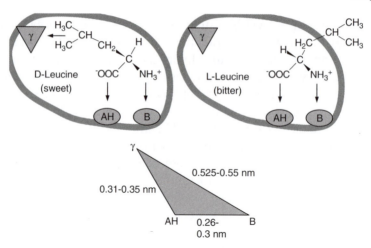

Figure 7.2 The third binding site (γ) proposed by Kier in 1972, as it affects the binding of D- and L-leucine. Opinions vary as to the exact dimensions of the triangle formed by AH,B and γ It is assumed that other elements of the receptor site obstruct the L-leucine side-chain. Glycine, which has a weak sweet taste, has just a single hydrogen atom in place of the $(CH_3)_2CHCH_2$ group in leucine and can therefore gain access to the receptor site even though it does not bind to the γ site.

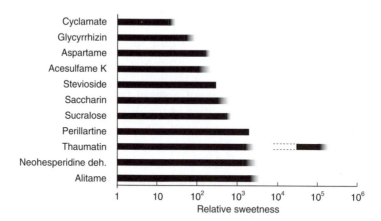

Figure 7.3 The relative sweetness of a range of non-sugar sweeteners. Sweetness is expressed on a logarithmic scale relative to sucrose (= 1). Data are obtained by diluting a solution of the test substance until its sweetness is judged to be equal to that of a dilute sucrose solution, and then relating its concentration to that of the sucrose standard. The blurring of most of the bars in this chart reflects the variations between results from different laboratories. The apparent precision of the data for stevioside and perillartine is merely due to the shortage of published data. For the protein thaumatin the concentration ratios were also calculated on a molarity basis, giving the extremely high relative sweetness values plotted to the right.

despite the evidence provided by Birch in favour of C-3 and C-4, shown in Figure 7.1. The γ site appears to have much less significance in the binding of sugars than its importance in the intensely sweet non-sugar sweeteners, some of whose chemical structures are shown in Figure 7.4.

Sugar derivatives form an important class of sweeteners. Sugar alcohols such as xylitol (2.20), sorbitol (2.21), mannitol (7.3), lactitol (7.4) and maltitol (7.5) are obtained from their parent mono- or disaccharides by hydrogenation. By contrast, isomalt, an increasingly popular member of this group, is made by a novel process involving immobilised cells of the soil bacterium, *Protaminobacter rubrum*. These cells carry an enzyme which is able to convert sucrose to its α-1,6 isomer isomaltulose (7.6), which is then hydrogenated to a mixture of 6-α-D-glucopyranosyl-(1,6)-D-sorbitol (7.7) and the corresponding mannitol derivative, and sold as isomalt:

(7.3) (7.4) (7.5)

(7.6) transglucosidase H_2 (7.7)

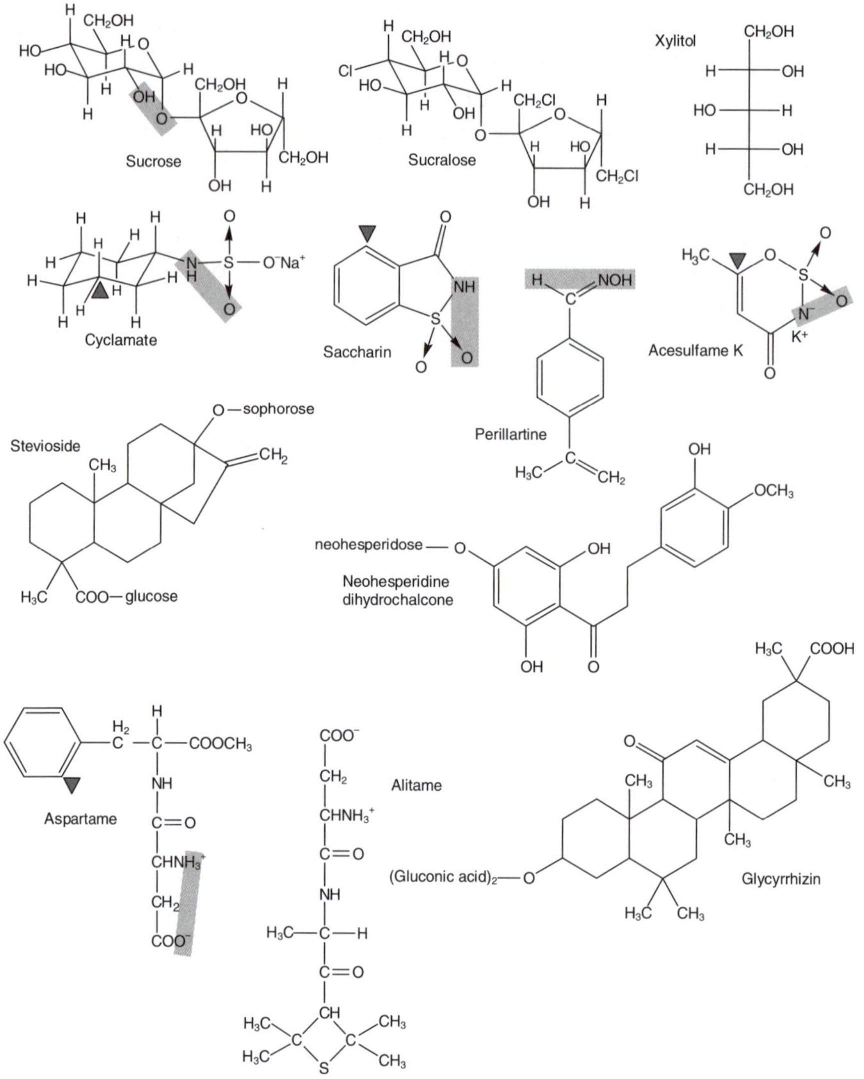

Figure 7.4   The chemical structures of sweeteners, showing the AH,B and γ sites, where these have been assigned with reasonable confidence. The AH and B sites are shown by the grey bars, the γ sites by grey triangles. Neohesperidose is α-L-rhamnose-1,2-β-D-glucose; sophorose is β-D-glucose-1,2-β-D-glucose.

Hydrogenated sugars are gaining importance as bulk low-calorie sweeteners. In general they are a little less sweet than sucrose. The term "bulk" refers to the fact that they can fill the other roles apart from sweetness, which sucrose fulfils in flour and sugar confectionery, but being largely unabsorbed by the small intestine they do not contribute to calorific intake. On the other hand, they also carry the water-binding properties of sugars into the large intestine and have the potential, if sufficient is consumed, to cause osmotic diarrhoea. They are frequently used in the formulation of commercial confectionery products marketed as being suitable for diabetics, although many medical authorities disapprove of this approach to diabetes management.

Non-sugar sweeteners have a much longer history than one might imagine. The attractive sweetness of saccharin was noticed within weeks of its first synthesis in 1878 and it was quickly exploited as a sweetening agent. By the turn of the century some 200 tons were being produced annually by a German company and in 1901 the Monsanto Chemical Company was created in the USA to manufacture saccharin. Shortages of sucrose during both World Wars led to its widespread use, both in the home and in commercial products such as soft drinks. The advantages of saccharin include its stability to processing procedures and the essentially clean bill of health it has received from toxicologists. It is also inexpensive—in terms of sweetening power only 2% the cost of sucrose. To many consumers it offers the advantage of a sweetener that contributes no "calories" either in the human diet or to the bacteria causing dental caries. Its main disadvantage is an unpleasant metallic aftertaste, which leads to it being blended with sucrose to reduce rather eliminate the calorific value of a product.

In more recent years a number of other non-sugar sweeteners have been introduced, some the products of chemical synthesis but others more natural in origin (Table 7.2). Two synthetic sweeteners, sodium cyclamate and acesulfame K (*see* Figure 7.4), share some structural features with saccharin. Both have a superior taste profile to saccharin, lacking its metallic aftertaste, and retain all the other dietary advantages. However, cyclamate fell foul of a single carcinogenicity test in 1969 and was immediately banned in the USA and many other countries. Subsequent biological testing did not support this conclusion, however, and it has been reinstated in most countries around the world, and the ban in the USA is currently being reconsidered.

Like saccharin, the sweet taste of aspartame, L-aspartyl-L-phenylalanine methyl ester (*see* Figure 7.4), was discovered by chance. In the case of aspartame it followed a lapse in laboratory hygiene on the part of a chemist working on a quite different problem. Aspartame has a clean, sucrose-like flavour which has made it extremely popular with soft drink manufacturers. Its main drawback is its instability outside the pH range 3–6, especially at temperatures above 70 °C. Its decomposition follows two distinct pathways: (i) simple hydrolysis to its components, aspartic acid, phenylalanine and methanol, and (ii) methanol loss, followed by cyclisation to a diketopiperazine (7.8).

Table 7.2 Non-sugar sweeteners permitted in Europe. Most of these are subject to restrictions in the levels allowed and the types of food in which they may be used. In particular, none are permitted in infant foods. The legal status of these and other sweeteners varies widely around the world. The Acceptable Daily Intake (ADI) figures, published by the Scientific Committee on Food of the European Union (SCF), are broadly similar to the figures produced by the Joint FAO/WHO Expert Committee on Food (JECFA).

Sweetener	E number	ADI (mg/kg)	Sweetener	E number
Acesulfame K	E950	<9	Sorbitol	E420
Aspartame	E951	40	Mannitol	E421
Cyclamate	E952	<7	Isomalt	E953
Saccharin	E954	<5	Maltitol	E965
Sucralose	E955	<15	Lactitol	E966
Thaumatin	E957		Xylitol	E967
Neohesperidine DHC	E959			

(7.8)

Fortunately the diketopiperazine proved to offer no toxicological problems, but aspartame itself caused some concern. The initial attraction of aspartame was that its natural hydrolysis in the gastrointestinal tract would not result in products whose "naturalness" at reasonable levels could be in doubt. However, the possibility arose that people suffering from the inherited disease phenylketonuria might be at risk, since their capacity to cope with an excess of phenylalanine is seriously impaired. In fact the dietary regime for sufferers from this disease is so exacting that avoiding food and drinks containing aspartame is only a minor tribulation, and the risk to symptomless carriers of the defective gene has been shown to be insignificant. Alitame (*see* Figure 7.4) emerged from the intense research into potentially sweet peptides which followed the discovery of aspartame. It is permitted in some countries around the world and is under consideration for approval in the USA and by the EU.

Sucralose is an intense sweetener derived from sucrose which is becoming increasingly popular in both soft drinks and confectionery as well as for domestic use. As with many other intense sweeteners it was discovered by accident. In 1976 chemists were exploring the possibility of using chlorinated sugar derivatives as intermediates in the synthesis of novel pesticides. Misheard instructions led to what is now known as sucralose being *tasted* rather than *tested*, and its intense sweetness came to light. Although derived from sucrose by selective chlorination, in strictly chemical terms it is a galactose derivative formed by isomerisation of the glucose ring, *see* Figure 7.4. It is now approved for use in foods in many countries, including the USA and the EU. It has the great advantage of being completely stable under all the conditions of temperature and pH likely to be encountered in culinary or food processing situations.

A number of intense naturally occurring sweeteners have attracted attention. Although neohesperidine dihydrochalcone is not known to occur in nature, close chemical relatives have been found to be the source of the strong sweet taste of a number of traditional food plants throughout the world. It is obtained by chemical conversion of naringin (7.9) (source of the bitterness of grapefruit, *Citrus paradisii*) to neohesperidine, followed by hydrogenation to the chalcone. Neohesperidine can also be obtained directly from the peel of the Seville orange (*Citrus aurantium*). It is not normally used on its own in food products since it is somewhat expensive, but it has been found to enhance the potency of many other types of sweetener. One unusual property is that it is able to mask the bitterness of other substances, making it an attractive ingredient for some pharmaceutical products. Glycyrrhizin (*see* Figure 7.4) is the natural sweet component of liquorice root (*Glycyrrhiza glabra*). Extracts containing glycyrrhizin are widely used in the Far East as flavourings and sweeteners for foods and drinks, medicines and tobacco. Nowadays liquorice-based confectionery products seem to be less popular in Britain than they once were. Unfortunately glycyrrhizin is also known to have less welcome biological activities. While its anti-allergenic, anti-inflammatory and expectorant properties are beneficial, its effects on kidney function and blood pressure are certainly not. For this reason it is widely permitted as a flavouring in confectionery and soft drinks only where its total consumption is unlikely to be high, but not as a sweetening agent. The well-known laxative action of liquorice may or may not be attributable to glycyrrhizin, but this factor does ensure that its level of consumption is to some extent self-limiting.

Stevioside (*see* Figure 7.4) is the principal terpenoid glycoside extracted from the leaves of *Stevia rebaudiana*, a member of the sunflower family cultivated in some parts of the Far East and South America. Extracts containing stevioside are approved for food use in some countries, but reservations about its safety (like glycyrrhizin it possesses a wide range of biological activities) mean that it is not permitted in the USA or the EU.

Sweetness is not the prerogative of relatively small molecules. Thaumatin is the best known example of a sweet-tasting protein. It is intensely sweet, especially compared to sucrose on a molar basis (*see* Figure 7.3). It is obtained from the fruit of a West African tree, katemfe (*Thaumatococcus danielli*), and is marketed under the name Talin™. Since it is unusually resistant to thermal denaturation, it is suitable for cooked products, but is totally hydrolysed to amino acids by the human digestive system. It has been approved for use in a number of countries, including the USA and the EU, but is more likely to find application as a flavour enhancer than as a sweetener.

Bitterness

The apparently contrasting nature of bitterness and sweetness have been linked together by the poetically and romantically inclined for centuries. However, this linkage at the emotional level is also supported at the chemical level. For the food scientist the two are often regarded as opposite sides of the same coin. Several types of substance which taste bitter have a close structural relationship to sweet-tasting substances, as we shall see. Inorganic salts are a notable exception; most of these have a very bitter taste and, as discussed in the next section, relatively few taste salty.

Bitterness is often a warning of potential toxicity. Microbial breakdown of food materials frequently results in the generation of bitter substances, and many plant toxins have a bitter taste. Although it may be unwise to generalise, it is significant that the human taste threshold for naturally occurring bitter substances is usually much lower than for naturally occurring sweet-tasting substances. For instance, the threshold in the case of quinine (7.9) is only $25\,\mu\text{mol}\,\text{dm}^{-3}$, compared with $10\,000\,\mu\text{mol}\,\text{dm}^{-3}$ for sucrose.

(7.9)

Bitterness is also a defining characteristic of the alkaloids, whose toxicity is discussed in Chapter 10. The ability to perceive bitterness has almost certainly evolved in order to protect man, and presumably other animals, from the dangers posed by the alkaloids present in many plants.[iii] Alkaloids are basic compounds containing nitrogen in a heterocyclic ring. Although we have little idea of the function of alkaloids in plants, many have profound pharmacological effects on animals, in addition to an extremely bitter taste; nicotine, atropine and emetine are all alkaloids. Thanks to its medicinal use quinine is one of the best known, and it is widely used as a bittering agent in soft drinks such as bitter lemon and tonic water.

Phenolic substances in the form of flavonoids are important sources of bitterness in fruit, particularly in citrus juices. The best known is naringin (7.10), a glycoside of the flavanone naringenin with the disaccharide neohesperidose, which occurs in grapefruit and Seville oranges. Its bitterness is such that it can be detected at a dilution of 1 in 50 000. Another bitter component in citrus juices is limonin (7.11), sometimes formed from tasteless precursors during commercial juice extraction.

α-L-rhamnose-1,2-β-D-glucose

(7.10)

(7.11)

[iii] The significance of the bitter taste capability in human evolution has recently been explored in studies of the geographical and ethnic distribution of genetic variations in the genes coding for one of the key proteins involved (*see* N. Soranzo, *et al., Curr. Biol.*, 2005, **15**, p. 1257). It appears that reduced sensitivity to bitterness may make certain foods rich in cyanogenic glycosides (*see* page 387) more palatable, allowing more to be eaten and thus leading to measurable, but sublethal, concentration levels of cyanide (CN) ions in the bloodstream. It is suspected that these provide a degree of protection against the malarial parasite.

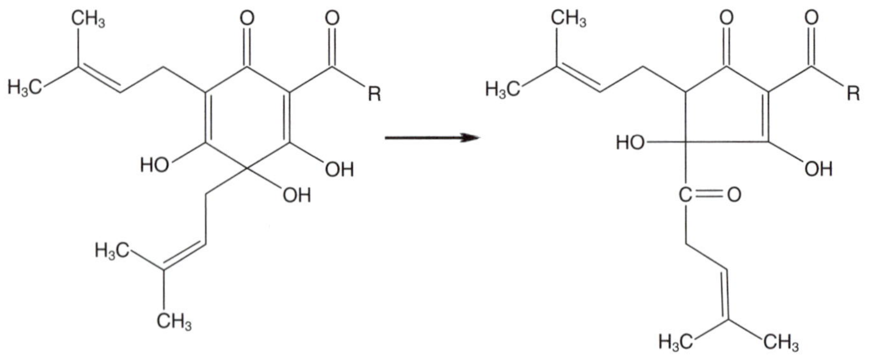

Figure 7.5 Isomerisation of the α-acids in hops. R is $(CH_3)_2CHCH_2$ in humulone, $(CH_3)_2CH$ in cohumulone, and $C_2H_5(CH_3)CH$ in adhumulone.

Bitterness is a sought-after characteristic in many types of beer, particularly those brewed in Britain. Bitterness is achieved by adding hops to the wort, the sugary extract from the malt, before it is boiled and then cooled during the stage preceding the actual fermentation. Hops are the dried flowers of the hop plant, *Humulus lupulus*, and they are rich both in the volatile compounds which give beer its characteristic odour, and in resins which include the bitter substances. The most important bittering agents in beer are the so-called α-acids. As shown in Figure 7.5, these isomerise during the "boil" to give forms which are much more soluble in water and much more bitter. The various varieties of hops contain different proportions of the three α-acids, and the structurally related but less important β-acids.

Until recently the bitter taste of many amino acids and oligopeptides was of only academic interest. However, the need for more efficient use of the proteins in whey, waste blood, *etc.,* has focused attention on the properties of the peptides obtained by partial hydrolysis of these proteins by enzymes or acid. These protein hydrolysates are being developed as highly nutritious food additives, with many useful effects on food texture. Examination of the tastes of the amino acids in protein hydrolysates shows that bitterness is exclusively the property of the hydrophobic L-amino acids valine, leucine, isoleucine, phenylalanine, tyrosine and tryptophan. Whether a peptide is bitter or not depends on the average hydrophobicity of its amino acid residues. A great deal of work on the conformation of peptides in relation to their taste has demonstrated that the structural requirements for bitterness mirror those required for sweetness in sugars and other compounds. There is a similar requirement for a correctly spaced pair of hydrophilic groups (AH and B in sugars), one basic and one acidic, together with a hydrophobic group, in the correct spatial relationship to the hydrophilic groups. The ability to predict the probable taste of a peptide can be linked to our knowledge of amino acid sequences, to predict the appearance of bitter, and therefore unwelcome, peptides in the hydrolysates of particular proteins. For example, caseins and soya proteins are rich in hydrophobic amino acids and therefore tend to yield bitter peptides. One possible solution is to ensure that the hydrolysis is limited to producing fragments with molecular weights above about 6000, which are too large to interact with the taste receptors.

In terms of chemical structure, close relatives of many non-sugar sweeteners are found to taste bitter. One example is the isomer of aspartame in which the L-phenylalanine residue is replaced by D-phenylalanine. A comparison of the structures of three other intense sweeteners and their bitter-tasting analogues is shown in Figure 7.6. Since there is at best only a limited market for novel bittering agents, the most likely outcome of these observations is a greater insight into the structure of the sweet taste receptor proteins.

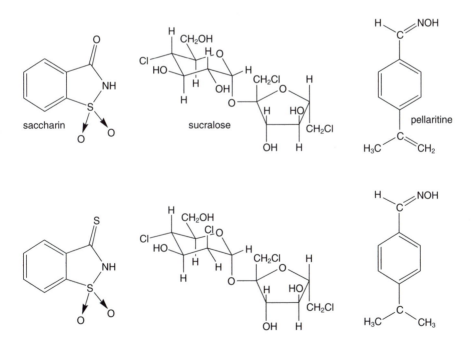

Figure 7.6 The chemical structures of three intense sweeteners, compared with their bitter tasting analogues (unnamed).

Saltiness

Saltiness is most readily detected on the sides and tip of the tongue, and is elicited by many inorganic salts apart from sodium chloride. However, it is only the saltiness of sodium chloride itself which has relevance to food. The contribution of sodium chloride to food flavour is often underestimated, as anyone who has experienced the awful taste of bread made without salt will testify. We frequently add salt to both meat and vegetables to enhance their flavour, but the sodium ions naturally present in many foods play a major role in their flavour. For example, when sodium ions were eliminated from a mixture of amino acids, nucleotides, sugars, organic acids, and other compounds known to successfully mimic the flavour of crab meat, the mixture was totally lacking in crab-like flavour, despite the fact that its composition corresponded closely to the water-soluble components of crab meat. In addition to its ability to stimulate meaty flavours, salt tends to decrease the sweetness of sugars, to produce the richer, more rounded flavour required for many confectionery products.

The relationship between molecular size and the saltiness or bitterness of salt was mentioned when bitterness was discussed. Studies with the alkali metal halides have suggested that the structural criterion which distinguishes saltiness from bitterness is simply a matter of size. Where the sum of the ionic diameters is below that of potassium bromide (0.658 nm), a salt which tastes both salty and bitter, the salty taste predominates. Sodium chloride, at 0.556 nm, is an obvious example; potassium iodide, however, at 0.706 nm, merely tastes bitter. Magnesium salts, such as the chloride, $MgCl_2$ (0.850 nm), also have a bitter taste.

Other aspects of the role of salt in food are dealt with elsewhere in this book, as an antimicrobial preservative in Chapter 9, and its contribution to elevated blood pressure in Chapter 11.

Sourness

Sourness is always assumed to be a property of solutions of high acidity (low pH), but it appears that the hydrogen ion,[iv] H^+, is in fact less important for taste than the undissociated forms of the organic acids which occur in acidic foodstuffs. One recent suggestion, based on studies

---

[iv] In aqueous solutions hydrogen ions are hydrated to form hydronium ions, H_3O, and strictly speaking it is these whose role in sour taste is under discussion.

with pickled cabbage (*sauerkraut*) and pickled cucumbers, is that sour taste intensity of a solution is linearly related to the sum of:

(a) the molar concentrations of all the of organic acids containing at least one protonated carboxyl group (*e.g.* ethanoic acid in the form CH_3COOH, but not that as CH_3COO^-), *plus*
(b) the molar concentration of hydrogen ions.

The value of (a) is readily calculated using the Henderson–Hasselbalch equation, using the pH, the total concentration of each organic acid and its respective pK_a value. The value of (b) is related to the pH by the expression, $pH = -\log[H^+]$.

In most fruit and fruit juices, citric acid (7.12) and malic acid (7.13) account for almost all the acidity. Tartaric acid (7.14) is characteristic of grapes, blackberries are especially rich in isocitric acid (7.15), and rhubarb is of course well known for the presence of oxalic acid (7.16), which occurs together with substantial amounts of malic and citric acids. Typical organic acid contents of citrus and grape juices are shown in Table 7.3.

(7.12)	(7.13)	(7.14)	(7.15)	(7.16)

Ethanoic (acetic) acid, at levels around 5%, is responsible for the sourness of vinegar and products derived from it. In many pickled

Table 7.3 The organic acid content of some fruit juices. The values here are typical, but wide variations occur between different varieties and are particularly influenced by the degree of ripeness. Grapes grown in relatively cold climates such as northern Europe, and the wines produced from them, have much higher acid content than those from warmer, sunnier climates.

	Acid content $(mmol\, dm^{-3})$		
	Malic	*Citric*	*Tartaric*
Orange	13	51	–
Grapefruit	42	100	–
Lemon	17	220	–
Grape	7	16	80

products, however, including pickled cabbage and cucumbers, it is lactic acid derived from the sugar in the vegetable by bacterial fermentation which is responsible for low pH and sourness. One should not overlook the contribution that acetic acid makes to the odour of foods, for example when fried fish and chips are dowsed in vinegar. Lactic acid, derived from lactose by fermentation, occurs in cheese at around 2%, contributing some of its sharpness. In the form of its sodium salt, lactic acid is also added to give the vinegar taste to "salt and vinegar" potato crisps.

Astringency

Astringency is a sensation clearly related to bitterness, but is registered throughout the oral cavity as well as on the tongue. We tend to register astringency in terms of "mouth feel", and although it may therefore be incorrect to regard this sensation as a *taste* we should not let semantics spoil our appetite. The reaction between polyphenolic substances and enzyme proteins was mentioned in the previous chapter (page 237) and it is believed that the sensation of astringency results from similar reactions involving proteins in the saliva, or possibly on the surface of the tongue, the palate, *etc.*

Although an undesirable characteristic in many food situations, astringency is usually regarded as a desirable characteristic in fruit, and in cider. It is in red wine and tea that it is most important, however. In both these beverages astringency is associated with a high content of the polyphenolics involved in giving them their colour, and described in Chapter 6. In wine and tea the polyphenols responsible in part for flavour are collectively described as tannins. In black tea the polyphenolics are considered to contribute to the astringency,[v] but it is the gallyl groups of the epigallocatechin gallate (*see* Figure 6.12) which are primarily responsible. Epigallocatechin itself is merely bitter, without discernable astringency, whereas the astringency of epicatechin gallate becomes apparent at concentrations above $50\,mg$ per $100\,cm^3$. There is evidence that it is interaction between the gallyl groups and caffeine (which alone is merely bitter) which produces astringency. When we drink black tea we normally feel the need to moderate the astringency. Two techniques are available, the choice being determined by custom and geography. In the USA, and in other countries where coffee rather than tea is the primary source of caffeine, lemon juice may be added. This lowers the pH, ionisation of the gallate carboxyl groups is repressed

---

[v] In the trade, the astringency of tea is referred to as "briskness".

and the astringency reduced. An alternative approach, deeply ingrained in the culture of the British Isles, is to add milk. The milk proteins successfully compete for the polyphenolics with the proteins lining the mouth. This technique is limited by the availability of pasteurised milk. Milk that has been subjected to more aggressive heat treatments, such as UHT or sterilisation, itself contains flavouring compounds, presumably generated by the Maillard reaction, and these clash with the delicate flavour of tea with unpleasant results. Coffee, whose flavour is more dependent on the Maillard reaction, is less sensitive to the heat treatment of the milk added to it.

In wine the situation is more complex, as there are many more elements in the flavour, but all the evidence suggests that it is flavonoids similar to those found in tea which are responsible for the astringency. In addition to their obvious anthocyanin content, which does contribute a little to the taste, red wines may contain up to $800 \, mg \, dm^{-3}$ of catechins (as in Figures 6.12 and 6.18), compared with no more than $50 \, mg \, dm^{-3}$ in white wine.

Pungency

Pungency, sometimes referred to as "hotness", is another sensation experienced in the entire oral cavity—and elsewhere, notably the anal mucosa. It is common nowadays to link pungency with astringency, together with the cooling effect resulting from certain compounds, under the general heading "chemesthesis". This is analogous to "somesthesis", the sensory system in the skin which responds to mechanical stimuli such as touch. The effect of chemesthetic stimulants may be better described as chemical irritation, and has much in common with the perception of pain.

Pungency is the essential characteristic of a number of important spices, but it is also found in many members of the family *Cruciferae*. The genus *Capsicum* includes the red and green chillies, those from the larger *C. annuum* being much less pungent than from the ferocious little fruit of *C. frutescens*. The active components in the *Capsicum* species are known as capsaicinoids. The most abundant and the most pungent of these are capsaicin (7.17), and also dihydrocapsaicin, which has a fully saturated side-chain.

$$H_3C-CH-C=C-(CH_2)_4-C-N-CH_2-\text{(aryl-OCH}_3\text{)}-OH \quad (7.17)$$

Pepper, both black and white, comes from *Piper nigrum*. The pungent active component is piperine (7.18), amounting to about 5% of the peppercorn, together with traces of less pungent isomers containing one or both of the double bonds in the *cis* form. From time to time concern has been expressed about the possible carcinogenicity of piperine. This is due to its close structural similarity to safrole (7.19), a major component of sassafras oil, and which is now recognised to be carcinogenic. In the past sassafras oil was used as a flavouring in beverages such as the so-called root beer, a non-alcoholic beverage popular in the USA.

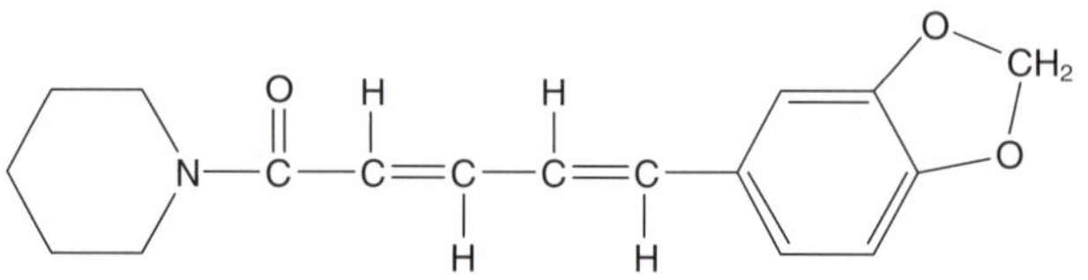

(7.18)

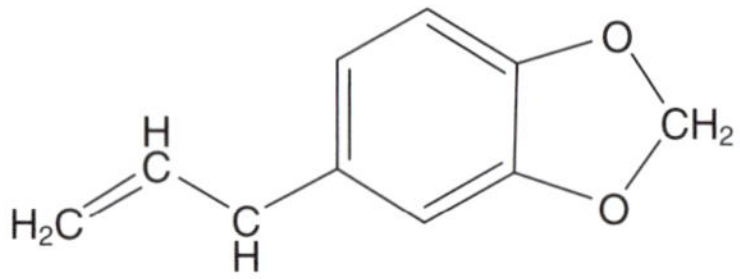

(7.19)

The active components of another pungent spice, ginger (*Zingiber officinale*), show a striking structural similarity to capsaicin and piperine. These are the gingerols (7.20) and shogaols (7.21), which predominate in versions where n = 4, but higher homologues are present in small amounts. Different types of ginger, *e.g.* green root ginger and dry powdered ginger, contain various proportions of gingerols and shogaols, resulting from the readiness with which gingerols dehydrate to the corresponding shogaols. The vanillyl side-chain is also found in eugenol (7.22), the pungent component of cloves.

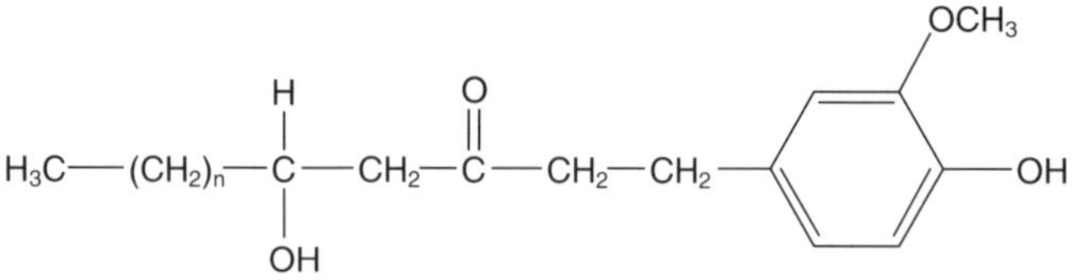

(7.20)

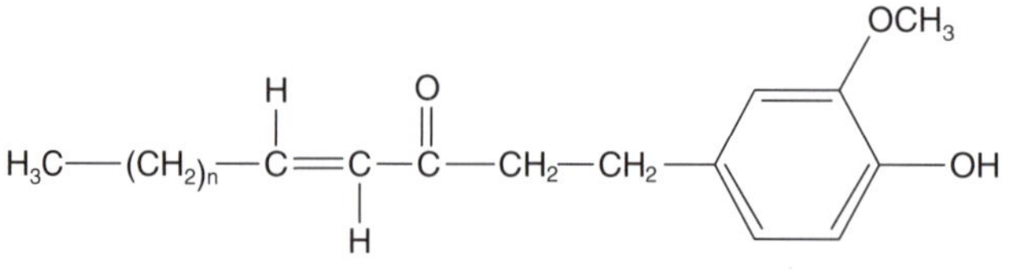

(7.21)

$$H_2C{=}CH{-}CH_2{-}\text{(aromatic ring with OCH}_3\text{ and OH)} \qquad (7.22)$$

The final group of pungent food components we should consider are the glucosinolates, found in plants belonging to the family *Cruciferae*, including the brassicas (cabbage, Brussels sprouts, broccoli) and many plants renowned for their pungency in the raw state, such as radish, black and white mustards, and horseradish. Total glucosinolates are normally present at up to 2 mmol (about 2 g) kg^{-1} fresh weight, with higher amounts in the most pungent varieties. The glucosinolates are not themselves pungent, but all the plants in which they occur also contain the enzyme, *myrosinase*. This is present in the cytoplasm of the cells, but it only encounters its natural substrates, the glucosinolates, at the time the plant tissue is disrupted when we or grazing animals chew it. This also happens if it suffers insect or other damage in the field, or when it is chopped up in the kitchen. As shown in Figure 7.7, myrosinase catalyses the hydrolysis of the unusual carbon–sulfur bond, liberating thiohydroxamate-*O*-sulfates. These compounds are unstable and depending on the conditions spontaneously form isothiocyanates or nitriles. The volatile isothiocyanates predominate, and it is these which provide the pungent taste.

Isothiocyanates are collectively referred to as mustard oils. The glucosinolates found in a range of plants are shown in Figure 7.8. In the cases of sinigrin and glucotropaeolin the thiocyanates are formed by enzymic isomerisation of isothiocyanates.

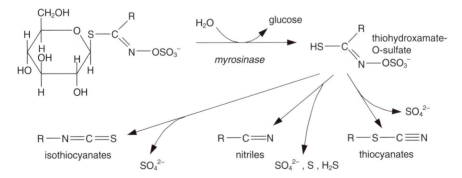

Figure 7.7 The breakdown of glucosinolates to form isothiocyanates and other products. Nitrile production is favoured by low pH, as for example in the production of *sauerkraut*.

Principal or characteristic glucosinolates

Cabbage	*Brassica oleracea*	sinigrin (7.23), glucobrassicin (7.24), progoitrin (7.25)
Brussels sprouts	*B. oleracea* var. *gemmifer*	glucobrassicin, neoglucobrassicin (7.26)
Broccoli	*B. oleracea* var. *italica*	glucobrassicin, glucoraphanin, (7.27)
Cauliflower	*B. botrytis* var. *cauliflora*	sinigrin, glucobrassicin, neoglucobrassicin
Black mustard	*B. nigrum*	sinigrin
White mustard	*B. alba*	sinalbin (7.28)
Rape	*B. napus*	progoitrin, glucobrassicin, neoglucobrassicin
Horseradish	*Amoracia lapathiofolia*	sinigrin
Radish	*Raphanus sativus*	glucoraphasatin (7.29)
Watercress	*Nasturtium officinalis*	gluconasturtiin (7.30)
Cress	*Lepidum sativum*	glucotropaeolin (7.31)

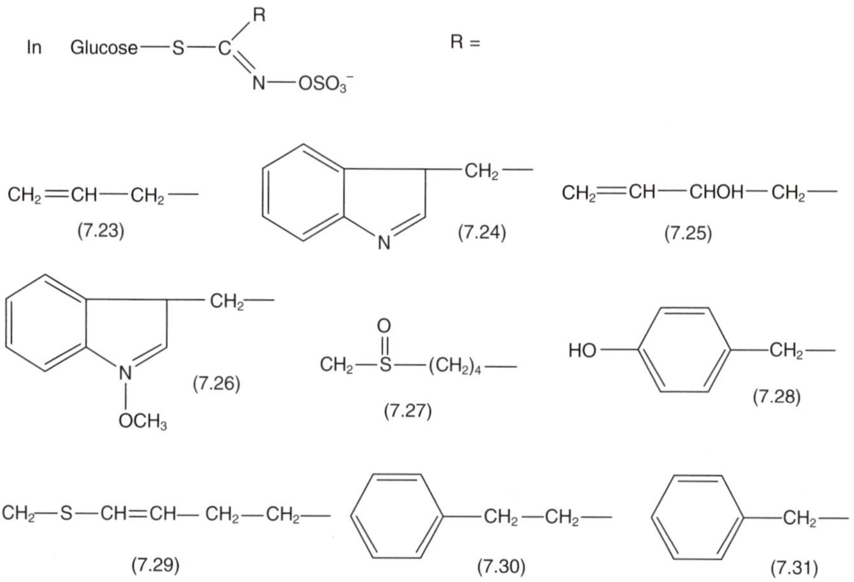

Figure 7.8 Glucosinolates found in members of the Cruciferae family.

Nitrile formation is a significant issue in the utilisation of rape-seed meal, the residue after the extraction of rape-seed oil. Another problem with rape-seed meal is the tendency of isothiocyanates having a hydroxyl group in the β-position, such as that derived from progoitrin (7.25), to cyclise spontaneously to oxazolidinethiones (7.32). These compounds, known as goitrins, interfere with iodine uptake by the thyroid gland, resulting in goitre, *i.e.* massive enlargement of the thyroid gland and deficiency of the hormone thyroxine. For these reasons, plant breeders have made great efforts to develop strains of rape with very low levels of glucosinolates. These efforts are not without their downside, however.

Breakdown products of glucosinolates are known to be important in defending plants against insect and fungal pathogens. The suggestion that milk from cattle fed largely on rape-seed meal and other brassicas might be harmful need not be taken seriously. The only time goitrins have proved to be a problem for humans was during the winter of 1916–17 in Germany. Food shortages caused by the First World War led to turnips (*Brassica rapa*) becoming a staple food, replacing both potatoes and cereals. Combined with other effects of malnutrition, the incidence of goitre rose sharply.

$$H_2C=CH-CHOH-CH_2-N=C=S \longrightarrow H_2C=CH-CH$$

(7.25)   (7.32)

The lower incidence of cancer in populations consuming large quantities of fruit and vegetables is now well recognised, and the beneficial effects, especially on cancers of the colon, can be related to some extent to the glucosinolate content of cruciferous vegetables. In experiments with laboratory animals, and using tissue cultures of human cells, allyl iso-thiocyanate (7.33) (from sinigrin) and sulforophane (7.34) (from glucor-aphanin) have been shown at realistic concentrations to kill cancerous cells. Sulforophane appears to increase the production of glutathione trans-ferases, a group of enzymes involved in the neutralisation of carcinogens.

$$CH_2=CH-CH_2-N=C=S$$

(7.33)

$$CH_3-\overset{\overset{\displaystyle O}{\|}}{S}-(CH_2)_3-CH_2-N=C=S$$

(7.34)

When vegetables belonging to the family *Cruciferae* are cooked, ideally by pouring boiling water over them followed by a few minutes at 100 °C, the myrosinase is destroyed before tissue disruption brings the enzyme and substrate into contact, and the formation of isothiocyanates does not take place. Instead, a wide range of other sulfur compounds arise and are responsible for giving over-boiled cabbage and similar vegetables their familiar unattractive odour. The difference in flavour after cooking frozen and fresh Brussels sprouts has been shown to be due to the different rates of myrosinase destruction caused respectively by commercial blanching and domestic cooking.

The sensation of cooling is a minor taste response which hardly justifies a section of its own. It is mentioned here simply because, like pungency, it is a manifestation of chemesthesis, and is also the natural opposite of heat. Cooling sensations are associated with some confectionery and products such as toothpaste and mouthwashes. Just as the burning taste of pungent substances is believed to be registered by similar nerve endings to those responding to high temperature, the cooling taste is assumed to invoke a response in the nerves which register low temperature. The best known cooling substance is (−)menthol (7.35). It is the dominant component of peppermint oil, extracted from *Mentha piperiita* and *M. arvensis,* which is used in confectionery, the other major component being menthone (7.36). Menthone and the optical isomers of (−)menthol all share the characteristic minty odour but it is only (−)menthol itself which produces a cooling sensation in the mouth. Certain other substances, notably xylitol (2.18), can also produce a cooling effect, but this is usually attributed to their large negative heat of solution. When the solid xylitol included in chewing gum dissolves in the saliva, its temperature briefly falls.

(7.35)   (7.36)

Meatiness

Of all the taste classes, the most recent to be investigated is meatiness. Although the flavour of meat is obviously an amalgam of our responses to both volatile and non-volatile compounds, a meaty taste has come to be associated with two particular compounds, inosine monophosphate (IMP, 7.37) and monosodium glutamate (MSG, 7.38). Both substances were first identified as active components in food ingredients used to enhance the flavour of traditional Japanese dishes. The Japanese word: うま味 (pronounced *umami*) is applied to the particular taste of these two compounds and others related to them. The literal translation of umami is "delicious taste", but it also carries the implication of meatiness, savoury, or broth-like. In recent years food writers have begun to familiarise the general public with the concept, and the word umami has found a place in the English language in its own right.

MSG was identified in 1908 as the source of the taste of the seaweed konbu, used to make a broth popular in Japan. Its commercial extraction and its use in food seasoning followed almost immediately and it is now manufactured by a bacterial fermentation process. A few years later IMP was isolated from a dried fish preparation (bonito flakes, made from skipjack tuna), also widely used in Japanese cooking. The fermented fish sauces used in Ancient Rome (known at the time as *garum* or *liquamen*) and still popular today in the Far East are essentially umami seasonings. Fish muscle was at one time the source of commercial preparations of all the 5'-nucleotides, but nowadays enzymic hydrolysates of RNA are the usual source.

(7.37)   (7.38)

On its own, neither of these substances has a particularly strong taste. They need to reach a concentration of about $300\,\text{mg dm}^{-3}$ before they begin to make an impact. However, when present together they show remarkable synergy. A mixture of equal proportions of IMP and MSG tastes some 20 times stronger than a similar weight of one alone. Although the mechanism of the synergy is not understood, its importance to food manufacturers is obvious. The addition of only a small amount of one of these substances (usually MSG) to a food product which itself contains even small amounts of the other can have a dramatic effect on the flavour. Thus MSG is used as a flavour enhancer rather than as a flavouring in dried soups and similar products, at levels giving about $1\,\text{g dm}^{-3}$ in the final dish. Although much more expensive, preparations of IMP, often blended with guanosine monophosphate (7.39), are now used by some food manufacturers.

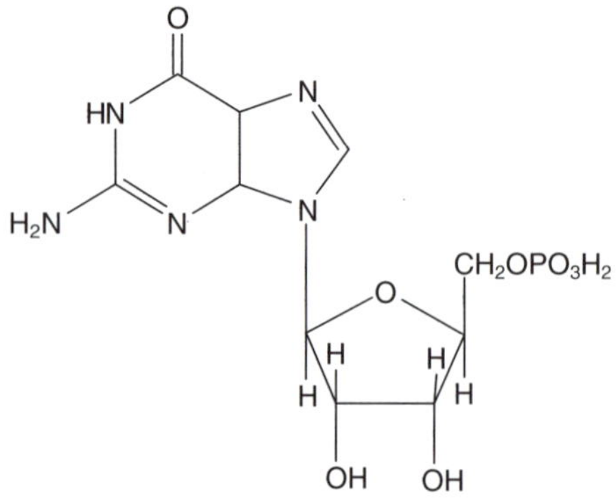

(7.39)

Glutamic acid in the form of MSG is present naturally at low levels in the muscle tissue of live animals, but in meat it is mainly derived from the protein breakdown occurring during the *post-mortem* ageing process. IMP is a breakdown product of adenosine monophosphate, AMP, which accumulates as adenosine triphosphate (ATP), utilised in the *postmortem* metabolism of muscle:

$$\text{ATP} \xrightarrow{-\text{phosphate}} \text{ADP} \xrightarrow{-\text{phosphate}} \text{AMP} \xrightarrow{-\text{NH}_3} \text{IMP}$$

Typical data for the levels of umami substances are given in Table 7.4. It is not difficult to correlate these data with one's personal impression of the flavour of many of these foods. The use of tomatoes to bring out the flavour of meat, and parmesan as a garnish for minestrone soup, are readily supported by these data. The difference in MSG level between human and cow's milk is also interesting, and one could speculate that it reflects the different dietary priorities of the herbivorous cow and the omnivorous human. By the time it is eaten any protein-rich food will have degraded sufficiently to accumulate detectable levels of umami substances, and our taste for these provides a good test for the likely presence of protein. Results obtained by the petfood industry suggest that cats, being obligate carnivores and using high protein foods to supply their nutritional needs, are much more sensitive to umami substances than are humans.[vi]

While IMP and MSG provide the basic meatiness to meat and meat products, the subtle difference in flavour between different meats depends on variations in the proportions of the two. The differences also involve

---

[vi] This argument also provides an explanation for the apparent indifference of cats to the sweetness which provides a human indicator of the presence of carbohydrates, our major energy supply.

Table 7.4 Levels of umami substances found in a range of foodstuffs. Levels found in different laboratories at different times can vary considerably. The gaps in the table indicate absence of data rather than zero concentration. Nori is the chopped seaweed pressed to form the dark strips wrapped around many types of sushi, the Japanese fish delicacy. Most of these data are from Nimomiya, 2002 (see Recent Reviews, page 311), the remainder from the publications of the Umami Information Centre, Tokyo.

	MSG	*IMP*	*GMP*
		mg/100 g	
Konbu (seaweed)	2240		
Nori (seaweed)	1378	8.5	12.5
Bonito flakes	26	687	
Fresh tuna		188	
Sardines	280	193	
Prawns	43	92	30
Cod	9	44	
Salmon	20		
Pork	9	260	2
Chicken	22	283	5
Parmesan cheese	1200		
Emmental cheese	308		
Shiitake mushroom (dried)	1060		150
Tomato juice	260		
Human milk	22		
Cow's milk	2		
Used as a food additive	20–80		

trace amounts of a number of other flavour compounds. For example, the IMP contents of beef and pork are roughly similar, but beef contains significantly more free amino acids than pork and twice as much MSG. Lamb is particularly rich in MSG but contains much lower levels of dipeptides such as carnosine (7.40), which also contribute to the meat flavour. The difference in flavour between veal and beef has been suggested to be the result of increasing levels of free amino acids as the animal matures.

(7.40)

Although there are some dishes containing raw meat (*e.g.* steak tartare), cooking is obviously the origin of the innumerable volatile compounds which are important to our appreciation of meat. These are discussed in the second half of this chapter.

ODOUR

While our sense of taste enables us to make a broad assessment of the nature of a potential foodstuff, it is odour that supplies the detail. Counting the number of different substances that we are able to detect is impossible, as is any attempt to list all the different odours available. It is also impossible to draw a line between food odours and the host of other odours, more often referred to as smells, that we encounter every day and have nothing to do with food. Yet another complication is that natural odours are hardly ever generated by single pure substances. The complexity of Scotch whisky was mentioned at the beginning of this chapter, and a typical fruit may similarly have as many as 200 different volatile components each making a contribution to its overall odour, but even in total these may comprise only a few parts per million in the fruit.

Attempts have been made to classify odours and to relate the different classes to elements of chemical structure. Seven primary odours were described by Amoore in 1972: camphoraceous, ethereal, musky, floral, minty, pungent and putrid. However, if these correspond to seven types of receptor in the nose, many substances must be able to interact with more than one receptor. Since most of the odours we experience in the real world outside the laboratory are caused by mixtures of a great many different substances, with several types of receptors responding simultaneously, the brain is left to make sense of an amalgam of nerve impulses, in much the same way as it does with colour vision. By analogy with what is now well understood in relation to taste, our olfactory organs must be equipped with cell membranes carrying receptor proteins able to discriminate between molecules of quite similar structure, including optical isomers. For example, of the two isochroman derivatives illustrated, one (7.41) has an intense musk odour, while the other (7.42) is odourless.

(7.41)   (7.42)

The organisation of the sections which follow is therefore based on a few broad groups of foodstuffs and makes no claim to be comprehensive.

Meat

The first half of this chapter concluded with an account of the taste of meat, and it seems reasonable to return to this food to emphasise the dependence of flavour on both taste and odour. There are three major groups of compounds that contribute to the flavour of meat and one of these, including the umami substances and dipeptides, was discussed in the previous section. The other two groups are:

- heterocyclic compounds derived from amino acids, nucleotides, sugars and thiamine *via* the Maillard reaction (*see* Chapter 2) during cooking; and
- unsaturated fatty acid breakdown products (*see* Chapter 4), also formed during cooking.

The Maillard reaction can yield a great diversity of carbonyl compounds from the residual sugars (from glycogen and nucleotides) in meat, including 2-oxopropanol (2.31), 2,3-butanedione (2.37), hydroxypropanone (2.38) and hydroxymethylfurfural (Figure 2.12). The Strecker degradation (Figure 2.15) of these with most amino acids yields the corresponding aldehydes but, as shown in Figure 7.9, cysteine gives rise to both hydrogen sulfide and ammonia, which can contribute to the formation of a variety of heterocyclic compounds. Thiazoles and oxazoles are important in the flavour of roasted meat. The reaction of hydrogen sulfide with aldehydes can give ring structures containing multiple sulfur atoms, and these are believed to be particularly important in the flavour of boiled (as oppose to roasted) meat. Several other sulfur compounds contributing to the flavour of cooked meat have been traced to the thermal decomposition of the vitamin thiamin.

Thermal decomposition of ribonucleotides (*e.g.* IMP, 7.37) has been shown to lead to compounds such as methyl furanolone (7.43), which not only have a pronounced meaty odour themselves but can also be the precursors of sulfur-containing compounds (the sulfur is derived from the breakdown of cysteine and methionine), for example methylthiophenone (7.44), which also have a strong meaty odour.

H_3C — O — HO — O

(7.43)

S — CH_3 — O

(7.44)

Fatty-acid breakdown, along the lines of the autoxidation reactions described in Chapter 4, is also an important source of the characteristic

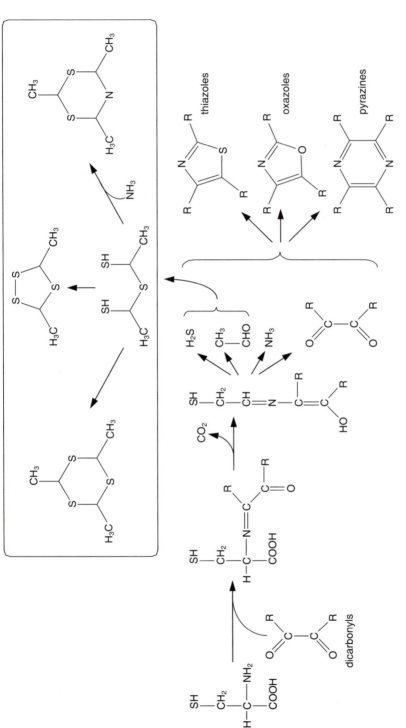

Figure 7.9 The formation of heterocyclic meat flavour compounds following the Strecker degradation of cysteine. For simplicity the complex sequence of condensation reactions postulated to lead to the ring compounds is not shown. R represents any of the terminal fragments of sugar molecules which can arise in the Maillard reaction. It is likely that aldehydes and dicarbonyls from the degradation of other amino acids will join the sequence and provide even greater diversity of end-products. The boxed reactions show the formation from acetaldehyde and hydrogen sulfide of the cyclic sulfides identified in the flavour of boiled meat.

odours of meat products. The low content of polyunsaturated fatty acids in the depot fat and intramuscular fat of beef, lamb and pork points to the polar membrane lipids of muscle cells as being the most likely source of the autoxidation products significant to flavour. Arachidonic acid has been shown to break down on heating to give four aldehydes which together provide the distinctive odour of cooked chicken:

$CH_3(CH_2)_4CH=CHCH_2CHO$	3-*cis*-nonenal
$CH_3(CH_2)_4CH=CH(CH_2)_2CHO$	4-*cis*-decenal
$CH_3(CH_2)_4CH=CHCH_2CH=CHCHO$	2-*trans*,5-*cis*-undecadienal
$CH_3(CH_2)_4CH=CHCH_2CH=HCCH=HCCHO$	2-*trans*,4-*cis*, 7-*cis*-tridecatrienal

Aldehyde fatty acid fragments such as these can react with hydrogen sulfide or ammonia to provide yet another source of odorous heterocyclic compounds, as illustrated (7.45).

(7.45)

Fruit

The flavour of fruit is very definitely a blend of taste and odour. The taste, an aggregate of sweetness due to sugars, the sourness of organic acids and sometimes the astringency of polyphenolics, provides the basic identity that confirms that one is consuming fruit. Even a simple solution of sucrose and citric acid tastes "fruity" to most people. However, to distinguish between different fruits we rely heavily on the distinctive odour of the volatile components. For instance if one has a cold, which effectively eliminates the sense of smell, it is extremely difficult to distinguish between raspberry and strawberry flavours.

A typical fruit may contain as many as 200 different volatile components, but even in total these may comprise only a few parts per million of the total fruit. In general it is observed that citrus fruit odours are dominated by terpenoids (*see* page 218, whereas in other fruit[vii] they

---

[vii] The term "berry-type" aroma is sometimes used to cover all non-citrus fruit, a satisfactory term when applied to raspberries and the like, but somewhat lacking when describing apples or bananas.

are much less in evidence. Long lists of compounds have been assembled for many fruit which show that acids and alcohols, and the esters resulting from their combination, and aldehydes and ketones, predominate. Considering apples as a typical example, we find:

- at least 20 aliphatic acids, ranging from formic to *n*-decanoic;
- at least 27 aliphatic alcohols, covering a corresponding range of chemical structures;
- over 70 esters, from combination of the predominant alcohols and acids;
- 26 aldehydes and ketones, also with structures corresponding to the acids and alcohols; and
- smaller numbers of miscellaneous ethers, acetals, terpenoids and other hydrocarbons.

This adds up to a total of 131 separate compounds! But the first point to remember is that by no means all of these substances make a significant contribution: volatility does not equate to aroma. Another point is that a very high proportion of these compounds are common to many different fruits. For example, of the 17 esters identified in banana volatiles only five are not found in apples. The particularly common occurrence of aldehydes, alcohols and acids which have linear or branched (*e.g.* isopropionyl) chains of up to seven carbon atoms, often with an occasional double bond, suggests some common origins. In fact there are two sources represented in many fruits. During ripening, cell breakdown is accompanied by oxidation of the unsaturated fatty acids in the membrane lipids, catalysed by the enzyme lipoxygenase. This converts fatty acids with a *cis*, *cis*-methylene-interrupted diene system (as in linoleic acid) into hydroperoxides. These break down either spontaneously or under the influence of other enzymes to give aldehydes, as shown later in Figure 7.10, which are important elements of vegetable flavour. The pattern of reaction products is very similar to that found in the non-enzymic autoxidation of fatty acids. The other source of aldehydes are amino acids. Enzymic transamination and decarboxylation of free amino acids often accompany ripening, for example leucine gives rise to 3-methylbutanal (7.46):

$$H_3C\text{---CHCH}_2\text{CHCOOH} \xrightarrow[\text{glutamate}]{\alpha\text{-ketoglutarate}} H_3C\text{---CHCH}_2\text{CCOOH} \xrightarrow{CO_2} H_3C\text{---CHCH}_2\text{CHO} \tag{7.46}$$

Aldehydes are not only important in fruit and vegetables. During alcoholic fermentation, yeast enzymes catalyse similar reactions with the amino acids from the malt or other raw materials. These aldehydes and the alcohols derived from them are important as flavour compounds in many alcoholic drinks, particularly in distilled spirits (where they are described as fusel oils) where they have a special responsibility for the headaches associated with hangovers. Other enzyme systems in plant tissues are responsible for the reduction of aldehydes to alcohols, or their oxidation to carboxylic acids, and the combination of the two to form esters.

In many cases the distinctive character of a particular fruit flavour is dependent on one or two fairly unique "character impact" substances. For example, *iso*-pentyl acetate is the crucial element in banana flavour, although eugenol (7.22) and some of its derivatives contribute to the mellow, full-bodied aroma of ripe bananas. Benzaldehyde is the character impact substance in cherries and almonds. It is difficult to establish to what extent hydrocyanic acid (HCN), which is also present at the level of a few ppm in cherries, contributes to their aroma. Hexanal ($CH_3(CH_2)_4CHO$) and 2-hexenal (7.47) provide what are described as green or unripe odours in a number of fruits and vegetables. In at least one variety of apple they have been shown to balance the ripe note in the aroma provided specifically by ethyl 2-methyl butyrate (7.48). The olfactory thresholds of these three substances are very low, 0.005, 0.017 and 0.0001 ppm, respectively! The distinctive character of raspberry aroma is due mostly to 4-(*p*-hydroxyphenyl)-2-butanone (7.49), often referred to as "raspberry ketone", but the fresh grassy aroma which is characteristic of the fresh fruit, but notably lacking in many raspberry-flavoured products, is principally due to *cis*-3-hexenol (7.50), supplemented by α- and β-ionone (*see* page 218). Furaneol, 2,5-dimethyl-4-hydroxy-2,3-dihydrofuran-3-one (7.51), is a major constituent of strawberry flavour.

Wine presents special challenges to the flavour chemist. Most varieties of red wine, *e.g.* Merlot, Grenache, *etc.*, share many of the same flavour compounds, merely in different proportions. Wine experts are well known for describing the aroma of a wine in terms of other fruit. For example, the presence of 2,6,6'-trimethyl-2-vinyl-4-acetoxytetrahydropyran (7.52) in a red wine is said to suggest a blackcurrant aroma. Similarly the peach and pineapple aromas of a number of red wines have been attributed to γ-decalactone (7.53) and "wine lactone" (3a,4,5,7a-tetrahydro-3,6-dimethyl benzofuran-2(3*H*)-one, 7.54), respectively. *cis*-Rose oxide (4-methyl-2-(-methyl-1-propenyl)tetrahydrofuran, 7.55) has been identified as the source of the lychee aroma characteristic of *Gewürztraminer,* a white wine.

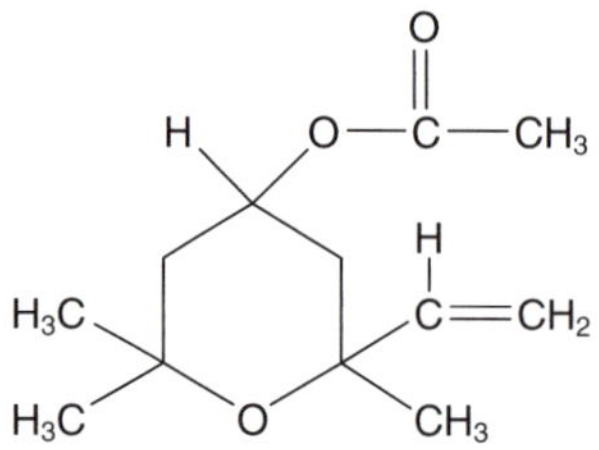

(7.52) 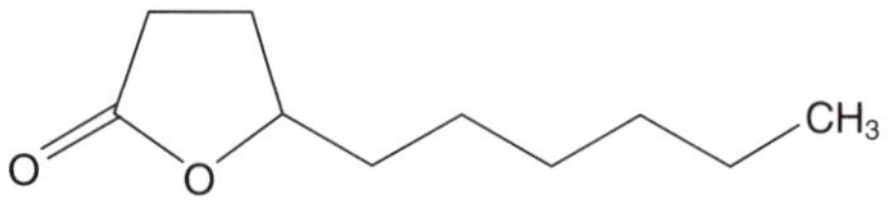   (7.53)

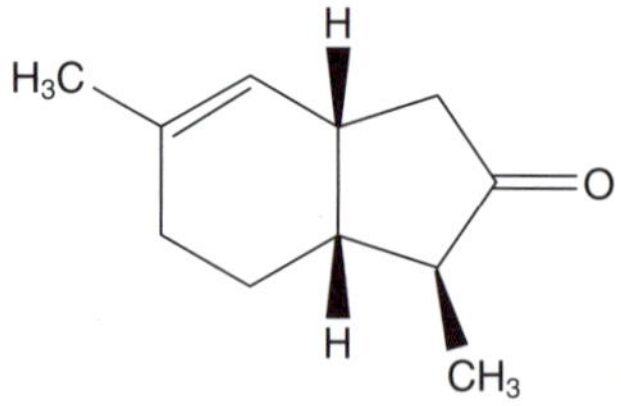

(7.54) 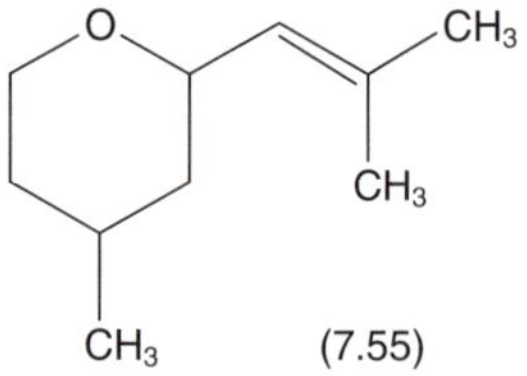   (7.55)

The character impact substances in many fruit, especially citrus fruits, are terpenoids. Terpenoids can often be isolated from plant materials by steam distillation. The oily fraction of the distillate is known as the "essential oil", and these are often used as flavourings and components of perfumes and toiletries. The essential oils of citrus fruits (*i.e.* of the entire fruit including the peel) consist mainly of terpenes, of which the monoterpene (+)-limonene (7.56) is usually at least 80%. The optical isomer of (+)-limonene, (−)-limonene (7.57), is reported to smell of lemons rather than oranges. However, (+)-limonene is not nearly as important a flavour component as the oxygenated terpenoids which occur in much smaller amounts in the oil. For example, fresh grapefruit juice contains about 16 ppm of (+)-limonene carried over from the peel during juice preparation, but the characteristic aromatic flavour of grapefruit is due to the presence of much smaller amounts of (+)-nootkatone (7.58). The characteristic bitterness of grapefruit juice was mentioned earlier in this chapter. The character impact substance in lemons is citral, which is more correctly described as a mixture of the isomers geranial (7.59) and neral (7.60). Most varieties of oranges do not appear to possess a clearly identified character impact substance, but the mandarin orange, *Citrus reticulata*, is characterised by α-sinensal (7.61). The various varieties of the so-called sweet or Valencia orange, *C. sinensis*, contain β-sinensal (7.62).

(7.56)   (7.57)   (7.58)   (7.59)   (7.60)

(7.61)   (7.62)

Vegetables

Compared with fruit, vegetables tend to have milder and more subtle flavours in which aroma is much less significant than taste. The aromas of fruit have evolved to entice insects, birds and other consumers into assisting with seed distribution and related activities. This hardly applies to the leaves of a cabbage or the tubers of a potato. The list of vegetables whose flavour is sufficiently desirable to prompt food manufacturers to seek out artificial alternatives to their flavours is a very short one—mushrooms and onions—and it certainly does not include either broccoli or beans. The character impact substance of the common mushroom, *Agaricus bisporus,* is 1-octen-3-ol, $CH_3(CH_2)_4CHOHCH = CH_2$. Many vegetables owe their flavour to the activity of the enzyme lipoxygenase. The action of this enzyme is illustrated in Figure 7.10. 2-*trans*-Nonenal, the end-product shown in Figure 7.10, is an important contributor to the aroma of cucumbers, although 2-*trans*-6-*cis*-nonadienal is recognised as the character impact substance. Lipoxygenase activity is not always a good thing, however. High levels of the enzyme occur in peas and beans, and if they are physically damaged during harvesting the enzyme is able to react with unsaturated fatty acids from the cell membranes, giving rise

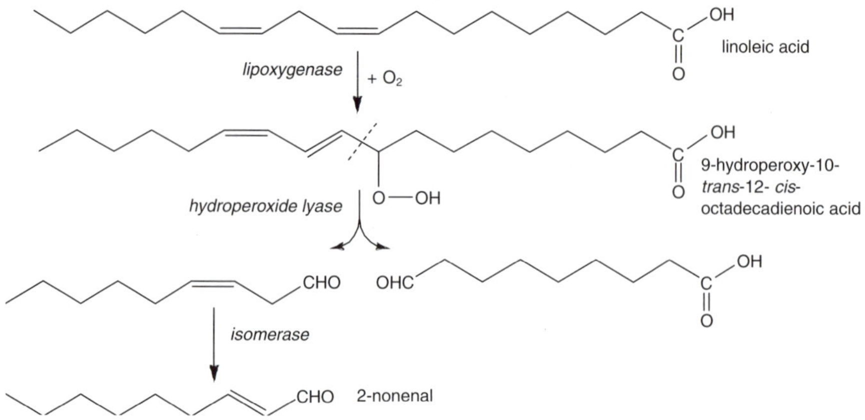

Figure 7.10 The formation of trans-2-nonenal, following the breakdown of linoleic acid initiated by lipoxygenase action. Lipoxygenases in some plant species react at the other end of the conjugated diene system giving, for example, 13-hydroperoxy-9-*cis*-11-trans-octadecadienoic acid, which inevitably leads to a quite different collection of end-products.

to off-flavours, notably "delay off-flavour" in peas. The delay in this case is the unintended few hours between the peas being mechanically removed from the pod by the vining machine, and the blanching process. Blanching, usually comprising a few seconds in a blast of steam, is designed to inactivate enzymes such as lipoxygenase.

Onion (*Allium cepa*), and garlic (*A. sativum*) are by far the most studied vegetables in terms of flavour. All members of this genus contain *S*-alkyl cysteine sulfoxides (*see* Figure 7.11). The 1-propenyl derivative is particularly associated with onions and the allyl derivative with garlic. Just as in the formation of isothiocyanates in cruciferous plants, the compounds present in the raw undamaged plant tissues are quite odourless. It is only when garlic is crushed, or the onion sliced, that the enzyme alliinase gains access to its substrates and the flavour appears. The fate of the unstable sulfenic acid produced by the action of the alliinase is dependent on the species. In the case of the onion a second enzyme, known as LF synthase (LF = lachrymatory factor), gives rise to thiopropionaldehyde-*S*-oxide (also known as propanethial-*S*-oxide). Until the discovery of this enzyme the reaction in onions was assumed to be spontaneous.[viii] This substance is the

---

[viii] Prior to the publication of the paper by Imai, *et al.*, *Nature*, 2002, **419**, p. 885, which describes the action of LF synthase, the accepted explanation for the different outcomes in onion and garlic was the difference in the position of the double bond in the side-chain. In onions the double bond was believed to be sufficiently close to interact with the sulfenic acid group, reducing its reactivity and facilitating isomerisation. By contrast, in garlic the reactivity of the sulfenic acid group is undiminished by interaction with the double bond.

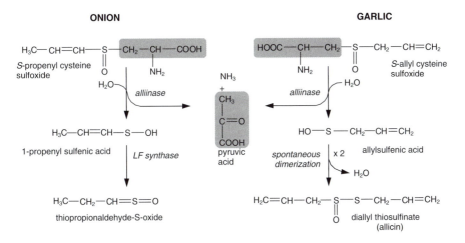

Figure 7.11 The breakdown of S-alkyl cysteine sulfoxides in onion and garlic. The pyruvate moiety is shaded.

source of the legendary lachrymatory effect of onions. In garlic, which is presumed to lack LF synthase, the reactivity of the sulfenic acid of the allyl sulfenic group leads to dimerisation to allicin, well known for its anti-bacterial properties.

The immediate biosynthetic precursors of S-alkyl cysteine sulfoxides are the γ–glutamyl derivatives (7.63), and considerable amounts of these compounds are present in the fresh plant material. They are not susceptible to the action of alliinase, but these, and any traces of the amino acid cysteine, are available to participate in the great variety of exchange reactions, dimerisations and condensations which occur at the time onions and garlic are cooked. Figure 7.12 illustrates some of the potential reactions that can lead to the formation of sulfonates and sulfides that can contribute to the flavour of cooked onion or garlic. Other, non-sulfur, flavour compounds, such as 2-methylbut-2-enal ($CH_3CH = C(CH_3)CHO$), and 2-methylpent-2-enal ($CH_3CH_2CH = C(CH_3)CHO$), are formed in similar reactions.

$$
\begin{array}{c}
R\!-\!S\!-\!CH_2\!-\!CH\!-\!COOH \\
\quad\ \|\quad\qquad | \\
\quad\ O\qquad\quad NH \\
\qquad\qquad\quad | \\
\qquad\qquad\quad CO \qquad\qquad (7.63)\\
\qquad\qquad\quad | \\
\qquad\qquad\quad (CH_2)_2 \\
\qquad\qquad\quad | \\
\qquad\qquad\quad CHNH_2 \\
\qquad\qquad\quad | \\
\qquad\qquad\quad COOH
\end{array}
$$

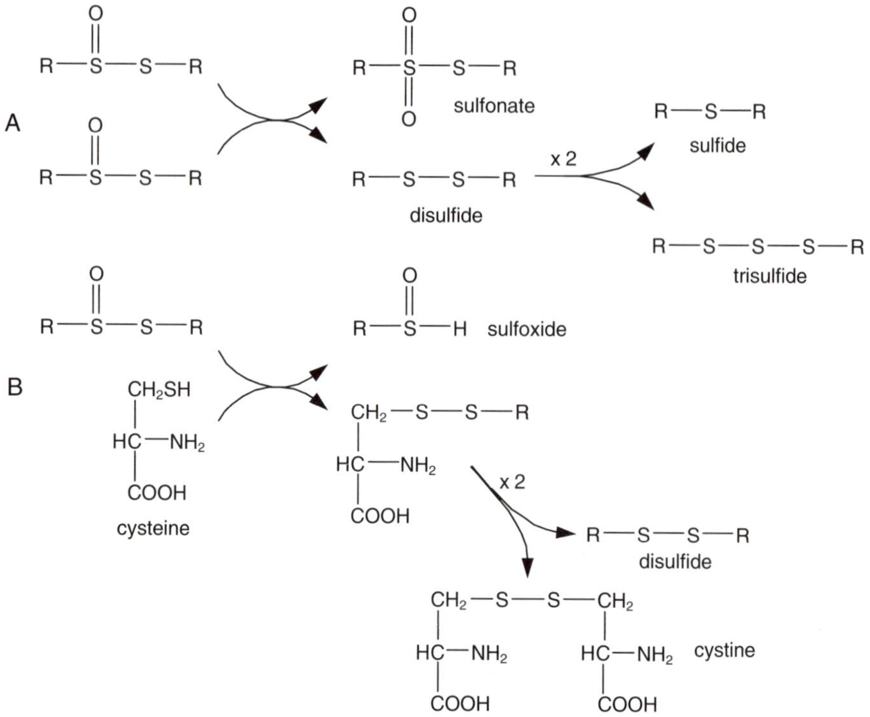

Figure 7.12   Exchange reactions taking place when onions and garlic are cooked involving, A, two alkyl thiosulfinate molecules and, B, an alkyl sulfinate with cysteine, both leading to the formation of sulfonates and volatile odorous sulfides. R may represent an allyl ($CH_2 = CHCH_2 -$), 1-propenyl ($CH_3CH = CH-$) or methyl (CH_3-) group.

Di- and trisulfides tend to be most significant when the cooking has taken place in water. However in fried dishes, including stir-fries, other compounds are more significant, including (*E*)-ajoene[ix] (7.64) and 2-vinyl-[4H]-1,2-dithiin (7.65).

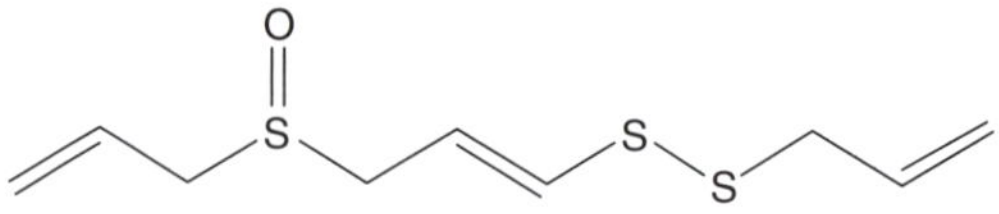

(7.64)

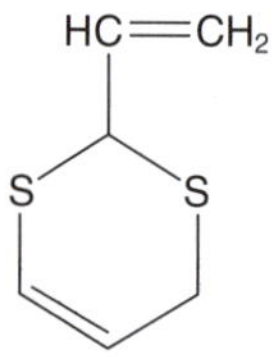

(7.65)

[ix] The (*E*) identifies the configuration of the central double bond as the *trans* form. (*Z*)-ajoene, with the double bond *cis*, also occurs. *Ajo* is the Spanish word for garlic.

Brief mention should be made of the medicinal properties of garlic. In spite of its long history of medicinal use, results in clinical trials with human subjects have been variable. Most studies show reductions in blood cholesterol levels of between 10 and 15% when 0.5–1.0 g of garlic powder (equivalent to 1.5–3.0 g fresh garlic) has been consumed daily for two or three months. In addition to its antibacterial properties allicin is a potent antithrombotic agent, inhibiting the aggregation of platelets. However it remains questionable whether the health benefits compensate for the social impact of this level of garlic consumption. The odour-free capsules of garlic oil available as alternative medicines are suspected to contain only small amounts of these biologically active substances.

Asparagus is another vegetable in which sulfur compounds make an important contribution to flavour. However, the aroma of the vegetable itself is of less interest than the unpleasant smell it imparts to the urine of consumers. The most likely compounds in urine causing the "rotten cabbage" smell are methyl mercaptan (methanethiol, CH_3SH), *S*-methylthioacrylate (7.66) and *S*-methyl-3-(methylthio)thiopropionate (7.67). These are assumed to be derived from asparagusic acid (7.68) which, unlike most of the other sulfur compounds in asparagus, is not encountered in any other vegetable. Only some 40% of asparagus eaters can detect this smell in their urine, but whether this is due to deficiencies in the sense of smell of the majority or their sulfur metabolism has not been established.

Herbs and Spices

These important sources of food flavour are usually lumped together, not least because the distinction between them is not very clear. Both terms describe plant materials used specifically as flavourings rather than for any nutrients they may contain. The culinary use of the word "herb" follows from its botanical meaning, a plant with no woody tissues, as in "herbaceous border". The word applies to leafy materials such as mint, basil, rosemary and sage. On the other hand, "spices" are usually derived from non-leafy plant tissues, such as seeds (nutmeg), flower buds (cloves), roots (ginger) or tree bark (cinnamon). The principal compounds responsible for the flavour of a selection of herbs and spices are

Principal aroma compounds

Basil	*Ocimum basilicum*	methylchavicol (estragole)(7.69)
Tarragon	*Artemisia dracunculus*	methylchavicol (estragole)(7.69)
Rosemary	*Rosmarinus officinalis*	verbenone (7.70), 1,8 cineole (7.71)
Sage	*Salvia officinalis*	α-thujone (7.72), camphor (7.73)
Oregano	*Origanum spp.*	carvacrol (7.74), thymol (7.75)
Peppermint	*Mentha piperita*	(-)-menthol (7.76), menthone (7.77), menthofuran (7.78)
Spearmint	*M. spicata*	(-)-carvone (7.79)
Caraway	*Carvum carvi*	(+)-carvone (7.80)
Nutmeg	*Myristica fragrans*	(+)-sabinine (7.81)
Cardamom	*Elettaria cardomomum*	α-terpinyl acetate (7.82)

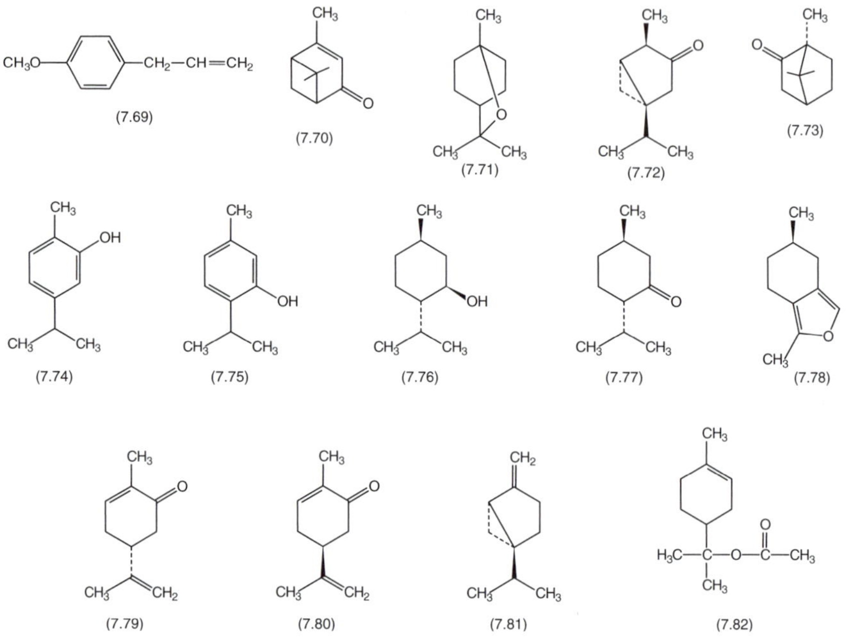

Figure 7.13 The principal aroma compounds in a selection of herbs and spices. The compounds shown are not necessarily the most abundant, but are those believed to impart the characteristic flavour notes. The carvones (7.79 and 7.80) demonstrate the common observation that optical isomers may differ markedly in aroma. (For clarity the −CH₃ groups carried by the bridging carbon atoms in 7.70 and 7.73 have been omitted.)

listed in Figure 7.13. Three important spices, ginger and black and white pepper, have been discussed in detail earlier in this chapter.

Synthetic Flavourings

A principal objective for the flavour chemist is the provision of synthetic flavourings for use by both the domestic cook and the food manufacturer. An important trend in recent years has been the replacement of simple esters, with their sweet and unsubtle odour (*e.g.* the amyl acetate in pear drops), with lighter and fresher (and often much more expensive) flavourings based on our knowledge of the natural flavour components. For commercial reasons, very little information about the composition of these products is released by flavour manufacturers, but the patent literature gives us some insights. Table 7.5 provides an example of a formulation designed for use in confectionery and similar products. Mixtures such as that described normally provide only the starting point for a commercial flavouring.

Flavour manufacturers are able to customise their products for individual food manufacturers, including retail chains offering own-label products, by making small changes to the proportions of the ingredients and by adding small amounts of other substances. Even though most of the components are pure synthetic compounds there is still a place for natural essential oils, as shown here, which may be complex mixtures in their own right. It is also important to remember that most of these compounds have been identified as components of the original fruit of which the flavour is being mimicked. Before arriving at a flavour formulation the flavourist will initially be guided by the identity and proportions of the volatiles in the original plant, determined by gas chromatography and mass spectrometry. However, the final composition will owe as much to the highly skilled nose of the flavourist as it does to science.

In recent years a demand has arisen in the food industry for flavourings beyond the range of the traditional fruit flavours. For example, the very high price of cocoa has stimulated demand for good synthetic chocolate flavourings. The natural flavour of chocolate has been

Table 7.5 A synthetic banana flavouring. The concentrations in the final product of even the most abundant of these components will only be of the order of parts per million.

Major components	*Minor components*	*Trace components*
Amyl acetate	Amyl valerate	Acetaldehyde
Amyl butyrate	Benzyl propionate	Butyl acetate
Ethyl butyrate	Cyclohexyl propionate	2-Hexenal
Isoamyl acetate	Ethyl caproate	Isoamyl alcohol
Isoamyl butyrate	Geranyl propionate	Lemon oil
Linalool	Heliotropin	α-Ionone
	Vanillin	Methyl heptanone
		Orange oil

Table 7.6 A synthetic chocolate flavouring
(US Patent No. 3619210). The
proportions of carriers, solvents,
etc. have been omitted.

	Parts
Dimethyl sulfide	1
2,6-Dimethylpyrazine	3324
Ethylvanillin	143
Isovaleraldehyde	100

extensively investigated, but not exhaustively. Two groups of compounds
have been implicated, sulfides (*e.g.* dimethyl sulfide and related sulfur
compounds) and pyrazines. No fewer than 57 different pyrazines have
been identified in cocoa volatiles! A patented formulation for a chocolate
flavouring containing a pyrazine and sulfide is shown in Table 7.6. In
view of the understandable desire of many people to minimise the
number of "chemicals" in their food, perhaps we should consider the
possibility of replacing natural ingredients with a few synthetic ones!

In recent years snack foods have appeared in a remarkable range of
synthetic savoury flavours—"smoky bacon" and "barbecue beef" are
typical examples. The synthesis of flavours such as these by replicating
the range of volatiles found in the original foods presents a number of
problems. Firstly, the flavour compounds are not usually present in the
original raw materials—they arise during cooking, mostly *via* the
Maillard reaction or the thermal degradation or oxidation of unsaturated fatty acids. Secondly, products of the Maillard reaction frequently
have highly complex molecular structures, placing them beyond the
range of commercial synthetic chemistry. A third issue is that such
compounds would in any case be unlikely to gain the necessary safety
clearance from the regulatory authorities. The solution has been the
development of what are known as "process flavours".

Process flavours are produced by heating precisely specified mixtures
of proteins, amino acids, sugars, and fats together under closely controlled conditions. The raw materials must be toxicologically acceptable
as food ingredients and heating conditions (time, temperature and pH)
must resemble those encountered in normal cooking. Comparison of the
composition of foodstuff raw materials with the range of flavour volatiles present after cooking will often enable the chemist to predict the
mixture of amino acids, *etc.*, most likely to produce an acceptable
process flavour. Another type of process flavour, "liquid smoke", is
discussed in Chapter 9.

SPECIAL TOPICS

1 Off-Flavours in Meat

Mutton, the meat from mature sheep, is often regarded as having an unattractive odour (often described as "muttony") compared with lamb. This has been attributed to the accumulation of traces of free branched-chain fatty acids in the depot fat triglycerides, the most important being 4-methyloctanoic (7.83) and 4-methylnonanoic acids (7.84).

$$CH_3(CH_2)_3\overset{\overset{\displaystyle CH_3}{|}}{CH}(CH_2)_2COOH$$

(7.83)

$$CH_3(CH_2)_4\overset{\overset{\displaystyle CH_3}{|}}{CH}(CH_2)_2COOH$$

(7.84)

Similarly, "boar taint" is a problem encountered with pig meat. Male animals tend to be leaner than females and in theory this should make them preferable for meat production, but this unfortunately presents a difficulty for pig farmers. As the uncastrated male pig reaches about 200 lb (90 kg) in live weight,[x] it accumulates 5-androst-16-ene-3-one (7.85) in its body fat. This by-product of normal steroid metabolism is the cause of boar taint. Being volatile, the steroid is driven off when the meat is cooked, and its odour may cause offence. Interestingly, not everyone is able to smell it. More than 90% of women but only about 55% of men are able to detect it at typical cooking concentrations, and of those that can, far more women than men find the smell unpleasant. One can only speculate about the implications for human behaviour or physiology of these gender differences. The odour is described as sharing common features with urine and sweat, and it is therefore hardly surprising that the meat industry would like to find a method of discouraging boars from synthesising it.

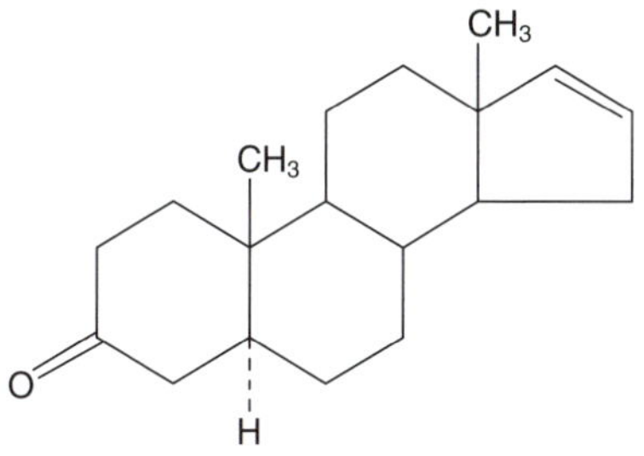

(7.85)

<hr>

[x] 200 lbs is the typical weight for a bacon pig; younger animals of around 140 lbs (65 kg) are used for pork.

2 Taints

Undesirable flavours that are found in foods as a result of contamination are known as "taints". An enormous diversity of substances can give rise to taint in food, and from a similar diversity of sources, that we only have space for a few examples here. The most important sources are storage areas, packaging materials, and the cleaning agents used on the processing plant. In many cases it can be extremely difficult to trace the source of the contamination. For example, a case has been reported[xi] in which a packaged foodstuff had an off-odour caused by trichloroanisole (7.86). This has a musty, mouldy odour detectable at 3 parts in 10^{16}! Some polyethylene packaging was found to be the source of the trichloroanisole, and it was eventually shown that a single polyethylene bag, containing polyethylene granules used in the manufacture of the packaging material, had been stored on a wooden pallet. The wood had been previously treated with chlorinated phenolic compounds as preservatives. Micro-organisms are known which are capable of converting chlorinated phenolics into the corresponding anisoles. Although this may be an extreme example, phenolic disinfectants are frequently the origin of such problems. Chlorophenol wood preservatives may find their way into the fibreboard used for cartons. Disinfectants applied to buildings, or to bedding for livestock or poultry, can ultimately reach milk and meat at low but detectable levels. *p*-Cresol (7.87), a common component of disinfectants, is detectable down to 2 parts in 10^7. Trichloroanisole contamination of the cork can also give rise to cork taint in wine.

(7.86)   (7.87)

FURTHER READING

C. Fisher and T. R. Scott, *Food Flavours: Biology and Chemistry*, Royal Society of Chemistry, Cambridge, 1997.

[xi] J. Ewender, *et al.*, in *Food and Packaging Materials – Chemical Interactions*, ed. P. Ackerman, M. Jägerstad and T. Olsen, Royal Society of Chemistry, Cambridge, 1995, p. 33.

Alternative Sweeteners, ed. L. O'Brien Nabors, Dekker, New York, 3rd edn, 2001.

G. Reineccius, *Flavor Chemistry and Technology*, Taylor and Francis, London, 2nd edn, 2006.

Flavor of Meat, Meat Products, and Seafoods, ed. F. Shahidi, Blackie, London, 2nd edn, 1998.

Food Flavourings, ed. P. R. Ashurst, Aspen, Gaithersburg, 3rd edn, 1999.

Flavour in Food, ed. A. Voilley and P. Etéivant, Woodhead, Cambridge, 2006.

Food Taints and Off-flavours, ed. M. J. Saxby, Blackie, London, 2nd edn, 1996.

R. S. Jackson, *Wine Science,* Academic Press, New York, 2000.

RECENT REVIEWS

P. A. Breslin and A. C. Spector, *Mammalian Taste Perception; Curr. Biol.*, 2008, **18**, p. R148.

M. Kroger, K. Meister and R. Kava, *Sweeteners and Other Sugar Substitutes: A Review of the Safety Issues; Comp. Rev. Food Sci. Safety*, 2006, **5**, p. 35.

A. G. Renwick, *The Intake of Intense Sweeteners – An Update Review; Food Addit. Contam.*, 2006, **23**, p. 327.

Drewnowski and C. Gomez-Carneros, *Bitter Taste, Phytonutrients, and the Consumer: A Review; Am. J. Clin. Nutr.*, 2000, **72**, p. 1424.

S. D. Johanningsmeier, R. F. Mcfeeters and M. Drake, *A Hypothesis for the Chemical Basis for Perception of Sour Taste; J. Food Sci.*, 2005, **70**, p. R44.

S. E. Ebeler, *Analytical Chemistry: Unlocking the Secrets of Wine Flavor; Food Rev. Internat.*, 2001, **17**, p. 45.

K. Srinivasan, *Black Pepper and its Pungent Principle, Piperine: A Review of Diverse Physiological Effects; Crit. Rev. Food Sci. Nutr.*, 2007, **47**, p. 735.

I. T. Johnson, *Glucosinolates in the Human Diet: Bioavailability and Implications for Health; Phytochem. Rev.*, 2002, **1**, p. 183.

Ninomiya, *Umami: A Universal Taste; Food Rev. Internat.*, 2002, **18**, p. 23.

M. Corzo-Martinez, N. Corzo and M. Villamiel, *Biological Properties of Onions and Garlic; Trends Food Sci. Technol.*, 2007, **18**, p. 609.

K. Srinivasan, *Role of Spices Beyond Food Flavoring: Nutraceuticals with Multiple Health Effects; Food Rev. Internat.*, 2005, **21**, p. 167.

CHAPTER 8

Vitamins

The vitamins are an untidy collection of complex organic nutrients which occur in the biological materials we consume as food (and in others we do not). In terms of chemical structure they have nothing in common with one another, and their biological functions similarly offer little help with their definition or classification. What does draw them together is that:

- they are essential components of the biochemical or physiological systems of animal life, and frequently plant and microbial life as well;
- as animals evolved they lost their ability to synthesise these substances for themselves in adequate amounts;
- they tend to occur only in tiny amounts in biological materials; and
- their shortage in the tissues, whether by absence from the diet or failure to absorb them, causes a specific deficiency syndrome.

This specification rules out:

- trace metals and other minerals, since these are not organic;
- essential fatty acids and amino acids, which are required in larger amounts;
- hormones, which are synthesised by the body as required and cannot be supplied in the diet; and
- substances which might well be beneficial in the treatment of some illness or condition, but whose absence does not invoke a disorder or disease in an otherwise healthy person.

Food: The Chemistry of its Components, Fifth edition
By T.P. Coultate
© T.P. Coultate, 2009
Published by the Royal Society of Chemistry, www.rsc.org

Animals are distinguished from most other forms of life by their dependence on other organisms for food. Either directly or indirectly, plants are a fundamental source of basic nutrients, and it is hardly surprising that animals have come to rely on plants for the supply of other substances as well. The risk of vitamin deficiency occurring in an otherwise adequate diet would have been slight at the time this dependence evolved, and was a small price to pay for no longer having to maintain the machinery for synthesising such a diverse range of complex substances. Several diseases around during classical times were caused by vitamin deficiency. For example, accurate descriptions of scurvy, a disease caused by deficiency of ascorbic acid, including its association with seamen living on preserved food, go back to 1150 BC (the Ebers papyrus) and Hippocrates (*c.* 420 BC). Similarly beriberi, caused by a deficiency of thiamin, was described in Chinese herbals dating back to around 2600 BC. Although the linkage to diet was noted, the idea that these conditions were caused by the *absence* of something from the diet was not widely accepted until the 20th century. For example, pellagra, caused by a deficiency of niacin, was correctly associated with maize but was assumed to be caused by the *presence* of a toxin.

Even after the association of a disease with lack of a dietary component had been confirmed, progress in isolation and identification of the substance could only be made if a suitable animal model was available. The discovery of thiamin is a classic illustration of this. In 1886 Christian Eijkman was sent to Indonesia to discover the cause of beriberi, presumed to be a bacterial infection. He failed to find a causative micro-organism, but he did notice a paralytic illness resembling beriberi in the chickens kept by the laboratory staff to supply their eggs. No connection with their nutrition or the human disease was made until the disease suddenly disappeared. This was quickly associated with a change in the chickens' diet—polished rice caused the disease, and brown rice prevented it.[i] Eijkman guessed that the symptoms shown by the chickens were very close to those of beriberi in humans, and from that moment Eijkman had his experimental model. He quickly confirmed the link when he found that the incidence of beriberi among Javanese prisoners fed a diet of polished rice was 2.8%, but only 0.09% in those on a diet of brown rice. In 1894 a Japanese naval surgeon named Takaki had also concluded that beriberi was diet-related, based on the observation that by adding milk and meat to the naval diet he was able to decrease the incidence of the disease. However, he believed the problem to be a manifestation of protein deficiency.

---

[i] Brown rice is the equivalent of wholemeal wheat, complete with the outer husk layers and the embryo, or germ. Polishing removes these, leaving only the white starchy endosperm.

Aqueous or alcoholic extracts of rice husks showed a similar effect, confirming the idea that the beriberi treatment might be related to a chemical substance. Subsequently, in 1912 Casimir Funk came close to isolating the anti-beriberi factor in rice husks, again using chickens to test the extracts. Suspecting that the factor was chemically an amine, he invented the term *vitamine*, from "vital amine". He later observed:

"I must admit that when I chose the word 'vitamine' I was well aware that these substances might later prove to not all be of an amine nature. However it was necessary for me to use a name that would sound well and serve as a 'catch word".

This has a modern ring to it. When it became apparent that other factors were not amines the final "e" was dropped.

Studies in the years up to 1914 on laboratory rats fed highly purified diets of proteins, fats, starch and minerals demonstrated that "growth factors" were missing from such diets. McCullom and Davis suggested that there were two of these, one associated with milk fat and egg yolk, described as "fat-soluble A", and another with wheat, milk and also egg yolk as "water-soluble B". The first of these became identified with the prevention of night blindness. However, in the 1920s a second fat-soluble factor was isolated and shown to prevent the bone disease, rickets. By this time the factor in fruit and vegetables which prevented scurvy was becoming known as vitamin C, although as yet unidentified. This meant that with A, B and C already spoken for, the anti-rickets (antirachitic) factor was named vitamin D.

Over the years, water-soluble B has been resolved into a host of different vitamins, referred to collectively as the "vitamin B complex", although they are not bound to one another as might be implied by the word "complex". Through the 1920s and 1930s plant and animal food-stuffs yielded an increasing number of water- and fat-soluble factors linked to the growth of laboratory animals or to their deficiency diseases. For a time this led to a proliferation of vitamins with suffixes extending through most of the alphabet. Once chemistry had advanced sufficiently to provide concise structural formulae, many of these disappeared. Some turned out to be identical to vitamins already identified, or to mixtures of other vitamins, and occasionally they were even found to lack the activity ascribed to them. Others were substances which were interesting but failed to meet the criteria set out above. Nowadays scientists try to use specific names wherever possible. Where a number of closely related substances, *e.g.* the tocopherols, have vitamin activity it is better to refer to them collectively, in this example as vitamin E, rather than suggesting that all tocopherols are equally potent as vitamins.

One should not underestimate the difficulties that the isolation and identification of vitamins presented to the original researchers. For example, the first isolation of pantothenic acid required a quarter of a ton of sheeps' livers as starting material! Although by the end of the 1930s the chemical structure of most vitamins had been established, that of cobalamin (vitamin B_{12}) was elucidated only in 1955, by Dorothy Hodgkin at Oxford University.

For the modern food chemist this brief detour into history is important to help appreciate how vitamin terminology evolved, and so that we can recognise when it is being misused. Even today it is common, especially among toiletry products and at the less responsible fringes of the food supplements industry, for obsolete terms to be resurrected and applied misleadingly.

The first task for chemists was the isolation of vitamins in a pure state and determination of their chemical structure. Once these were known, chemists were then expected to provide sensitive and accurate methods for the determination of vitamin levels in food materials. Initially the only measurements available were biological assays based on the relief of deficiency diseases in laboratory animals, but their high cost and the time they required made them unsuitable for routine measurements.[ii] Nowadays almost all vitamin analysis is carried out using chromatographic methods, particularly high-performance liquid chromatography (HPLC). A second objective for vitamin chemists was to devise commercially viable chemical syntheses for vitamins, to make these substances available for clinical purposes and as dietary supplements. Chemical routes are now available for the synthesis of most vitamins but in some cases, notably riboflavin and cobalamin, bacterial fermentation is a more economic method. Although the tocopherols can be chemically synthesised, they are more usually extracted from vegetable oils.

THIAMIN (VITAMIN B_1, ANEURINE)

Thiamin occurs in foodstuffs either in its free state (8.1a) or as its pyrophosphate ester (8.1b) complexed with protein. No distinction is made between the two forms by normal analytical techniques, or in tables of food composition. Although small amounts occur widely, only a few foodstuffs can be regarded as convenient sources. In general it is present in the greatest

---

[ii] In the early years of the 20th century the moral issue of the use of laboratory animals was less prominent than it is today.

amounts (0.1–1.0 mg per 100 g) in foodstuffs rich in carbohydrates or associated with a high level of carbohydrate metabolism in the original living material. Examples are legume seeds and the embryo component of cereal grains (the germ), which have high rates of carbohydrate metabolism during germination of the seed. Although in meat both muscle tissue and liver contain very little carbohydrate in the living animal, carbohydrate is the chief source of energy in these tissues. The reason why pork should contain about ten times the amount of thiamin (about 1 mg per 100 g) than in beef, lamb, poultry or fish has yet to be established.

(8.1a)   (8.1b)

The association of thiamin with carbohydrates arises from its role in metabolism. Thiamin pyrophosphate (TPP) is the prosthetic group[iii] of a number of important enzymes catalysing the oxidative decarboxylation of α-keto acids, including pyruvic acid (8.2) and α-ketoglutaric acid (oxoglutaric acid, 8.3), which occur as intermediates in the Krebs cycle, the key metabolic pathway of respiration. Details of the reactions involving coenzymes are not normally the province of food chemists but these can be found in the first of the Special Topics at the end of the chapter. There is some evidence that thiamin, probably as the pyrophosphate, has a quite separate role in nerve function, but the details have not yet been established.

(8.2)   (8.3)

[iii] A prosthetic group is a non-protein component of an enzyme which is involved in the reaction being catalysed and remains bound to the enzyme throughout the reaction sequence.

Although thiamin, like the other B vitamins, is nowadays usually estimated by HPLC, it is actually one of the few vitamins which can be determined using chemical methods. Even so, these methods involve the use of sophisticated techniques. First, the vitamin is extracted with hot dilute acid. Then the extract is treated with the enzyme phosphatase to convert any TPP into thiamine, and cleaned up by column chromatography. This extract is then treated with an oxidising agent such as a hexacyanoferrate[II] (ferricyanide), to convert the thiamin into thiochrome (8.4), the concentration of which is measured fluorimetrically.

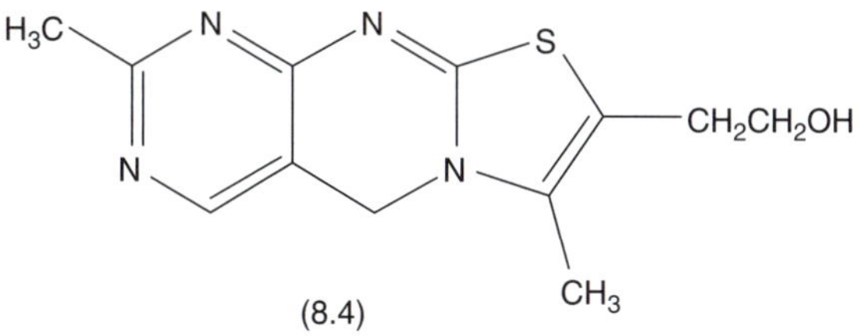

(8.4)

Thiamin is one of the most labile of vitamins. Only under mildly acid conditions, *i.e.* below pH 5, will it withstand heating. The reason for this is the readiness with which nucleophilic displacement reactions occur at the atom joining the two ring systems. Ions such as OH^-, and particularly HSO_3^-, split the molecule into two fragments (8.5):

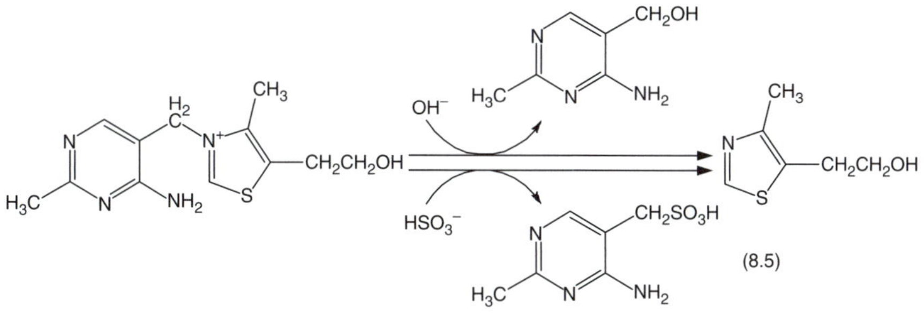

(8.5)

This means that sulfite added to fruits and vegetables to prevent browning will cause total destruction of the thiamin. Fortunately these foods are not important sources of dietary thiamin. Heat treatments such as canning cause losses of up to 20% of the thiamin, especially when the pH is above 6, but only in the baking of bread, where up to 30% may be lost, do these losses become really important. The greatest losses of thiamin, in domestic cooking as well as in commercial food processing, occur simply as a result of its water solubility rather than due to chemical attack. It is impossible to generalise about the extent of such losses, depending as they do on the amount of chopping, soaking and cooking time, but they can be minimised by making use of meat drippings and vegetable cooking water—perhaps gravy is nutritious after all!

Polishing rice, *i.e.* removing the outer layers of the grain and the germ, enhances its storage life by denying invading moulds of some of their nutrients. Although this practice is a major cause of the thiamin deficiency common in those parts of the world where rice is a staple, the difficulty of storing rice in warm, humid climates makes it essential. One way in which its effects can be minimised is by partial boiling, or "parboiling". The rice is steeped in water, steamed and dried. Ostensibly this is carried out to make polishing easier, but in fact it allows the thiamin to diffuse from the germ into the endosperm, so that it is not lost when the outer layers are removed. On the other hand parboiling does not have a beneficial effect on the availability of niacin. This vitamin's poor solubility means that it does not diffuse into the endosperm under these conditions.

The question as to how much of any particular vitamin is required to maintain health in various circumstances is more appropriate to textbooks of nutrition, but thiamin is a somewhat unusual case. Its involvement in carbohydrate metabolism has already been referred to, and it is recognised nowadays that determination of the thiamin content of a single staple foodstuff must allow for the energy that the foodstuff itself contributes. Authorities recommend a minimum thiamin intake of $96 \, \mu g \, MJ^{-1}$ $(0.4 \, mg \, kcal^{-1})$ for all humans. Armed with this figure, it can be easily demonstrated that vegetarian diets dominated by refined cereals (from which the germ has been removed) will automatically result in thiamin deficiency unless supplemented by legumes or whole-cereal products.

In some parts of the world where raw or fermented (but uncooked) fish is extensively consumed, including rural Thailand, an enzyme found in shellfish and the viscera of freshwater fish, thiaminase I, has been implicated in thiamin deficiency. The enzyme breaks the $-CH_2-$ link with the nitrogen atom of the thiazole ring. Fortunately the activity of the enzyme is destroyed by heating. Polyphenolic compounds, particularly those found in tea, coffee and many fruit and vegetables can also act as antithiamin factors and can lead to deficiency symptoms, notably in communities where chewing fermented tea leaves or betel nuts is common.

RIBOFLAVIN (VITAMIN B₂)

The structure of this vitamin is usually presented as riboflavin itself (8.6), the isoalloxazine nucleus with only a ribitol side-chain attached, but in most biological materials it occurs predominantly in the form of two nucleotides, flavin mononucleotide (FMN, 8.7), and flavin–adenine dinucleotide (FAD, 8.8). These occur both as prosthetic groups in respiratory enzymes known as flavoproteins, and also in the free state, when they are referred to as coenzymes.

The distribution of riboflavin in foodstuffs is very similar to that of thiamin, in that it is present at least in small amounts in almost all biological tissues. It is particularly abundant in meat (0.2 mg per 100 g), and especially in liver (3.0 mg per 100 g). In contrast to thiamin, cereals are not a particularly rich source, but milk (0.15 mg per 100 g) and cheese (0.5 mg per 100 g) are valuable sources. Dried brewer's yeast and yeast extracts contain large amounts of riboflavin and many other vitamins, but it is interesting that the thiamin does not leak out of the yeast during fermentation whereas riboflavin does (giving a concentration in beer of up to 0.04 mg per 100 g).

The flavoprotein enzymes of which riboflavin forms part of the prosthetic group are all involved in oxidation–reduction reactions. It is the iso-alloxazine nucleus that receives electrons from substrates which are being oxidised and donates electrons to substrates which are being reduced (8.9):

Most flavoprotein enzymes are involved in the complex respiratory processes which occur in the mitochondria of living cells, but some take part in other aspects of metabolism. The enzyme glucose oxidase, mentioned in Chapter 2, is an example. This enzyme transfers two hydrogen atoms from C-1 of the glucose to a molecule of oxygen *via* the FAD prosthetic group, resulting in the formation of hydrogen peroxide.

Riboflavin is one of the most stable vitamins. The alkaline pH conditions in which it is unstable are rarely encountered in foodstuffs. The most important feature of its stability is its sensitivity to light. While this is not significant in opaque foods such as meat, the effect can be dramatic in milk. It has been reported that as much as 50% of the riboflavin contained in the usual glass bottle of milk is destroyed by two hours' exposure to bright sunlight. The principal product of irradiation is known as lumichrome (8.10), and at neutral and alkaline pH some lumiflavin (8.11) is also produced. The breakdown of riboflavin has wider significance than simply loss of vitamin activity. Both breakdown products are much stronger oxidising agents than riboflavin itself and catalyse massive destruction of ascorbic acid (vitamin C). Even a small drop in riboflavin content of milk can lead to virtual total elimination of the ascorbic acid content. Fortunately milk is not an important source of vitamin C in most diets. The other important side-effect of riboflavin breakdown is that in the course of the reaction, which involves interaction with oxygen, highly reactive states of the oxygen molecule are produced, in particular singlet oxygen. Although this has only a transient existence it is able to initiate the autoxidation of unsaturated fatty acids in the milk fat, as discussed in Chapter 4. The result is an unpleasant off-flavour. When milk is to be sold in supermarkets, whose chilled cabinets are often brightly lit by fluorescent lights, it is obviously essential that opaque containers rather than traditional glass bottles are used.

(8.10)   (8.11)

Since vegetables are not an important source of riboflavin, losses by leaching during blanching or boiling in water are not of significance.

Other cooking operations, including meat roasting and cereal baking also have a negligible effect.

Riboflavin is not readily determined in the laboratory unless HPLC is available. The most commonly applied chemical technique requires initial treatment with dilute hydrochloric acid at high temperature to release the riboflavin from the proteins to which it is normally bound. After clean-up procedures, the extract is examined fluorimetrically. The absorbance by riboflavin at 440 nm is accompanied by an emission at 525 nm.

Before HPLC became widely available laboratories routinely determining the levels of many different vitamins in food frequently used microbiological assays. These depend on the existence of special strains of bacteria which have lost the ability to synthesise one particular vitamin for themselves. The special strains are obtained by subjecting more normal (*i.e.* wild-type) members of the species to irradiation or mutagenic chemicals and isolating mutants which possess the required characteristics. Specially prepared growth media are used, providing all the nutrients required by the test strain except for the vitamin being determined. The amount of bacterial growth in a culture using this medium depends on how much of the vitamin is present, either from a food sample extract or from a standard solution used for comparison. Although difficult to carry out, its sensitivity in experienced hands is impressive. The assay for riboflavin using a strain of *Lactobacillus casei* covers the range 0–200 ng per sample, and this is one of the least sensitive. The vitamin B_{12} assay using *L. leichmannii* has an assay range of 0–0.2 ng per sample!

Although a great deal is known about the biological functions of riboflavin, few distinctive deficiency symptoms have been identified. Those which have been recognised are also associated with deficiencies of other B vitamins, notably pyridoxine.

PYRIDOXINE (VITAMIN B_6, PYRIDOXOL)

The form of this vitamin which is active in the tissues, again as the prosthetic group of a number of enzymes, is pyridoxal phosphate (8.12). However, when the vitamin is identified in foodstuffs it is in one of three forms, all having lost the phosphate group: pyridoxine (or more correctly pyridoxol, 8.13), pyridoxal and pyridoxamine (8.14). In foodstuffs of plant origin the first two usually predominate, whereas in animal materials the last two predominate. At least two-thirds of the vitamin in living tissues is probably present as enzyme prosthetic groups, in other words tightly bound to protein. In muscle tissue one particular enzyme, phosphorylase, provides the major share of the total. Milk is one exception, only about 10% being protein-bound.

$$\text{(8.12)} \qquad\qquad \text{(8.13)} \qquad\qquad \text{(8.14)}$$

Pyridoxine (this name is used to describe all the active forms of the vitamin other than pyridoxol) is believed to be widely distributed among foodstuffs, but the problems posed by its assessment mean that reliable data are scarce. It has been found at least in small amounts (around 10 µg per 100 g) in almost all biological materials in which it has been sought. Meat and other animal tissues and egg yolk are particularly rich sources (around 500 µg per 100 g); milk and cheese contain about 50 µg per 100 g. The germ of wheat grains is another good source (around 500 µg per 100 g), accounting for over 90% of the total in the complete grain. In consequence a high proportion (75–90%) of the vitamin is lost in the manufacture of white flour. The wide distribution of this vitamin, however, means that any diet poor enough to be deficient in it will almost certainly also be deficient in most of the other B vitamins as well, and a specific clinical deficiency is rarely encountered in humans.

Compared with other vitamins in prosthetic groups, pyridoxal phosphate is quite versatile in terms of the type of reaction in which it is involved. Almost all the enzyme-catalysed reactions in which it participates have amino acids as substrates, phosphorylase being an exception.[iv] Details of a transamination reaction, one of the best known involving pyridoxal phosphate, are given in the first of the Special Topics at the end of the chapter.

The question of the stability of pyridoxine during food processing is complicated by the fact that its different forms differ in stability, and they also have a tendency to interconvert during some processing operations. Pyridoxol is very stable to heat within the pH range usually encountered in foodstuffs. The other two forms are slightly less stable but, as with the other water-soluble vitamins, leaching is the major cause of loss during cooking and processing. Although there is little loss of vitamin activity during milk processing, the more extreme treatments such as the production of evaporated milk cause extensive conversion of pyridoxal, the predominant form in raw milk, to pyridoxamine. A similar change occurs during the boiling of ham. When milk is dried, extensive losses of pyridoxal can occur due to interactions with the sulfhydryl groups of

---

[iv] Pyridoxine has recently been shown to have a role in the control exercised by steroid hormones over the synthesis of certain proteins.

proteins. Interaction with the free amino groups of proteins leads to pyridoxamine formation—which does not cause loss of vitamin activity.

The analysis of pyridoxine levels in foods is not practical using chemical techniques, and HPLC or microbiological methods must therefore be employed. A special source of difficulty with microbiological assay of the vitamin is that its different forms differ in their potency for microorganisms, although apparently they do not influence animal nutrition.

NIACIN (NICOTINIC ACID, NICOTINAMIDE)

Niacin is a collective name for nicotinic acid (8.15) and its amide, nicotinamide (8.16).[v] Niacin is included in the vitamin B group (as are all the water-soluble vitamins apart from ascorbic acid), but its previous designation as vitamin B_3 is now obsolete. In living systems the pyridine ring occurs as a component of nicotinamide adenine dinucleotide (NAD, 8.17), and its phosphate derivative (NADP, 8.18).[vi] NAD occurs as a prosthetic group in a few enzymes, *e.g.* glyceraldehyde-3-phosphate dehydrogenase, but most of the NAD and NADP in living cells functions as an electron carrier in respiratory systems.

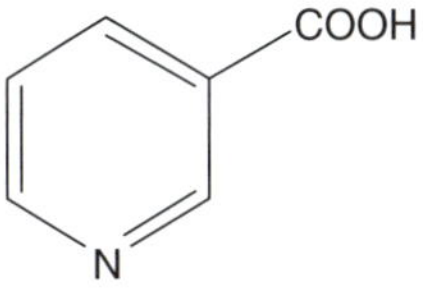

(8.15) 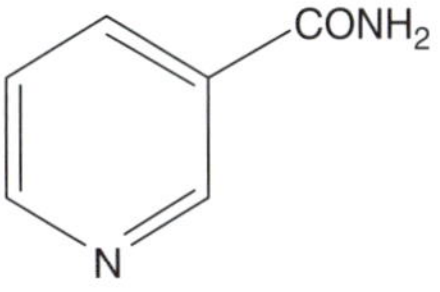   (8.16)

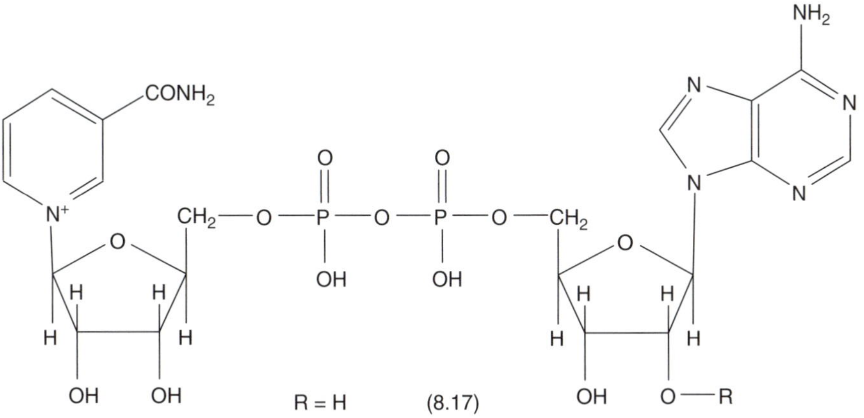

R = H   (8.17)
R = PO_3H_2   (8.18)

---

[v] In the USA niacin specifically refers to nicotinic acid rather to than the two forms of the vitamin.
[vi] In older literature NAD and NADP are referred to as di- and triphosphopyridine nucleotides, DPN and TPN, respectively.

For example, in most of the Krebs cycle reactions (the key metabolic pathway of respiration) the electrons removed from the organic acid substrates are transferred initially to NAD (8.19), and are then used to reduce oxygen to water in the oxidative phosphorylation system of the mitochondria. The route from NAD involves first flavoproteins and then porphyrin ring-containing proteins known as cytochromes. A striking feature of the reduction of NAD is the very high absorbance shown by NADH at 340 nm (molar extinction coefficient, $\varepsilon = 6.22 \times 10^6$), a wavelength at which the oxidised form has almost no absorbance at all.

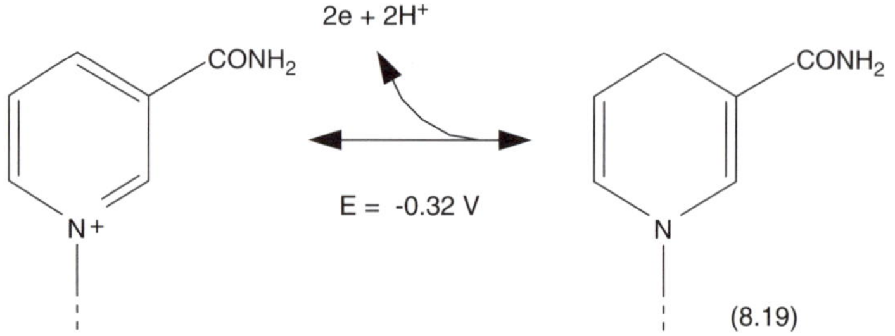

$$2e + 2H^+$$

$$E = -0.32 \text{ V}$$

(8.19)

In general the distribution of niacin in foodstuffs is similar to that of the other vitamins already discussed in this chapter. Although present in small amounts in all biological materials, it is meat (with 5–15 mg per 100 g) which provides the richest source of niacin in most diets. Dairy products such as milk, cheese and eggs are surprisingly poor sources, with no more than 0.1 mg per 100 g. On the other hand fruit and vegetables, particularly legume seeds, are quite useful sources (0.7–2.0 mg per 100 g). Green coffee beans contain trigonelline (*N*-methylnicotinic acid, 8.20), which is converted to nicotinic acid during the roasting process. The resulting levels of nicotinic acid range from 80 mg kg^{-1} when medium roasted, and up to 500 mg kg^{-1} in dark roasted beans.

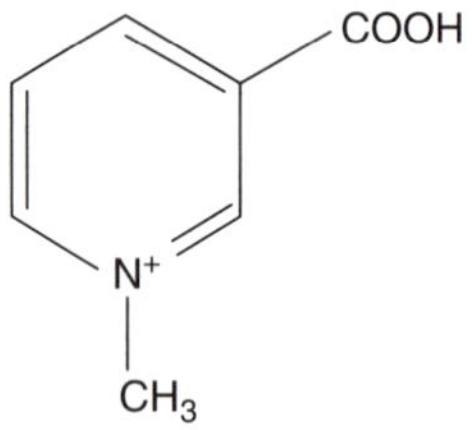

(8.20)

Many cereals contain apparently high contents of niacin (*e.g.* whole wheat flour, 5 mg, and brown rice 4.7 mg per 100 g). However, as with pyridoxine, milling can have a dramatic effect since a large proportion of the niacin is located in the germ; white bread and wholemeal bread contain 1.7 and 4.1 mg per 100 g, respectively.

To compound the problem, much of the niacin in cereals is tightly bound to other macromolecules in the grain. In wheat, rice and maize a small protein has been implicated, and in both wheat and maize a polysaccharide, probably a hemicellulose, is also involved. The normal digestive processes in the small intestine do not liberate the vitamin and it remains unabsorbed. This problem is especially important for consumers of maize. There is no record of pellagra, the disease caused by a deficiency of niacin, before 1753 but after this time it became increasingly prevalent in southern Europe. The increase is now clearly seen to be associated with the replacement of rye with maize as a dietary staple, but the connection was not made at the time. Pellagra was endemic in the USA, especially the south. Its absence from the Yucatan peninsular, Mexico, where maize was first cultivated was never noticed, however, and even when the link to maize was identified it was assumed that a microbial toxin present in spoilt maize was to blame. The reason for the apparent immunity to pellagra in Mexico is that the maize flour used for making tortillas, the dietary equivalent of bread, was traditionally ground from maize kernels initially treated with lime water. The purpose of the alkali was to weaken the husk of the maize kernel to make milling easier, but we now know that under alkaline conditions the bound unavailable niacin was liberated by heating! Consumers of "corn chips" and other Westernised versions of tortillas will be reassured that alkali treatment of maize is still a key step in their manufacture.

A further complication in the assessment of niacin contents of food-stuffs is that the human body itself has a limited but significant ability to synthesise the vitamin. By a rather tortuous metabolic pathway the amino acid tryptophan (*see* Figure 5.2) can be converted to nicotinic acid. The effectiveness of this route is inevitably low, due to the competing demands of protein synthesis for tryptophan. Experimental evidence suggests that about 60 mg of dietary tryptophan is required to replace 1 mg of dietary nicotinic acid. This explains the beneficial effect of milk and eggs in cases of pellagra, even though both are poor sources of the actual vitamin. It also illustrates the care needed in the interpretation of food composition tables when diets are being assessed.

Niacin is stable under most conditions of cooking and processing, leaching being the only significant cause of losses. During the baking of cereals, especially in products rendered slightly alkaline by use of baking powder, the availability of niacin may actually increase.

COBALAMIN (CYANOCOBALAMIN, VITAMIN B₁₂)

The chemical structure of this vitamin is by far the most complex of all
the vitamins. As shown in Figure 8.1 its most prominent feature is a
"corrin" ring system. This resembles the porphyrin ring system found in
chlorophyll (*see* Chapter 6) and haem proteins, *etc.* (*see* Chapter 5). It
differs in two respects—the ring substituents are mostly amides and, in
cobalamin, the coordinated metal atom is cobalt. The fifth coordination
position is occupied by a nucleotide-like structure based on 5,6-dime-
thylbenzimidazole, which is also attached to one of the amide groups in
the corrin ring. The sixth coordination position (R in Figure 8.1), on the
face of the corrin opposite the fifth position, can be occupied by a

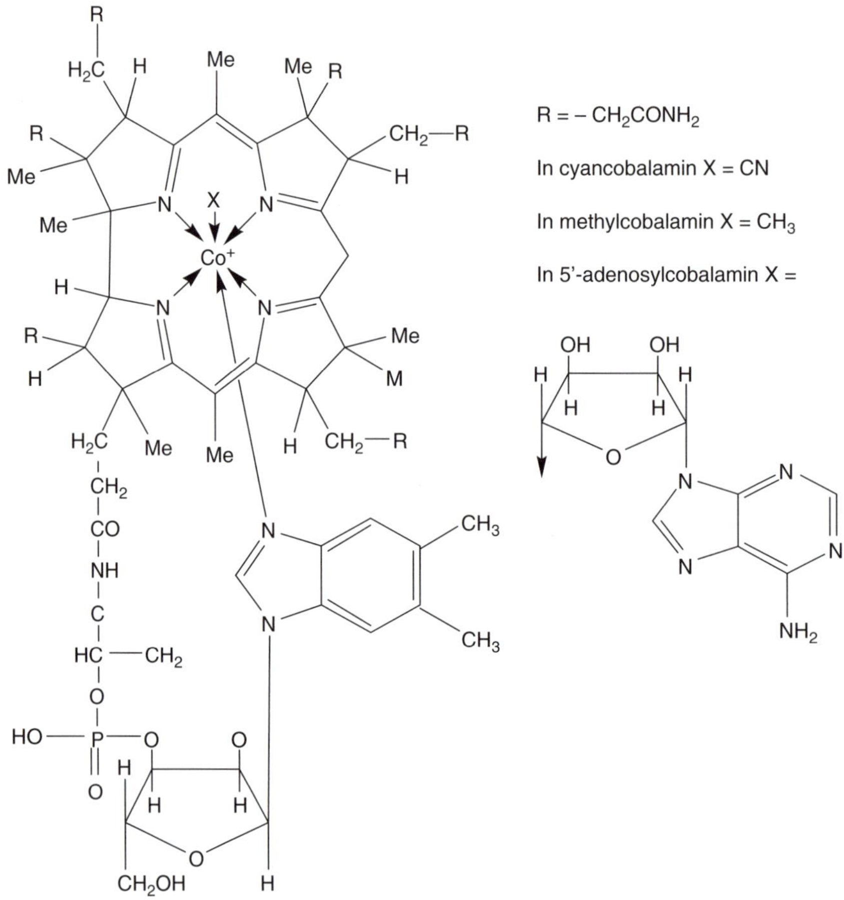

Figure 8.1 The structure of cobalamin and its derivatives. In the absence of these or
other ligands (i.e. R) a water molecule occupies the sixth coordination
position.

number of different groups. As normally isolated, the vitamin contains a CN group in the sixth position, hence the name cyanocobalamin. In its functional state, in which it is the prosthetic group of a number of enzymes, it may have either a methyl group or a 5′-deoxyadenosyl residue in the sixth position.

The distribution of cobalamin in foodstuffs is most unusual. Bacteria are the only organisms capable of synthesising it, and it is therefore never found in higher plants. When traces are detected in vegetables the source is almost always surface contamination with faecal material used as fertiliser. The bacterial resident in the colon collectively secrete a total of about 5 µg daily, but since there is no absorption from the colon this is wastefully excreted. The daily requirement is about 1 µg. The vitamin occurs in useful quantities in most foodstuffs of animal origin (per 100 g: eggs 0.7 µg, milk 0.3 µg, meat (muscle tissue) 1–2 µg, liver 40 µg, kidney 20 µg), but in all these cases the vitamin has arrived either from micro-organisms or animal tissues consumed in the diet, or from synthesis by intestinal micro-organisms high enough in the alimentary tract for absorption to occur.

As a result of the low requirement for this vitamin and its abundance in many foodstuffs deficiency is extremely rare, except among dietary extremists such as vegans. It has even been suggested that a delay in the onset of deficiency symptoms in the young children of vegan parents is caused by the lack of toilet hygiene inevitable in young children, and this is providing a source of the vitamin from unwashed hands. A less picturesque explanation is that it may take as long as 10 years after ceasing consumption of the vitamin for an otherwise healthy individual to show deficiency symptoms, since the vitamin is lost from the body at an extremely slow rate. In contrast to the other water-soluble vitamins the body is able to conserve vitamin B_{12} by storing it in the liver.

The classic disease of cobalamin deficiency is pernicious anaemia. This is almost always caused by failure to absorb the vitamin rather than lack of it in the diet. Patients with pernicious anaemia are unable to synthesise a carbohydrate-rich protein known as the "intrinsic factor". This is normally secreted by the gastric mucosa and complexes the vitamin. Only in this complexed form is the vitamin readily absorbed from the small intestine. In the 1920s pernicious anaemia was successfully treated by having patients consume large daily amounts (100–200 g) of calves' liver (apparently raw!) which would provide doses of the vitamin sufficient to overcome its inefficient absorption.

In mammals the catalytic activities of cobalamin enzymes involve one-carbon units such as methyl groups, an element of metabolism in which coenzymes based on the vitamin folic acid are also involved. The details are really outside the scope of food scientists but will be touched on here to show how a deficiency of cobalamin leads to pernicious anaemia.

The synthesis of DNA is obviously dependent on a supply of the purines and pyrimidines (adenine, guanine, cytosine and thymine) which their component nucleotides contain. The synthesis of purines and pyrimidines depends on the supply of a particular form of tetra-hydrofolate (THF), the biochemically active form of folic acid. The enzyme reaction which maintains the supply of this utilises methyl cobalamin as a coenzyme. A deficiency of cobalamin thus inhibits DNA synthesis, and the rapidly dividing cells of the bone marrow which generate red blood cells (erythrocytes) are among the first to die. The resulting shortage of erythrocytes is known as a megaloblastic anaemia. Not surprisingly, a deficiency of folic acid also causes anaemia, but pernicious anaemia caused by cobalamin deficiency is specifically accompanied by neuropathy, or damage to the nerve cells, which starts in the peripheral nervous system but eventually reaches the spinal chord and brain, with catastrophic results.

Despite the complexity of its structure cobalamin is fairly stable to food processing and cooking conditions, and the losses which do occur are not sufficient to cause concern to nutritionists. As with all water-soluble vitamins, leaching is the major cause of loss. Microbiological assay (*see* page 321) is the only practical method of determining the vitamin in foodstuffs.

FOLIC ACID (FOLACIN)

Folic acid is the name given to a group of widely distributed compounds whose vitamin activity is related to that of the cobalamins. The structure of folic acid itself is shown in Figure 8.2.

The form active as a coenzyme is tetrahydrofolate, which has the pteridine nucleus in its reduced state and several ($n = 3$ to 7) glutamate residues carried on the *p*-aminobenzoate moiety. Accurate determination of the folic acid content of foodstuffs is difficult, even using microbiological assays, since it is necessary to remove all but one of the glutamyl residues before consistent results are obtained. This can now be achieved by enzymic methods, but it has meant that earlier estimates of the folic acid content of foodstuffs could be orders of magnitude too low. Only recently has measurement of particular glutamyl–folic acid conjugates become possible. In animal tissues, in which the pentaglu-tamyl conjugate predominates, liver is a particularly good source (300 µg per 100 g); muscle tissues contain much less (3–8 µg per 100 g). Leafy

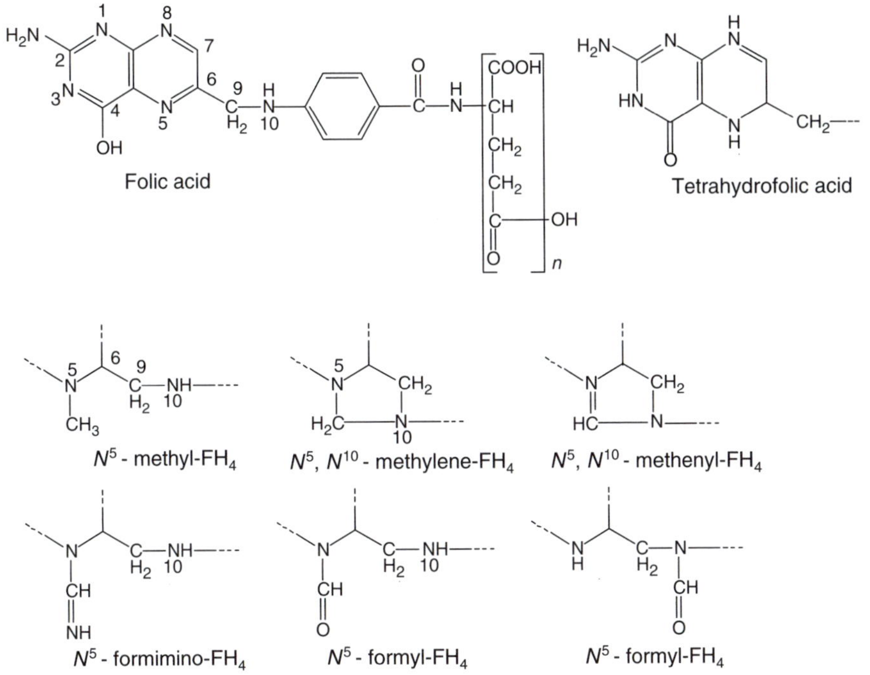

Figure 8.2 The structure of folic acid, tetrahydrofolic acid and the one-carbon units carried by tetrahydrofolate.

green vegetables, in which the heptaglutamyl conjugate predominates, have around 20–80 µg per 100 g) and probably constitute the most important source in modern diets.

As a coenzyme, tetrahydrofolate (FH_4) is concerned solely with the metabolism of one-carbon units, which it carries at N-5 or N-10. While bound to the coenzyme, the one-carbon unit may be oxidised, reduced or undergo other changes, giving the variety of structures shown in Figure 8.2. These are intimately involved in the metabolism of many amino acids, purines and pyrimidines (components of nucleotides and nucleic acids), and indirectly haem. In animals the metabolism of methyl groups, which are mainly supplied in the diet by choline, is particularly complex. Both folic acid and cobalamin-requiring reactions are involved, and it is not surprising that there is some overlap in the deficiency symptoms of these two vitamins.

There is growing support for the view that deficiency of folic acid is widespread, even in the comparatively well fed populations of Europe and North America. It is becoming clear that a deficiency of folic acid in the very early stages of pregnancy, described as the periconceptual period, can lead to neural tube defects such as spina bifida in the foetus.

An average diet may not supply sufficient folic acid, and the UK authorities now recommend vitamin supplements to ensure that all young women consume 0.4 mg daily to ensure adequate folate reserves in advance of pregnancy. An intake ten times this figure is recommended if a neural tube defect has occurred during a previous pregnancy. The importance of supplementation is increased by the fact that the bio-availability of naturally occurring folates is actually lower than that from folate supplements. There is also considerable support for the view that improving the levels of folates in the diet can reduce the incidence of strokes and other cardiovascular problems.

In the past routine dietary fortification has not been recommended. This is because extra folic acid can alleviate, and therefore mask, the obvious anaemia caused by cobalamin deficiency. However, it has no effect on the neurological damage accompanying cobalamin deficiency, and by the time the damage is detected this cannot be reversed. In spite of this reservation, the authorities in the USA in 1998 made folic acid fortification mandatory in specified grain products, including bread-making flour and corn meal, at a level of 140 mg per 100 g. Since adopting this policy, rates of neural tube defects in the USA have almost halved. In the UK between 700 and 900 pregnancies are affected each year by neural tube defects, and it has been estimated that some 13 million people in the UK have a low folate intake. Canada and Chile have also introduced similar fortification and in 2007 the Food Standards Agency recommended that it should be adopted in the UK.

Folic acid and its derivatives are considered to be reasonably stable under the conditions in most cooking and food processing operations. Stability to heat and acidity depend both on the number of glutamyl residues and the nature of any attached one-carbon units. Oxidation can be a problem, but ascorbic acid provides a valuable protective effect. As with other water-soluble vitamins, most cooking and processing losses are the result of leaching.

BIOTIN AND PANTOTHENIC ACID

These two vitamins have little in common in terms of structure or function. What brings them together is the total absence of reports of deficiency symptoms in man or animals other than under the artificial conditions of the laboratory. Biotin (8.21) occurs as the prosthetic group of many enzymes which catalyse carboxylation reactions, such as the conversion of pyruvate to oxaloacetate. Most foodstuffs contain at least a few µg per 100 g (assayed microbiologically), and there appears to be a more than adequate amount present in a normal diet. The only known deficiency of biotin occurs when extremely large quantities of raw egg are eaten. Egg white contains a protein, avidin, which complexes biotin, but avidin is denatured when eggs are cooked. Biotin is apparently quite stable to other cooking and processing procedures.

(8.21)

As with biotin, much more is known about the biochemical function of pantothenic acid (Figure 8.3) than about its status as a food component. It occurs in two forms in living systems, either as part of the prosthetic group of the acyl carrier protein (ACP) component of the fatty acid synthetase complex, or as part of coenzyme A, the carrier of acetyl (ethanoyl) and other acyl groups in most metabolic systems. In each of these roles the acyl group is attached by a thioester link (*see* Figure 8.3) to the sulfhydryl group, which replaces the carboxyl group in the free vitamin. Pantothenic acid is extremely widely distributed, usually at around 0.2–0.5 mg per 100 g, a more than adequate dietary level. As with many vitamins, liver contains especially high levels, in this case about 20 mg per 100 g. Quite high losses of pantothenic acid have been reported during food processing, mostly due to leaching, but some decomposition is also known to occur during heating under alkaline or acid conditions.

The term "provitamin" is used to describe substances which are direct precursors of vitamins, and their presence in the diet can be a substitute for the vitamin itself. The most obvious example is β-carotene, whose relationship to retinol, vitamin A, was described in Chapter 6 and will be further discussed later in this chapter. However, the term "provitamin B_5" has been adopted for certain ingredients of cosmetics and shampoos. For reasons which seem to be connected more with marketing than nutrition, this name is applied to panthenol (8.22) and its ethyl ester. While panthenol can be absorbed by the skin (at least by the skin of laboratory rats) and converted to pantothenic acid which then

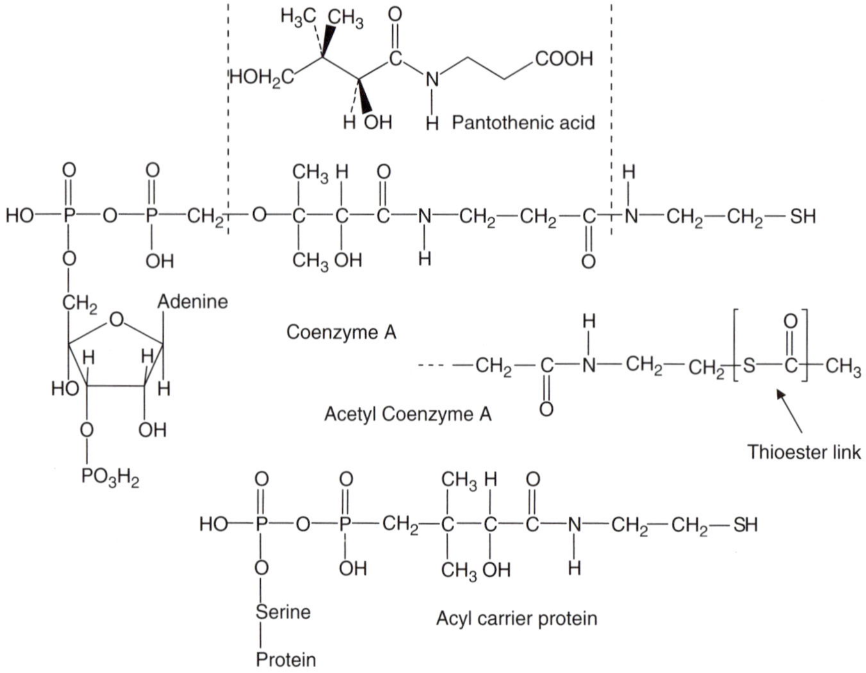

Figure 8.3 Pantothenic acid and its relationship to coenzyme A and acyl carrier protein.

appears in the urine, the relevance of this to the nutrition of the skin or the appearance of one's hair is doubtful.

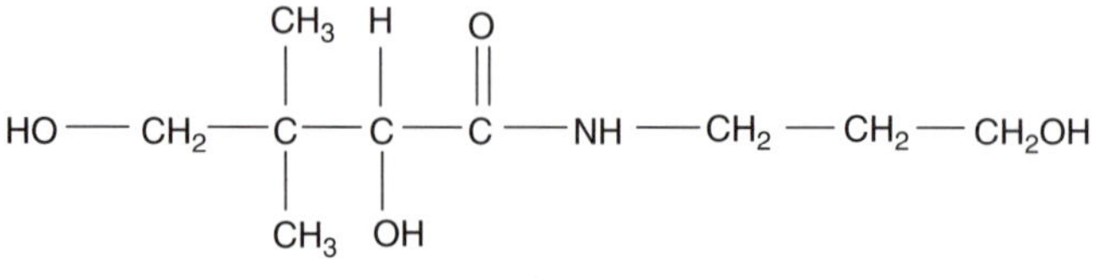

(8.22)

ASCORBIC ACID (VITAMIN C)

Ascorbic acid is a rather curious vitamin. It occurs widely in plant tissues, and is also synthesised by almost all mammals, and is not therefore a vitamin for them, the exceptions being the primates, the guinea-pig and some fruit-eating bats. The only discernible common feature of this otherwise disparate collection of mammals is a liking for fruit, a rich source of the vitamin. The structure of L-ascorbate (pK_a 4.04 at 25 °C) and some related substances are shown in Figure 8.4. The oxidised form, dehydro-L-ascorbic acid (DHAA) (an unfortunate name, since the

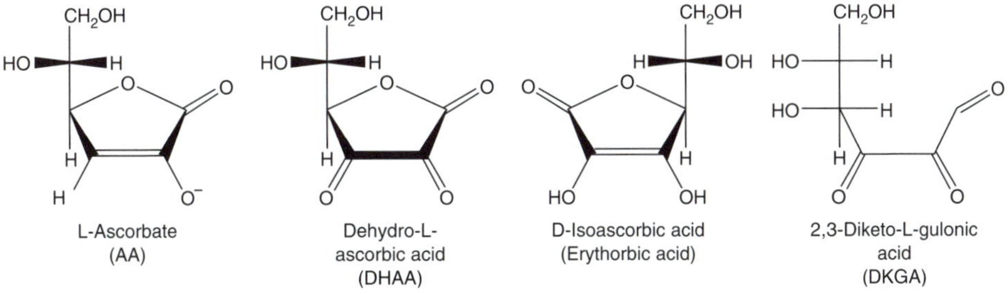

Figure 8.4 The structure of ascorbic acid and related substances.

Table 8.1 Typical ascorbic acid content of some fruit and vegetables. Values for particular samples may vary considerably from these figures, which represent that present in the edible part when fresh.

	Ascorbic acid (mg/100 g)		Ascorbic acid (mg/100 g)
Plums	3	Carrots	6
Apples	6	Old potatoes	8
Peaches	7	Peas	25
Bananas	11	New potatoes	30
Pineapple	25	Cabbage	40
Tomatoes	25	Cauliflower	60
Citrus fruit	50	Broccoli	110
Strawberries	60	Horseradish	120
Blackcurrants	200	Sweet red peppers	140
Rose hips	1000	Parsley	150

molecule has no ionisable groups), is distributed with the reduced form in small, rarely accurately determined proportions. DHAA has virtually the same vitamin activity as L-ascorbate. The mirror image isomer, most commonly referred to as erythorbic acid, has no vitamin activity but does behave similarly to ascorbate in many food oxidation–reduction systems, and it therefore finds some similar applications as a food additive.

Much more is known about the distribution in foodstuffs of ascorbic acid than any other vitamin. The reason is not that ascorbic acid is inherently more interesting or more important, but simply because it is much easier to determine, so that when a research worker or student decides to do a project "on vitamins" this is the one that tends to be chosen. The richest sources are fruits, but as the data in Table 8.1 show, wide variations do occur. It has to be remembered that a foodstuff that is eaten in large quantities will be an important dietary source even if it contains only modest levels of the vitamin. The very high levels to be found in parsley are nutritionally insignificant when compared with the quantity in the new potatoes it is used to garnish. Cereals contain only traces, and of the animal tissue foods only liver (30 mg per 100 g) is a valuable source. Human milk provides enough ascorbic acid to prevent scurvy in breast-fed infants, but cow's milk and other dairy products contain insignificant amounts due to oxidative losses during processing.

The role of ascorbic acid in mammalian physiology and biochemistry is far from fully understood but is quite different from its role in plants. In plants its function is as an antioxidant (*see* page 123), particularly in helping to protect the photosynthetic machinery of the chloroplasts from the effects of singlet oxygen and other highly reactive oxygen species (*see* page 153). Glutathione is also involved in this activity (*see* page 207). In animals, it is involved in hydroxylation reactions. It has been implicated in a number of reactions, particularly in the formation of collagen, the connective-tissue protein. After incorporation into the collagen polypeptide chain, proline (*see* Figure 5.2) is hydroxylated to hydroxyproline in a complex reaction in which ascorbic acid is one of the essential components (8.23).

$$(8.23)$$

Many of the symptoms of scurvy, an ascorbic acid deficiency disease, are to be expected from a failure of normal connective-tissue formation, for example, wounds fail to heal, internal bleeding occurs and the joints become painful. Another important reaction requiring ascorbate is the hydroxylation of 3,4-dihydroxyphenylethylamine (dopamine, a derivative of tyrosine) to noradrenaline (8.24).

$$(8.24)$$

This may account for the disturbances of the nervous system which accompany scurvy. There is also evidence that the uptake of non-haem iron (*see* page 447) from the duodenum is enhanced by the presence of high levels of ascorbic acid. Ascorbic acid reduces iron from its normal ferric (Fe^{III}) state in food to the ferrous (Fe^{II}) state. Compared with ferrous iron very little ferric iron is absorbed, as it forms the insoluble hydroxide. Unfortunately the amounts of additional ascorbic acid required to double the iron uptake from a typical meal are quite high, around 100 mg. Ascorbic acid also has a role in the transport of iron in the blood and its mobilisation from reserves.

The diversity of roles played by ascorbic acid in human physiology and biochemistry probably account for its adoption as a wonder cure-all for a number of conditions and diseases, particularly the common cold. The actual evidence for the value of extra-high doses (up to 5 g per day, more than 100 times the recommended dietary intake) is extremely flimsy, something that cannot be said for the profits to be made by selling ascorbic acid to health food enthusiasts.

There are many different routes available for the chemical synthesis of ascorbic acid, but the most usual commercial route is that shown in Figure 8.5. This is an interesting combination of traditional synthetic chemistry with biotechnology.

Almost all of the numerous methods devised for the determination of ascorbic acid depend on measurement of its reducing properties by titration with an oxidising agent, the most popular being 2,6-dichlorophenolindophenol (DCPIP, 8.25). In its oxidised form DCPIP is blue, colourless when reduced. Highly coloured plant extracts such as blackcurrant juice obviously present problems, because the end-point of the titration is obscured. There are numerous more or less satisfactory techniques for circumventing the problem. One example is titration with N–bromosuccinimide (NBS) in the presence of potassium iodide and ether. At the end-point of the titration, when all the ascorbic acid present has been oxidised to DHAA, the excess NBS liberates iodine from the potassium iodide, which dissolves in the ether phase to give a visible brown colour. Very few of the methods commonly used determine the DHAA present; one that does not discriminate between the two compounds is the fluorimetric method using o-phenylenediamine.

(8.25)

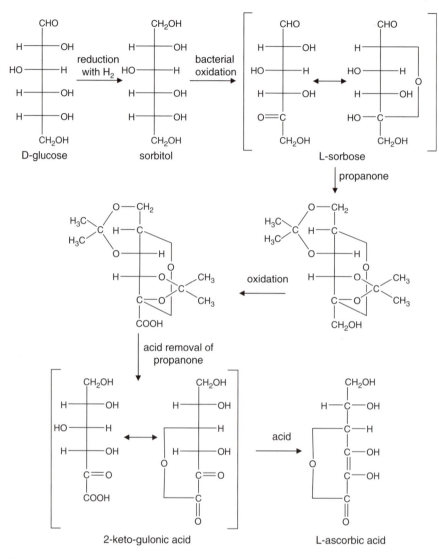

Figure 8.5   The commercial synthesis of ascorbic acid from glucose.

Information resulting from a variety of studies shows that ascorbic acid is one of the least stable vitamins. Losses during processing, cooking or storage can occur by a number of different routes, but few experiments have been devoted to discovering which routes are involved in the case of particular foodstuffs. Leaching is obviously important in vegetable preparation if the cooling or processing water is discarded. Green vegetables may lose more than 50% of vitamin C in this way if they are boiled for prolonged periods. Losses from root vegetables are usually smaller, simply because they present a smaller surface area in relation to their weight.

In the presence of air, most degradation of the vitamin occurs through the formation of the less stable DHAA. Once formed, DHAA rapidly undergoes irreversible ring-opening to give 2,3-diketo-L-gulonic acid (DKGA), which has no vitamin activity. Oxidation to DHAA can occur by a variety of mechanisms. In plant tissues stored without blanching, especially if they have been sliced, peeled or otherwise damaged, the enzyme ascorbic acid oxidase is very active, catalysing the reaction:

$$\text{ascorbate} + \tfrac{1}{2} O_2 \rightarrow \text{DHAA} + H_2O_2$$

The enzyme phenolase is also responsible for loss of ascorbic acid when, as discussed in Chapter 6, ascorbic acid reduces *o*-quinones to the original *o*-diphenols. Metal cations, particularly Fe^{3+} and Cu^{2+}, catalyse the oxidation and may cause serious losses in food products. Even in the absence of catalyst the oxidation occurs quite rapidly at elevated temperatures. Formation of DKGA is virtually instantaneous at alkaline pH, rapid at neutrality, and slow under acidic conditions. The reason now becomes clear for not adding sodium bicarbonate to preserve the colour of green vegetables during cooking.

The vitamin is quite stable in the acid environment of fruit juice processing. Its stability is enhanced by the presence of citrates and flavonoids, both of which can complex metal cations, but the juice should still be kept deaerated as far as possible. Under anaerobic conditions at elevated temperatures ascorbic acid will undergo breakdown in the same way as other sugars (*see* Chapter 2), except that in this case CO_2 is evolved.

The low cost of ascorbic acid and its obvious acceptability as a nutrient make it a valued food additive, for technological rather than nutritional purposes. In fruit and vegetable products, notably dehydrated potato, it is used as an antioxidant to prevent the browning catalysed by phenolase. In the manufacture of cured meats it is used as a reducing agent to lower the concentration of nitrite needed to give a good pink colour (*see* page 365), and in modern bread making it is used as a flour improver (*see* page 205).

At the beginning of this chapter the author denied any need to describe how each vitamin was discovered. In spite of this, no apology is required for the inclusion of the following paragraphs, a reminder that scientists have not always taken themselves as seriously as they sometimes do today.

Chemists who in exasperation may at times have referred to an unidentifiable sugar on their chromatogram as "@*!-knows" may be comforted to know that this brand of terminology has an honourable history. In 1928 Albert Szent-Györgyi set out to publish the details of a sugar acid with strong reducing properties he had isolated, initially from the adrenal cortex of cattle and later from orange juice and cabbage water. In the draft of his paper he pencilled in the name "ignose" (Latin, *ignorare*, not to know). This, and his alternative suggestion, "Godnose", were each rejected by the editor of the *Biochemical Journal* on the grounds that they didn't publish jokes. Szent-Györgyi reluctantly had to settle for "hexuronic acid". His paper did hint at the possibility that this substance might actually be the antiscorbutic factor (*i.e.* scurvy preventing) already designated "vitamin C", but this was not confirmed until 1932.

Once the structure was firmly established by Haworth in Birmingham, using the kilogram quantities Szent-Györgyi was able to isolate from a local vegetable, red pepper, the name ascorbic acid was adopted in recognition of its antiscorbutic properties. Food chemists should forgive Szent-Györgyi's later remark that *"...vitamins were, to my mind, theoretically uninteresting. 'Vitamin' means that one has to eat it. What one has to eat is the first concern of the chef, not the scientist."*

RETINOL (VITAMIN A)

This vitamin is the first of the fat-soluble vitamins to be considered in this chapter. Its active form in mammalian tissues is the alcohol retinol (8.26), but most of the vitamin occurs in the diet as precursors, described earlier in this chapter as provitamins, of which there are several.

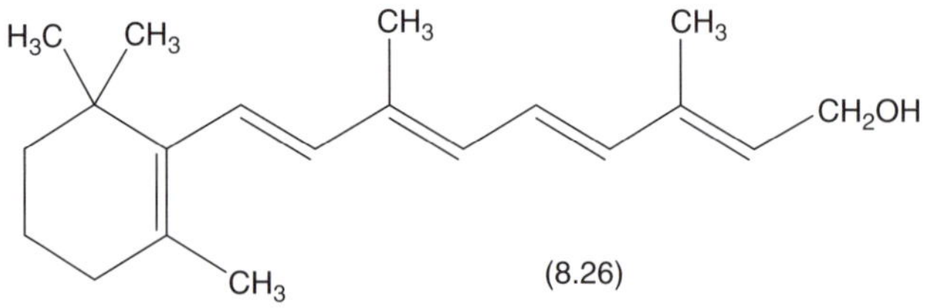

(8.26)

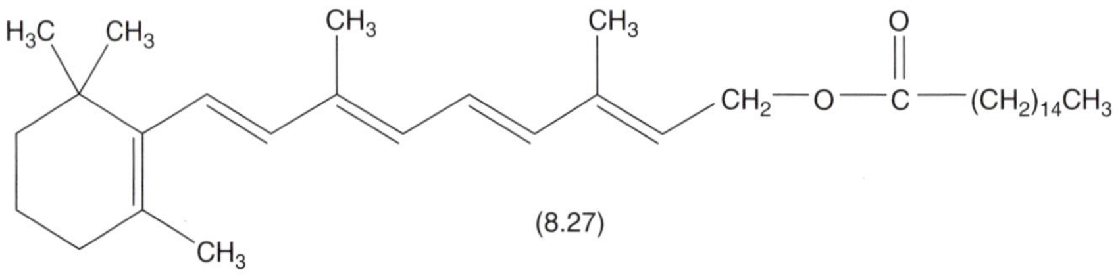

(8.27)

Plant tissues contain the vitamin in the form of carotenoids. As was indicated in Chapter 6, any carotenoid that contains the β-ionone ring system, as do β-carotene (at both ends of the molecule), α- and γ-carotene and β-*apo*-8′-carotenal (at one end only), is converted by enzymes in the mucosa of the small intestine to retinol. The absorption of carotenoids and their conversion to retinol do not occur with total or uniform efficiency, which makes estimation of the vitamin activity of different foodstuffs complicated. At the present time the equivalence figures recommended by the FAO/WHO[vii] in 1967 are in general use, *i.e.* 6 μg of β-carotene or 12 μg of other active carotenoids being regarded as the equivalent of 1 μg of retinol. It is increasingly recognised that this approach is over-simplified, firstly because the efficiency of conversion is inversely related to the level of β-carotene intake, giving a range of 4 μg to 10 μg equivalent to 1 μg of retinol. Ideally allowances also need to be made for the source of dietary carotenoids since less than 10% may be absorbed from raw vegetables whereas around 50% is absorbed from cooked vegetables. On cooking, dietary carotenoids tend to dissolve in any lipids present and in this state their absorption into the blood stream is much more efficient.

In 2001 the US Institute of Medicine recommended that this system of recording vitamin A activity in terms of retinol equivalents (mg RE) should be replaced by one based on retinol activity equivalents (mg RAE), in which the retinol activity equivalency ratio for β-carotene was set at 12 : 1 and that for other provitamin A carotenoids at 24 : 1. In other words, 1 mg of retinol was to be regarded as nutritionally equivalent to 12 mg β-carotene, or 24 mg of other provitamin A carotenoids. These figures were based on the assumption that the subject was healthy and consuming a good mixed diet. Studies of people living on very poor diets, *e.g.* in the Far East, which tend to include a low proportion of fats or oils, indicate that a factor of around 21 : 1 should be applied to the total carotenoid content of the diet when its vitamin A status is assessed. At the present time these new equivalence factors are yet to be adopted in the UK or EU and have not been applied to any of the data reported in this book.

Leafy green vegetables contain between 100 and 3000 retinol equivalents (*i.e.* an amount equivalent to 1 μg of retinol per 100 g), mostly as β-carotene. Carrots, the classic source, contain about 2000 retinol equivalents per 100 g. The other major plant source is red palm oil, whose β-carotene content is sufficient to give some 20 000 retinol equivalents per 100 g.

[vii] FAO = Food and Agriculture Organisation (of the United Nations); WHO = World Health Organisation.

In animal tissues the vitamin, always derived originally from plant sources, is stored and transported as the retinyl ester of long-chain fatty acids, mostly palmitate (8.27) and stearate. Dietary retinyl esters are first hydrolysed and then re-esterified by the intestinal mucosal cells during the process of absorption. A specific protein binds the retinyl esters in the liver, with the result that liver and liver oils are particularly good sources in the diet. Cod and halibut liver oils are well known for their vitamin A content, around 10^5 and 10^7 retinol equivalents per 100 g, respectively. The vitamin A content of mammalian liver varies widely, from 3000 in pig's liver and 17 000 in sheep's liver—to 600 000 retinol equivalents per 100 g in polar bear liver! The relationship between the diet of these animals and the vitamin A content of their livers is obvious. Meat (*i.e.* muscle tissue) contains very little of the vitamin. Although milk itself has only a low concentration, dairy products in which the lipid phase of the milk is concentrated, such as butter and cheese, contain 500–1000 retinol equivalents per 100 g. Margarine sold in Britain is required by law to be supplemented with vitamin A, usually in the form of synthetic retinyl acetate or red palm oil, to give a level of 800–1000 retinol equivalents per 100 g, comparable with butter.

The chemical synthesis of vitamin A was first achieved in 1947. In the form of retinyl acetate it is now produced commercially on a large scale in a simple process using β-ionone as starting material. As one might expect from its chemical structure, ensuring the resistance of vitamin A preparations to autoxidation is a major issue, especially in the presence of trace metals.

The function of vitamin A in the visual process is well known, and the sequence of events in the "visual cycle" of the rod cells in the retina is shown in Figure 8.6. The inefficiency of the cycle inevitably results in losses, and the supply of all *trans*-retinol is replenished from the circulation. The absence of adequate supplies results in loss of vision in dim light, known as "night blindness". This was linked to a deficiency of the vitamin as early as 1925.

Deficiency of vitamin A can ultimately lead to death—the result of effects quite unconnected with night blindness. These include abnormal bone development, disorders of the reproductive system and xerophthalmia, a drying and degenerative condition of the cornea of the eye, which untreated causes total blindness. There are indications of a role for retinol in the synthesis of the glycoproteins occurring in mucous membranes, a deficiency is known to lead to keratinisation of these membranes. This can lead to increased susceptibility to infection following the degeneration of the epithelia which line the digestive and respiratory tracts.

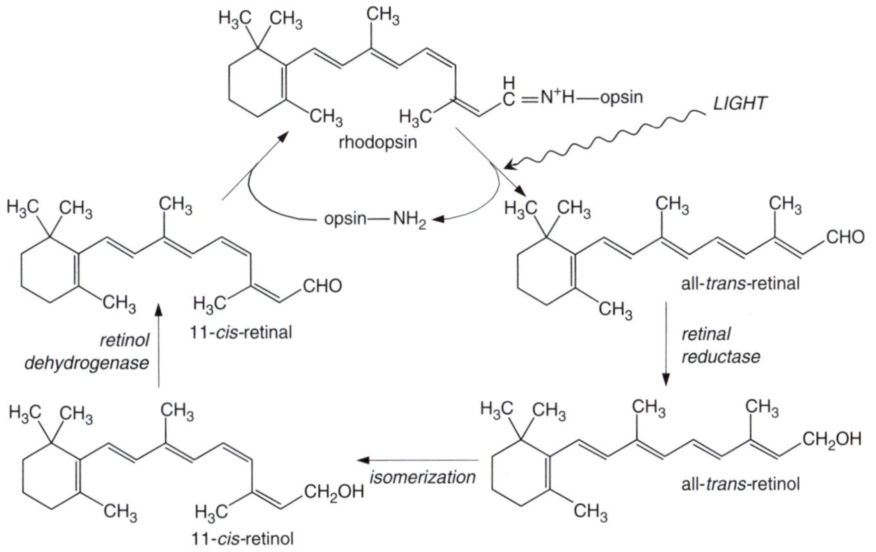

Figure 8.6 Vitamin A in the visual cycle. In the dark, 11-cis-retinal combines with the protein opsin to form rhodopsin. When this absorbs light (λ_{max} 500 nm) the *cis* bond isomerises to *trans* and ultimately the all-*trans* retinal detaches from the protein. During the course of this isomerisation the complex processes which lead to the generation of the nerve impulse are initiated. The 11-*cis*-retinal is then regenerated to recombine with opsin in darkness.

It is now recognised that retinol has even greater involvement in mammalian physiology than the preceding paragraph might imply, including control of protein synthesis resembling the action of the steroid hormones. Retinol circulating in the blood stream is taken up by target cells and converted to retinoic acid, which binds to specific receptors in the cell nuclei and influences the process of cell differentiation, and hence the growth and development of embryos.

Only since the 1960s has it been fully appreciated that vitamin A deficiency afflicts millions of people in the developing countries of Asia. In these areas neither dairy products nor green vegetables feature much in the diet of the poor, and a variety of methods are being introduced to supplement diets with synthetic retinyl acetate. The possibility of introducing "golden rice" varieties, in which the natural content of β-carotene has been enhanced by genetic modification, has been rejected since the contribution of vitamin A this would make to the diet was too small to be useful.

Considerable attention has been focused in recent years on the possibility that vitamin A might have an anticancer function. Laboratory experiments with unnaturally high doses of the vitamin have shown promise but it has not proved possible to apply these to therapeutic effect. Large doses simply lead to an increase in the liver reserves without action on the target tissues. The beneficial effect of vitamin A in this context is in its antioxidant activity. Both retinol and β-carotene are weak antioxidants (*see* page 123), but β-carotene also undergoes a specific reaction with singlet oxygen (*see* page 153). This quenching reaction may help to eliminate the dangerous by-products resulting from reactions involving oxygen in the body. There is growing support for the view that vitamin A, vitamin E (*see* page 346), selenium (as a component of the enzyme glutathione peroxidase), and ascorbic acid function together as the body's antioxidant system, and should therefore be considered collectively when the relationship between health and the vitamin content of the diet is assessed.

An unusual feature of retinol is its toxicity in excess. Most victims of hypervitaminosis have been health food fanatics, but it has been reported that polar explorers have also suffered—following the ill-advised consumption of a polar bear. As the recommended daily intake for the vitamin is only about 750 retinol equivalents, one modest portion of polar bear liver was equivalent to over two years' supply of retinol! While the excessive consumption of polar bear liver is unlikely to present much of a hazard for most of us, it is now recommended that pregnant women should guard against excessive consumption of vitamin A by not eating liver. Fortunately there is no risk of hypervitaminosis from excess β-carotene consumption: as its level in the diet increases, the efficiency of its conversion to retinol falls.

Both retinyl esters and β-carotene are fairly stable in food products. Most of the information on stability concerns β-carotene. High temperatures in the absence of air, such as in canning, can cause iso-merisation to neocarotenes (*see* Chapter 6), which only possess vitamin activity if one end of the molecule remains unaffected. In the presence of oxygen breakdown can be rapid, especially in dehydrated foods in which a large surface area is exposed to the atmosphere. The hydroperoxides resulting from the autoxidation of polyunsaturated fatty acids will bleach carotenoids and can be presumed to have a similar destructive effect on retinyl esters.

The determination of the amount of vitamin A in a foodstuff is complicated by the diversity of forms in which it occurs. An indication can be gained from the absorbance of a solvent extract at 328 nm, the absorption maximum of retinol. Elaborate procedures are available to compensate for the interfering absorbance of other substances at this wavelength, but nowadays the problem can be avoided using preliminary chromatography to separate the different carotenoids and retinyl esters.

CHOLECALCIFEROL (VITAMIN D, CALCIFEROL)

Only one form of this vitamin, known as cholecalciferol (8.28) or vitamin D₃, occurs naturally in the diet. A different form of the vitamin, ergocalciferol (vitamin D_2, 8.29), is used for pharmaceutical applications and for food supplementation, since it can be synthesised on the industrial scale from ergosterol (8.30), a cheap starting material.

There is an increasing tendency for physiologists—but not nutritionists—to regard cholecalciferol as a hormone rather than a vitamin. As we shall see, this is because it functions in the body in the same way as a hormone, rather than as an enzyme prosthetic group like most other vitamins. Furthermore, humans and other mammals under the right circumstances have a more than adequate capacity for synthesising it themselves.

Animals are able to synthesise cholesterol and other steroids, one intermediate in the pathway being 7-dehydrocholesterol. The epidermal cells of the skin contain this substance, which is converted to cholecalciferol by the action of the UV component of sunlight. The course of the reaction is shown in Figure 8.7.

When human skin is adequately exposed to sunlight the body is able to provide sufficient vitamin D for its needs (2.5–10 μg per day, infants,

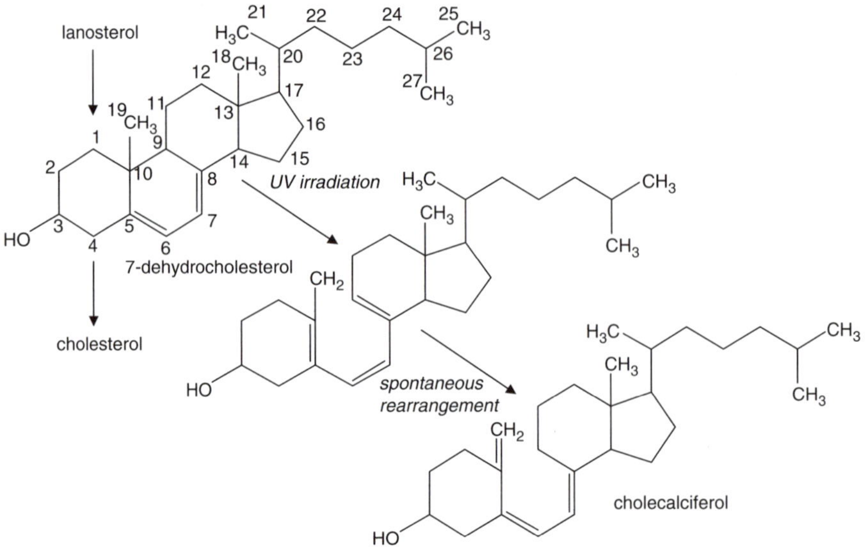

Figure 8.7   The formation of cholecalciferol in the skin. The standard numbering of the carbon atoms of steroids is used in 7-dehydrocholesterol, the immediate biosynthetic precursor of cholesterol. The optimum wavelength for the first reaction is 300 nm. The structure shown for cholecalciferol appears to differ from that in the text (8.30), but this is only due to the single bond between carbons 6 and 7 being rotated through 180° to illustrate more clearly the relationship between cholecalciferol and its parent compounds.

children, and pregnant or lactating women having the highest requirement). The body is able to maintain useful stores of the vitamin, but unlike the other fat-soluble vitamins it is not only stored in the liver but is found in the lipid elements of most tissues, particularly in adipose tissue. These reserves provide some protection against its reduced formation in the skin during the winter months, but without dietary supplies many people would still have a vitamin D deficiency for part of the year. A supply in the diet is naturally important in parts of the world without much sunshine, especially since such regions are usually too cold to encourage sunbathing. The heavy skin pigmentation associated with ethnic groups originating in tropical regions prevents a high proportion of the ultraviolet light from reaching the epidermal cells in which cholecalciferol is formed. It may well be that the pale complexion of the northern European races is an adaptation to ensure the most efficient use of the restricted level of sunlight available.

    The pattern of distribution of cholecalciferol in foodstuffs is strikingly similar to that of retinol. As with retinol, the fish liver oils contain spectacular levels, *e.g.* mackerel liver oil has 1.5 mg per 100 g. The muscle tissues of fatty fish such as salmon, herring and mackerel are also valuable dietary sources (5–45 µg per 100 g), but mammalian tissues, including liver, provide less than 1 µg per 100 g. Milk contains as little as 0.1 µg per 100 g, but fatty dairy products such as butter and cream can contain useful amounts (1–2 µg per 100 g) and egg yolk has about 6 µg per 100 g. An unfortunate contrast with vitamin A is that plant materials do not contain significant amounts of vitamin D.

It is nowadays common practice to enhance the vitamin D contents of certain foods, including breakfast cereals, milk (in the USA), and margarine. In the UK there is a legal requirement for the vitamin D content of margarine to be between 7.05 and 8.82 µg per 100 g,[viii] a level some ten times greater than that present in butter. These levels are normally expressed in International Units; 1 IU of vitamin D equals 0.025 µg of cholecalciferol or ergocalciferol. The growing unpopularity of fatty dairy foods (*see* Chapter 4) may eventually lead to a requirement for vitamin D supplementation of milk in the UK.

Rickets, a childhood disease closely linked to vitamin D deficiency, is a failure of proper bone development. First described in 1645, it was always looked upon as a disease of the urban poor. By the late 19th century 80% of children in London showed symptoms. Although some authorities guessed correctly that it was a "sunlight deficiency" disease, syphilis and hereditary factors were also blamed. The fact that poverty and nutritional deprivation could in some senses be inherited escaped attention. Cod liver oil was being used as cure as early as 1848 (at Manchester Infirmary) but most of the medical profession remained sceptical until the 1920s.

Whether obtained from the diet or formed in the skin, cholecalciferol is converted in the liver to the physiologically active compound 1,25-dihydroxycholecalciferol (calcifetriol, 8.31). This is one of the three hormones (the others are calcitonin and parathormone) which together control calcium metabolism. Calcifetriol promotes the synthesis of the proteins which transport calcium and phosphate ions through the cell membranes. A lack of calcifetriol prevents the uptake of calcium from the intestine, and the resulting shortage of calcium for bone growth is manifested as rickets.

[viii] This curious degree of precision has no nutritional significance; it is merely a mathematical quirk arising from the conversion of ounces to grams.

(8.31)

Another feature of this vitamin is that it shares with retinol its toxicity in excess. Babies and young children were at one time the usual victims, a result of confusion in the dosages of halibut liver oil and cod liver oil required (3000 µg and 250 µg per 100 g, respectively). Massive overdoses cause calcification of soft tissues such as the lungs and kidneys, but in babies even modest overdoses can cause intestinal disorders, weight loss and other symptoms.

At the levels occurring in most foodstuffs the determination of cholecalciferol is extremely difficult. Biological assays using rats reared on a rickets-inducing diet were essential until the development of gas chromatography (GC) and high-performance liquid chromatography (HPLC). As a result very little work has been done on the behaviour of cholecalciferol in food. In general it is believed to be fairly stable to heat processing and only liable to oxidative breakdown in dried foods such as breakfast cereals.

VITAMIN E (α-TOCOPHEROL)

Of all the vitamins, this one and the one concluding this chapter, vitamin K, have defeated efforts to replace their old alphabetical designations with more scientific terminology. In each case this is because no single chemical name is both sufficiently comprehensive and sufficiently exclusive.

The E vitamins are a group of derivatives of 6-hydroxychroman carrying a phytyl side-chain. As shown in Figure 8.8, tocopherols have a

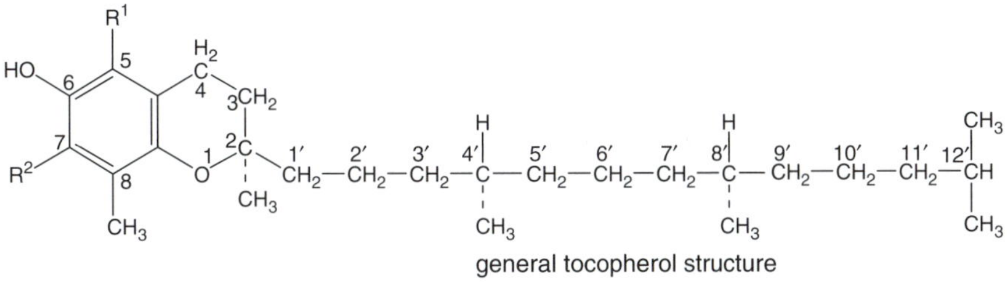

general tocopherol structure

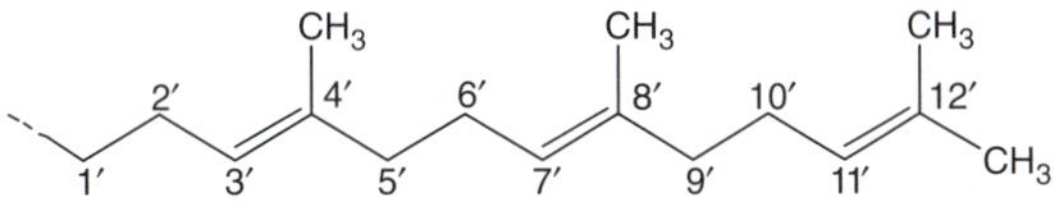

tocotrienol side chain

Figure 8.8 Vitamin E structures. The naturally occurring tocopherols and tocotrienols all have the configurations of their asymmetric centres (2, 4' and 8') as shown here and are distinguished by the prefix *d* included in the name, e.g. α-(*d*)-tocopherol. In the absence of *d* or *dl*, *d* is assumed. Synthetic tocopherols are mixtures of all eight possible isomers and the prefix dl is used instead.

fully saturated side-chain whereas in tocotrienols the side-chain is unsaturated. Variations in the degree of methylation of the chroman nucleus give rise to the α-, β-, γ- and δ-members of each series. All eight compounds shown in Figure 8.13 are found in nature, but of these only α-, β- and γ-tocopherols and α- and β-tocotrienols are widespread.

The evaluation of much of the data on the occurrence of vitamin E is complicated by the variations in vitamin activity between the different tocopherols and tocotrienols. Expressions of total tocopherol content are of no value. Taking α-(*d*)-tocopherol as 1.0, the vitamin activities of the β-, γ- and δ-(*d*) forms are 0.27, 0.13 and 0.01, respectively. The *dl*-racemic mixtures have about three-quarters of the activity of the corresponding pure *d*-isomers. Of the tocotrienols only the α form has significant vitamin activity (about 0.3). Data such as these make any statement of total tocopherol content in mg in a given weight of the foodstuff nutritionally meaningless. In situations like this we must resort to the International Unit (IU). In the case of vitamin E this is defined as the average minimum daily amount of the vitamin necessary to prevent sterility in a female rat, which actually turns out to be an amount close to 1 mg of α-tocopherol. It is common, as for example in Table 8.2, to use "mg α-tocopherol equivalents" rather than IU. Although α-tocopherol is the most potent, these data show that the less potent but often more abundant γ-tocopherol is at least as important in nutritional terms.

Table 8.2   The vitamin E, tocopherol and tocotrienol content of vegetable oils. These figures are based on data from *McCance and Widdowson* (*see* Chapter 1, Further Reading) and *Handbook of Vitamins*, ed. L. J. Machlin, Dekker, New York, 2nd edn, 1990. They should be regarded as typical rather than absolute values.

	Vitamin E as α-(d)-tocopherol	d-Tocopherols				Total tocotrienols
		α	β	γ	δ	
			mg/100 g			
Cottonseed oil	43	39	–	39	–	–
Olive oil	5	5	–	trace	–	–
Palm oil	33	26	–	32	7	46
Soybean oil	16	8	–	59	26	–
Wheatgerm oil	137	119	71	26	27	21
Maize (corn) oil	17	11	5	60	2	–

The most important sources of vitamin E in the diet are the plant seed oils (50–200 mg α-tocopherol equivalents per 100 g). Most other plant tissues contain less than 0.5 mg per 100 g. Among the vegetables the leafy green varieties are the richest sources, *e.g.* spinach ∼1.7 and cabbage ∼7 mg α-tocopherol equivalents per 100 g. The older dark green outer leaves of cabbages contain much higher levels (the figure quoted here) than the almost white inner tissues, from which the vitamin is virtually absent.

Animal tissues, including liver, milk and eggs, contain similarly low levels. Unlike other fat-soluble vitamins, vitamin E is not found in particularly large amounts in fish liver oils; cod liver oil, for example, has only ∼25 µg per 100 g.

The tocopherols are the natural antioxidants of animal tissues, and possibly also those of plants. Vegetable oils such as those listed in Table 8.2 are also protected by their tocopherol content. Antioxidants block the free-radical chain reactions of lipid peroxidation (*see* Chapter 4). The tocopherols have a special affinity for the membrane lipids of the mitochondria and endoplasmic reticulum of animal cells. It has been suggested that the shape of the tocopherol side-chain permits the formation of a complex with the arachidonic acid component of the phospholipids in these membranes. Mitochondrial membranes from animals reared on diets deficient in vitamin E are exceptionally prone to peroxidation. In view of the antioxidant role of vitamin E it would be expected that dietary supplements would be beneficial in reducing the likelihood of cardiovascular disease. Unfortunately this attractive hypothesis has not been borne out by recent research, which has tended to show that the intake levels required before measurable effects on clinical markers of cardiovascular disease are detected are extremely high, of the order of 1000 IU per day!

A deficiency of dietary selenium can cause symptoms in farm and laboratory animals, similar to those arising from vitamin E deficiency. The only known physiological role of selenium is in the active site of glutathione peroxidase, the enzyme responsible for the breakdown of hydrogen peroxide and organic peroxides. This enzyme therefore constitutes a second line of defence against the consequences of lipid autoxidation.

In spite of the tendency of some health food enthusiasts to regard vitamin E as beneficial in all aspects of human sexual behaviour,[ix] no such deficiency condition has ever been identified in man in normal circumstances. It is now well established that in order to maintain satisfactory levels of the vitamin in the tissues, the minimum intake is related to the quantity of polyunsaturated fatty acids in the diet. A diet high in vegetable oils which are rich in vitamin E will also unfortunately be rich in the polyunsaturated fatty acids that these oils contain (*see* Chapter 4).

Determination of vitamin E in foodstuffs is complicated by the diversity of its various forms, each with different vitamin activity. Straightforward chemical tests for total tocopherols can be carried out, but they are clearly valueless. Chromatographic methods, GC and HPLC, can however distinguish between the different forms and are now well established.

It is difficult to draw generally applicable conclusions about the stability of vitamin E in foodstuffs during storage or processing. The most significant loss of tocopherols occurs when food materials which also contain polyunsaturated fatty acids are exposed to air, and the vitamin then exercises its antioxidant function. An example is in the manufacture of breakfast cereals, when drum-drying or extrusion cooking may be involved. In extracted vegetable oils the vitamin is quite stable, unless conditions arise which permit autoxidation of the polyunsaturated fatty acids. Its presence in the oil will delay the onset of rancidity, but inevitably the build-up of peroxides in the oil will eventually overcome the effect of the tocopherols and cause oxidation to compounds lacking antioxidant and vitamin activity. Vitamin activity is rapidly lost from the oil content of commercial frozen products such as potato chips which have been deep fat-fried. Low-temperature storage does not, however, prevent the loss of much vitamin activity.

[ix] The characteristic symptom of vitamin E deficiency in laboratory animals is reproductive failure caused by foetal resorption or testicular degeneration. The name tocopherol is derived from the Greek, τοκοσ (offspring), and φερειν (to bear).

VITAMIN K (PHYLLOQUINONE, MENAQUINONES)

This is the last of the fat-soluble vitamins to be considered. As has already been mentioned, the alphabetic designation has been maintained in view of the lack of suitably concise chemical names for the various substances with vitamin K activity. All the K vitamins are chemically menadione (2-methyl-1,4-naphthoquinone, 8.32) or its derivatives. Phylloquinone (8.33), known as vitamin K_1, has a phytyl side-chain. The vitamin K_2 series, the menaquinones, have side-chains of varying lengths of up to 13 isoprenoid units, but those most commonly encountered have between 4 and 10, in particular vitamin K_2 itself, which has seven (8.34).

(8.32)   (8.33)

(8.34)

Vitamins K are widely distributed in small amounts in biological materials, but reliable data are hard to find. Leafy green vegetables such as spinach and cabbage are particularly rich in the vitamin ($> 100 \,\mu g$ per 100 g), but most other vegetables such as peas or tomatoes contain 50–100 μg per 100 g. Animal tissues, including liver, contain similar levels. Potatoes, fruit and cereals contain less than 10 μg per 100 g. Cow's milk and human milk contain small amounts, 6 and 1.5 μg per 100 g, respectively. Although animals are unable to synthesise the K vitamins, humans are not particularly dependent on dietary supplies because considerable quantities are synthesised by the bacteria of the large intestine. Vitamin K_2 was first isolated from putrefying fish meal. Deficiency of vitamin K is an important issue in poultry farming but it is not encountered in otherwise healthy adult humans. Very occasionally, new-born infants may show deficiency symptoms.

The only clearly identified physiological role for vitamin K is in blood clotting. The process of clot formation depends on the conversion of a number of different protein factors from inactive to active forms. The best understood of these is the conversion of the inactive prothrombin to the active proteolytic enzyme, thrombin. This conversion requires the carboxylation of several glutamate residues (8.35). A vitamin K molecule in its reduced form, a hydroquinone (8.36), is an essential component of the carboxylation reaction.

(8.35)   (8.36)

Vitamin K deficiency thus results in failure of the blood clotting mechanism. Many substances which antagonise the activity of vitamin K are now identified. Most are derivatives of coumarin, such as dicoumarol, 3,3'-methylene-bis-4-hydroxycoumarin (8.37), and warfarin, 3-(α-acetonylbenzyl)-4-hydroxycoumarin (8.38). Dicoumarol is produced when clover for animal feeding is spoiled by fermentation and causes disease in cattle. Warfarin is used as a rat poison, and at rather lower doses in human medicine to prevent blood clotting in patients with cardiovascular disease.

(8.37)   (8.38)

SPECIAL TOPICS

1. Details of the Coenzyme Functions of Thiamine and Pyridoxine

The role of TPP in oxidative decarboxylation is shown in Figure 8.9. Although the details of enzyme reaction mechanisms are not essential for food chemists, this is a valuable illustration of the reason vitamin structures are so often complex. It is also worth noting that the essential component of coenzyme A, pantothenic acid, is also a vitamin.

The reactions involving pyridoxal phosphate are a further example of the complexity of the role of vitamins in metabolism. The reaction

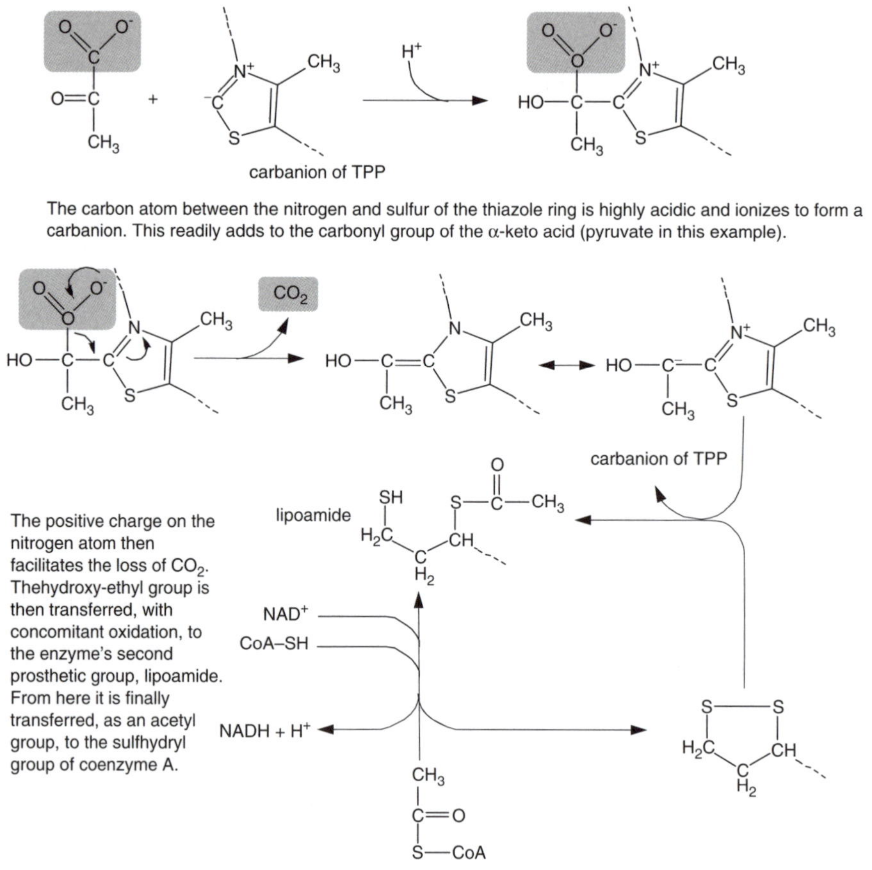

carbanion of TPP

The carbon atom between the nitrogen and sulfur of the thiazole ring is highly acidic and ionizes to form a carbanion. This readily adds to the carbonyl group of the α-keto acid (pyruvate in this example).

carbanion of TPP

The positive charge on the nitrogen atom then facilitates the loss of CO_2. Thehydroxy-ethyl group is then transferred, with concomitant oxidation, to the enzyme's second prosthetic group, lipoamide. From here it is finally transferred, as an acetyl group, to the sulfhydryl group of coenzyme A.

Figure 8.9   The mechanism of oxidative decarboxylations involving thiamin. Lipoamide is the other prosthetic group required in this type of decarboxylase. NAD^+ and NADH are, respectively, the oxidised and reduced forms of coenzymes based on the vitamin niacin, described on pages 323–325.

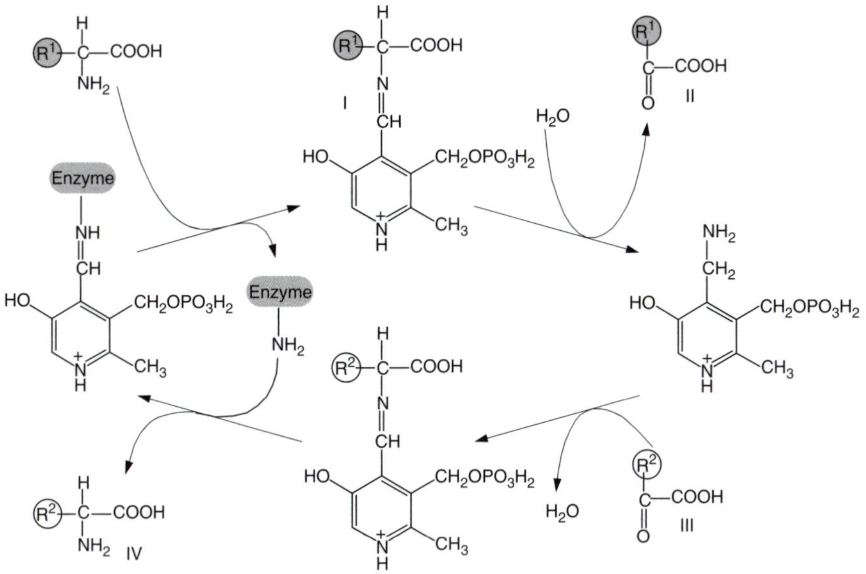

Figure 8.10 The mechanism of a transamination reaction. The ε-amino group of lysine in the active site of the enzymes carries the pyridoxal phosphate. This is displaced by the incoming amino acid to form a Schiff base (*I*). The hydrolysis of the weakened N–C bond then liberates the α-keto acid (*II*). The sequence is then reversed; as the second α-keto acid (*III*) enters, the second amino acid (*IV*) is formed. The overall result is a reaction of the type:

$$\text{Glutamate} + \text{oxaloacetate} \rightleftharpoons \alpha\text{-ketoglutarate} + \text{aspartate}$$

sequence is a transamination, a typical pyridoxal phosphate-requiring reaction, and is shown in Figure 8.10.

2. Further Details of the Role of Ascorbic Acid

While a full treatment of the reactions in which ascorbic acid partici-pates are more properly the concern of biochemists, it is instructive for food chemists to have a systematic overview of the full range of potential activities of vitamin C.

Ascorbic acid is an effective scavenger of the various free radical forms of oxygen which arise in the tissues, notably as by-products of oxygen metabolism or the autoxidation of unsaturated fatty acids (*see* Chapter 4). As shown in Figure 8.11, the unpaired electrons in the free radicals are in effect transferred to the ascorbic acid. Ascorbic acid is

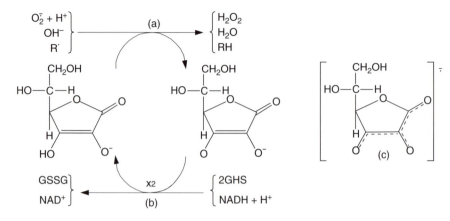

Figure 8.11   The reaction of ascorbic acid with free radicals (*a*) and regeneration by reducing agents such as reduced glutathione (**GSH**) or reduced nicotinamide dinucleotide (**NADH + H⁺**) (*b*). The stability of the ascorbic acid free radical is the result of extensive delocalisation, as indicated in the alternative structure (*c*).

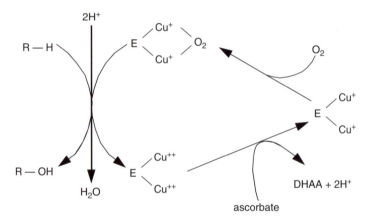

Figure 8.12   The role of ascorbic acid and copper in monooxygenase reactions.

then regenerated by reaction with other reducing agents such as glutathione or NADH (*see* page 324).

Ascorbic acid is an important cofactor in many reactions catalysed by oxygenases. These are reactions in which an oxygen atom is incorporated into a substrate in the form of a hydroxyl group. In monooxygenase reactions, shown in Figure 8.12, one of the two oxygen atoms emerges from the reaction as water. However, in dioxygenase reactions the second oxygen atom is utilised in the oxidation of the auxiliary substrate, α-ketoglutarate, to succinate:

$$R-H + O_2 + \text{α-ketoglutarate} + \text{ascorbate} \rightarrow R-OH + \text{succinate} + DHAA$$

The hydroxylation of proline side-chains in collagen (8.23) referred to earlier in this chapter is a dioxygenase reaction, whereas the conversion of dopamine to noradrenaline (8.24) is a mono-oxygenase reaction. Two essential stages in the biosynthesis of carnitine (8.39) from the amino acid lysine are catalysed by dioxygenases. Carnitine is an essential component of the system for transporting long-chain fatty acids across mitochondrial membranes, and in vitamin C deficiency its depletion may well be the basis of some of the symptoms of scurvy.

$$H_3C-\overset{\displaystyle CH_3}{\underset{\displaystyle CH_3}{N^+}}-CH_2-\overset{\displaystyle H}{\underset{\displaystyle OH}{C}}-CH_2-COOH$$

(8.39)

3. Non-Vitamins

A number of substances not mentioned so far in this chapter are sometimes described as vitamins or listed with them. As in the case of vitamin F, these are sometimes terminological hangovers from the 1930s, but there are others with even less credibility.

Carnitine (mentioned in the previous section) is widely sold as a "vitamin-like supplement". Humans and most other animals synthesise it and, although it occurs naturally in meat, fish and dairy products, humans (except vegans) obtain further supplies from their diet. The lower levels of carnitine in the plasma of strict vegetarians are not believed to be of any clinical significance. p-Aminobenzoic acid (PABA, 8.40) is another popular "supplement" often and erroneously described as a member of the vitamin B complex. Although it is a growth factor for a number of micro-organisms there is no evidence that it is required by humans. *myo*-Inositol (8.41), the only one of nine isomers of inositol occurring widely in nature, has been shown to be a growth factor necessary for a few types of human cells growing in culture, but there is no evidence of its requirement *in vivo*. There is therefore no reason to regard any of these three substances as vitamins, nor for them to be adopted as dietary supplements.

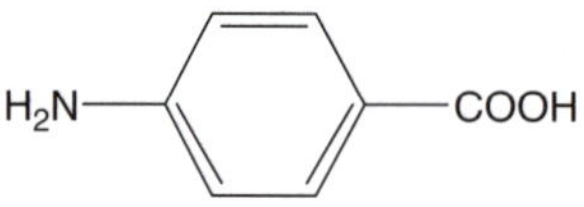

H_2N—⟨benzene ring⟩—COOH

(8.40)

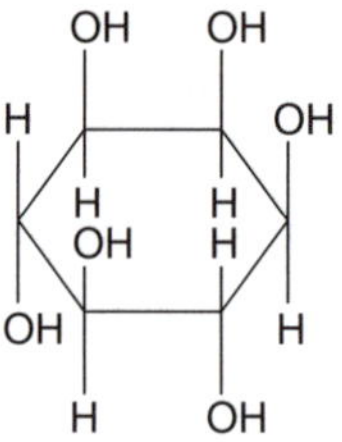

(8.41)

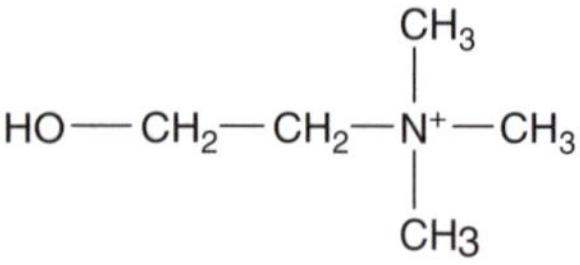

HO—CH_2—CH_2—N^+—CH_3 (with CH_3 above and CH_3 below)

(8.42)

Choline (8.42) has been listed as a vitamin from time to time but its status has now been clarified. The body uses considerable quantities of choline in the form of phosphatidyl choline or lecithin (*see* page 140) as a component of cell membranes. Although humans are able to synthesise choline by the addition of methyl groups to phosphatidyl ethanolamine, the process still requires a source of methyl groups. These can be supplied by the essential amino acid methionine, but a normal diet is unlikely to provide a surplus of methionine beyond that required for synthesising protein. As adults we require at least 0.5 g of choline per day. Growing infants and children may require somewhat higher levels in relation to body weight. Fortunately any normal diet supplies at least 1 g of choline per day. The amounts required and its function as a tissue component mean that it can be considered as an essential nutrient alongside the essential fatty acids (Chapter 4) and essential amino acids (Chapter 5) – but not as a vitamin.

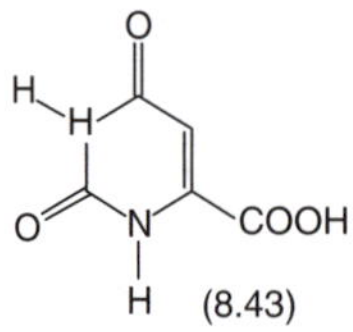

(8.43)

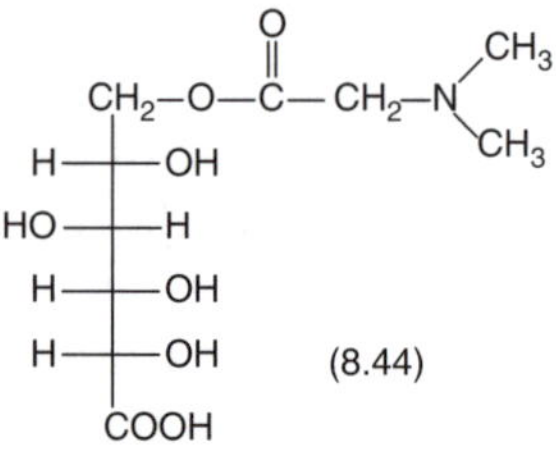

(8.44)

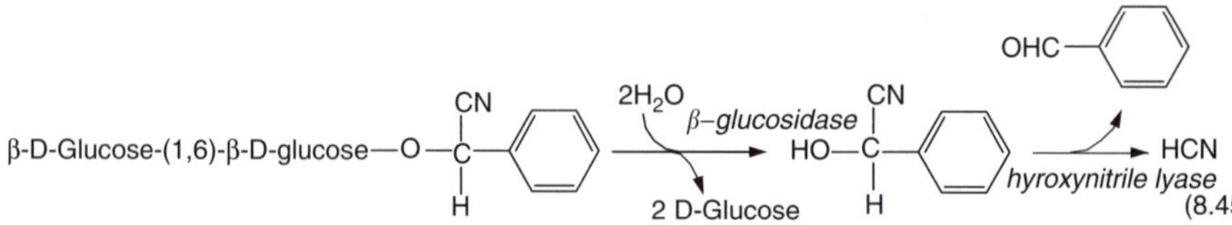

(8.45)

Three other substances are often referred to as vitamins, although beyond the realms of conventional nutrition. "Vitamin B_{13}" is actually orotic acid (8.43), a normal intermediate in the biosynthesis of pyrimidines. The supposedly scientific name pangamic acid is otherwise known as "vitamin B$_{15}$" but both these names have been applied to a variety of substances, most notably D-gluconodimethylaminoethanoic acid (8.44). "Vitamin B$_{17}$" was encountered in Chapter 2 as amygdalin, and under the name laetrile it has achieved notoriety as a supposed cancer cure. The theory, unsupported by any experimental evidence, was that cancer cells are especially rich in the β-glucosidases which can cause its breakdown, ultimately releasing cyanide (8.45), which could supposedly kill the cancer cells selectively. In actual fact, the levels of cyanide released would normally be detoxified by conversion to thiocyanate by the enzyme rhodanase, which is widely distributed in the tissues.

FURTHER READING

G. F. M. Ball, *Vitamins in Foods: Analysis, Bioavailability, and Stability*, CRC Press, Boca Raton, Florida, 2006.

G. F. M. Ball, *Vitamins: Their Role in the Human Body*, Blackwell, Oxford, 2004.

T. Sanders and P. Emery, *Molecular Basis of Human Nutrition*, Taylor and Francis, London, 2003.

C. D. Berdanier, *Advanced Nutrition: Micronutrients*, CRC Press, Boca Raton, Florida, 1998.

T. K. Basu and J. W. Dickerson, *Vitamins in Health and Disease*, CAB International, Wallingford, Oxon., 1996.

The Technology of Vitamins in Food, ed. P. B. Ottaway, Blackie, Glasgow, 1993.

Nutritional Aspects of Food Processing and Ingredients, ed. C. J. K. Henry and N. J. Heppell, Aspen, Gaithersburg, 1998.

Dietary Reference Intakes for Thiamine, Riboflavin, Niacin, Vitamin B$_6$, Folate, Vitamin B$_{12}$, Pantothenic Acid, Biotin and Choline, Institute of Medicine Standing Committee on the Scientific Evaluation of Dietary Reference Intakes, National Academy of Sciences, Washington D.C., 1998.

Dietary Reference Intakes: Calcium, Phosphorus, Magnesium, Vitamin D, and Fluoride, ibid., 1997.

Dietary Reference Intakes for Vitamin C, Vitamin E, Selenium, and Carotenoids, ibid., 2000.

RECENT REVIEWS

S. P. Stabler and R. H. Allen, *Vitamin B12 Deficiency as a Worldwide Problem; Annu. Rev. Nutr.*, 2004, **24**, p. 299.

A. J. A. Wright, P. M. Finglas and S. Southon, *Proposed Mandatory Fortification of the UK Diet with Folic Acid: Have Potential Risks been Underestimated? Trends Food Sci. Technol.*, 2001, **12**, p. 313.

A. Szent-Györgyi, *Lost in the Twentieth Century*, in *Annu. Rev. Biochem.*, 1963, **32**, p. 1. (Hardly "recent" but nevertheless recommended reading!)

CHAPTER 9
Preservatives

Micro-organisms, particularly bacteria, yeast and moulds, have nutritional requirements remarkably similar to our own. Unless they use photosynthesis (as some bacteria do), their energy is obtained by the oxidation or fermentation of organic compounds.[i] Similarly, organic compounds are the source of carbon, nitrogen and other elements for the biosynthesis of cell material. It is therefore hardly surprising that, given the opportunity, few types of micro-organisms will be unwilling to utilise the same nutrients that make up a typical human diet. The attractiveness of human foodstuffs to micro-organisms is enhanced by two factors, their tendency to be at moderate pH and temperature and, secondly, that they offer a wide enough range of oxygen concentrations to tempt both obligate anaerobes and aerobes.

It is therefore to be expected that food technology has always had a preoccupation with the control of food-borne micro-organisms. The word "control" rather than "eliminate" acknowledges the fact that the elimination of micro-organisms from food is often not even attempted, and indeed in some foodstuffs it would be undesirable.

There are four quite distinct aspects of food science to which microbiologists can contribute. The most important is the issue of safety.

---

[i] There are some bacteria, not relevant to human food, which obtain their energy by oxidation of inorganic compounds.

Food: The Chemistry of its Components, Fifth edition
By T.P. Coultate
© T.P. Coultate, 2009
Published by the Royal Society of Chemistry, www.rsc.org

Foodstuffs are ideal carriers of pathogenic bacteria, *e.g. Salmonella* sp., as well as being ideal substrates for the growth of bacteria and fungi which secrete toxins, *e.g. Clostridium botulinum* and *Aspergillus flavus*. A second aspect of food microbiology is that of biodeterioration. If food materials were not vulnerable to degradation by the extracellular enzymes secreted by invading bacteria and moulds, there would be little prospect of success for the parallel processes of digestion that occur in our own alimentary canal.

The other two aspects of food microbiology are more positive in nature, and concern the use of microbial activity in food production. Many processes, often under the blanket description of fermentation, have been practised for thousands of years. Examples of man's exploitation of biodeterioration include pickling and cheese making. Microbial action reduces the raw material to a more stable product in which there is much less scope for further breakdown. This usually happens because microbial activity has been accompanied by the loss of water and/or the accumulation of ethanol, lactic acid or acetic acid (ethanoic acid).

The fourth aspect of food microbiology, which attracts less interest nowadays than previously, is the use of the micro-organisms themselves as food. While we have always eaten fungi such as mushrooms, scientists have more recently explored the use of single-celled micro-organisms. These could be utilised directly as human food, but with one notable exception the plan was for the protein to be extracted and used as a food supplement for livestock. Strains of micro-organisms were developed which could feed readily on methanol and other by-products of the oil industry. Subsequent increases in world oil prices have ruined the economics of most single-cell protein (SCP) production. However, one notable exception is the mycoprotein known commercially as Quorn™ (*see* Chapter 5 page 200, and R. Angold, *et al., Food Biotechnology*, in Further Reading). The carbon source for the production of Quorn is the starch generated as a by-product of the commercial production of wheat gluten.

Of these various aspects of food microbiology it is the use of chemical substances to deter unwanted microbial activity which is of greatest interest to chemists. Even with unwanted micro-organisms, it is often only the inhibition of growth which can be achieved. As we have seen elsewhere in this book, the heat-processing methods required to obtain sterility, in other words the complete elimination of viable vegetative cells and spores, are seldom compatible with maximum nutritive value. Furthermore, the essential similarity between the metabolic processes of micro-organisms and those of man means that chemical sterilants (*e.g.* high concentrations of chlorine or phenol) will never be acceptable as food ingredients. The advantage of a chemical additive approach to the reduction of microbial activity is that many products need to remain stable for a period after the package has been opened. A jar of jam or a bottle of tomato ketchup will be subjected to repeated and massive recontamination once it comes under the influence of junior members of the household.

It is actually quite unusual for a single antimicrobial procedure to be used on its own to protect a food product. Wherever possible several relatively mild procedures are combined to maximise the inhibition of microbial activity, at the same time minimising their adverse effects on nutritional value. For example, the long-term storage of meat, fish or vegetables at ambient temperature requires their canning with the temperature in the centre of the can maintained above 115 °C for over an hour. Otherwise the surviving spores of *C. botulinum* would find the anaerobic, nutrient-rich, neutral pH environment perfect for germination, growth and production of toxins. On the other hand the natural acidity of fruit eliminates the risk of botulism, and the more modest heat treatments we associate with home bottling are adequate. As we shall see in our discussion of individual food preservatives, these are very often used in combination with other antimicrobial procedures. For example, the safety of cooked ham depends on (i) the salt content, to maintain low water activity, (ii) cooking, to destroy most vegetative bacterial cells and some spores, and (iii) the inclusion of nitrites, to inhibit spore germination and bacterial growth.

The use of preservatives other than sodium chloride and smoke as recognised food additives is covered by legislation in Europe and elsewhere. The principal preservatives approved for use in the EU are listed in Table 9.1.

SODIUM CHLORIDE

Common salt, sodium chloride, was undoubtedly the first antimicrobial substance to be used in foodstuffs. One can be confident that in early civilisations it was regarded as a preservative rather than for flavouring. Salting is the traditional method of preserving meat, often in combination with smoking and drying. Modern technology has provided more rapid methods of getting the salt into the meat, but the essentials have remained unchanged for centuries. Solutions containing 15–25% salt

Table 9.1   Antimicrobial preservatives permitted in Europe and mentioned in
the text. There are restrictions on the range of foods and the concentrations in which particular preservatives may be used. Sodium
chloride, alcohol and sugars, which exert their antimicrobial action
by reducing the availability of water, are not treated as food
additives in the context of the regulations. In the cases of the
asterisked preservatives individual E-numbers are applied to the
free acids and their individual salts. For example benzoic acid is
E210, and its sodium, potassium and calcium salts are E211, E212
and E213, respectively.

Preservative	E number
*Sorbic acid and sorbates	E200–E203
*Benzoic acid and benzoates	E210–E213
*p-Hydroxybenzoate esters	E209, E214–E219
Sulfur dioxide, *sulfites, *hydrogen sulfites and *metabisulfites[ii]	E220–E228
Nisin	E234
Pimaricin (*Natamycin*)	E235
*Nitrites	E249, E250
*Nitrates	E251, E252
*Acetic acid and acetates	E260–E266
Lactic acid	E270
*Propionic acid and propionates	E280–E283

are used to bring the water activity, a_W, down to about 0.96.[iii] This has
the effect of retarding the growth of most micro-organisms, including
the majority of those responsible for meat spoilage. With the advent of
other preservative methods, notably canning and refrigeration, the
importance of salting has diminished, but not without some salt-pre-
served meat products such as ham and bacon remaining firmly established in our diet. As the necessity for high salt content for preservation
has decreased, our taste has adapted to bacon and ham which is less
salty and less dry.

The health implications of the sodium chloride content of our diet are
discussed in Chapter 11.

NITRITES

Nowadays very little meat is preserved using common salt alone. At
some distant point in history it was realised that it was the unintended

---

[ii] In UK legislative documentation sulfur compounds continue to be spelt with *ph* rather than *f*. The
interrelationships between these preservatives are detailed in Figure 9.1.

[iii] The water activity of an aqueous solution is equal to its vapour pressure divided by the vapour
pressure of pure water at the same temperature. This concept is considered in greater detail in
Chapter 12.

presence of saltpetre (sodium nitrate) as an impurity in crude salt which gave an attractive red or pink colour to preserved meat. Subsequently nitrates and/or nitrites have become an almost indispensable component of the salt mixtures (known as "pickles") used for curing bacon and ham. In spite of the antiquity of the process, it is only very recently that the special antimicrobial properties of nitrites and their mechanism of action have been recognised. There are many subtle variations in curing procedures, but the well-known Wiltshire cure is typical. Sides of pork are injected with 5% of their weight of a pickling brine containing 25–30% sodium chloride, 2.5–4% sodium or potassium nitrate, and sometimes a little sugar. The sides are then submerged in a similar solution for a few days. After removal they are stored for 1–2 weeks to mature. If prolonged storage is called for a second preservative, smoke, is applied. By the time the maturation stage is completed, the salts will have become evenly distributed throughout the muscle tissue and a complex series of reactions will have produced the characteristic red colour of uncooked bacon and ham. The essential features of these reactions are as follows.

(i) Some of the nitrate present is reduced to nitrite:

$$NO_3^- + 2[H] \rightarrow NO_2^- + H_2O$$

This is due either to salt-tolerant micro-organisms in the brine or the respiratory enzymes of the muscle tissue.

(ii) The nitrite oxidises the iron of the myoglobin in the muscle to Fe^{III} iron (*see* Chapter 5):

$$Fe^{II} + NO_2^- + H^+ \rightarrow Fe^{III} + NO + OH^-$$

In other words, myoglobin (Mb) is converted to metmyoglobin (MMb) and nitrogen oxide (*i.e.* nitric oxide) is formed.

(iii) The nitrogen oxide reacts with the MMb to form nitrosyl met-myoglobin (MMbNO).

(iv) The MMbNO is immediately reduced by the respiratory systems of the muscle tissue to nitrosyl myoglobin (MbNO), the red pigment present in uncooked bacon and ham.

The distribution of electrons around the iron atom of MbNO is similar to that in oxymyoglobin (MbO_2), hence the similarity in colour. When bacon is grilled or fried, or when ham has been boiled, the nitrosyl myoglobin is denatured and a bright pink pigment, nitrosylhaemochromogen, is formed. There is some uncertainty about its structure, but it is believed that the denaturation of the globin allows a second nitrogen oxide molecule to bind to the iron in place of the histidine side-chain.

In recent years the use of nitrates in food has been regarded with increasing suspicion owing to the possible formation of nitrosamines, which are potent carcinogens. Secondary amines react readily with nitrous acid to form stable *N*-nitroso compounds (9.1):

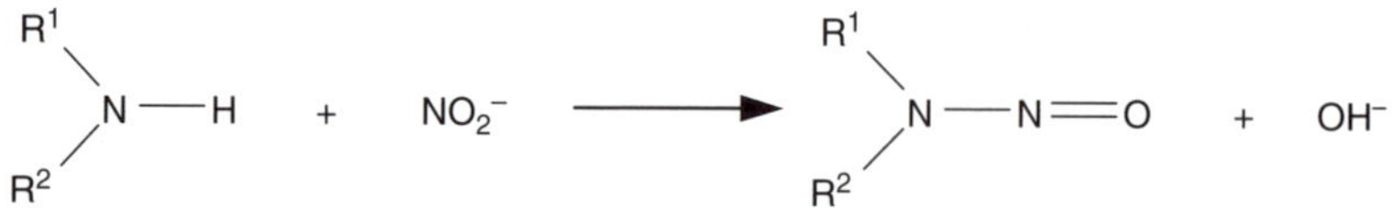

$$R^1R^2N-H \; + \; NO_2^- \; \longrightarrow \; R^1R^2N-N{=}O \; + \; OH^-$$

(9.1)

The reaction with primary amines, which are of course abundant in meat as free amino acids, leads simply to deamination:

$$RNH_2 + NO_2^- \rightarrow N_2 + OH^-$$

The reaction with amino acids is important since it eliminates some of the excess nitrite in cured meat. Secondary amines are much less abundant in meat but are assumed to arise as a result of microbial action, especially by anaerobes. The decarboxylation of proline, which occurs spontaneously at the high temperatures involved in frying, leads to the formation of the very important nitrosamine, *N*-nitrosopyrrolidine (NOPyr, 9.2):

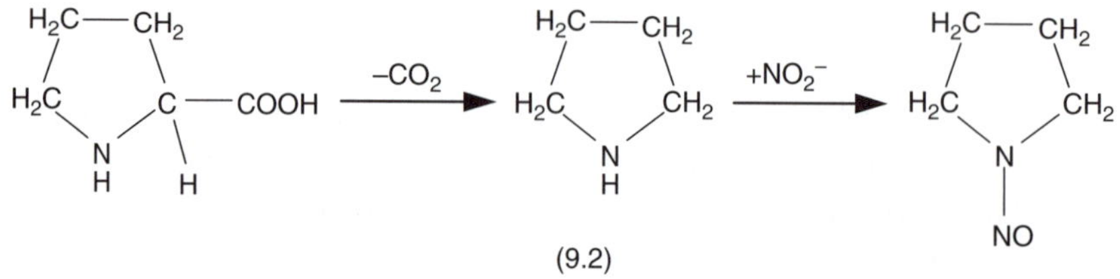

$$\text{proline} \xrightarrow{-CO_2} \text{pyrrolidine} \xrightarrow{+NO_2^-} \text{N-nitrosopyrrolidine}$$

(9.2)

Although nitrosamines have been detected in many cured meat products such as salami and frankfurters, it is clear that their presence is most significant in cured meat cooked at high temperature, such as fried bacon. Levels of NOPyr in fried bacon are consistently around $100 \, \mu g \, kg^{-1}$, along with somewhat lower levels of *N*-nitrosodimethylamine (NDMA). Although the carcinogenicity of these volatile nitrosamines has been confirmed in laboratory animals, there is little evidence that the consumption of cured meats is actually responsible for causing cancer in man. Studies in Finland have however shown that over a long period there is a positive correlation between NDMA intake and the incidence of colorectal cancer. In this study a high intake of smoked or salted fish was however found to be a much more significant factor, and there was no correlation with consumption of cured meat. The potential hazard is enough to ensure that steps remain in place to reduce the consumption of *N*-nitrosamines.

Beer is another product in which the nitrosamine content has received attention. Studies in the late 1970s showed that levels of NDMA around 2–3 ppb were common. These arose when the malted barley was kilned, *i.e.* heated and dried to prevent continued germination once the starch-degrading α- and β-amylases had been synthesised in the grain. It was usual to expose the malt directly to the flue gases from an oil- or solid fuel-fired furnace. Nitrogen oxides were generated in the flames of the burners and were transferred to the germinated barley by direct exposure to the flue gases. It was discovered that malt kilning was to blame. Changes in the construction of the dryers, and the incorporation of sulfur dioxide into the brewing process to inhibit nitrosation, have effectively eliminated the NDMA problem.

Improvements in processing controls are making a steady reduction in residual nitrite levels possible, but the most important measure is the inclusion of ascorbic acid in curing-salt mixtures. Ascorbic acid is beneficial in two ways. Firstly, being a reducing agent, it enhances either directly or indirectly the rate of the key reducing reactions in the formation of MbNO, thereby allowing lower levels of nitrites or nitrates to be used in the pickles. Secondly, it actually inhibits the nitrosation reaction.

If the only value of nitrite was as a colouring agent, even a remote possibility of a hazard in its use for curing would warrant its prohibition. (The question of whether nitrite actually contributes to the flavour of cured meat remains unresolved.) However, the antimicrobial properties of nitrites are extremely valuable, and would justify their inclusion even if they had no effect on colour. It had been known for some years that the growth of many types of anaerobic bacteria, including the organism causing botulism, *Clostridium botulinum*, is prevented in cured meat by an unidentified product of the interaction between the residual nitrite and the meat during the time the meat is cooked. If a ham were given sufficient heat treatment to ensure that all spores of *C. botulinum* had been killed, it would be unacceptably overcooked. Canned stewing beef, which is given this degree of heat treatment, provides an illustration of the kind of texture the ham would have. This effect of the residual nitrite in ham was termed the "Perigo effect" after its discoverer, and a great deal of work was done to identify what became known as the "Perigo inhibitor". It is now well established that during cooking much residual nitrite is broken down to nitrogen oxide. This is not liberated from the meat but becomes loosely associated with some of the exposed amino acid side-chains and iron atoms of the denatured meat proteins. From this reservoir nitrogen oxide is available for the inhibition of sensitive bacteria. It appears that nitrogen oxide is a specific and potent inhibitor of at least one enzyme (pyruvate:ferredoxin oxidoreductase) which plays an essential role in the energy metabolism of anaerobes such as *C. botulinum*.

The use of nitrites therefore presents those concerned with food safety with a paradox. On the one hand we have nitrite the preservative, in use for centuries and now known to be essential for microbiological safety, and on the other hand we have nitrite the colouring agent, suspected of giving rise to carcinogens. Apart from the obvious but unacceptable solution of eliminating cured meats from our diet, the most we can do is to ensure that residual nitrite levels are kept below the minimum, about $50 \, \mu g \, g^{-1}$, at which toxins are produced. It is also worth remembering that nitrates naturally present in other foods, particularly vegetables, and even in drinking water can be reduced to nitrite by our intestinal bacteria. Nitrates therefore also have the potential to cause problems, even for those who do not eat cured meat (*see* Chapter 10).

SMOKE

Smoke is another preservative traditionally associated with meat and fish. We can be fairly sure that the flavour produced by wood smoke was regarded initially as merely a valuable side-effect of drying over a wood fire. The preservative action of smoke almost certainly went unnoticed. Nowadays meat and fish are rarely preserved by drying alone and refrigeration has made the preservative action of smoke less important than the flavour it gives. When meat is dried for long-term preservation, as for example in the production of South African biltong, the strips of meat (beef, ostrich or antelope) are marinated in vinegar and salted before drying, each of them effective anti-bacterial procedures in their own right.

Smoke consists of two phases, a disperse phase of liquid droplets and a continuous gaseous phase. In smoking, the absorption of gas phase components on the food surface is considered to be much more important than the actual deposition of smoke droplets. The gas phase

of wood smoke has been shown to include over 200 different compounds, including formaldehyde (methanal, HCHO), formic acid
(methanoic acid, HCOOH), short-chain fatty acids, vanillic (9.3) and
syringic (9.4) acids, furfural (9.5), methanol, ethanol, acetaldehyde
(ethanal), diacetyl (butanedione), acetone (propanone), and 3,4-benz-
pyrene (9.6). Of these the most important antimicrobial agent is almost
certainly formaldehyde.

(9.3)   (9.4)   (9.5)   (9.6)

The detection of known carcinogens such as 3,4-benzpyrene (benzo
[*a*]pyrene, or BaP) and many other polynuclear aromatic hydrocarbons
(PAHs) in wood smoke has caused concern about the safety of smoked
foods. It has been suggested that the high incidence of stomach cancer in
Iceland and the Scandinavian countries might be related to the large
quantities of smoked fish consumed. Although there is no suggestion that
the amount of smoked food consumed elsewhere in the world is sufficient
to cause similar problems, liquid-smoke preparations which contain much
lower concentrations of polynuclear aromatics are being increasingly
adopted. Liquid smokes are prepared from wood smoke condensates by
fractional distillation and extraction with water. The carcinogenic polynuclear hydrocarbons are insoluble in water and remain behind. The
value of liquid smoke preparations is illustrated by current EU legislation.
Although it may represent only a small proportion of the total PAHs, the
BaP level is the accepted marker for these carcinogens. The maximum
permitted level of BaP in traditional smoked foods, including meat, fish
and derived products, is 5 µg per kg. However in foods flavoured by liquid
smoke preparations the corresponding figure is as low as 0.03 µg per kg.

SULFUR DIOXIDE

Sulfur dioxide (SO_2) has been used in wine-making for hundreds of years
to control the growth of unwanted micro-organisms. It was originally

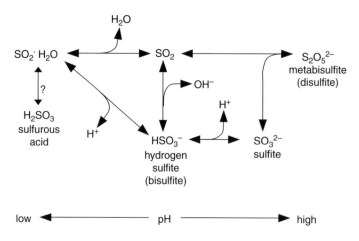

Figure 9.1 Structural relationships between sulfur dioxide-generating compounds.

obtained by the rather haphazard process of exposing the "must", the unfermented grape juice, to the fumes of burning sulfur. Nowadays the gas itself is rarely used, being replaced by a number of SO_2-generating compounds, particularly sodium sulfite (Na_2SO_3), sodium hydrogen sulfite[iv] ($NaHSO_3$) and sodium metabisulfite (sodium disulfite, $Na_2S_2O_5$). The relationship between these, sulfur dioxide, and sulfurous acid are outlined in Figure 9.1. The antimicrobial activity of such compounds increases dramatically as the pH falls, and it has therefore been assumed that the antimicrobial activity is due to undissociated sulfurous acid. However, in recent years the presence of undissociated sulfurous acid in aqueous sulfur dioxide has been called into question, the sulfur dioxide being present simply as a complex of formula $SO_2.H_2O$. It is therefore now thought that sulfur dioxide itself is probably the effective antimicrobial agent.

Total sulfur dioxide levels around 100 ppm are added to wine musts to achieve a differential effect. The desirable wine yeast *Saccharomyces cerevisiae* is able to grow and ferment the sucrose to ethanol, but some undesirable yeast species such as *Kloeckera apiculata* and lactic acid-producing bacteria are suppressed. Sulfur dioxide inhibits many of the NAD^+-dependent dehydrogenases in yeast and bacteria but is particularly effective against glyceraldehyde dehydrogenase in yeast, and malate dehydrogenase in *Escherichia coli*. Both of these enzymes play a central role in energy metabolism. About two-thirds of the total sulfur dioxide

---

[iv] Sodium hydrogen sulfite is still frequently referred to by its now obsolete name, sodium bisulfite.

in a must or wine is bound to anthocyanins or other flavonoids (*see* Chapter 6), sugars (9.7) and other aldehydes:

(9.7)

In addition to the added sulfur dioxide, many strains of wine yeasts themselves produce sulfur dioxide by the reduction of sulfates present in the grape juice. When the wine is bottled, further sulfur dioxide is added to prevent secondary fermentation of residual sugar in the bottle; concentrations of sulfur dioxide up to 500 ppm are employed.

A wide range of other foodstuffs, particularly fruit and vegetable-based products, contain sulfites either added as a preservative or as a residue from processing operations. The use of sulfite as an antioxidant has been discussed in Chapter 6.

The use of sulfur dioxide has generally been regarded as being without any toxicity hazard at the usual levels. Of all the preservatives, sulfur dioxide and its relatives are consumed in the largest amounts. Although there are no more recent data than 1986, UK consumption cannot be expected to have changed significantly since then. At that time average consumption per person was estimated to be 18.2 mg per day, with alcoholic drinks, sausages, hamburgers and dried fruit being the major sources. The margin between this level of intake (and more especially intakes at the top end of the range) and the Acceptable Daily Intake (ADI, *see* page 416) is not large. The ADI of $3.5\,mg\,kg^{-1}$ body weight corresponds to 245 mg for a typical 70 kg adult. Nevertheless, the UK authorities do not at present consider levels of sulfur dioxide intake to be a cause for concern.

Recently, levels at the upper end of the normal range in food have been shown in a small proportion of asthmatics to cause symptoms similar to a severe allergic response. It remains to be seen whether this will be sufficient for a reappraisal of the GRAS (Generally Recognised as Safe) status of sulfur dioxide. The use of sulfur dioxide presents a more significant problem. The sulfite ion reacts destructively with the vitamin thiamin (*see* Chapter 8). Vegetables such as potatoes are often stored in sulfite solutions during the intermediate stages of processing and can lose considerable proportions of their thiamin content. It is for this reason that many countries prohibit the use of sulfur dioxide in foodstuffs which provide an important source of thiamin in that country's diet.

A further drawback to the use of sulfur dioxide and sulfites is the taste they can produce. Above 500 ppm most people are aware of the dis- agreeable flavour sulfur dioxide causes, and some can detect it at much lower levels. Some white wines are actually characterised by their slight sulfur dioxide flavour, notably *Gewurztraminer*, from the Alsace region of France. Very little is known for certain about how sulfur dioxide inhibits microbial growth. The most probable explanation is that it forms addition compounds with the aldehyde groups in key metabolic intermediates or coenzymes, but these have not been identified.

Sulfur dioxide is one of the easier preservatives of which to measure the concentration in foodstuffs and drinks, and one which has not yet been taken over by the chromatographers. In the most popular version of the Monier–Williams method, the sample is placed in a distillation flask and acidified to convert disulfites *etc.* to sulfite/SO_2 and to liberate the sulfur dioxide from complexes with aldehydes. The volatile sulfur dioxide is then removed by distillation under reflux, the volatile organic acids being retained. The sulfur dioxide passing over is trapped in hydrogen peroxide solution:

$$SO_2 + H_2O_2 \rightarrow SO_4^{2-} + 2H^+$$

and this allows titrimetric estimation. Sulfur dioxide can also be con- veniently measured enzymically using a combination of the two enzymes, sulfite oxidase and NADH peroxidase:

$$SO_3^{2-} + O_2 + H_2O \xrightarrow{\text{sulfite oxidase}} H_2O_2 + SO_4^{2-}$$

$$H_2O_2 + NADH + H+ \xrightarrow{\text{NADH peroxidase}} 2H_2O + NAD^+$$

A concentration of 0.10 mM sulfite gives a fall in the absorbance at 340 nm (due to NADH, *see* page 325) of 0.622, providing ample sensi- tivity for food analysis purposes.

BENZOATES

Benzoic acid (9.8) occurs naturally in small amounts in many foods, particularly in fruit, and notably the cloudberry (at about 0.8%), cranberries and prunes. Whether this qualifies benzoates produced synthetically to be described as "natural" is another matter.[v] The sodium salt is much more soluble in water than the free acid, and this is the form usually used as a food preservative.

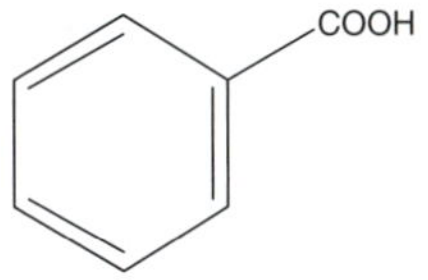

(9.8)

On the other hand it is the undissociated acid which is active against micro-organisms, particularly yeasts and bacteria, through its inhibitory effect on the activities of the enzymes of the tricarboxylic acid cycle (*i.e.* the Krebs or citric acid cycle), α-ketoglutarate dehydrogenase and succinate dehydrogenase. Since it is active only in the protonated form, use of benzoic acid is restricted to acidic foods (pH 2.5–4.0), such as fruit juices and similar beverages. At the levels used, 0.05–0.1%, no deleterious effects on humans have been detected. Benzoates do not accumulate in the body but are converted by condensation with glycine into hippuric acid (*N*-benzoylglycine, 9.9), and excreted in the urine.

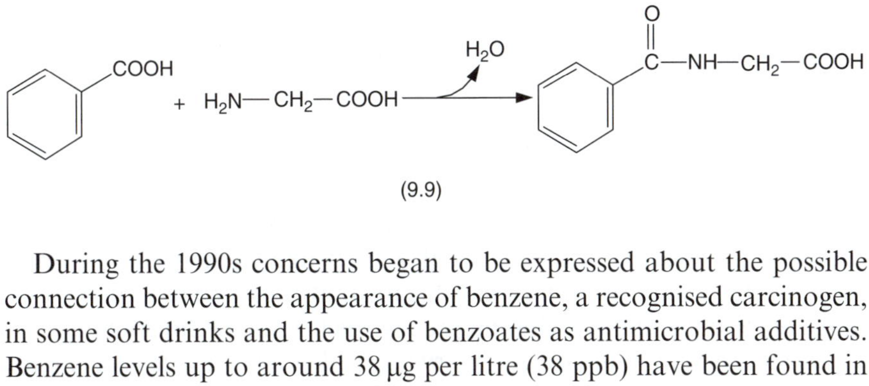

(9.9)

During the 1990s concerns began to be expressed about the possible connection between the appearance of benzene, a recognised carcinogen, in some soft drinks and the use of benzoates as antimicrobial additives. Benzene levels up to around 38 μg per litre (38 ppb) have been found in drinks such as colas containing ascorbic acid and benzoates. Although it was assumed initially that the benzene originated from contaminated raw materials or during processing, the real source has now been identified. Under acidic conditions, which are normally the case in soft drinks, low concentrations of ascorbic acid in the presence of oxygen and traces of transition metal ions (*e.g.* Fe^{III} or Cu^{II}) can give rise to

---

[v] Cases such as this exemplify the futility of the "naturalness equals wholesomeness" debate.

hydroxyl radicals (OH·). These can react with benzoic acid, resulting in its decarboxylation to benzene.

Although some concern has been expressed in the popular press in the USA and the UK, the authorities have pointed out that the benzene emitted in automobile exhausts causes about 200 µg daily to be routinely inhaled by city dwellers. Cigarette smokers may inhale as much as 7000 µg each day! These levels are vastly in excess of the amounts that could possibly be consumed in soft drinks. It must also be recognised that many fruit-based drinks contain significant quantities of both benzoates and ascorbic acid naturally.

Esters of *p*-hydroxybenzoic acid with methanol, propanol (9.10) or other alcohols, are known collectively as "parabens" and are also commonly used in many of the same situations as benzoic acid, and present similarly few toxicity problems.

$$HO-\underset{}{\bigcirc}-\overset{\overset{O}{\|}}{C}-O-(CH_2)_2-CH_3 \qquad (9.10)$$

OTHER ORGANIC ACIDS

A number of other organic acids are useful preservatives, despite their rather unpromising chemical structures. Sorbic acid (9.11) is a particularly effective mould inhibitor against a wide range of enzymes important in intermediary metabolism, including enolase and the tricarboxylic acid cycle dehydrogenases. As with benzoates, it is the undissociated form rather than the anion which is effective. However, the higher pK_a value (4.8) of sorbic acid compared with benzoic acid (4.2) makes it useful at pH values up to 6.5. This means that it can be used in a wider range of food products, including processed cheese, low-fat spreads such as margarine, and flour confectionery (*e.g.* cakes and pastries). It cannot be used in bread, however, due to its inhibitory effect on the yeast. Since it does show some activity against bacteria, there is growing interest in using it to replace at least some of the nitrite in cured meat products.

At the levels normally used (up to 0.3%) no toxic effects have been detected. It has been suggested that it is readily metabolised by mammals by a similar route to naturally occurring unsaturated fatty acids.

$$CH_3-CH=CH-CH=CH-COOH$$
$$(9.11)$$

It might be assumed that the association of acetic acid with preserved foods such as pickles was a consequence of its acidic character, but in fact it is an effective inhibitor of many types of spoilage bacteria and fungi at concentrations too low (0.1–0.3% in the undissociated form) to have any effect on the pH of a foodstuff. Apart from its inevitable presence in vinegar-based food products, acetic acid is becoming increasingly popular as an inhibitor of moulds and the formation of "rope"[vi] in bread. One of its special virtues, and the one which has led to it superseding propionic acid and its salts, is that it can be added as vinegar, which can then be listed as a simple ingredient on the packaging and not described as an "added chemical".

Like acetic acid, propionic acid ($CH_3CH_2CH_2COOH$) is found naturally in many foodstuffs. Some Swiss cheeses contain as much as 1%. It has a similar spectrum of antimicrobial activity to that of acetic acid and has many applications similar to those of sorbic and acetic acids.

There is a surprising lack of information about the mechanisms by which organic acid preservatives actually inhibit microbial growth. The most probable site of action is the cell membrane, where it is envisaged that such molecules will form close associations with the polar membrane lipids. In doing so, they could disrupt the normal processes of active transport into the cell.

NISIN AND NATAMYCIN

These two antibiotics are unlikely preservatives. The better known of the two, nisin, is a polypeptide comprising 34 amino acids in a chain, and containing a number of small loops formed by sulfhydryl bridges involving unusual amino acids such as lanthionine. Though fascinating to biochemists, the structure of nisin, beyond the simplified diagram shown in Figure 9.2, need not concern us further.

Nisin was identified in the 1930s as a problematic inhibitory substance occurring naturally in many dairy products. It was found to be produced by some strains of the bacterium, *Streptococcus lactis*. This organism commonly occurs in milk, and was found to inhibit the growth of other bacteria (but not fungi or yeasts) important in cheese making, including both the desirable lactobacilli and the undesirable clostridia. The origins and activity of nisin mean that it could strictly be regarded as an antibiotic, but since it has never been used in the treatment of disease the term is not normally used.

[vi] Rope occurs in bread when spores of *Bacillus mesentericus* survive the baking temperatures in the centre of the loaf and germinate once the loaf cools. The bacteria then attack the starch to produce a foul-smelling mass of sticky dextrins along the centre of the loaf—the "rope".

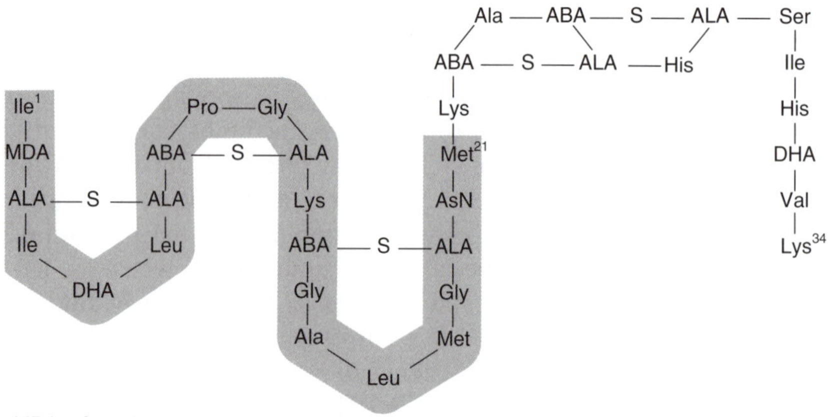

MDA = β-methyldehydroalanine, $CH_3CH=CNH_2COOH$
DHA = dehydroalanine, $CH_2=CNH_2COOH$
ALA-S-ALA = lanthionine, $S(CH_3CH=CNH_2COOH)_2$
ABA-S-ALA = β-methyl-lanthionine, $HOOCCH(NH_2)CH_2SCH(CH_3)CH(NH_2)COOH)$

Figure 9.2   The amino acid sequence in nisin. The amino acids are numbered from the NH$_2$ terminal in the conventional way. The usual abbreviations for amino acids are used, plus those shown. The first 21 amino acids (shaded) are predominantly hydrophobic and form a region of the molecule able to bind with the membrane lipids of susceptible micro-organisms.

Nowadays nisin is manufactured by growing selected strains of *S. lactis*. It is likely to prove valuable in a wide range of foodstuffs in which spoilage is caused by gram-positive bacteria, but at present it is mainly used in processed cheese. In this it prevents the growth of bacteria such as *Clostridium butyricum*, which has spores able to survive the 85–105 °C melting temperature involved when the cheese is processed. In common with antibiotics and enzyme preparations, it is usual to quantify nisin in units of activity rather than absolute units of weight: 1 g of pure nisin is equivalent to 4×10^7 international units (IU). For most applications, levels of nisin giving 250–500 IU g^{-1} are required.

Nisin functions by binding to a specific protein in the cell membrane of susceptible bacteria, and in doing so it forms a pore in the membrane, allowing ions to leak across. Nisin has highly specific requirements for binding to target cells, and this is clearly the reason for its lack of unwanted effects on humans. It has a very long history of natural occurrence in dairy products.

Natamycin (9.12) is an antifungal antibiotic (it is a polyene) which has a limited clinical application in opthalmics and a similarly restricted application as a food preservative. It is permitted as a surface treatment for cheese to deter unwanted mould growth. It is produced by a number of *Streptomyces* species, including the splendidly named *S. chattanoogensis*.

(9.12)

IRRADIATION

It might be argued that the use of ionising radiation to reduce the microbial content of foodstuffs belongs within the province of microbiology or physics. However, chemists are concerned with the effects of irradiation on nutrients and other components of foodstuffs and should therefore have some appreciation of the process. The side-effects of another physical method of food preservation, heating, have been mentioned in the preceding chapters of this book.

The possibility that food might be preserved by irradiation was first recognised in the early 1950s, but in spite of varying degrees of acceptance by the authorities in many countries it remains a potential rather than a practical process. Irradiation has been seen as a method of killing contaminating pests and micro-organisms without damaging the physical structure, flavour or nutritive value of food.

A range of levels of irradiation can be used to achieve different objectives. Novel names have been given to some of these processes, as shown in Table 9.2, but for the sake of the English language we must hope that they do not become widely adopted.

Although X-rays or electron beams from accelerators could provide the necessary radiation, the ideal penetration into food material is given by the γ-rays emitted by the radioactive decay of either ^{60}Co (1.17 and 1.33 MeV) or ^{137}Cs (0.66 MeV). The mechanism by which irradiation

Table 9.2   Levels of irradiation treatment.

Process	Treatment per kGy[a]	Effect
Inhibition	0.02–3.0	Delayed sprouting and ripening of vegetables and fruit
Disinfestation	0.1–1.0	Elimination of insect pests from stored food materials
Radurisation	1–5	Reduction of numbers of major spoilage bacteria and fungi
Radicidation	2–8	All vegetative microbial cells (but not spores) eliminated
Decontamination (hygienisation)	3–30	Reduction of total microbial load in food ingredients, and diet sterilisation for specialised clinical applications
Radapperisation	~50	Effective sterilisation

[a]1 kilogray (kGy) = 1000 gray (Gy). 1 Gy corresponds to 1 joule kg^{-1} of energy absorbed from ionising radiation. 1 Gy = 100 rad (1 rad = 100 erg g^{-1}).

kills micro-organisms or induces other biological effects depends in the first instance on the absorption of energy by the atoms of the irradiated material. In the case of food materials the vast majority of the atoms present are of low atomic number. In a typical "wet" biological tissue, hydrogen (~10%), carbon (~12%), nitrogen (~4%), and oxygen (~73%) account for 99% of the elements present. None of these elements is converted into unstable radioactive isotopes by radiation at the energy levels used. It follows that the induced radioactivity will be at a very low level, much lower in fact than that resulting from naturally occurring radioisotopes of heavier elements.

In the majority of foods the energy of irradiation is absorbed by the water present, according to the Compton effect. The incident photon of the γ-ray transfers some of its energy to an orbital electron, which is ejected from the absorbing atom. The residual energy is emitted as a photon of reduced energy. When water is the absorbing substance the loss of an electron leads to a range of free radicals, ions and other products, including ·OH, H·, H_2, H_2O_2 and H_3O^+. The ejected electron is immediately hydrated (indicated by e_{aq}^-) and gives rise to further free radicals. It is these various free radicals which account for the chemical and biological effects of irradiation. The distinction between irradiation and processes such as heating for delivering energy to food materials, is that irradiation focuses a high proportion of the energy into a small number of chemically reactive molecules. In contrast, heat energy is dispersed throughout all the molecular species present in the material.

The lethal effect of irradiation on the reproductive capability of micro-organisms is attributed to the effect of free radicals on the vulnerable purine and pyrimidine bases of the nucleic acids. The dosage required to kill vegetative bacteria lies between 0.5 and 1.0 kGy; bacterial spores require doses in the range 1.0–5.0 kGy. The greater complexity of higher organisms leads them to be affected by much lower lethal doses: 5–10 Gy for mammals, and 10–1000 Gy for insects. A more detailed consideration of radiation effects on micro-organisms is the concern of microbiologists.

Other important biological effects of irradiation are those which affect the development of fruit and vegetables. Depending on the fruit and the precise dosage, small amounts of radiation can either accelerate or retard the onset and speed of ripening. Sprouting in stored potatoes can also be inhibited. The actual mechanism of these effects is obscure, which is hardly surprising when one considers how little is known of the biochemical control of plant development in general, even under conventional conditions.

It is the side-effects of irradiation on the chemical components of food which are our primary concern. In general, the effect on carbohydrates resembles that of heating under alkaline conditions (*see* Chapter 2). Attack by free radicals causes fragmentation to compounds of low molecular weight, such as formaldehyde and glyoxal. The glycosidic bonds of oligo- and polysaccharides are also vulnerable; in model systems amylose solutions have been shown to lose most of their viscosity with radiation doses of 20 kGy. The total amount of carbohydrate breakdown products obtained is small, and they do not appear to include any which are not also associated with heating effects.

Much the same situation applies to amino acids and proteins. Losses of amino acids are small, particularly in the dry state, when direct interaction of the γ-ray photons with the target molecules occurs; this is not the case with water-derived products. As with polysaccharides, some depolymerisation occurs in aqueous systems. While this can have a deleterious effect on some physical properties, such as the whippability of egg white, nutritional value is largely unaffected.

Lipid materials, particularly unsaturated fatty acids, are much more vulnerable to irradiation. The susceptibility of lipids to free radical attack was described in Chapter 4, and experiments have shown that irradiation can initiate autoxidation along classic lines with all the expected end-products. If oxygen is abundant, very high levels of fatty acid hydroperoxides are formed. When fatty food materials are irradiated in an anaerobic environment and this is subsequently maintained, the extent of autoxidation and the development of rancidity are greatly reduced.

With the notable exceptions of thiamin, retinol and tocopherols, loss of vitamins during irradiation of foods is only slight. Many vitamins, *e.g.* ascorbic acid, are quite unstable in simple aqueous solutions but are stabilised by close association with food components such as sugars and proteins. Substantial losses of thiamine occur on irradiation due to attack by both $^{\bullet}$OH radicals and e_{aq}^{-} at the C=S and C=O double bonds. In addition to retinol, carotenoids also suffer from irradiation-induced autoxidation in fatty foods but are more stable in vegetable foodstuffs. Similar autoxidation reactions can overwhelm tocopherols.

The conclusion to be drawn is that treatments of up to 5–10 kGy can greatly reduce the rate of microbial spoilage, and have little adverse effect on all but fatty foods. In general, irradiation of foods while they are frozen minimises any deleterious effects on food components without significant reduction in the anti-microbial effect. High doses have a similar effect on the chemical components of food to that of sterilisation by heat. In spite of the reassurance these observations might provide for consumers, there is considerable opposition to the use of irradiation of food preservation. Much of this opposition is based on the unfounded fear of induced radioactivity. However, legislators and consumer organisations are more concerned about the risk that unscrupulous food processors may attempt to describe foods as "fresh" when the popular understanding of this adjective does not apply. Successful legislation will require a reliable test which can indicate whether a foodstuff has been irradiated. Several methods have been developed in recent years but none is universally applicable to all foods.

In spite of the observation that up to 300 mg kg^{-1} radiolytic products may be produced per 10 kGy of radiation, the nature of these products, as discussed in the preceding paragraphs, varies with the composition of the food. Heat processing can also generate so many similar compounds that the number qualifying as "unique radiolytic products" (URPs) is in fact very small. A failure to detect URPs cannot be taken as evidence that food has not been irradiated. A valuable screening technique is to examine the microbial flora present in a suspect food. Too few bacteria, or more precisely, many more dead bacterial cells detected by the direct epifluorescence filter technique (DEFT) than those remaining alive, strongly suggests that irradiation has been used.

Free radicals generated by irradiation can remain trapped in foods containing dry rigid matrices, such as bone (including fragments), shell (of shellfish), and certain nuts and seeds. Their unpaired electrons can be detected by electron spin resonance (ESR) but the instrumentation involved is expensive and the technique is highly skilled. Most plant-based foods carry traces of the silica present in soil particles, even after washing, and irradiation energy is trapped by the crystalline matrix of silica. Heating releases this energy as light, and the thermoluminescence produced can be measured and related to the irradiation dose.

When foods contain a certain amount of lipid it is worthwhile to look for fatty acid breakdown products, provided these can be distinguished from those arising from heat treatment or oxidative rancidity (*see* Chapter 4). The amounts of long-chain unsaturated fatty acids, such as tetradecene ($C_{14}H_{28}$) and hexadecadiene ($C_{16}H_{30}$), have been shown to rise in relation to the irradiation dose. They can be separated from the foodstuff by vacuum distillation and then identified and quantified by gas chromatography linked to mass spectrometry (GC/MS).

Another type of fatty acid/triglyceride breakdown products which qualify as URPs are the alkylcyclobutanones, whose formation is shown in Figure 9.3. These too can be detected by gas chromatography.

One effect of irradiation on DNA is to create breaks in the polymer chain, leading to the formation of small fragments. These fragments can be detected by a technique known as microelectrophoresis. Although repeated freezing and thawing also damages DNA, the pattern of

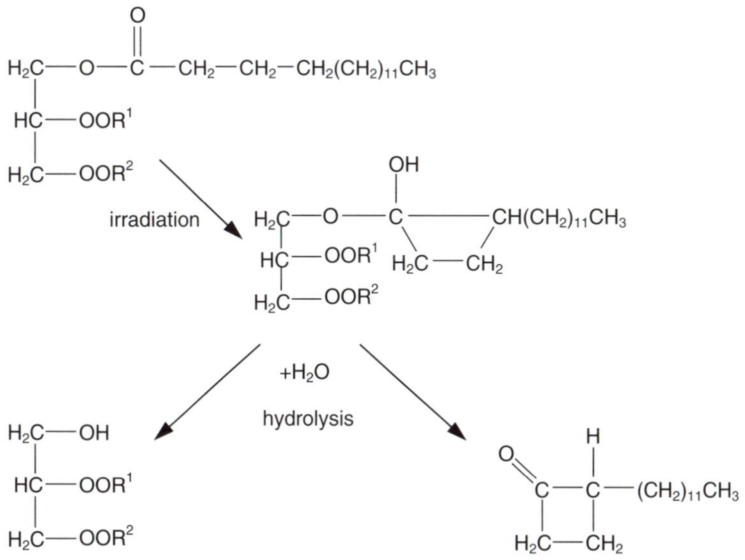

Figure 9.3 The formation of 2-alkylcyclobutanones from triglycerides during the irradiation of food. The example shown is the formation of 2-dodecylcyclobutanone from a triglyceride containing palmitic acid.

fragments produced by irradiation is sufficiently distinctive for this to be a valuable procedure for detecting irradiation.

Despite exhortations from the authorities and the food industry that losses of flavour and nutritional value caused by low doses of irradiation are trivial, these have not been sufficient to quell public unease. Although the use of irradiation is increasing in many parts of the world, *e.g.* irradiated fresh meat is on sale in the USA, in Britain the food industry is possibly more cautious of public opinion. It has been said that in bringing irradiated products to the marketplace the major food retailers in Britain are in a desperate race "to be the first to be second". At the present time herbs and spices are the only irradiated products on sale in Britain. One source of consumer concern, alluded to above, is that unscrupulous food suppliers might pass off less than fresh foods by using irradiation to eliminate their high microbial population without reversing the other signs of excess age, such as nutrient losses and the formation of bacterial exotoxins (*see* Chapter 10), or off-flavours.

FURTHER READING

R. Angold, G. Beech and J. Taggart, *Food Biotechnology*, Cambridge University Press, Cambridge, 1989.

J. M. Jay, J. Loessner and D. A. Golden, *Modern Food Microbiology*, Springer–Verlag, New York, 2005.

M. Doyle, *Food Microbiology: Fundamentals and Frontiers*, American Society for Microbiology, 3rd edn, 2007.

Antimicrobials in Food, ed. P. M. Davidson, J. N. Sofos and A. L. Branen, CRC Press, Boca Raton, Florida, 3rd edn, 2005.

M. J. Waites, N. L. Morgan, J. S. Rockey and G. Higton, *Industrial Microbiology: An Introduction*, Blackwell Science, Oxford, 2001.

RECENT REVIEW

J. Farkas, *Irradiation for Better Foods; Trends Food Sci. Technol.*, 2006, **17**, p. 148.

CHAPTER 10
Undesirables

The previous chapters of this book will have left the reader in no doubt that the foods we eat contain many substances which are neither directly nutritive nor of value in other ways. While we can assume that our bodies are indifferent to many minor food components, the rule that substances which are not friends are likely to be enemies is a helpful guide. This chapter discusses a number of substances present in our food which we would probably be better off without. Obviously there are many other substances which may have some adverse effects but also play an important nutritional role, such as the saturated fatty acids.

So far in this book substances have been classified either in terms of their chemical structure, (Chapters 2 to 5), or according to their function (Chapters 6 to 9). Neither of these approaches is acceptable when we consider undesirable components of food. The enormous diversity of their chemistry, and the nature of their physiological effects, makes it better to classify them by origin. This gives us four groups of toxins which will be discussed in this chapter:

- Endogenous toxins: substances which are natural components of food materials;
- Microbial toxins: substances secreted by contaminating moulds or bacteria;
- Toxic residues: substances which are carried over into food from procedures applied by man to the living plants or animals used as food; and
- Toxic contaminants: substances resulting from food processing, preservation or cooking operations.

Food: The Chemistry of its Components, Fifth edition
By T.P. Coultate
© T.P. Coultate, 2009
Published by the Royal Society of Chemistry, www.rsc.org

ENDOGENOUS TOXINS IN FOODS DERIVED FROM PLANTS

It is widely assumed that if a foodstuff is "natural" it will be safe. If we follow this assumption it follows that afflictions of mankind which can be traced back to diet will disappear if we avoid "unnatural" foods, and if we stop processing natural foods in unnatural ways. For the author there is no clearer demonstration of the naivety of this attitude than the abundance of naturally occurring toxins in plant food materials. In the face of the evidence it is hard to maintain the view that plants were placed on the planet with the primary function of sustaining human life. The physiological similarity between humans and the animals we eat means that endogenous toxins are likely to be encountered much more often in plants than in animals, and more frequently in invertebrates or fish than in mammals. In this section a representative selection of plant toxins are examined.

One of the best known plant toxins is solanine, which occurs in potatoes. This steroidal glycoalkaloid (10.1) is not only found in potatoes, along with smaller amounts of related alkaloid derivatives, but also in other members of the Solanaceae family such as the aubergine and the highly poisonous nightshades. Potatoes normally contain 2–15 mg solanine per 100 g fresh weight. When potatoes are exposed to light and turn green, the level of solanine may reach 100 mg per 100 g, concentrated mainly just beneath the skin. The growing popularity of commercially prepared potato products in which the skin is intended to be eaten is a cause for concern, since even without obvious signs of greening the total concentration of glycoalkaloids in the skin can reach 60 mg per 100 g. Other metabolically active tissues, such as sprouts on potatoes, may contain even higher concentrations. Solanine is not one of the more potent toxins, although over the years a number of fatalities have been reported in the scientific literature.

Solanine has been confirmed as an inhibitor of the enzyme acetyl choline esterase, a key component of the nervous system, and signs of neurological impairment have been recorded after ingestion of the toxin at approximately 2.8 mg kg^{-1} of body weight. The relatively high doses needed to produce toxic effects are believed to be due to its poor absorption from the gastro-intestinal tract, since laboratory animals show a much stronger reaction if the toxin is injected directly into the bloodstream.

The widespread awareness of the general public to the dangers of eating greened potatoes has restricted solanine poisoning to a low level. It has been reinforced by the observation that potatoes with solanine above about 20 mg per 100 g have a bitter taste. As solanine is heat stable and fairly insoluble in water it is not lost during normal cooking procedures. The usually accepted safety limit for total glycoalkaloids is 200 mg kg^{-1}, but lower limits have been suggested. This is partly due to the wide variations found even in samples from the same batch of potatoes, and also since the milder symptoms of solanine poisoning such as abdominal pain and vomiting may go unreported.

Caffeine (10.2), already mentioned in Chapter 6, is a purine alkaloid. Another important member of this group, known as the methyl-xanthines, is theobromine (10.3). It could be argued that these substances, which are found in tea, coffee, cocoa and cola beverages, should not be classified as toxins since they are normally regarded as stimulants. In the author's opinion there are two good reasons for treating them as toxins. Firstly, whether their physiological effects are beneficial or not depends on the amount consumed and, secondly, they could never be looked upon as nutrients.

(10.2)   (10.3)

Roasted coffee beans contain between 1 and 2% caffeine, but the level in the beverage itself is highly dependent on the method of preparation and its strength. Values between 50 and 125 mg per cupful are normal. Black leaf tea (*see* Chapter 6) contains some 3–4% caffeine, giving around 50 mg per cupful, and rather less theobromine, about 2.5 mg per cupful. Cocoa powder contains around 2% theobromine and 0.2% caffeine. The levels in plain chocolate are about a quarter of this, and in milk chocolate about one-tenth. Nowadays the caffeine content of cola drinks is restricted to $200 \, mg \, dm^{-3}$, although the average value is nearer to $65 \, mg \, dm^{-3}$. In recent years "energy drinks" containing caffeine have become popular. These may also include extracts of the South American plant, guarana (*Paullinia cupana*), which supplies additional caffeine to give a total concentration in the $200–300 \, mg \, dm^{-3}$ range.

Nearly all physiological studies on methylxanthines have concentrated on caffeine. The stimulant action of caffeine is the result of its promotion of the synthesis or release of the catecholamine hormones epinephrine and norepinephrine (better known as adrenalin and noradrenalin, respectively) into the bloodstream. Most of the other effects of coffee on the level of glucose, triglycerides and cholesterol in the blood can be traced back to release of these hormones. The diuretic effect of caffeine is well known, of course, but the details are beyond the scope of this book. Although the various effects of caffeine tend to lie within the normal range, even for quite heavy coffee drinkers, it remains questionable whether they will be beneficial in the long term.

There have been too few cases of death attributable to caffeine consumption to provide useful data about its acute toxicity, but doses from 150 to $200 \, mg \, kg^{-1}$ are known to have proved fatal in humans. This level of dosage will hardly ever be achieved, even by obsessive coffee drinkers. A vast amount of work has been carried out, both using animal experimentation and epidemiological methods, in an attempt to establish whether caffeine consumption can have adverse effects on human reproduction. At the present time caffeine has been exonerated from causing either birth defects or low birth weight, but its stimulant effects are not regarded as being beneficial during pregnancy.

For many consumers the answer to the possible adverse effects of caffeine is to consume instant coffee in its decaffeinated form. In this case the caffeine is removed by solvent extraction before the beans are roasted. Water is first added to the beans to bring the moisture content up to about 40%. They are then extracted in a counter-current system with the solvent methylene chloride at between 50 and $120 \, °C$, conditions fairly selective for caffeine. Residual solvent and moisture are then driven off and the beans roasted. An alternative extremely selective solvent for caffeine is liquid, or supercritical, carbon dioxide at a pressure of the order of 120–180 times atmospheric. The beans must still be moist, and the caffeine is removed from the circulating supercritical CO_2 by adsorption on to activated charcoal. This process has the advantage that removal of desirable flavour elements is minimised and there is no question of possible toxicity from residual solvent.

Another group of alkaloids, the pyrrolizidine alkaloids (PAs), are associated with herbal remedies rather than foods. They would barely deserve a mention in this book apart from their very occasional appearance in milk or honey when the cows or bees concerned have fed on unsuitable plants. However, there is one plant material, comfrey, which is often used in herbal tea, a product lying on the borderline between food and medicine. Comfrey contains at least twelve different pyrrolizidine alkaloids. Symphytine (10.4) is one of the most abundant, and illustrates a chemical feature characteristic of these compounds, fused 5-membered rings with a shared nitrogen atom. A cup of herbal tea made from comfrey root could contain as much as 9 mg total PAs. With so many different compounds occurring in such a wide range of plants, reliable data on the toxicity of PAs are difficult. Symphytine is thought to be possibly carcinogenic, and PAs are usually assumed to be to blame in cases of liver damage associated with the use of herbal preparations.

(10.4)

Substances which have adverse effects when consumed by humans often turn out to be natural insecticides, in other words substances synthesised by plants to discourage insect pests. A good example of this phenomenon is the psoralens found in celery. (*Apium graveolens* var. *dulce*), Psoralens are a group of substances, also known as fur-ocoumarins, occurring widely in the plant kingdom. Their significance as components of edible plants came to notice in the USA in 1986 when numerous reports of severe skin rashes and related conditions among farm workers, packers and shop workers were traced back to the handling of a new variety of celery bred to resist insect pests. Psoralens, notably 8-methoxypsoralen (10.5), were found in the green leafy parts of this type of celery at levels around 6200 ppb, compared with 800 ppb in other varieties. The psoralens from the plant are absorbed by the skin and enter the nuclei of the skin cells. Here they interact with the DNA, finding their way into the space between neighbouring pairs of thymidine residues, a process known as *intercalation*. Subsequent irradiation with UV light, as experienced by a person working outdoors, causes the formation of a link between the psoralen molecule and the pyrimidine ring of the thymidine. Although human systems for the repair of damaged DNA can recognise psoralen's effects, some damage still results. Suffice it to say that this variety of celery is no longer grown.

(10.5)

Another food plant which has presented a similar problem is the lime. Limes (*Citrus aurantiifolia*) can also produce UV-induced photo-dermatitis. In this case the usual victims are sunbathers who have whiled away their time on the beach drinking margaritas,[i] leaving traces of fresh lime juice on their skin.

Chocolate is also known for its ability to cause migraine headaches in certain individuals. This is related to its content of phenylethylamine, a vasopressor (*i.e.* it raises blood pressure) which occurs widely in foods, some of the more notable being shown in Figure 10.1. They may be normal products of amino acid metabolism in the plant or, in the case of cheese and wine, the micro-organisms involved in fermentation. Normally the body is well equipped to deal with these amines. The activity of the enzyme monoamine oxidase, situated in the mitochondria of many tissues, converts them to the corresponding aldehydes. It is clear that some individuals are either particularly sensitive to these amines or have difficulty oxidising them. Inhibitors of monoamine oxidase are important drugs (*e.g.* in the treatment of depression), and patients taking them are advised to avoid this type of food. The usual effect of these amines is constriction of the blood vessels in various parts of the body. The general result is hypertension, but in the brain vasoconstriction causes intense headaches. There is also some support for the suggestion that the prevalence of a certain type of heart disease in West Africa may be caused by large amounts of plantain, a relative of the banana, in the diet. A typical plantain-based diet might easily give a daily intake of 200 mg serotonin.

Another African staple food, cassava (manioc, *Manihot esculenta*), poses a quite different toxicological problem, that of cyanogenesis.

---

[i] A margarita is a cocktail of lemon or lime juice with tequila.

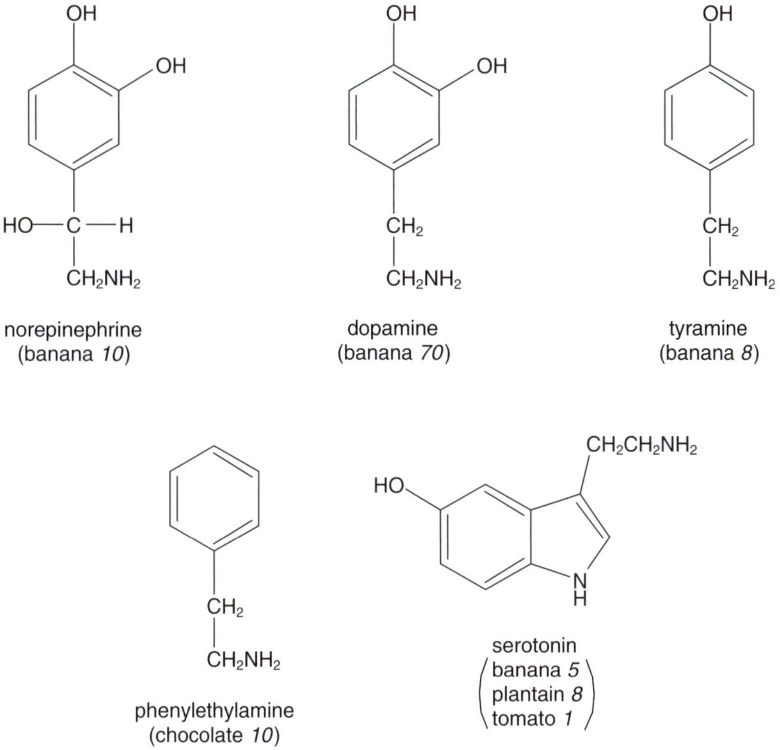

Figure 10.1 Vasopressor amines. The figures in italics are typical concentrations of the amine in the food material indicated, expressed in mg per 100 g fresh weight. Phenylethylamine is also found in significant amounts in many cheeses and in red wines.

In Chapter 2 the glycoside amygdalin was briefly described. This is one of a group of glycosides which occur in many plant families, but particularly in the Rosaceae and Leguminoseae (*see* page 356). Although the presence of amygdalin in the stones of some fruit does not present much of a hazard, the higher levels of α-hydroxyisobutyronitrile and related compounds in cassava do. Fresh cassava may give rise to 50 mg hydrogen cyanide 100 g due to the activity of two enzymes, β-glucosidase and hydroxynitrile lyase (8.46). Both occur in the plant tissue and become active (*cf.* phenolase, Chapter 6) when the tissue becomes damaged during harvesting or prior to cooking. Although animals do not themselves produce either of these enzymes, there is evidence that intestinal bacteria are able to release hydrogen cyanide from cyanogenic glycosides. It is a well established tradition in West Africa for those who depend on cassava to allow a fermentation period before cooking, which allows the endogenous enzymes to release the HCN, which being volatile is lost during cooking.

In spite of this food preparation procedure it is becoming clear that chronic cyanide poisoning, caused by ingestion of low levels of HCN over prolonged periods, is widespread in parts of the world dependent on cassava. The possibility that low levels of serum cyanide may afford a degree of protection against the malarial parasite was discussed in the context of the bitter taste of cyanogenic glycosides in Chapter 7. Two common diseases, one a degenerative neurological condition (ataxia) and the other a form of blindness, tropical amblyopia, have been attributed to the combination of cyanide and a deficiency of the vitamin cobalamin. Another problem is caused by the body's use of the enzyme rhodanase to detoxify small amounts of cyanide by converting it to thiocyanate, CNS^-. Where the cyanide intake is high, *e.g.* in those parts of eastern Nigeria where dried unfermented cassava is popular, high levels of thiocyanate are found in the blood. Thiocyanate is known to interfere with iodine metabolism, resulting in the goitre common in eastern Nigeria.

Another important food crop which can contain dangerous levels of cyanogenic glycosides is the lima bean (*Phaseolus lunatus*), which is grown in many parts of the world. The difficulty of preparing the beans to overcomes their toxicity has led to the breeding of bean varieties with reduced levels of glycosides.

Cyanogens are not the only toxins found in legumes. Another group are the protease inhibitors. Legume seeds, including peas, beans, soya and peanuts, all contain proteins which can inhibit the proteases in the mammalian digestive tract, notably trypsin and chymotrypsin. The inhibition is the result of one-to-one binding which blocks the active site of the enzyme. These proteins fall into two groups, those with molecular weights around 22 000 which are specific to trypsin, and those with molecular weights between 6000 and 10 000 which can inhibit chymotrypsin as well. Experimental studies with animals fed on diets of little apart from unheated soya bean meal show reduced efficiency in utilising proteins (*see* Chapter 5). If the diet is persisted with, the pancreas, the organ which synthesises trypsin and chymotrypsin, shows a marked increase in size, presumably due to some type of compensatory mechanism.

When legumes have been heated, the efficiency with which they are utilised as sources of dietary protein rises, but experiments with soya have shown that prolonged heating does not allow the theoretical efficiency to be reached. This is because the loss of essential amino acids caused by the Maillard reaction becomes significant.

Any examination of the effects of protease inhibitors in legumes is complicated by the presence of a second group of toxic proteins, the lectins. The lectins are characterised by their ability to bind to the surface of certain animal cells, particularly red blood cells, causing them to clump together—hence their alternative name, haemagglutinins. Because individual lectins show a high degree of specificity for particular types of animal cells, *e.g.* red blood cells of different blood groups, they have become popular laboratory tools with animal biochemists and physiologists. Injected directly into the bloodstream some are extremely potent poisons, with toxic doses in the region of $0.5\,\mathrm{mg\,kg^{-1}}$.

Lectins are proteins of molecular weight around 10^5. Due to their usefulness to biochemists some have been studied in great detail, but until recently the results have not had much relevance to food science. Successful experiments have now been conducted to insert the genes of particular lectins into other plant species by recombinant DNA techniques. The idea is that this could provide a non-chemical method of enhancing the pest resistance of valuable agricultural crops. Lectins may represent a significant fraction of the total protein in legumes, amounting to 3% in the case of soya beans. Experimental animals fed diets containing as little as 1% unheated soya beans show impaired growth, but some other legumes can be lethal at this level. The differences appear to be caused by the variable susceptibility of different lectins to breakdown by digestive enzymes. The lectin in kidney beans is one of the more resistant. The toxic effect of orally administered lectins is reduced uptake of nutrients from the digestive tract, believed to be caused by damage to the intestinal mucosa.

The relevance of lectin toxicity to humans is debatable. From time to time there are bursts of enthusiasm in the media for eating raw vegetables, which as we have seen can be a potential hazard. As long as the contribution of uncooked legumes to total protein intake remains small the danger of lectin-induced illness is likely to be slight. Fortunately for advocates of the culinary value of barely cooked sprouted beans, massive breakdown of lectins occurs at the time the seeds germinate.

Soya beans are the best known source of another group of biologically active substances which occur quite widely in plants, the phytoestrogens. It is debatable whether they should be considered as undesirables, but in the absence of a specific discussion of what are nowadays referred to as "nutraceuticals" this is as good a home for them as any. They were discovered in the 1940s as a cause of fertility problems in Australian sheep fed a diet rich in a species of clover which is now recognised as a

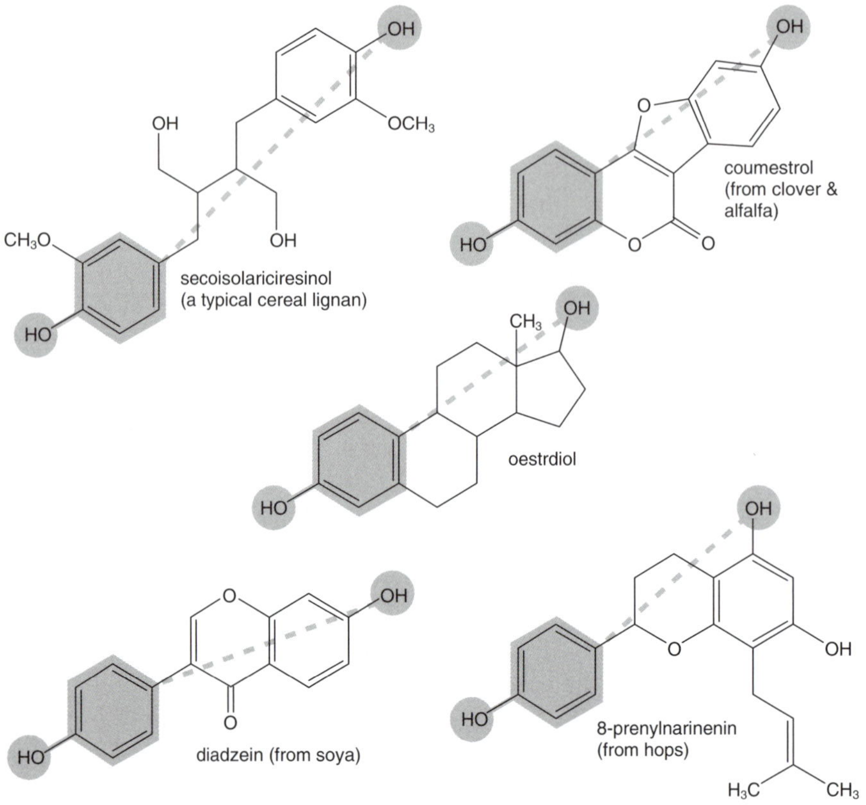

Figure 10.2 The structure of phytoestrogens in relation to that of the human hormone, oestradiol.

major source of a phytoestrogen known as coumestrol. This substance is unusual in that it binds to the oestrogen receptors in the body with an affinity comparable to that of oestradiol, whereas the soya phytoestrogens, the only ones of significance in human diets, have a binding affinity between one thousandth and one ten-thousandth that of oestradiol. As illustrated by the structures shown in Figure 10.2, phytoestrogens have their critical phenolic and hydroxyl groups in a similar structural relationship to those in oestradiol.

The legume phytoestrogens are classed as isoflavones.[ii] The structures of the principal soya bean phytoestrogens are shown in Figure 10.3.

[ii] Isoflavones differ from other flavonoids in that their B ring (*see* Formula 6.9) is attached at the 2 rather than the 3 position of the central oxygen-containing ring.

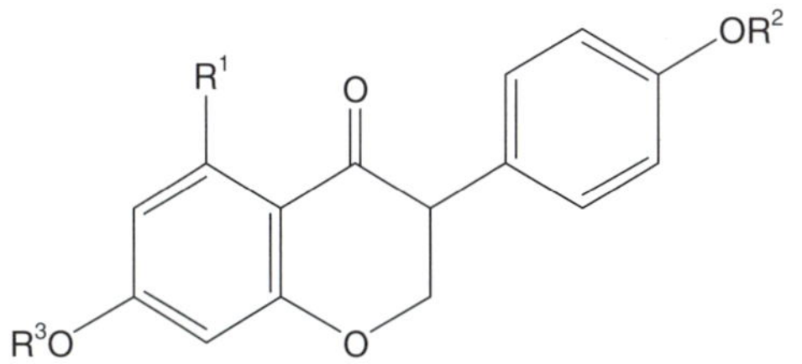

	R^1	R^2	R^3
daidzein	H	H	H
genistein	OH	H	H
glycetein	H	OCH_3	H

Figure 10.3 The principal isoflavanoid phytoestrogens of soya beans. In the raw beans there is usually a glucose residue attached at position 3. These glycosides are known as daidzein, genistein and glycitein, respectively. Glycitein is not considered to be oestrogenic.

In Europe most soya-based foods are consumed as part of vegetarian diets, whereas in the Far East and among communities of Far Eastern origin in North America and elsewhere, fermented soya products such as soy sauce, tofu, miso and tempeh are components of virtually everyone's diet. Amounts of total phytoestrogens found in soya beans can vary widely, between 20 and 300 mg kg^{-1}, mostly as genistein or daidzein, as shown in Table 10.1. Levels in foods other than legumes rarely exceed 1 mg kg^{-1}.

Total phytoestrogen levels in soya-based infant formula milk products vary, but are normally in the region of 9 mg dm^{-3} when made up. Adults consuming omnivorous European-style diets have been estimated to consume up to 0.8 mg of phytoestrogens daily, although the figure for adult vegetarians may be 100 times this figure. The daily phytoestrogen intake of Chinese and Japanese adults is between 20 and 200 mg daily.

Concern is regularly expressed about the possible effect on subsequent reproductive function of high consumption of soya milk in

Table 10.1 The phytoestrogen content of soya-based food products. Adapted from data given in *Phytoestrogens and Health* (Committee on the Toxicity of Chemicals in Food Consumer Products and the Environment, Food Standards Agency, 2003).

Food	*Total phytoestrogen content (mg/100 g)*
Soya beans	14–300
Soya milk	5–10
Soya flour	130–200
Soya protein-based burgers, sausages, *etc.*	8–15
Fermented foods: tempeh, tofu, miso	15–70
Soy sauce	0.1–1.5
Non-dairy infant formula	up to 20

infancy (providing phytoestrogen intakes in the region of $4\,mg\,kg^{-1}$ of body weight per day), but there has been no substantial evidence published to support this concern. Potential benefits of phytoestrogen appear in the scientific literature and popular press at regular intervals. The authors of the report referred to in Table 10.1 considered a wide range of possible phytoestrogen influences reported in the literature, including thyroid function, fertility and development, the central nervous system, the skeleton (with regard to osteoporosis), coronary heart disease, interaction with endogenous hormones, diabetes, and cancer of the breast and prostate. They concluded that epidemiological data confirm that populations consuming relatively large quantities of soya do in fact show higher bone mineral density and lower incidence of breast and prostate cancer, but there is little evidence to support the other effects. The actual mechanism by which phytoestrogens exert their effects are only beginning to be understood.

It remains to be seen whether the apparent benefits of a soya-rich diet will have much effect on western diets. The other significant issues for European soya consumers is the increasing dependence on genetically modified soya varieties grown in North America, or Amazon rain forest clearance for soya grown in South America.

As an aside to the problems with soya, this is an appropriate point to mention the contamination of some soya-based products with 3-monochloropropane-1,2-diol (3-MCPD,10.6). This substance is found from time to time at very low levels ($\sim 10\,\mu g\,kg^{-1}$) in a wide range of food products. The UK authorities have concluded that at this level 3-MCPD is unlikely to present a carcinogenic risk to man, bearing in mind that the "no observable effect" level (NOEL, *see* page 417) has been established at $1.1\,mg\,kg^{-1}$ of body weight per day. However, in some products containing acid-hydrolysed soya protein, notably soy sauce, 3-MCPD has been found at higher levels ($>20\,\mu g\,kg^{-1}$) and gives cause for concern. The 3-MCPD arises from reaction between the hydrochloric acid used for the hydrolysis and the inevitable residue of soya fatty acids.

$$Cl-CH_2-\overset{\displaystyle H}{\underset{\displaystyle OH}{C}}-CH_2OH$$

(10.6)

Another legume, the fava bean, *Vicia faba*,[iii] is also well known for its toxicity. In some parts of the world consumption can lead to a potentially fatal haemolytic anaemia (*i.e.* premature destruction of red blood cells) known as *favism*. The association of fava beans with attacks of what we now call favism was known to the doctors of ancient Greece, and also to the mathematician Pythagoras, who published a "rule" or "consultation" stating, κυάμων ἀπέχεσθαι which roughly translates as "abstain from fava beans".

The problem is caused by two pyrimidine glycosides, vicine (10.7) and convicine (10.8). These occur respectively at around 700 and 200 mg per 100 g in the bean cotyledons, the main edible portion. During the normal process of digestion these are broken down by micro-organisms in the gut to yield the corresponding aglycones, divicine (10.9) and isouramil (10.10). Although fava beans are widely grown in many of the temperate zones of the world, favism is mainly encountered in territories bordering the Mediterranean such as Greece, southern Italy, Turkey, Cyprus, Sardinia, Algeria, Lebanon and Egypt. Favism also occurs in the Near East (Iraq and Iran) and parts of India.

A common feature of these regions is the prevalence of malaria. A significant proportion of the population has an inherited condition which leads to reduced levels of the enzyme glucose-6-phosphate dehydrogenase

---

[iii] The species *V. faba* is normally divided into the subspecies *V. faba minor*, the fava bean proper, and *V. faba major*, the broad bean better known in the UK. Both subspecies contain similar levels of pyrimidine glycosides.

(G6PD) in their red blood cells (erythrocytes). This reduced enzyme activity results in the cells providing an unfavourable environment for the development of the malarial parasite, giving these individuals considerable resistance to the disease. The downside of the protective effect is that their erythrocytes become less stable to attacks of favism. It is now established that the fava bean glycosides have the same effect on the erythrocytes' hospitality to malarial parasites as G6PD deficiency does.

One might expect the incidence of favism to lead vulnerable populations to stop growing fava beans, but for two reasons this has not occurred. Firstly, these beans are an excellent source of protein and can be produced successfully under difficult environmental conditions. Secondly, the presence of divicine and isouramil provides a useful degree of protection against malaria, especially in individuals with reduced G6PD. The risks of favism in individuals having the lowest levels of G6PD can be assumed, in evolutionary terms, to be a price worth paying.

The actual mechanism of the effect of divicine and isouramil on the metabolism of erythrocytes is far from simple and is examined further in the Special Topics at the end of this chapter.

The concluding toxin in this section is myristicin (10.11). This occurs in significant levels in nutmeg and in smaller amounts in black pepper, carrots and celery. There is sufficient myristicin in 10 g of nutmeg powder to induce similar symptoms to a heavy intake of ethanol: initial euphoria, hallucinations, and finally narcosis. Higher doses induce symptoms also similar to those of ethanol poisoning, including nausea, delirium, depression and stupor. Of course, the amount of myristicin in the quantities of nutmeg used as a flavouring are well below those affecting the brain, but it is interesting that the consumption of nutmeg-flavoured foods during pregnancy has traditionally been regarded as ill-advised.

(10.11)

ENDOGENOUS TOXINS IN FOODS OF ANIMAL ORIGIN

The reason for the relative shortness of this section has already been mentioned. Most animal toxins of significance to human eating habits are found in taxonomic groups distant from man, such as the crustaceans, but one of the most interesting occurs in a number of fish species, notably the infamous puffer fish. The toxin tetrodotoxin (10.12) occurs in various organs of the puffer fish, including the liver and ovaries (at around 30 mg per 100 g) and to a lesser extent in the skin and intestines. The minimum oral lethal dose of tetrodotoxin is estimated to be between 1.5 and 4.0 mg.The muscles and testes are generally free from the toxin, and it is these which form a delicacy popular in Japan. Unfortunately, great skill is required by the cook to separate the deadly parts of the fish from those which are merely risky. Stringent licensing of expert cooks has reduced the risk, but fatalities still occur.

(10.12)

The toxin is absorbed rapidly and the first symptoms may become apparent as soon as ten minutes after eating the toxic fish, although an interval of an hour or so is more usual. The effect of the toxin is to block movement of sodium ions across the membranes of nerve fibres. The result is that nerve impulses are no longer transmitted and the victim suffers a range of distressing nervous symptoms, developing into total paralysis and respiratory failure, and leading to death within 6–24 hours. No effective treatment is available and, in the light of continued enthusiasm for eating these fish, all that can be done is to ensure the best possible training of Japanese chefs.

Of the other toxins associated with marine foods the best known are difficult to classify unequivocally as being of animal origin. For example there is a group of toxins which occur under certain circumstances in edible shellfish such as mussels, cockles, clams and scallops. At certain times of year the coastal waters in the hotter parts of the world develop a striking reddish colour due to the proliferation of red-pigmented dinoflagellates, a type of plankton. These "red tides" are alluded to in the Old Testament (Exodus 7:20–21) and may have given the Red Sea its name. Problems also occur in cooler regions such as the coasts of Alaska and Scandinavia, and the dinoflagellates are not always red-pigmented. Some species of dinoflagellates contain toxins that quickly find their way into the food chain *via* shellfish for human consumption. There are a number of different classes of shellfish toxins, but the best known are those which cause paralysis of the nervous system, resulting in rapid death after consuming only a few mg. Saxitoxin (10.13) is one of the best known of the toxins causing paralytic shellfish poisoning (PSP). Most of the others differ chemically only in minor variations in the groups around the rings. A number of different toxins may be responsible for a single outbreak of PSP. The relationship between red tides and the toxicity of shellfish appears to have been recognised for many centuries in coastal communities, and as a result the number of outbreaks of poisoning is limited. However, the number of fatalities occurring in a single outbreak of PSP is such that the authorities in regions such as the Pacific coast of North America maintain routine checks on toxin levels in shellfish catches. A serious outbreak of PSP, fortunately not fatal, occurred in 1968 when 78 people were taken ill after eating mussels from sites on the north-east coast of England. Since that time shellfish in this area have been regularly monitored and on only one occasion, during ten weeks in the summer of 1990, were levels above 80 g per 100 g of mollusc flesh recorded and warnings issued.

(10.13)

Diarrhetic shellfish poisoning (DSP), involving a rapid onset of diarrhoea, nausea and vomiting lasting up to three days, occurs mostly in Japan but cases have been reported world-wide. Although no cases have been recorded in Britain, the toxins concerned have been detected in shellfish, *e.g.* queen scallops, in British waters. The best known of the DSP toxins is okadaic acid (10.14), again produced by species of dinoflagellates. Cases of amnesic shellfish poisoning (ASP) were first identified in Canada. ASP can cause severe neurological symptoms which can lead to permanent brain damage. The toxin involved, domoic acid (10.15), is produced by various species of diatom.

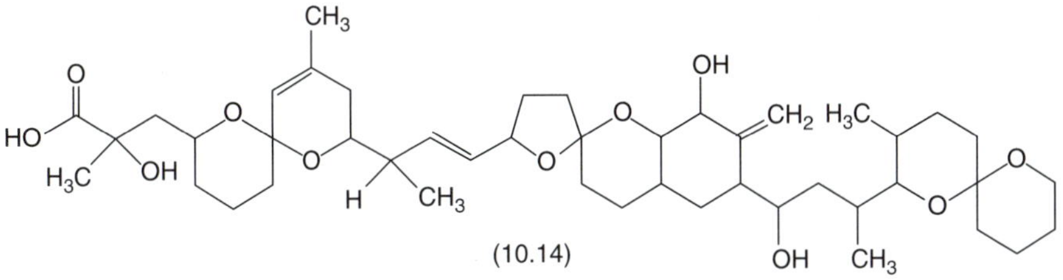

(10.14)

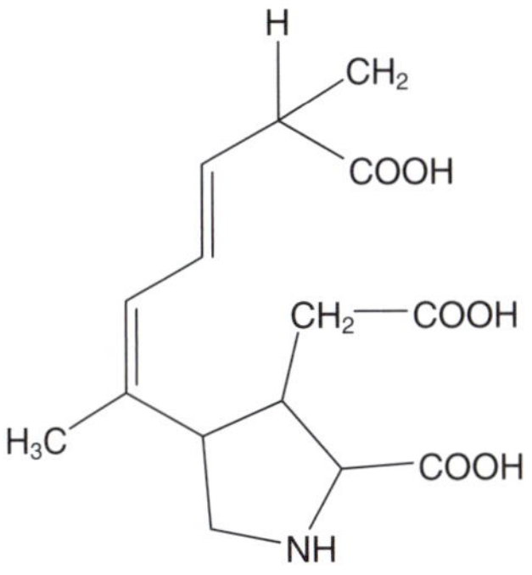

(10.15)

A quite different type of poisoning is associated with fish of the family Scombridae. This family includes many commonly eaten fish, including mackerel, tuna and sardines. When freshly caught the fish are not toxic, but if held at temperatures above 10 °C for several hours high levels of histamine (10.16) may sometimes be found. Histamine is formed from the amino acid histidine, which occurs naturally in the muscle. The reaction is catalysed by a decarboxylase produced by a number of common bacteria, but that of *Proteus morganii*, an otherwise innocuous marine bacterium, has been particularly incriminated. In spite of levels of histamine above 100 mg per 100 g being produced by the bacteria, obvious putrefaction of the fish may not have taken place. Consumption of large doses of histamine from contaminated fish does not necessarily produce adverse effects, since histamine is poorly absorbed from the intestine. Nevertheless the collection of symptoms seen from time to time, including severe headaches, palpitations, gastrointestinal upsets, skin flushes and erythaema, correspond closely to the effects of histamine injection into the bloodstream. In addition, the symptoms are typically relieved by antihistamine drugs. It is assumed that victims of scombrotoxic fish have some minor intestinal lesion or disorder that has allowed the histamine through.

$$\text{histidine} \longrightarrow \text{(10.16)} + CO_2$$

Histamine is classified as a biogenic amine, *i.e.* a low molecular weight amine having recognised biological activity. Other biogenic amines, which occur in plant materials and are grouped together by their tendency to raise blood pressure in humans, were described on page 386. Such biogenic amines are not uncommon in foods of animal origin when micro-organisms have become involved, as in the spoilage of meat and the maturation of cheese. Histamine, tyramine, cadaverine [$H_2N(CH_2)_5NH_2$, derived from the amino acid lysine] and putrescine [$H_2N(CH_2)_4NH_2$, from glutamine] are all abundant in cheese, typically at levels between 100 and 500 mg kg^{-1}. The levels of amines in meat are generally around one tenth of these figures.

MYCOTOXINS

In the preceding chapter the susceptibility of human and animal food-stuffs to contaminating micro-organisms was explained. While the value to man of the fermentative micro-organisms involved in dairy products is generally recognised, most people instinctively reject food which looks mouldy. This rejection may seem to be an aesthetic reaction, but in actual fact the growth of many species of fungi on foods and food raw materials is accompanied by the production of toxins that can be extremely dangerous.

The best known of these diseases is ergotism. In the Middle Ages this reached epidemic proportions throughout much of Europe, particularly where rye was an important cereal crop. During the 20th century smaller outbreaks did occur in Britain and France. It is now established that the cause of the disease is the infection of rye and other cereals in the field by the commonly occurring mould *Claviceps purpurae*, or ergot. At one stage during its life cycle the mould produces structures known as sclerotia, which are hard purplish-black masses of dormant cells. About the same size as a cereal grain, these may inadvertently be harvested along with the grain and unless removed will be milled into the flour. The sclerotia have been found to contain at least 20 different toxic alkaloids, the most abundant of which are peptide derivatives of lysergic acid, shown in Figure 10.4.

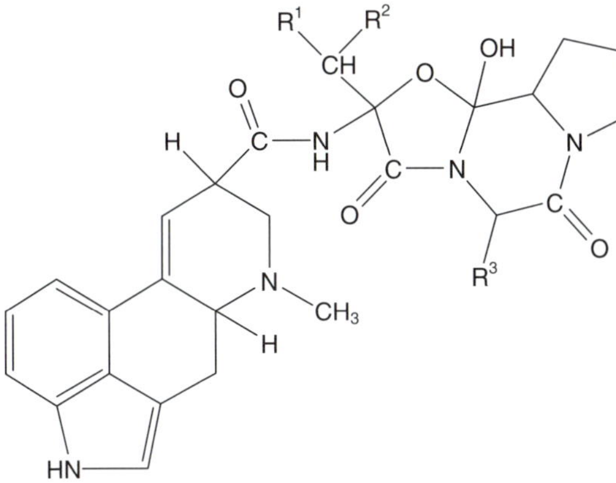

Figure 10.4 Ergot alkaloids. In ergotamine, one of the most abundant ergot alkaloids, R^1, R^2 and R^3 are H, CH_3 and C_6H_5, respectively. Others have
various combinations of H or CH_3 at R^1, R^2 and R^3, and $C_6H_5CH_2$,
$(CH_3)_2CHCH_2$ or CH_3CH_2 at R^3.

The biological effects of the ergot alkaloids appear to be due to the
effect of individual alkaloids on different features of human physiology,
which means that the symptoms will vary from outbreak to outbreak
depending on the range of toxins present. The onset of the disease is
usually slow but insidious, and is generally catastrophic in outcome.
Through poorly understood effects on the nervous and vascular systems
the victim initially suffers general discomfort, followed by intense burning
pains in the hands and feet. These give way to a total loss of sensation in
the limbs, followed firstly by gangrenous withering and blackening, and
finally loss of the limbs. These symptoms are often accompanied by
mental derangement and gastrointestinal failure before the victim finally
succumbs. It is hardly surprising that in the Middle Ages this awful disease was given the picturesque name, "St Anthony's Fire".

The comparative rarity of the disease today is the result of a number
of factors. Modern fungicides reduce the incidence of infected plants to a
minimum, and modern grain drying systems largely prevent post-harvest
spread of the disease. Nowadays millers inspect all grain shipments for
telltale signs of the black sclerotia, and in addition modern grain
cleaning methods are able to separate the heavier grains from the light
sclerotia. Despite the success of these preventive measures there is no
room for complacency. One should be cautious of "amateur" millers,
especially as these are also likely to favour cereals grown without the
benefit of fungicide treatment.

To date some 150 different mould species have been shown to produce
toxins when they grow on human or animal foodstuffs. Cereals are the

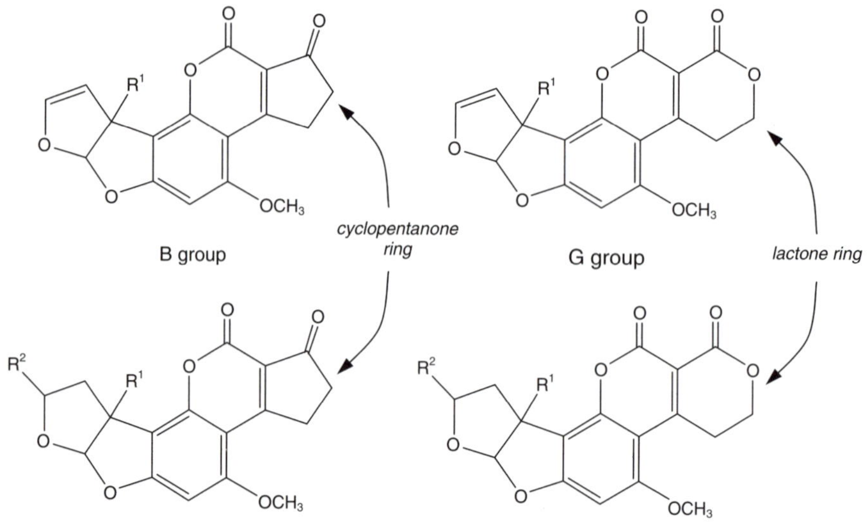

Figure 10.5 The mycotoxins of *Aspergillus flavus*. The aflatoxins are bifur-
anocoumarins fused in the B group to a cyclopentanone ring, or in the G
group to a lactone ring. Similar variations in the substitution pattern of
the bifuranocoumarin system occur in both groups, as shown in the inset
table. Hydroxylation at R^2 eliminates toxicity.

B group	G group	R^1	R^2
B_1	G_1	H	
M_1	GM_1	OH	
B_2G	G_2	H	H
B_2a	G_2a	H	OH

most popular targets. Most moulds are members of one of the three
genera *Fusarium*, *Penicillium* or *Aspergillus*. It is impractical to consider
even a small selection in the detail we have given to ergot, but there is
one *Aspergillus* species, *A. flavus*, which deserves mention.

In 1960 thousands of turkeys were killed in England following an
outbreak of what became known as Turkey X disease. The cause was
traced to Brazilian groundnut (peanut, *Arachis hypogaea*) meal which
had become infected with *A. flavus*, and it was not long before the group
of toxins shown in Figure 10.5 was identified. Although there are differences in the relative toxicity of specific aflatoxins towards individual
animal species, it is now clear that aflatoxin B1 is one of the most potent
liver carcinogens known. A diet containing as little as 15 ppb over a few
weeks can produce tumours in experimental animals.

As might be expected, data on the effects of these toxins on humans are scarce, but they are recognised as being acute hepatotoxins. They are carcinogenic in many animal species, but not all. In humans, long-term exposure is a probable cause of liver necrosis and it plays a part in liver cancer. There are also links to the occurrence of hepatitis B. A further complication is that the mechanism of aflatoxin-induced carcinogens is quite different in humans from that in rodents. In the tropics numerous cases have now been recorded of fatalities due to acute hepatitis and related disorders linked to the consumption of mouldy cereals, especially rice, later shown to contain between 0.2 and 20 mg kg^{-1} of aflatoxins. Of much greater importance are the epidemiological studies now being carried out in many parts of the world, notably in Africa and the Far East. These studies have shown a positive correlation between the incidence of liver cancer within a community and the level of aflatoxins in its staple foodstuffs. Although most plant grains appear to be vulnerable to aflatoxin contamination, there is no doubt that peanuts and derived products such as peanut butter can present particular hazards. Levels of aflatoxins in peanut butter produced in some Third World countries may routinely reach 500 μg kg^{-1}. As a result of such widespread contamination it is not uncommon to find the total daily aflatoxin intake in some Third World communities estimated at some 5 ng kg^{-1} body weight.

Since the climate of Western Europe tends not to favour production of aflatoxins, the regulatory authorities in Britain focus attention on a small range of imported products, including peanuts and peanut butter, dried figs, animal feed, and other types of nuts known to be vulnerable to contamination. The legal limit for aflatoxin contamination in the UK is 4 μg kg^{-1}. The problem facing the authorities in detecting contaminated foodstuffs is the wide variability of samples. For example, some years ago an analysis of 36 separate 1 kg samples selected at random from a single consignment of imported dried figs showed that the majority were below this limit, but three were above 50 μg kg^{-1} and one of these had over 2 mg kg^{-1}.[iv] Wherever contaminated animal feedstuffs are used, aflatoxins can be expected in milk. Aflatoxins B1 and G1 are converted to M1 and GM1 (M indicates for milk), respectively, by liver enzymes, which then appear at levels of around 0.05–0.5 μg dm^{-3} in liquid milk. It is remarkable how much of the aflatoxin in a cow's diet subsequently appears in its milk; the concentration of aflatoxin is generally found to be around 1% of the level in the feed. This may not seem much, but it is sufficient for aflatoxins to present a potentially serious problem for the

---

[iv] Quoted in Ministry of Agriculture, Fisheries and Food, Food Surveillance Paper No 36, *Mycotoxins: Third Report*, HMSO, 1993.

dairy industry. The concentration of aflatoxins found in meat is less than one-thousandth of the feedstuff level.

The accumulation of detailed knowledge of aflatoxin distribution achieved by the authorities in many parts of the world has been greatly facilitated by the relative ease with which aflatoxins can be detected and quantified. Aflatoxins fluoresce strongly in ultraviolet light. At around 365 nm the B-types give a blue and the G-types a green fluorescence. Solvent extraction has to be followed by a fairly elaborate clean-up procedure before TLC or HPLC can be used for detection at the μg level. Determinations down to 10 ppb (based on the original food) are now possible using techniques such as enzyme-linked immunosorbent assay, ELISA.

Food scientists now recognise the potential hazard caused by a number of other mycotoxins. Ochratoxins, notably ochratoxin A (10.17), were first identified as products of *Aspergillus ochraceus*, which occurs mostly in warm climates. In temperate climates other aspergilli and penicillia, notably *P. verrucosum,* in cereals and cereal products are more important. Ochratoxins have been implicated in kidney damage in pigs but the toxin is destroyed by the bacteria in the rumen of cattle. A serious form of kidney disease affecting humans is endemic in parts of the Balkans, and this also has been associated with ochratoxins. The total intake of ochratoxins from a typical UK diet is very low, coming either directly from cereal products, or from pigs' kidneys or black pudding made from pigs' blood, the two tissues in which the toxin accumulates.

(10.17)

In the USA the maize (*i.e.* corn) fungi, *Fusarium verticilloides* and *F. proliferatum* are a cause for concern. Contamination with these pathogens can lead to the presence of mycotoxins known as fumonisins, the best known of which, fumonisin B_1 is illustrated (10.18). They are unusual in being water-soluble and can be carried through into corn-based food products. They are known to cause a wide range of symptoms in livestock, and are also suspected of being human carcinogens. In recent years the levels of fumonisins detected in corn samples has however fallen considerably, possibly due to subtle climate changes.

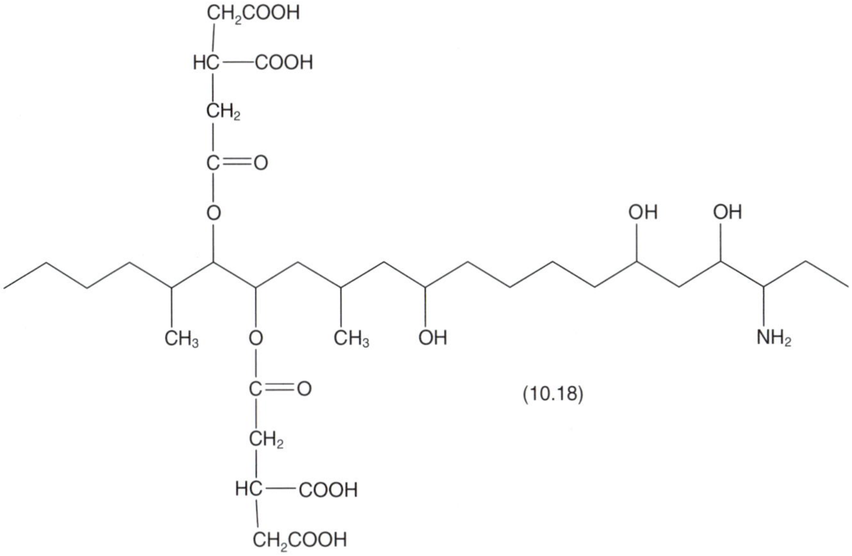

(10.18)

Patulin (10.19) is classified as a mycotoxin but is not considered dangerous. Although it has been found to cause mutations in test bacteria in the laboratory, animal studies have not shown it either to be carcinogenic or teratogenic. The possibility has even been explored that it might have some useful antibiotic activity. Patulin is particularly associated with apple juice, in which its presence tends to be regarded as a marker for detecting mouldy fruit rather than as a health hazard in its own right. It is also found in home-made jam which has mould growing on its surface. Patulin is produced by a number of species of *Aspergillus* and *Penicillium* (particularly *P. expansum*) which cause rotting of stored fruit. Most clear apple juice sold in the UK has less than $10\,\mu g\,kg^{-1}$ but cloudy apple juice frequently contains patulin above this level. The difference is caused by the tendency of patulin to associate with the particulate material in cloudy juice and is removed when the juice has been filtered.

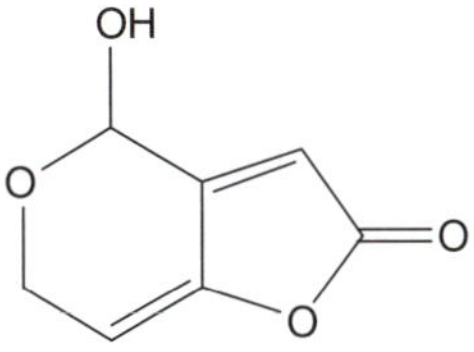

(10.19)

The increasing awareness of the health problems mycotoxins can cause has led to food manufacturers and processors taking a number of preventive steps. The need to discard mouldy cereals and nuts is obvious, but there are many food products which traditionally require moulds to ripen. These include blue cheeses and fermented sausages such as salamis. The approach now being adopted with these products is to avoid adventitious contamination from factory walls, employees or machinery, and to add pure starter cultures of mould strains which are known not to produce toxins.

BACTERIAL TOXINS

The growth of moulds on food is usually obvious to the naked eye. The growth of bacteria, on the other hand, is far from obvious and the first we may know of toxins being produced is when we start to suffer their effect. When considering the diseases caused by food-borne bacteria it is essential to distinguish between infection and intoxication. *Infection* results when harmful bacteria contained in our food are ingested. Once in the gastrointestinal tract they proliferate and produce toxins which cause the symptoms of a particular disease. Diseases caused by *Vibrio cholerae* and the innumerable species and subspecies of the genus *Salmonella* are among the most important of food-borne infections.

Our concern in this section is with *intoxication*, which results from growth of harmful bacteria in food before it is consumed. Although the toxins secreted may well be deadly it is curious, although of little consolation, that the conditions encountered in the intestines of the victim may render the bacteria quite harmless, while the toxins they secrete are unaffected.

The most feared of the bacterial food intoxications is botulism, caused by the toxin secreted by *Clostridium botulinum*.[v] The causative organism is commonly found in soil throughout the world, although being an obligate anaerobe it proliferates only in the absence of oxygen. Unlike the spores, its vegetative cells are not particularly resistant to heat, but even the least resistant *C. botulinum* spores are able to withstand heating at 100 °C for 2–3 minutes. The heating regime required for food preservation by canning is designed to eliminate all reasonable risk of *C. botulinum* spores surviving. Experience has shown that for safety the entire contents of a can must be held at 121 °C for at least 3 minutes. Times approaching half an hour are required if the temperature is only 111 °C. To achieve such temperatures at the centre of a large can may

---

[v] *Botulus* is Latin for sausage.

require prolonged heating of the can as a whole. Acidic foods such as fruit or pickled vegetables present no hazard, since the pH is too low for the growth of the bacteria and production of the toxin. Obviously the home bottling of non-acidic foods, including vegetables such as beans or carrots, is risky unless salt is added at high concentration in the form of a pickling brine containing at least 15% w/v NaCl. Other factors which prevent growth are reduced water activity, below 0.93 (*see* Chapter 12), or the presence of nitrites (*see* Chapter 9). If heating is inadequate and the other conditions are favourable, any spores carried into the can with the raw food materials will eventually multiply and secrete the toxin.

The species *C. botulinum* is subdivided into a number of types (A to G) which can be distinguished serologically, but more importantly which differ in the lethality of their toxins and in their geographical distribution. For example, in the USA west of the Rocky Mountains type A predominates, whereas to the north in Canada and Alaska type E, and elsewhere in North America type B, predominate. In soil samples collected in Western Europe type B alone is found. Throughout the world type E is associated with intoxication from seafoods. Types C and D are most often connected with botulism in domestic animals rather than man. Ducks are particularly vulnerable due to the rotting vegetation they consume from the anaerobic depths of ponds.

As with the mycotoxins, and for similar reasons, data are scarce on the amounts of botulinum toxin which produce adverse effects in humans. Estimates of the minimum lethal dose are around 1 μg for an adult, *i.e.* approximately $1.4 \times 10^{-2}\,\mathrm{g\,kg^{-1}}$. This amount of toxin is small, both in absolute terms and also relative to the productivity of the bacteria. Levels of toxin around $50\,\mathrm{g\,cm^{-3}}$ have been encountered in an infected can of beans. Laboratory cultures using ideal media may contain concentrations 100 times greater than this. Of the most common toxin types (A, B and E), B appears to be the most toxic and E the least. The variations in toxicity are principally due to differences in the stability of the toxins in the intestine and different rates of absorption into the bloodstream. By direct injection the toxicity becomes 10^3–10^6 times higher!

Botulinum toxin comprises a complex of two distinct protein components: a haemagglutinin and a neurotoxin of molecular weight 5×10^5 and 1.5×10^5, respectively. After absorption the toxin is transported throughout the body by the blood and the lymphatic system. It reaches the nerve endings, where it binds apparently through the haemagglutinin component and blocks the transmission of nerve impulses to the muscles. The result is that some 12–36 hours after ingestion of the toxin the first signs of neurological damage appear: dizziness and general weakness, followed over a day or so by general paralysis, and death by respiratory and cardiac failure. When the victim is known to have consumed food contaminated with the toxin but before extensive symptoms appear, it is possible to retrieve the situation by the administration of an antitoxin for the specific toxin involved. Unfortunately the difficulty in rapidly identifying the causative toxin type means that analysing the geography and similar clues may be the best indicator of which antitoxin type to use.

The intractability of botulism means that it is much better to prevent the disease rather than attempt to cure it. Although rarely seen in the UK, there are typically 40 outbreaks per year in the USA. In the main this difference can be ascribed to the greater enthusiasm for home-based food preservation in rural America, rather than to any lack of diligence on the part of the authorities or food manufacturers. Although the spores of *C. botulinum* are resistant to heat, the toxins themselves, in common with most other biologically active proteins, are quickly denatured at high temperatures. Any low-acid, high water-activity home-bottled or canned food can be boiled for 10 minutes before eating, to ensure freedom from this deadly toxin. On rare occasions cases of botulinum *infection* are encountered in infants where the conditions in the infant's intestines have allowed its growth to take place.

In contrast to the rarity and severity of botulism, the consequences of consuming food contaminated by the toxin from *Staphylococcus aureus* are rarely fatal, but will be experienced by almost everyone at one time or another. The staphylococci are a ubiquitous group of micro-organisms detected in air, dust and natural water. One species, *S. aureus,* is particularly associated with humans and is to be found on the skin and in the mucous membranes of the nose and throat of a high proportion of the population. In the ordinary way we remain unaffected by the presence of this microbial guest. However, human carriers represent an extremely efficient distribution system. Flakes of skin shed naturally from the hands can transfer the organism to anything touched, and the microdroplets from a sneeze will ensure aerial distribution.

One particular strain of *S. aureus*, known as phage type 42E, presents special problems. On arrival on a food material which offers a favourable environment, this strain will start to grow and at the same time secrete an enterotoxin. There are at least five different types of enterotoxin produced by various strains of *S. aureus*. All are proteins with molecular weights around 32 000. Although very similar in their clinical effects, the different toxins show wide variability in their stability to heating. The least stable, type A, loses 50% of its activity after 20 minutes at 60 °C, whereas the most stable, type B, still retains over 50% activity after 5 minutes at 100 °C. All these enterotoxins show considerable resistance to the proteolytic digestive enzymes of the stomach and small intestine. At the molecular level little is known of the mechanism of the action of the staphylococcal enterotoxins. What is known is that 1 µg is sufficient to cause the development of symptoms within an hour or so. The victim suffers vomiting and sometimes diarrhoea, accompanied by a selection of symptoms which may include sweating, fever, hypothermia, headache and muscular cramps. Rarely has the toxin proved fatal, and the symptoms usually abate after a few hours.

The physiological requirements of *S. aureus* indicate which foods it is likely to be found in. Nutritionally it requires most types of amino acids and some of the B vitamins. Oxygen enhances growth but is not essential. Although growth and toxin production are optimal at around 36 °C, there are indications that toxin can be produced at temperatures as low as 15 °C and as high as 43 °C. Toxin is not produced below pH 5 or in salt concentrations above 10% w/v. These requirements make cooked meats such as sliced ham, prepared custard, and hamburger patties (prior to grilling) prime targets. Any food not consumed straight after cooking is vulnerable, especially if it is not adequately refrigerated. Studies of food items incriminated in outbreaks of *S. aureus* poisoning have shown numbers of organisms in the range 10^7 to $10^9 \, \text{g}^{-1}$. The discrepancy between the requirements for optimal growth and optimal toxin production, and also between the thermal stability of the micro-organism and its toxin, mean that there is no correlation between cell count and toxin level. Responsibility for the prevention of *S. aureus* poisoning thus lies with food handlers, retailers and caterers rather than with food processors or manufacturers.

As might be expected, the determination of enterotoxin levels in foodstuffs is not straightforward. Traditionally animal assays have been necessary, often using kittens since they are especially sensitive and show symptoms similar to humans. Serological methods can be applied, but although they are sensitive (down to $0.05 \, \mu\text{g} \, \text{cm}^{-3}$) they take several days to complete.

ALLERGENS

Most of the undesirable substances so far discussed in this chapter do not discriminate between their victims, except that the youngest, the oldest or the frail may be most at risk. However, a small but not insignificant number of individuals can suffer from certain food components to which the majority are totally indifferent. These are the food allergens. In view of our natural tendency to apply the term allergy to any negative response, especially in the young, to something the rest of us regard as valuable,[vi] it is as well to remind ourselves that when doctors and scientists talk about an allergy they have something very specific in mind. Recent surveys have shown that as many as one in five people in Britain consider themselves allergic (in the sense of experiencing an unpleasant reaction) to certain foods. However, similar surveys have shown that when proper diagnostic tests are applied to these individuals, only one in ten actually demonstrates a measurable reaction (*i.e.* 2% or less of the total adult population). Up to 8% of infants may show a genuine food allergy. Although most of those who believed they suffered from a food intolerance identified food additives as the culprits, the incidence of reactions to food additives is in fact much rarer and is estimated to affect only 1 in 10 000 people. To be regarded as valid, any diagnosis of a food allergy must be based on double-blind placebo-controlled oral food assessment. A major reason for the current interest in food allergy is that the proportion of allergy sufferers in the population appears to be increasing.

Allergy (sometimes referred to as hypersensitivity) is defined as an abnormal reaction of the immune system to a foreign (but not infectious) material (the allergen), leading to reversible or irreversible injury to the body. Immune reactions can be divided into four types. Reactions involving food components are almost always Type I, involving IgE antibodies. Types II and III also involve antibodies, but are not concerned with reactions to food components. Type IV reactions develop slowly and are caused by migrating lymphocytes (white blood cells). The only Type IV reaction of interest to us is coeliac disease, in which the presence of wheat gluten causes damage to the lining of the small intestine.

Food allergy is associated with a wide range of symptoms, including skin reactions such as eczema (atopic dermatitis) and urticaria, and gastro-intestinal responses such as vomiting and diarrhoea. In anaphylaxis, an exaggerated immune response resulting from re-exposure to an allergen, these symptoms can be very severe and also involve the respiratory system, with tissue swelling causing obstruction to the airways. These allergic reactions follow the release of histamine and other inflammatory mediators, provoked by the presence of the allergen. In food allergy this immune response is directed against a specific protein in the diet. It is important to recognise that for an allergic reaction to take place the allergen must penetrate the lining of the digestive tract. A combination of digestive enzymes, the mucous membrane lining and lymphoid tissue provides the barrier which normally protects us from the allergens and invasive micro-organisms that we consume. Most of

---

[vi] "Hard work" is a classic example.

the common types of food allergy are shown only in infants and young children, but "growing out" of the problem is more likely to be the result of a gradual strengthening of these barriers rather than the body becoming accustomed to the allergen.

In general, allergens need to have a molecular mass greater than 5000 and below 70 000. Molecules below this range only provoke a reaction when they have become bound to a larger protein; above this they are too large to reach the immune system. In most cases the allergenic effect of proteins is not lost when the protein is denatured (*see* page 166). This suggests that the immunological reaction is directed towards relatively short sections of the molecule containing a particular amino acid sequence, rather than the tertiary structure or overall shape of a section of the molecule. In the cases of many allergenic proteins, *e.g.* those in cow's milk, the location and amino acid sequences of many of these sections, known as epitopes, have been established.

Cow's milk allergy (CMA) is well known, affecting up to 7% of infants, but very few older children. CMA should not be confused with cow's milk intolerance (CMI). CMI is caused by the inability to absorb lactose due to a lack of the enzyme lactase, as described in Chapter 2. Any of the milk proteins could be the culprit in a particular case of CMA. However, studies of large numbers of CMA sufferers have revealed that the most abundant milk proteins, α-lactalbumin, β-lacto-globulin and the caseins (*see* Chapter 5), are the major allergens.

There are considerable similarities between the amino acid sequences of the versions of these proteins which occur in cows, sheep and goats, which means that victims of cow's milk allergy are likely to react to these milks as well. Babies consuming only breast milk may not be immune if their mother consumes a lot of cow's milk. Some cow's milk proteins can be absorbed through the mother's gut mucosa more or less intact and are secreted in the milk. The proportions of these potentially allergenic proteins varies considerably between the three species, offering the possibility that replacing cow's milk with that from another species might provide relief from the symptoms of CMA. In fact the only practical alternative is goat's milk, not least because of its availability. The protein involved in a significant proportion of CMA cases is α_{s1}-casein. An unusual characteristic of the genetics of goats is that individual goats (apparently regardless of breed) fall into one of two categories in terms of the proportion of α_{s1}-casein in the total casein content of their milk, approximately 2% or 18%. Either figure compares favourably with the corresponding figure of 35–40% for cow's milk. This is the protein involved in a significant proportion of CMA cases and many children who suffer from CMA can tolerate goat's milk. The food industry's use of milk proteins, in the form of sodium caseinate or whey proteins, as functional ingredients in a wide range of food products can make the avoidance of CMA much more difficult than simply not consuming obvious dairy products such milk and cheese.

Soya milk, a preparation of proteins extracted from soya beans, does not cause such cross-reaction but infants who have suffered badly from CMA often go on to become allergic to soya proteins. This is because the damage to the gut lining caused by the CMA can allow further interaction between the soya proteins and the immune system.

Coeliac disease is the clinical name given to an allergy to wheat proteins, more specifically the gliadin fraction (*see* page 201) of the gluten proteins. Victims also react to the corresponding proteins in barley and rye, and sometimes also oats. Coeliac disease occurs throughout the world wherever wheat is the staple cereal, and tends to cluster in families, suggesting that there is a strong genetic component. The disease begins when babies are weaned and wheat-based foods begin to appear in their diet. However, the effects of the disease do not normally become evident until after the child's first birthday. More rarely, it may not be properly diagnosed until the patient in much older. Along with the appearance in the blood of antibodies to gliadin proteins, there is severe damage to the cells lining the jejunum (part of the small intestine). This damage causes greatly impaired digestion and absorption of nutrients, resulting in a general failure to thrive, together with anaemia, lethargy, and gastro-intestinal and other problems. The precise sequence of events in coeliac disease is still not well understood, but it appears to have much in common with other inflammatory autoimmune diseases.

Complete removal of wheat from the diet, and the other cereals mentioned above, is the only satisfactory remedy. As with the allergens in cow's milk, the allergenicity of the gliadins is unaffected by cooking. Fortunately both maize and rice are safe for coeliac sufferers, and this makes it possible to provide gluten-free cereal products which can take the place of the bread, cakes and pastry that the rest of us can enjoy. The use of wheat flour in many processed food products which are not obviously cereal in character poses special problems and emphasises the importance of accurate food labelling.

A number of other foodstuffs are known for their allergenicity, although the number of sufferers is small. Egg proteins (notably ovo-mucoid, in the white, *see* page 185) can produce similar symptoms to those given by milk. The allergenicity of ovomucoid survives cooking. Traces of two other egg proteins, ovalbumin and ovotransferrin, have been found in raw chicken meat, but fortunately their allergenicity does not survive cooking.

For some time nut allergies, especially to peanuts (also known as groundnuts, *Arachis hypogaea*) have attracted particular attention, both within the scientific and medical communities and with the general public. Far more cases of peanut allergy occur compared with allergies to "tree" nuts, such as almonds, brazil nuts and cashews, but much greater quantities of peanuts and peanut butter are consumed. The edible part of the peanut, the cotyledon, contains around 25% protein. When the nut geminates this protein provides nutrients for the growing seedling. It includes a number of members of an important family of plant storage proteins, the cupins,[vii] which are particularly associated with legumes such as peanuts, soya beans, almonds, walnuts and cashews. In the case of peanuts, two proteins, known as arachin and conarachin, dominate the edible portion of the nut and are the usual culprits in peanut allergy. Immunologists refer to conarachin and arachin as the *ara h 1* and *ara h 3* allergens respectively. They are particularly heat resistant and can be detected in products containing only small amounts of peanut flour.

Peanut allergens can suddenly produce serious symptoms in susceptible individuals, including anaphylaxis, and these may be life threatening. Studies among the general population have suggested an incidence of peanut allergy of 5 to 6 per 1000 in the UK and USA. Peanut allergy accounts for a major proportion of food-induced anaphylactic reactions and more than 50% of food allergy fatalities. It has been estimated that approximately 95% of peanut allergy cases involve reaction to the *ara h 1* allergen.

The increase in peanut sensitivity during the last decade or so has been ascribed to the growing popularity of peanut products within the population, and especially the introduction of peanut-containing items to children's diets from an early age. Another possible explanation for the increase is that peanut (*i.e.* groundnut) oil is nowadays a common ingredient of ointments and creams, and if these are applied to the broken skin of infants this may provide an entry route for traces of allergenic proteins.

Peanut allergic reactions may be provoked by doses of as little as 100 mg of peanut protein, equivalent to no more than 400 mg of peanut meal. One recent study has suggested that sufficient amounts of peanut protein may remain in the saliva of someone who has recently eaten a peanut butter sandwich to make it unwise for a peanut allergy sufferer to be kissed by them!

The potential problems posed by peanut allergens have made their detection and quantification in foods a major objective for food

---

[vii] The name "cupin" comes from the Latin for barrel, and refers to the shape of the subunits that make up these complex proteins. The other major family are the prolamins, to which the glutenins and gliadins of cereals belong.

analysts. Two methods are in regular use. The most popular is sandwich ELISA, a type of enzyme-linked immunosorbent assay. The other, which detects fragments of peanut DNA which correspond to the allergenic protein, utilises the polymerase chain reaction (PCR). This is an exactly similar technique to that used in forensic science to identify the source of DNA residues left behind at crime scenes. Both methods are remarkably sensitive, with a lower detection limit below 5 mg of peanut protein per kg of foodstuff.

TOXIC AGRICULTURAL RESIDUES

Modern agricultural methods make use of innumerable products from the modern chemical and pharmaceutical industries. Whether we approve of them or not, the high yields demanded of agriculture are nowadays dependent on these chemicals, and the food scientist should therefore be familiar with them.

The most worrying agricultural chemicals are obviously pesticides. It was found in the 1960s that the insecticide DDT, dichlorodiphenyltrichloroethane (10.20), had penetrated virtually every food chain studied. This was the result of a combination of its extreme stability in the environment and the vast scale on which it was used from the time of World War II until the early 1960s. Not only was DDT cheap to manufacture, it was also extremely effective in controlling both the plagues of mosquitoes carrying malaria and yellow fever, as well as a wide range of agricultural pests.

(10.20)

It was the accumulation of DDT in birds of prey and its catastrophic effects on their numbers which first came to public attention, and once it was realised that human fat and milk were also contaminated, restrictions on its use followed. In 1972 in Britain DDT levels of 2.5 ppm were found to be generally present in human fat. The mean levels of DDT detected in human milk in Britain have fallen sharply since the 1960s, as indicated in Figure 10.6. Fortunately there is no indication that DDT at the levels normally encountered in our diet contributes to the incidence of cancer.

In general, the levels of DDT in cow's milk were found to be around 1% of these figures. The insecticide had travelled *via* the fat of beef and

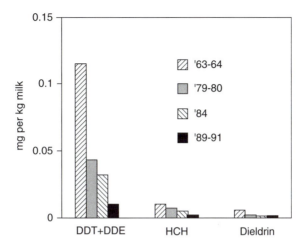

Figure 10.6   The decline in mean levels of DDT and related compounds in human milk in the UK during the period 1963 to 1991. HCH is hexachlorocyclohexane and DDE is 2,2-bis(4-dichlorophenyl)-1,1-dichloroethylene, the major metabolite of DDT. (Based on data from the Ministry of Agriculture, Fisheries and Food, Food Surveillance Paper No 34, *Report on the Working Party on Pesticide Residues, 1988–90*, 1992.)

other meat, and had also reached man through dairy products. Now that the use of DDT is either banned or greatly constrained in Europe, North America and many other parts of the world, the levels to be found nowadays in human tissues in these areas are much reduced. On the other hand, in Africa and other tropical regions its cost effectiveness for controlling insect-borne disease makes its use a much more complex issue. The 1970 report from the National Academy of Sciences (USA) stated that DDT had prevented 500 million otherwise inevitable deaths from malaria, and that this fact had to be balanced against its effect on wild life such as eagles and ospreys. Since that time the impact of banning DDT on the incidence of malaria has become only too apparent. For example, once DDT was banned in the mid-1990s the incidence of malaria rapidly increased to ten times the pre-ban levels. Fortunately, limited spraying with DDT was resumed in 2001, and malaria rates have now begun to fall. It has become recognised that the key factor in protecting vulnerable human populations is to spray the DDT *indoors*. This means that much lower levels of DDT are required than had to be used in outdoor spraying in the agricultural context. What makes it particularly valuable is that DDT used in this way not only kills mosquitoes, it repels them.

From the difficulties encountered in the use of DDT, the authorities learned the need for continual monitoring of pesticide levels in all types of human foodstuffs. The disturbing result is that we now realise just how many foods are contaminated with a wide range of pesticides and their breakdown products. Many countries, including Britain and the USA, maintain intensive surveillance programmes to monitor the level of a large number of different pesticides in items of everyday food. A summary of the British data for 2006, the most recent full year for which figures were available at the time of writing, are shown in Table 10.2.[viii] The authorities in each country have their own policies regarding the choice of raw materials and commodities to be tested. In Britain the major contributors to the national diet in which any residues might be

Table 10.2 A summary of results of pesticide residue analysis of foods sold in Britain during 2006, including imported foodstuffs. The italicised data are examples selected from the major sections. Source: *Annual Report of the Pesticide Residues Committee 2006*, Department for Environment, Food and Rural Affairs, UK (Crown copyright).

Commodity	Total samples analysed	Total with no detectable residues	Total with detectable residues	Number above MRL[a]
Baby foods (fruit or vegetable based)	72	70 (97.2%)	2 (23.8%)	none
Fruit and vegetables	1791		845 (47.1%)	59 (3.3%)
potatoes	*139*	*86*	*53*	*none*
apples	*119*	*30*	*87*	*2*
grapes	*298*	*102*	*196*	*5*
lemons	*72*	*12*	*60*	*none*
lettuce	*95*	*52*	*43*	*4*
Starchy foods and grains	404	158 (39.3%)	245 (60.5%)	1 (0.2%)
bread	*145*	*54*	*91*	*none*
wheat flour	*70*	*17*	*53*	*1*
Animal products	1007	952 (94.5%)	55 (5.5%)	none
milk (cow's and goat's)	*300*	*300*	*none*	*none*
lamb	*120*	*93*	*27*	*none*
Others (including dried and canned fruit and fruit juice)	288	255 (88.5%)	33 (11.5%)	none
TOTAL	3562	2322 (65.2%)	1179 (33.1%)	61 (1.7%)

[a]MRL: Maximum Residue Level.

[viii] Similar data covering all the member states of the EU are published by the EU Commission and can be accessed at: http://www.ec.europa.eu/food/fvo/specialreports/pesticides_index_en.htm

expected to make a major contribution to a consumer's total exposure, such as wheat and potatoes, apples, milk and margarine, are monitored more or less continuously. Among other fruit and vegetables, where there are vast numbers of different species, varieties and sources that need to be monitored, two selection policies operate in parallel. On the one hand, fruit and vegetables which are particularly liable to contain pesticide residues, *e.g.* celery and carrots, are subjected to continuous and intensive monitoring. On the other hand there is a rolling programme of tests for more rarely consumed items every few years, for example persimmons and pomegranates were last examined in 1995, and swedes and kohlrabi were screened for the first time in 1999. The substances tested for are not all pesticides or fungicides, *e.g.* tecnazene is used to inhibit sprouting in stored potatoes. Depending on the food being examined, the total number of different residues determined in a single sample can range from as few as eight to more than 90.

Samples are reported as containing an undesirable substance if the quantity exceeds the detection limit of the analytical procedure. The sensitivity of modern instrumentation, especially gas chromatography, is such that in most cases the detection limit is rarely higher than 50 µg per kg (50 ppb), and often one tenth of this figure. The legal limit for residues is described as the Maximum Residue Level (MRL). In UK legislation these limits are not only arrived at from toxicity considerations but also from an assessment of what is achievable by good agricultural practice. Perhaps surprisingly, we are assured that this actually results in lower limits for most types of residue than would otherwise be the case.

The degree of pesticide contamination varies from year to year and also from country to country, due to differences in agricultural practice. For example, the number of bread samples in which malathion was detected by the UK authorities fell from about 14% in previous years to zero between 1988 and 1990, a period of unusually hot dry summers. It has been suggested that the fine weather might have meant that a greater proportion of home-grown wheat was suitable for bread making, reducing the use of wheat imported from countries which commonly treated wheat with malathion prior to shipping.

A special problem in the UK is that so much of our food is imported, especially fruit and vegetables. Some of the problems of sampling imported produce were mentioned earlier in this chapter. Data of this type give an overall picture for particular foodstuffs, but naturally cannot tell us about the pesticide residue exposure of a typical consumer from their diet as a whole. Surveys to gather this information are massive undertakings requiring the correlation of data on the overall

Table 10.3 Pesticide residues in total food intake. Acceptable Daily Intake (ADI) figures are those assigned by the FAO/WHO, expressed per kg of body weight (the typical adult male is taken as weighing 70 kg). Intake covers all dietary sources, based on data published by the FDA in the USA and MAFF in the UK. The 1987 US data discriminate between different sections of the population; the data given are for males aged 14–16. The UK data refer to the average of the whole population in either 1984–85 or 1989–90. The MRLs indicated were those featuring in contemporary UK legislation at the time. Different foodstuffs have different MRLs, with cereals at the upper end of most of the ranges shown. (nd: no data available).

Pesticide	ADI	US intake	UK intake	UK MRL
		µg/kg body weight/day		*mg/kg*
Dieldrin	0.1	0.0045	<0.001	0.006–0.2
Carbaryl	10	0.0173	nd	0.5
Chlordane	0.5	0.0018	nd	0.002–0.05
Chlorpyrifos-methyl	10	0.0066	0.01	0.01–10
DDT	20	0.0192	0.0017	0.04–1
Diazinon	2	0.0123	nd	0.02–0.7
Endosulfan	8	0.0206	nd	0.1
Fenitrothion	3	0.00051	nd	0.05–0.5
Heptachlor	0.5	0.0018	nd	0.01–0.2
Malathion	20	0.119	<0.0017	8
Pirimiphos-methyl	10	0.0016	0.05	10
Quintozine	7	0.00112	0.000029	(no MRL)
Tecnazene	10	0.0001	0.055	(no MRL)

content of diets and the pesticide residue present in each component of the diet. Data from such studies in the USA and UK are shown Table 10.3. These are referred to as a total diet survey, or shopping basket survey. Interpretation of this type of data is difficult since an individual's actual intake may differ considerably from the mean, and it is affected by age, gender, economic and ethnic factors, as well as by personal preferences. In spite of the obvious reservations due to the date of the surveys, they do give a valuable impression of the overall situation, especially since the data gathered meanwhile on the levels of pesticide residues in individual commodities show that the incidence of pesticide residues has not increased significantly.

ADI figures have been established internationally by the combined efforts of the Food and Agriculture Organisation and the World Health Organization and represent the maximum daily exposure which offers no appreciable risk. They are arrived at by dividing by 100 the highest level of exposure which has not resulted in any detectable toxicity in the most sensitive animal species tested, the No Observable Effect Level (NOEL).[ix] Such tests normally are not simply based on a single test but involve continuous exposure over the natural lifetime of the animal concerned. NOELs are not readily established for carcinogens, as current understanding of carcinogenicity implies that even one molecule of a carcinogen has the potentiality for causing cancer. In this case risks are calculated on the basis of the number of additional cancers that lifetime exposure to a pesticide at particular level are expected to cause. In the USA the standard of "negligible risk" for a pesticide is one additional cancer arising per million of the population. This should be set against the overall incidence of all cancers of approximately one in four of the population, *i.e.* 250 000 per million. For suspected carcinogens the term ADI is being replaced by the Tolerable Daily Intake (TDI), since the NOEL is no longer relevant.

One obvious inference to be drawn from these data is that pesticides are now so universally distributed through our environment that for the foreseeable future they will remain unavoidable components of our diet. Although little work appears to have been done on the subject, it seems likely that restricting one's diet to items claimed to be produced "without the use of chemicals" will probably have only a marginal effect on one's personal pesticide intake. This does not rule out the influence that may be exerted on suppliers through the workings of the market. There is, for example, a much greater willingness nowadays to accept that the fungicides used on peanut crops are much less hazardous than the aflatoxins those fungicides prevent. One must remember that the levels that we normally encounter in our diet are extremely low, usually two orders of magnitude below the maximum ADI levels established by studies based on experimental animals. That we are able to measure such low levels at all is a credit to the skill of the analytical chemists. It must also be pointed out that the recorded cases of human poisoning by pesticides have tended to arise from relatively massive intake following accidental cross-contamination of food containers, or from factory accidents. These observations should not be taken to imply complacency. What is required is constant vigilance and a willingness to accept that there are times when the dangers of using pesticides can be outweighed by the benefits.

Another view of this balance is arises from the use of hormones in animal husbandry. For example, it is well established that castration of male animals leads to a higher proportion of fat in the carcass compared with lean muscle tissue and to less efficient conversion of feed to meat.

[ix] These are sometimes reported as No Observable Adverse Effect Level, NOAEL, values.

Early research showed that the use of male sex hormones, androgens, as additives in animal diets was expensive and relatively ineffective and that there was a serious risk of overdosage. In the 1950s it was discovered that the female hormones, the oestrogens, had a similar effect to androgens in increasing the output of growth hormone. It rapidly became the norm to supply synthetic oestrogens, hexoestrol (10.21), or diethylstilboestrol (DES,10.22), to meat animals, particularly cattle and poultry. The hormone was either added to the animals' feed or implanted as a pellet under the skin.

(10.21)   (10.22)

Concern about the use of these hormones in food arose in the 1970s when an unusual number of cases of vaginal cancer in young women began to appear. It was realised that the mothers of all these young women had been treated with DES for various complications during their pregnancy. Subsequently, prenatal exposure to this type of hormone was shown to cause a wide range of disorders in the reproductive organs of children of both sexes. The uproar generated by these findings made the banning of DES inevitable. However, what was overlooked was that clinical doses of DES had ranged from 1.5 to as much as 150 mg per day. Even allowing for poor transmission across the placental barrier, the actual levels found in meat were well below those which had caused problems. In the mid-1970s only a very small proportion of beef livers from treated cattle were found to contain DES levels above $0.5 \, \mathrm{g \, kg^{-1}}$. Levels in muscle tissue were estimated to be about one tenth of this.

Some other hormones were regarded as less hazardous and have continued to be permitted in Britain until recently. At the present time hexoestrol is still permitted for use in poultry. A pellet containing the hormone is implanted under the skin at the top of the neck of young cockerels to simulate the effects of castration, producing a capon.

Modern agricultural practices can present more insidious problems than hormones. In Chapter 9 the problems posed by nitrosamines in cured meats were discussed. However, the formation of nitrosamines in the human digestive tract from nitrates in our food is no less important. Out of a typical daily intake of nitrates from food of around 90 mg, vegetables

Table 10.4 Typical nitrate levels in fresh vegetables. The nitrate content of individual samples of vegetables can vary between one-third and three times these values. Cooking in boiling water can reduce the figures by up to 75%.

	Nitrate as NaNO₃ (mg/100 g)
Beetroot	121
Celery	230
Spinach	163
Lettuce	105
Potatoes	16
Tomatoes	2

including potatoes contribute over 75%. As shown in Table 10.4 some vegetables contain spectacular amounts of nitrates, but wide variations are found between different regions, reflecting local soil types, fertiliser usage and levels present in the water supply. Solid food normally accounts for more than 80% of total nitrate intake. However if the nitrate level in drinking water is high (above $100 \, mg \, dm^{-3}$) this takes over as the major source, and total intake can be increased by 2–3 times. The average intake of nitrate from drinking water in Britain is around 13 mg daily, but the figure for South-East England is nearly three times this level and that for North Wales only one-twentieth. The UK limit for nitrate in drinking water is $80 \, mg \, dm^{-3}$, averaged over a period of three months.

In Chapter 9 (*see* page 365) the carcinogenicity of nitrosamines was considered. It is now recognised that the reaction of nitrites and secondary amines to form nitrosamines can occur *in vivo* and is probably an important cause of gastric cancer. Nitrate entering the gastro-alimentary tract can be converted into nitrite at two points:

$$NO_3^- + 2[H] \rightarrow NO_2^- + H_2O$$

Nitrates, first absorbed from food and drink by the small intestine, find their way into the saliva and become exposed to the nitrate-reducing bacteria present in the natural flora of the mouth. On the other hand, once the saliva reaches the stomach the acidity is normally sufficient to prevent this continuing. However, reduced hydrochloric acid secretion, achlorhydria, can lead to the stomach being at a sufficiently high pH for bacterial nitrate reduction to occur. Like stomach cancer, achlorhydria is associated with both old age and poor levels of nutrition over a long period. The incidence of stomach cancer in the elderly is significantly higher in regions of Britain where high levels of nitrate in the water coincide with economic deprivation. The frequent absence of an association between nitrate intake and stomach cancer in other parts of the world may well be the result of good nutrition, or a failure to include in medical statistics the incidence of stomach cancer in old age.

The final type of agricultural residues to be considered are antibiotics. Antibiotics are used in agriculture for two quite separate purposes. Their most obvious use is in the treatment of bacterial disease in farm animals. Good agricultural practice demands that milk from a cow treated for mastitis is discarded, and there are also regulations governing the maximum permitted levels of antibiotics such as penicillin in milk. An occasional lapse will have the potential for causing allergic reactions in sensitive individuals, but the major problem lies with manufacturers of dairy products such as yoghurt and cheese. These require lactobacilli to generate their lactic acid content, and even very low levels of antibiotics will impair the growth of this unusually sensitive group of bacteria.

Less obvious is the practice of including antibiotics in animal feedstuffs to act as growth promoters. A number of different antibiotics have been used for this purpose over the past 30 years. Antibiotics are added to the feed at the rate of 5–$50\,g\,ton^{-1}$, depending on the antibiotic and the animal species. Although the mechanism by which they can give an increase in growth rate of up to 50% and feed conversion of 15% is obscure, the benefits are clear. The carry-over of antibiotic residues from the animal's diet into either meat or dairy products is insignificant, but this does not imply that the practice is without risk.

The problem is that an animal receiving an antibiotic over a prolonged period can become a breeding ground for strains of bacteria which are resistant to the antibiotic. This is a particular problem with enteric bacteria such as *Escherichia coli* and the salmonellae. It is the ability of these bacteria to exchange genetic material, including the R-factors which carry antibiotic resistance, which creates havoc. Resistance developed within a harmless or even a beneficial bacterial strain may be transferred to a dangerous pathogen, with serious long-term results for health. Some years ago it was realised that the appearance of strains of disease-causing bacteria which were resistant to a wide range of antibiotics could be blamed on the use of antibiotics in agriculture. In consequence, the use of antibiotics for growth promotion is now heavily restricted in many parts of the world. Within the EC seven antibiotics are permitted, but none of these are used in the treatment of human disease. Current UK regulations list the maximum levels of residues of a long list of antibiotics, sulfonamides and other medicinal products allowed in meat and meat products.

TOXIC METAL RESIDUES

Discussion of metallic residues requires us to depart from our classification of undesirables by origin. Toxic metals may reach our food from a variety of sources. The more important of these are:

- the soil in which human or animal feedstuffs are grown;
- sewage sludge, fertilisers and other chemicals applied to agricultural land;
- the water used in food processing or cooking;
- contamination from dirt, *e.g.* soil on unwashed vegetables; and
- equipment, containers and utensils used for food processing, storage or cooking.

The contribution of these different sources to the undesirable metals in our diet will become apparent as we consider the individual metals involved. We shall need to distinguish between metals which are desirable and required for health in the same way that vitamins are, and those which are toxic and undesirable. As we shall see, the distinction is frequently one of amount rather than nature, and where a metal is beneficial at the levels normally encountered in food it will be discussed in the next chapter.

Lead

Lead is undoubtedly the metal that first springs to mind when the issue of undesirable metals in food is concerned. The ancient Greeks were aware of the health hazards it posed to miners and metal workers, and since that time the need for legislation to protect both lead workers and the general public has been recognized—although not necessarily provided. On a worldwide basis at least 5 million tons of lead are used annually, so it is hardly surprising that a certain amount of this finds its way into our lives. Although organic compounds such as the tetraethyl lead used in motor fuels are nowadays probably the most worrying source of lead in our environment, we shall restrict our consideration here to the lead compounds reaching us through food and drink. Repeated examination of crops from fields close to busy roads has failed to confirm contamination from vehicle exhaust fumes. Similarly, the milk from cows grazing near busy roads shows no increase in lead content from this source.

The level of lead in untreated water supplies varies widely, depending on the lead content of the rock the water has encountered; most supplies contain no more than around 5 ppb. The WHO recommended maximum for drinking water is 50 ppb. The use of lead piping and lead-lined tanks for domestic water supply can lead to much higher levels if the water is particularly soft, such as the acidic moorland water collected in many parts of upland Britain. The problem of acidic water dissolving lead from household pipes is exacerbated if the water has remained in contact with the household pipework overnight, as it may contain in excess of 100 ppb of lead. In the past the use of lead pipes and tanks in breweries and cider factories often gave rise to lead poisoning, and it was blamed in 1767 for what was described as the "endemic colic of Devonshire". Nowadays lead contamination of beverages is restricted to illicitly distilled spirits such as "moonshine" whisky and "bathtub" gin in the USA, and "poteen" in Ireland. Lead levels in excess of $1 \, \text{mg} \, \text{dm}^{-3}$ are still reported in American moonshine. Home-made stills are often made with lead piping, or from car radiators with metal joints sealed with lead-rich solder.

Another source of lead in beverages is pottery glaze. If earthenware pottery is glazed at a temperature below $120 \, ^\circ\text{C}$, lead compounds in the glaze are not rendered sufficiently insoluble. There is then a serious risk that lead salts will be leached into acidic materials stored in such pottery. There are numerous cases in the literature of fatal or near-fatal poisonings resulting from the storage of pickles, fruit juices, wine, cider and vinegar in amateur pottery. In one case, the lead content of apple juice was stated to have reached $130 \, \text{mg} \, \text{dm}^{-3}$ after three days' storage in poorly glazed earthenware.

Canning is another potential source of lead contamination in food. The original containers for canning were made of tinplate—steel coated with tin—and this was first used about 160 years ago. Lead-based solder was used to seal the joins in the cans, and the relative position of tin and lead in the electrochemical series ensured that the tin component was the first to dissolve. Unless solder was splashed into the can during manufacture, only a small surface area was normally exposed to the contents of the can. An illustration of the leaching effect is shown by the analysis of the contents of a can of veal a century after it was prepared for use in the Arctic expedition of 1837. The veal was found to contain $3 \, \text{mg} \, \text{kg}^{-1}$ lead, $71 \, \text{mg} \, \text{kg}^{-1}$ iron and $783 \, \text{mg} \, \text{kg}^{-1}$ tin.

Nowadays most tinplate used for canning is lacquered. The lacquers are based on thermosetting resins which are polymerised on to the surface of the tinplate at high temperature before the can is fabricated. Unfortunately lacquering is not always as successful as might be desired. Blocking the reaction between the tin and the contents of the can apparently enhances the loss of lead from the solder. The lead content of canned foods is also affected by the time and temperature of storage, as well as the acidity of the food. It is therefore difficult to be precise about typical lead contents in canned food, but most analyses indicate figures between 100 and $1000\,g\,kg^{-1}$.

Another source of lead in the diet, although affecting only a small section of the population, is the lead shot present in game birds such as pheasants and wild ducks, and in rabbits and hares. The possibility of inadvertently swallowing a lead pellet lodging unnoticed in a roast pheasant breast is a minimal hazard, since lead in metallic form is poorly absorbed. The problem arises from the dispersion of the metal in the mildly acidic muscle fluids of the bird or animal. This can give rise to levels as high as $10\,mg\,kg^{-1}$ in the meat and in products such as game pâté. Unless the consumption of game were to be banned completely (not a likely prospect when one considers the lifestyles of some of our legislators), this is one source of dietary lead we will have to live with. However, it may be provide a toxicological element to add to the economic argument for discouraging young children to eat such expensive food. A more valid objection to game consumption is that in its pursuit a vast amount of lead is fired off into the environment without ever hitting anything edible.

Most unprocessed food, and foods in which the processing has not added to the lead content, have lead levels in the range 50–500 $\mu g\,kg^{-1}$. In the UK most foods are subject to regulations restricting their lead content to $1.0\,mg\,kg^{-1}$. Food products specifically sold for infants must be below $0.2\,mg\,kg^{-1}$. In recognition of their inevitably high lead content both game birds and shellfish are subject to a limit of $10\,mg\,kg^{-1}$. The total lead intake from food and beverages has been estimated for adults in various industrialised countries to be in the region of 250–300 μg per day. The low water solubility of most lead compounds is a major factor in its poor absorption from the gastrointestinal tract, indeed 90% of intake is unabsorbed and is excreted in the faeces. Unfortunately, young children may absorb a much higher percentage than this.

The inevitable result of all this lead in our diet is that the human body contains a measurable amount of lead even before birth. The lead content of the body builds up as we get older and adults typically contain between 100 and 400 mg, all but about 10% bound tightly in the bones, but not irreversibly. The remaining lead is held in equilibrium in the soft tissues of the body and the blood. Three-quarters of the lead absorbed by the body is excreted in the urine, also *via* the sweat, hair and nails. It is widely believed that serum lead levels below 40 μg per $100\,cm^3$ are not deleterious to health. Between 40 and 80 μg per $100\,cm^3$, there are biochemical indications that the body is failing to cope with the excessive lead burden. Levels higher than 80 μg per $100\,cm^3$ indicate chronic lead poisoning. Clinical studies suggest that an adult male needs to assimilate over $100 \mu g$ daily before his serum lead rises above $40 \mu g$ per $100 \, cm^3$. This corresponds to a dietary intake of about 1 mg daily. Data from the UK 2000 Total Diet Study indicate an average dietary exposure of $7.4 \mu g$ per day, less than a third of the figure reported only a few years earlier. More detailed information on the dietary sources of the UK population of the metals mentioned in this section is given in the second of the Special Topics at the end of this chapter.

Ordinary consumption in the diet will not lead to acute lead poisoning. Furthermore, the obvious symptoms of chronic lead poisoning such as anaemia are rarely associated with intake from food or drink. More worrying are the effects on young children of serum levels approaching the "safe" level quoted above. There are indications that a number of neuropsychological indicators, including IQ test performance and aspects of social and learning skills, show definite negative correlations with serum lead level once other variables have been allowed for. It is difficult at the moment to assess the effect of dietary lead to the damage suffered by some children. There appears to be a particular problem with children reared in an inner-city environment, and it may well be that tetraethyl lead in vehicle exhaust fumes is the prime culprit rather than food.

Mercury

Mercury presents food scientists with problems similar to those of lead, but on a different scale. Like lead, it occurs in three different forms: the free metal (Hg^0), inorganic mercuric (Hg^{2+}) salts, and alkyl mercury compounds. From time to time liquid mercury has found its way into food, for example from broken thermometers, but in this form mercury cannot be regarded as particularly toxic. There are reports of people drinking 500 g and suffering no more than occasional diarrhoea! Mercuric salts are more hazardous (the lethal dose of Hg^{2+} is about 1 g), though still very rare as food contaminants. It is however much harder to dismiss the problem posed by organic mercury compounds.

Alkyl mercury compounds reach our food from two sources. On occasion there have been serious outbreaks of mercury poisoning caused by consumption of cereal grains intended for planting which had been dressed with antifungal mercurial compounds. The practice of dyeing treated seeds lurid colours has not prevented this happening, and many countries now ban the use of mercurial seed dressings altogether. The most important source of alkyl mercury compounds is industrial pollution of coastal waters, leading to contamination of fish and other seafoods. The problem of marine pollution by mercury compounds first came to world attention in Japan in the mid 1950s when a number of cases of mercury poisoning were traced to the contamination of Minimata Bay by methylmercury, CH_3Hg^+. In spite of the source of the poisoning being identified, the pollution continued until the 1970s, by which time there had been over 700 cases of poisoning including 46 deaths. Fish and shellfish in the bay had mercury levels of up to $29\,mg\,kg^{-1}$, which pushed the average local daily intake to over $300\,\mu g$ per head. A more normal daily intake in Japan would be unlikely to exceed $10\,\mu g$, and data from the 2000 Total Diet Study indicate an average dietary exposure in Britain of only $1.5\,\mu g$ per day.

Subsequently, even higher levels were found in fish from some polluted Swedish and North American lakes, but local fishing bans have prevented further tragedies.

Dangerous levels of methylmercury can accumulate even if the pollutant is an inorganic mercury salt or the free metal. Various anaerobic bacteria which occur in muddy sediments are able to generate methane in the following manner. The methyl groups may be provided by methylcobalamin. There is evidence that similar systems for methylating mercury occur with many types of anaerobic bacteria.

$$Hg^{2+} \xrightarrow{+2e} Hg^0 \xrightarrow{+2[CH_3]} (CH_3)_2Hg \xrightarrow{H^+} CH_3Hg^+ + CH_4$$

Pollution is not always to blame for high mercury levels in seafood. Some deep-sea fish such as tuna appear to accumulate mercury, as CH_3Hg^+, to around $0.5\,mg\,kg^{-1}$. Fish from the most polluted of Britain's coastal waters have similar levels. The widespread occurrence of mercury in fish leads inevitably to its accumulation in the tissues of marine mammals. However, these animals are apparently able to detoxify the mercury by converting it into an uncharacterised complex involving mercury and selenium. The formation of a similar complex in man may offer some degree of protection to those exposed to mercury over long periods.

It is generally recognised that the clinical signs of mercury poisoning are seen in adults once the intake of organic mercury exceeds $300\,\mu g$ per day, and a safe limit for adults has therefore been adopted at approximately one-tenth of this. There is every reason to suppose that children may be more sensitive. The symptoms of mercury poisoning are variable, but all point to damage to the central nervous system. In the case of the Minimata poisonings, many mothers who themselves showed no clinical signs of poisoning nevertheless gave birth to children who went on to suffer symptoms resembling cerebral palsy.

Arsenic

Arsenic is another element which occurs at low levels in food as a consequence of its widespread distribution in nature. Ordinary drinking water in most parts of the world contains around 0.5 ppb, but some thermal spring and spa waters may have as much as $1 \, mg \, dm^{-3}$. In Bangladesh serious problems of arsenic contamination of the drinking water from shallow pipewells were first identified in 1993. It is currently estimated that over 30–40 million people in Bangladesh and the neighbouring West Bengal region of India regularly consume drinking water whose arsenic content exceeds $50 \, \mu g \, dm^{-3}$. Health problems characteristic of arsenic poisoning, such as skin problems including cancer, are widespread. The amount in food rarely exceeds $1 \, mg \, kg^{-1}$, with the exception of seafoods. The tendency of fish and crustaceans to accumulate toxic metals is difficult to explain but, as we have seen in the case of mercury, it can cause problems. In British coastal waters prawns have been caught with arsenic levels as high as $170 \, mg \, kg^{-1}$! Data from the UK 2000 Total Diet Study indicated an average dietary exposure of 55 μg per day, of which fish accounted for almost 75%. An important factor in the elevated levels of arsenic in fish is that most of it is present as the organic derivative arsenobetaine, $(CH_3)_3As^+CH_2COO^-$, a compound which is not metabolised by man. Published safety guidelines, such as the Provisional Tolerable Weekly Intakes (PTWIs) set by the Joint FAO/WHO Expert Committee on Food Additives (JECFA), refer only to inorganic forms of arsenic.

Most cases of arsenic poisoning *via* food have occurred following large-scale accidental contamination. A classic incident occurred in 1900 when 6000 cases of arsenic poisoning, including 70 deaths, occurred among beer drinkers in Manchester and elsewhere. The cause was "brewing sugar", a glucose-rich syrup produced from starch by hydrolysis with crude sulfuric acid, which in this instance was contaminated with arsenic. In 1955 12 000 infants were poisoned in Japan by dried milk in a bottle-feeding formula, and at least 120 died. Sodium phosphate had been used as a stabiliser in the milk powder, but it was contaminated with arsenic trioxide.

It is unclear at what level arsenic in food becomes hazardous. At present British legislation sets a limit of 1 ppm for most foods, apart from seafood, for which no limit is specified.

Cadmium

Had it not been for one outbreak in Japan in the late 1960s, cadmium would hardly warrant a mention as a toxic metal contaminant in food.

The amount in different foodstuffs varies widely, but in most diets it averages around 50 ppb. Data from the UK 2000 Total Diet Study indicated an average dietary exposure of 9 μg per day. The WHO suggests that 50 μg daily should be the maximum for an adult, and normal diets do not therefore present any difficulty. The Japanese outbreak occurred over a period of twelve years during which rice paddies were irrigated with water contaminated with cadmium from a mining operation. The contamination of the rice resulted in daily intakes of cadmium between 100 and 1000 μg. The cadmium caused an extremely painful demineralisation of the skeleton, especially in post-menopausal women. The disease, known as *itai-itai* from the Japanese equivalent of "ouch-ouch", was exacerbated by dietary calcium deficiency.

Two other sources of cadmium should be noted. One is the use of cadmium-plated components in food-processing machinery. While specialist manufacturers of such machinery can be relied upon to avoid cadmium plating, there can be a risk if non-food machinery is adapted for foodstuff use. Another possible source of cadmium is from zinc plating, or galvanising. Zinc usually contains some cadmium. If galvanised containers are used for storing acidic foods, some zinc, in itself harmless, may dissolve and carry traces of cadmium into solution.

Tin and Aluminium

The occurrence of tin in canned foods has already been mentioned in connection with lead. Most studies of tin toxicity have concluded that even quite high intakes are harmless; the minimum toxic dose is around 400 mg for an adult.

Another important packaging metal is aluminium. There is no evidence that ingested aluminium is harmful in any way. We consume it in considerable quantities from a number of sources including toothpaste, baking powder and alkaline indigestion remedies. A number of aluminium compounds are used as food additives. For example sodium aluminium phosphate is a slow-acting leavening agent in combination with sodium bicarbonate, and sodium and calcium aluminium silicates are used as anti-caking agents. The general level in fresh unprocessed food averages below $10 \, mg \, kg^{-1}$ (many foods contain less than $1 \, mg \, kg^{-1}$), but when levels above this are encountered in ome processed foods it is almost invariably due to the use of aluminium-based additives such as those mentioned. Tea is often quoted as an exception, but in spite of levels around $500 \, mg \, kg^{-1}$ present in the leaves, the aluminium remains undissolved and tea infusions contain only about $3\,mg\,dm^{-3}$. The level of aluminium in beer supplied in aluminium casks and cans is no higher than that from wooden casks or glass bottles.

Concern was expressed some years ago about the higher levels of aluminium in infant formulae based on soya protein, compared with cow's milk-based formulae, averaging $0.98\,mg\,dm^{-3}$ and $0.11\,mg\,dm^{-3}$, respectively. Reformulation of soya formulae has succeeded in halving this figure, but the natural aluminium content of soya beans limits what is achievable. The amount of aluminium leached from aluminium cooking utensils is low. Eliminating the use of aluminium cooking utensils and foil reduces the total daily intake by only 2 mg.

The amounts of aluminium consumed as the hydroxide in indigestion remedies and buffered aspirin preparations can be massively higher than the intake from food, as much as 1 g per day. The only question mark over the safety record of aluminium is a recent report that it can accumulate in the diseased brain tissue found in Alzheimer's patients. Although several epidemiological studies have hinted at an association between the incidence of Alzheimer's disease and the concentration of aluminium in drinking water, the small contribution of water to total intake makes the results difficult to interpret. We need to know much more about the absorption of the metal from the gut before useful conclusions can be drawn. At the present time there is every reason to believe that this accumulation is a secondary and unimportant effect of this distressing condition rather than a cause.

TOXINS GENERATED DURING HEAT TREATMENT OF FOOD

A substantial proportion of the food we eat is heated before consumption. The benefits of this are clearly considerable. In addition to the physical effect of tenderisation, for example, cooking provides a measure of protection against food-borne micro-organisms and their toxins. Unfortunately we cannot automatically assume that the application of heat to food is in itself entirely blame-free. Polyaromatic (or polynuclear) hydrocarbons (PAHs) were mentioned in the previous chapter as components of smoke. It is now becoming clear that this type of carcinogen can be present in other types of food. When carbohydrates or, more significantly, fatty acids are heated to above $500\,^{\circ}C$, they can give rise to unsaturated fragments such as the following types:

$$CH_2{=}CH_2 \quad CH_2{=}CH{-}CH{=}CH_2 \quad CH_2{=}C(CH_3)CH{=}CH_2$$

and their radicals. These condense to form progressively larger poly-aromatic systems, until compounds such as 3,4-benzpyrene (9.6) are formed. While smoked fish, bacon and sausages typically have levels of 3,4-benzpyrene below 5 ppb, barbecued meats, hamburgers, chops, steaks, *etc.*, frequently have much higher levels, up to 100 ppb. These elevated levels appear to be caused by fat dripping from the meat on to the burning charcoal of the barbecue and decomposing in the familiar puff of smoke and flame. Designs for barbecues and grills which prevent fat dripping into the fire would clearly be of benefit. "Clean" fuels such as charcoal are also to be preferred to wood or coal.

There is little doubt as to the carcinogenicity of the PAHs, at least in laboratory animals. There is considerable argument, however, as to whether the levels actually ingested in barbecued or grilled meat are sufficient to cause tumours in humans. Much higher levels of them are found in the tar of tobacco smoke and in diesel exhaust fumes, and eliminating these would do much more for human health than outlawing the barbecue. Studies published by the UK Food Standards Agency in 2002 indicated estimated average levels of benzpyrene consumption from food to be $1.6\,\text{ng}\,\text{kg}^{-1}$ body weight per day, approximately one-fifth of the level found in 1979.

In recent years several new classes of potential carcinogens have been identified in cooked protein-rich foods such as meat, fish, soya products and cereals. These substances are identified as potential carcinogens on the basis of their ability to promote mutations in special strains of *Salmonella typhimurium* in the Ames test. At the present time there is little information as to the actual carcinogenicity of these substances in animal systems. Some of them appear to be the result of pyrolytic breakdown of amino acids, particularly tryptophan, and others arise from amino acids caught up in the ramifications of the Maillard reaction. It is too early to speculate about whether these substances are responsible for the elevated incidence of cancer of the gastrointestinal tract in countries where a lot of meat is eaten.[x] Moreover, the likelihood that these substances may contribute to the attractive flavour of roast meat will not help to resolve the issue.

The unexpected appearance of benzene as a contaminant in some types of soft drinks was mentioned in Chapter 9 (*see* page 371).

---

[x] Recently published research (M. H. Lewin, *et al., Red meat enhances the colonic formation of the DNA adduct* O-*carboxymethyl guanine: implications for colorectal cancer risk; Cancer Res.,* 2006, **66**, p. 1859) has suggested that the myoglobin of red meat interacts with naturally present nitric oxide (NO) in the intestine leading to the generation of potentially carcinogenic *N*-nitroso compounds.

PACKAGING RESIDUES

The plastics used to package our food can give rise to two types of contaminant. The plastics themselves are polymers of extremely high molecular weight, and there is therefore little possibility of traces of the polymer itself contaminating food products. On the other hand, polymeric materials of this nature may contain traces of low molecular weight material. For example, vinyl chloride monomer (VCM), $H_2C=CHCl$, can be leached out of PVC (polyvinyl chloride) and this is a potential issue for food stored in PVC containers. Although there have been occasional incidents when much higher concentrations have been detected, UK legislation requires that food packaged in PVC contains less than $10\,g\,kg^{-1}$ of VCM. Materials such as cooking oils would be much more vulnerable to contamination with VCM than non-fatty materials such as mineral water, but fortunately PVC bottles are rarely used in this role.

Of greater concern are plasticisers. These are substances incorporated into plastics such as PVC to render them flexible. At one time attention was focused on the cling films used for wrapping food. These were originally made of PVC, in films around $10\,\mu m$ thick. Their clinging property made them ideal for wrapping irregular food items and covering glass vessels, and they also had an acceptable degree of permeability to oxygen and water vapour. The plasticiser most commonly used in PVC cling film, and the one which has received most attention from the authorities, is dioctyl adipate (di-2-ethylhexyl adipate, DOA or DEHA,10.23). A wide variety of plasticisers has been used in PVC in the past, but we will concentrate on DEHA.

(10.23)

Extensive surveys conducted in the late 1980s showed that significant levels of DEHA were present in a wide range of the foods consumed in Britain. Taking into account typical consumption of these foods, the maximum daily intake of DEHA in 1988 was calculated to be 8.2 mg per person.[xi] This figure was half that recorded in 1986. In response to the problems of posed by PVC manufacturers turned to the use of low density polyethylene (LDPE) for cling films. These are undoubtedly inferior to PVC based films in terms of both clinging ability and permeability, but they have helped to sustain a continued fall in the intake of packaging residues. Phthalates, in particular di(2-diethylhexyl)phthalate (DEHP,10.24), were also widely used as plasticisers in food packaging plastics. They have now been almost entirely eliminated from food packaging, but they remain an important issue of which food chemists should be fully aware. Although these do not appear to be carcinogenic, the possibility is now recognised that in some cases they may mimic mammalian female sex hormones. This could implicate them in a fall in male fertility and an increase in other male reproductive disorders reported over recent decades, in both humans and other species. Regardless of the outcome, this is a valuable reminder that toxicologists need to maintain vigilance in areas beyond the current preoccupation with carcinogenicity.

(10.24)

The complex issue of food packaging is one that is outside the normal scope of food chemists, and has only received brief attention in this book. A major cause of the complexity is the enormous variety of different polymers now in use and their tendency to turn up unexpectedly. An innocent looking glass jar will probably have a lid that includes a plastic liner to ensure that it seals properly. All that can be expected of the readers of this book is an awareness that the problem exists; a real understanding of the situation demands specialised expertise beyond that of the food scientist.

[xi] Ministry of Agriculture, Fisheries and Food, Food Surveillance Paper No. 30, *Plasticisers: Continuing Surveillance*, HMSO, 1990.

ENVIRONMENTAL POLLUTANTS

Some environmental pollutants command the attention of food chemists, despite their origins having little to do with food. The dioxins, the
closely related dibenzofurans (often referred to simply as furans), and
the dioxin-like polychlorinated biphenyls (PCBs), are the most important. Their structures are shown in Figure 10.7. In view of the widely
differing toxicity of the possible variants, analytical data on the dioxins,
furans and PCBs are reported in the form of Toxic Equivalents (TEQs),
i.e. 2,3,7,8-TCDD Toxic Equivalents. By analogy with the retinol
equivalents used for vitamin E measurements (*see* page 347), 1 g of
TEQs consists of a quantity of these substances of overall toxicity
equivalent to that of 1 g of 2,3,7,8-TCDD. The values to be applied to
UK survey data on the relative toxicities of dioxins, PCBs, *etc.* (the
Toxic Equivalency Factors, or TEFs) are periodically reviewed in the
light of current research by the WHO, most recently in 2005. Where
appropriate the data on dietary intakes are then adjusted. TEFs range
from 1 to as low as 0.00003.

PCBs were manufactured for use as the dielectric fluid in transformers
and capacitors. In the light of their toxicological problems and

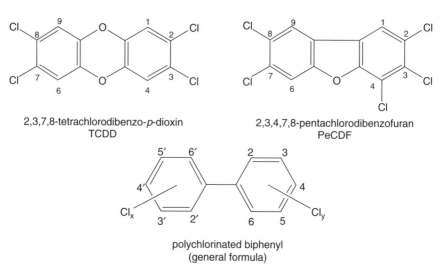

2,3,7,8-tetrachlorodibenzo-*p*-dioxin
TCDD

2,3,4,7,8-pentachlorodibenzofuran
PeCDF

polychlorinated biphenyl
(general formula)

Figure 10.7 Dioxins, dibenzofurans and polychlorinated biphenyls. The dioxin
shown is just one of 210 possible congeners, differing in their pattern of
chlorine substitution. A similar abundance of dibenzofuran congeners is
possible. Polychlorinated biphenyls are not normally referred to specifically in terms of the numbers or the location of substituent chlorine
atoms. When commercial PCBs were supplied to the electrical industry
they were identified merely by the average number of carbon atoms and
chlorine atoms per molecule present in a batch.

persistence in the environment, manufacture was ceased by 1980 and their use is now banned. Existing electrical equipment containing PCBs is being gradually replaced and its PCB content destroyed. Dioxins and dibenzofurans arise from numerous sources. By-products of the industrial production of chlorinated chemicals account for about a third of the total. Bleaching of wood pulp with chlorine can also generate them, but the major source is the combustion of materials which contain chlorinated compounds. This occurs in waste incinerators, metal refining, domestic and industrial coal burning, and in petrol and diesel engines. Replacing the dumping of sewage sludge at sea by incineration or spreading it on the land may be an important future source. There are also suggestions that wholly natural events such as forest fires also may increase the level of dioxins in the environment.

With the exception of workers in the chemical industry, particularly the victims of accidents such as that at Seveso in Italy in 1973, human exposure to dioxins is almost entirely through food. However, their ubiquitous character and their persistence in the environment make it impossible to identify particular routes by which they reach our food. Once in the food chain, their solubility in fats and their total insolubility in water causes them to accumulate in the fatty tissues of animals, fish and humans. Although levels of dioxins and PCBs have fallen steadily over recent decades, the incidence of both of these groups in human milk remains considerably higher than in cow's milk. This is a reflection of the difference in the diet of cows and humans, but the advantages of feeding babies with human milk are considered to far outweigh this isolated disadvantage.

Total Diet Studies of dioxins and PCBs, similar to those mentioned earlier with reference to pesticide residues, are carried out from time to time by the UK authorities. Results from the 2001 survey are shown in Table 10.5. The authorities currently accept a value for the TDI of 2 pg TEQ per kg body weight per day. Table 10.5 also includes details of the proportion of the UK population estimated to have dioxin and PCB intake in excess of the TDI. These data show that an uncomfortably high proportion of young children exceed the TDI. However, the TDI relates to exposure over the full natural lifespan.

The implications of exceeding the TDI in childhood are not clear. What is clear, however, is that there is an even greater incentive for authorities worldwide to maintain the policies which have resulted in the sharp decline in the level of these pollutants over the past two decades. Reducing the level of dioxins in the diet is regarded as a priority for governments in Europe and elsewhere, but it is more likely to be

Table 10.5   Total Diet Studies of dioxins and PCBs in the diet of adults and
                  children in Britain in 2001. Estimated levels of dioxins plus
                  dibenzofurans (dioxins) and dioxin-like polychlorinated biphenyls
                  (PCBs) are expressed as pg of Toxic Equivalents (TEQs) using
                  2005 TEFs, as discussed in the text. (Based on data published by
                  the Food Standards Agency in *Food Surveillance Information
                  Sheet 4/00*, 2000.).

	Average dietary intake (pg TEQ/kg body weight/day)		*Percentage estimated to exceed the TDI*
	Dioxins	*PCBs*	
Children (1.5–2.5 years)	1.0	0.9	34.0
Children (4–6 years)	0.7	0.7	14.0
Children (11–14 years)	0.4	0.3	0.0
Adults	0.3	0.3	0.03

achieved through controls on environmental pollution rather than by
the action of farmers and food processors.

SPECIAL TOPICS

1. Favism

As implied earlier in this chapter a full appreciation of the actual
mechanism by which the *Vicia faba* toxins damage red blood cells
requires specialised knowledge. The following additional details may be
of help.

Red blood cells (RBCs) are unusual in lacking the mitochondria which
provide reduction potential in other mammalian cells. In consequence
they rely on the pentose phosphate pathway (also known as the hexose
monophosphate shunt) which provides reducing power in the form of
NADPH (*see* Chapter 8). This is generated in two steps in the initial
oxidative phase of the pathway, catalysed in turn by glucose-6-phosphate
dehydrogenase (G6PD) and 6-phosphogluconate dehydrogenase:

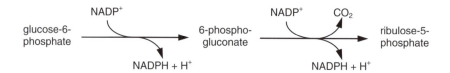

The NADPH is utilised in a number of processes, including synthesis
of the haem component of haemoglobin. However, we are most

concerned with its reaction with glutathione (5.13), which we first encountered in the context of the sulfhydryl reactions in dough (*see* page 208), a reaction catalysed by the enzyme glutathione reductase:

$$GSSG + NADPH + H^+ \xrightleftharpoons{\substack{\text{glutathione} \\ \text{reductase}}} 2GSH + NADP^+$$

The ratio of the reduced form (the dimer GSSG) to the oxidised form (the monomer GSH) is normally about 500 : 1. The reduced glutathione has a number of crucial roles in erythrocytes. One of these is in maintaining the iron atom of haemoglobin in the ferrous, Fe^{II}, state in which it carries out its normal function of reversibly binding oxygen. (Chapter 5, page 193, gives a detailed account of the parallel function of myoglobin.) It is also responsible for disposing of hydrogen peroxide, an inevitable but dangerous by-product of oxygen metabolism:

$$2GSH + H_2O_2 \xrightarrow{\substack{\text{glutathione} \\ \text{peroxidase}}} GSSG + 2H_2O$$

Reducing conditions and a plentiful supply of reduced glutathione are essential for the development of malarial parasites in the erythrocytes. This means that mutations reducing the activity of G6PD in erythrocytes would make them an inhospitable environment for the parasites and could confer a degree of resistance to the disease.

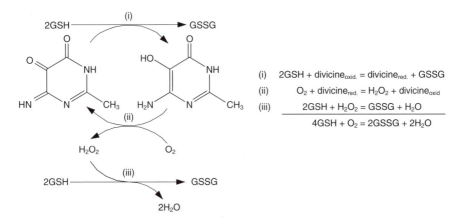

Figure 10.8   The removal of reduced glutathione by the spontaneous "futile cycle" reactions involving divicine. The other fava bean pyrimidine, isouramil, participates in the same reactions. The "futility" of this sequence of reactions is best demonstrated by arranging the equations for all three in a sum, as on the right. In the best algebraic tradition, any terms appearing on both sides of the equals signs are cancelled out.

The impact of the fava bean toxins on this situation is now well understood. As shown in Figure 10.8, their aglycone forms react spontaneously with oxygen, with concomitant production of hydrogen peroxide. The oxidised product is then reduced in a spontaneous reaction with reduced glutathione. The reaction scheme in Figure 10.8 shows how these reactions become what is known as a "futile cycle", in which the erythrocytes' normal balance between the reduced and oxidised forms of glutathione is disrupted. The result is a similarly inhospitable environment for the malarial parasite as results from reduced levels of G6PD.

The problem of favism arises when an individual, whose supply of GSH is already rendered marginal by an inherited low level of erythrocyte G6PD activity, consumes a quantity of fava beans. An attack of haemolytic anaemia is the inevitable outcome of the twin assaults on the GSH economy of the erythrocytes.

2.   Total Intake of Undesirable Metals

The concept of the Total Diet Survey was introduced on page 416 in connection with the total exposure of an individual to pesticide residues in the diet. In 2000 the UK Food Standards Agency published similar data regarding the intake of a wide range of metals, based on the food consumption data taken from the National Food Survey in 1997, and ongoing surveys of the toxic metal content of food conducted by the UK Committee on the Toxicity of Chemicals in Food, Consumer Products and the Environment. Some of the results for metals discussed in this chapter are shown in Table 10.6.

Table 10.6   Undesirable metals in the total UK diet. The daily safety guidelines shown are the Provisional Tolerable Weekly Intakes (PTWIs) set by the Joint FAO/WHO Expert Committee on Food Additives (JECFA), divided by a factor of 7.

	Mean estimated dietary exposure (μg/kg body weight/day)				
	Children (1.5–4.5 years)	Young people (4–18 years)	Adults	Elderly[a]	Safety guidelines
Aluminium	165	121	68	82	1000
Arsenic	2.7	1.7	1.6	1.6	2.14
Cadmium	0.32	0.22	0.12	0.14	1.0
Lead	0.25	0.17	0.10	0.12	3.5
Mercury	0.07	0.05	0.04	0.04	0.23[b]
Tin	70	38	20	17	2000

[a]Institutionalised.
[b]Methylmercury.

With the obvious exception of arsenic, the margins between the mean actual and the tolerable intakes are reassuringly large. This also remains true when the calculated intake for those at the upper end of the intake range, not included here, are considered.

It is not easy in routine analysis to distinguish between the dangerous inorganic forms of arsenic and the relatively benign organic forms such as arsenobetaine. The intake data shown refers to the total arsenic content of the diet, whereas the tolerable level refers only to the inorganic forms. Since the bulk of dietary arsenic is in the organic form the data given here are less alarming than they appear at first sight.

FURTHER READING

Food Safety and Food Quality, ed. R. E. Hester and R. M. Harrison, Royal Society of Chemistry, 2001.

S. T. Omaye, *Food and Nutritional Toxicology*, CRC Press, Boca Raton, Florida, 2004.

Natural Toxicants in Food, ed. D. H. Watson, Sheffield Academic Press, Sheffield, 1998.

Food Toxicology, ed. W. Helferich and C. K. Winter, CRC Press, Boca Raton, Florida, 2000.

Nitrates and Nitrites in Food and Water, ed. M. J. Hill, Ellis Horwood, 1991.

Toxins in Food, ed. W. M. Dabrowski and Z. E. Sikorski, CRC Press, Boca Raton, Florida, 2005.

Environmental Contaminants in Food, ed. C. F. Moffat and K. J. Whittle, Sheffield Academic Press, Sheffield, 1999.

Food Chemical Risk Analysis, ed. D. R. Tennant, Chapman and Hall, London, 1997.

Pesticide, Veterinary and Other Residues in Food, ed. D. H. Watson, CRC Press, Boca Raton, Florida, 2004.

G. W. Ware, *Pesticide Book*, Meister Publishing Co., Danvers, Massachusetts, 7th edn, 2007.

C. Reilly, *Metal Contamination of Food: Its Significance for Food Quality and Human Health*, Blackwell, Oxford, 3rd edn, 2002.

Detecting Allergens in Food, ed. S. J. Koppelman and S. L. Hefle, CRC Press, Boca Raton, Florida, 2006.

Pesticide Residues in Food and Drinking Water: Human Exposure and Risks, ed. D. Hamilton and S. Crossley, Wiley, Chichester, 2004.

RECENT REVIEWS

M. C. M. Rietjens, *et al., Molecular Mechanisms of Toxicity of Important Foodborne Phytotoxins; Mol. Nutr. Food Res.*, 2005, **49**, p. 131.

D. Siritunga and R. Sayre, *Engineering Cyanogen Synthesis and Turnover in Cassava (*Manihot esculenta*); Plant Mol. Biol.*, 2004, **56**, p. 661.

J. V. Higdon and B. Frei, *Coffee and Health: A Review of Recent Human Research; Crit. Rev. Food Sci. Nutr.*, 2006, **46**, p. 101.

P. G. Bradford and A. B. Awad, *Phytosterols as Anticancer Compounds; Mol. Nutr. Food Res.*, 2007, **51**, p. 161.

A. Faraj and T. Vasanthan, *Soybean Isoflavones: Effects of Processing and Health Benefits; Food Rev. Internat.*, 2004, **20**, p. 51.

E. Dittmann and C. Wiegand, *Cyanobacterial Toxins: Occurrence, Biosynthesis and Impact on Human Affairs; Mol. Nutr. Food Res.*, 2006, **50**, p. 7.

M. E. van Apeldoorn, H. P. van Egmond and G. J. I. Bakker, *Toxins of Cyanobacteria; Mol. Nutr. Food Res.*, 2007, **51**, p. 7.

C. Ruiz-Capillas and F. Jiménez-Colmenero, *Biogenic Amines in Meat and Meat Products; Crit. Rev. Food Sci. Nutr.*, 2004, **44**, p. 489.

P. A. Murphy, *et al., Food Mycotoxins: An Update; J. Food Sci.*, 2006, **71**, p. 51.

L. Monaci, V. Tregoat, A. J. van Hengel and A. Anklam, *Milk Allergens, their Characteristics and their Detection in Food: A Review; Eur. Food Res. Technol.*, 2006, **223**, p. 149.

M. J. Scotter and L. Castle, *Chemical Interactions between Additives in Foodstuffs: A Review; Food Addit. Contam.*, 2004, **21**, p. 93.

I. S. Arvanitoyannis and L. Bosnea, *Migration of Substances from Food Packaging Materials to Foods; Crit. Rev. Food Sci. Nutr.*, 2004, **44**, p. 63.

G. Lack, *Epidemiological Risks for Food Allergy; J. Allergy Clin. Immunol.*, 2008, **121**, p. 1331.

CHAPTER 11
Minerals

In the previous chapter we saw that there are many inorganic elements whose presence in our diet is undesirable. The food chemist has a special responsibility to ensure that these elements are not consumed in harmful amounts and therefore needs to understand how they reach our diet, from the environment, raw materials, processing operations, *etc*. On the other hand, when we turn to nutritionally desirable elements, the food chemist's task becomes somewhat simpler and can be examined under three headings:

- The analysis of foodstuffs to determine how much of the various inorganic nutrients they contain in relation to our nutritional needs. For each mineral element there are a number of laboratory techniques available, depending on the analytical objectives and the resources available.[i]
- Studies of the form in which certain inorganic elements occur in foodstuffs where this affects our ability to absorb them from the gastrointestinal tract.
- Studies of the actual behaviour of inorganic elements in food in cases where this has an influence on other aspects of food quality.

The list of bulk and trace elements required in the diet is a long one but, as Figure 11.1 shows, it is largely restricted to elements of low

---

[i] This aspect of food chemistry is beyond the scope of this book.

Food: The Chemistry of its Components, Fifth edition
By T.P. Coultate
© T.P. Coultate, 2009
Published by the Royal Society of Chemistry, www.rsc.org

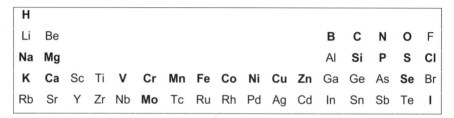

Figure 11.1   The elements essential for animal life. This section of the period table
shows in bold the elements regarded as essential; those almost certainly
not required are in lighter type.

atomic number. As we have observed previously, the fact that our diet
consists almost exclusively of materials that were once living organisms,
coupled with the broad similarity of the biochemistry of all forms of
animal and plant life, means that we can expect any reasonably balanced
diet to supply our mineral needs in about the right proportions.

THE BULK MINERALS

Sodium

The alkali metal sodium has a well established role in animal physiology
and is abundant in both human tissues and human diets. It occurs
at levels between 50 and 100 mg per 100 g in most foods of animal
origin, including milk, meat and fish, but eggs contain around 150 mg
per 100 g. In their raw state vegetables and cereals have very much less
sodium, between 1 and 10 mg per 100 g, but the almost universal
practice of adding sodium chloride during cooking or at the table
brings their sodium content, as eaten, up to similar levels to those in
animal foods.

It is difficult to state with certainty the minimum sodium intake
necessary for good health. Special low-sodium diets prescribed for
patients suffering from heart disease and certain other conditions can
bring the daily intake of salt down to around 2 g, but this requires
considerable distortion of normal eating habits. A strict vegetarian diet
without added salt could approach this figure, but at the cost of low
palatability. One should not overlook the fact that our taste buds are
able to distinguish sodium chloride from other salts (*see* Chapter 7), and
that there is a sound physiological basis for the other responses provided
by our sense of taste.

The importance of sodium chloride as a preservative was discussed in
Chapter 9, and in foods in which it has this role, *e.g.* bacon, kippers and
butter, levels ten times those quoted are quite common. There is

considerable pressure from the medical profession to reduce the salt content in our diet. Until a few years ago the emphasis was on the use of low sodium diets in the management of patients with hypertension, *i.e.* high blood pressure. Hypertension is widely accepted as a major risk factor in the development of cardiovascular disease.[ii] Two major investigations in the USA, the "Intersalt" study published in 1988 and the DASH (Dietary Approaches to Stop Hypertension) study reported in 2003, confirmed that the relative risk of cardiovascular disease increases as blood pressure rises, even within the normal range of blood pressure. Another important point which emerged from the DASH study was that people vary widely in their response to reduced sodium intake. Identifying individuals as *sodium responders* is difficult, however, supporting the view that current recommendations for lower salt intake are more usefully directed at the population as a whole, rather than attempting to identify susceptible individuals.

In the UK the authorities now recommend a maximum intake for adults of 6 g of sodium chloride per day. This figure is much lower than the actual intake for a significant proportion of the population, especially males. (The median for adult males in the 2003 National Diet and Nutrition Survey was 8.3 g.) The figure of 6 g is also significantly above the current Reference Nutrient Intake figure for adult males. If the recommended target is to be achieved it will require considerable retraining of our palate.

Potassium chloride, often suggested as an alternative for culinary and table use, simply does not bear comparison with the real thing, especially in important products such as bread.

Potassium

In common with sodium the alkali metal potassium also has an important part to play in animal physiology. It occurs abundantly in both human tissues and human diets, in foodstuffs of both animal and plant origin. Almost all foods, except on the one hand the oils and fats, which have virtually no mineral content, and on the other hand the seeds and nuts, with 0.5–1.0% potassium, lie within the range 100–350 mg per 100 g. The character of one's diet therefore has little effect on total potassium intake. The only food in which the potassium level is of particular interest is jam. This has nothing to do with the nutritional role of potassium, but it does provide an insight into that most elusive quality

---

[ii] Although the first modern suggestions by the medical profession of this relationship were heard in the 1940s, Chinese physicians were warning their patients as early as 2500 BC that too much salt in their food would "harden their pulse".

parameter of jam, its fruit content. The essential similarity of the sugars, the organic acids, pectins, pigments, and many of the more easily determined flavour compounds, makes analytical confirmation of the jam-maker's assertions about his ingredients almost impossible. It is routine for acids and pectins to be included from sources other than the fruit named on the label. The level of potassium in a jam will provide a rough measure of the total content of fruit (and vegetable) tissues, provided potassium metabisulfite has not been added as a preservative.[iii] The weakness of this entire area of food analysis becomes apparent when one considers that counting and identifying the pips is still a routine procedure in the quality assessment of many types of jam.

Magnesium

Like potassium, magnesium is widely distributed in foodstuffs of all types (at 10–40 mg per 100 g), although foods derived from plant seeds such as wholemeal flour, nuts and legumes may contain more than 100 mg per 100 g. It is difficult to envisage an otherwise nutritionally adequate diet which would be deficient in either potassium or magnesium. Beer brewed in many parts of Britain (notably in Yorkshire) owes some of its bitterness, and perhaps some of its cathartic effect, to the naturally high levels of magnesium sulfate in the water.

Calcium

When we turn to calcium we come across a much more complex situation. The calcium content of different foodstuffs covers a very wide range. At the lower end are some fruits and vegetables, such as apples, peas and potatoes, with less than 10 mg per 100 g. Other vegetables contain a great deal more, *e.g.* broccoli and spinach have around 100 and 600 mg per 100 g, respectively. Meat and fish are generally low in calcium, around 10 mg per 100 g. Cereals are also relatively low, for example wholemeal flour has around 35 mg per 100 g, but white flours are fortified with additional calcium carbonate. Current UK regulations require that white flour for retail sale and most food manufacturing applications[iv] must contain between 235 and 390 mg of calcium carbonate per 100 g. Self-raising flour contains calcium carbonate in any event, as the source of carbon dioxide.

[iii] Patients with kidney problems are often advised to pursue a low potassium diet, whose only notable feature is abstinence from bananas! The potassium content of bananas, at around 400 mg per 100 g, is $1\frac{1}{2}$ to 2 times that of most other fresh fruit.
[iv] Flours for making communion wafers and matzos are exempt.

There is evidence from some parts of Britain that even in today's enlightened and affluent world many children's diets would contain dangerously low levels of calcium without this fortification. The question of adult calcium requirement is complex, and beyond the scope of this book. Mild calcium deficiency cannot be diagnosed, since the body tends to maintain serum levels at the expense of skeletal reserves. Similarly, the level of calcium in milk during lactation may be maintained by depletion of reserves. The uptake of calcium from the intestine and its movement between the blood, skeleton and other tissues are under hormonal control. When deficiencies in calcium levels in the body do occur, resulting in rickets and other disorders, they are most frequently caused by impaired alimentary absorption resulting from a deficiency in vitamin D.

Dairy products are traditionally our primary source of calcium. The calcium content of cow's milk varies, but on average is about 120 mg per 100 g. In hard cheeses such as Cheddar the calcium of milk is concentrated to around 800 mg per 100 g, but in soft cheeses levels in the region of 450 mg per 100 g are more typical. The biological reason for the presence of large quantities of calcium in milk is obvious: the newborn mammal has a skeleton to build. The skeleton's primary raw material is calcium phosphate, $Ca_3(PO_4)_2$, and this presents a problem. One of the features of calcium phosphate which is most valuable in skeleton construction is that it is insoluble at neutral pH. However, the phosphate is delivered in milk, and it is not easy to visualise how calcium and phosphate ions could share the same delivery system without forming insoluble crystals. Nature's answer lies in the subtle interactions of these two incompatible ions with the caseins discussed in Chapter 5. Cow's milk contains on average:

$0.065 \, mol \, dm^{-3}$ inorganic phosphate (0.62 g per $100 \, cm^3$, as PO_4^{3-});
$0.024 \, mol \, dm^{-3}$ organic phosphate (0.23 g per $100 \, cm^3$, as PO_4^{3-}), essentially the phosphoserine of α- and β-caseins;
$0.009 \, mol \, dm^{-3}$ citrate; and
$0.030 \, mol \, dm^{-3}$ calcium (0.12 g per $100 \, cm^3$ as Ca^{2+}).

At the normal pH of fresh milk, about 6.6, the inorganic phosphate is present almost entirely as HPO_4^{2-} and $H_2PO_4^-$, in the ratio 1:3. The virtual absence of PO_4^{3-}, the fact that a significant proportion of the HPO_4^{2-} and $H_2PO_4^-$ are protein-bound, and the intervention of the citrate (largely citrate^{3-} at pH 6.6), rule out the formation of the highly insoluble $Ca_3(PO_4)_2$. Instead we have a network of anions and cations which help to maintain the structure of the casein micelles, as discussed in Chapter 5.

The interaction of calcium with phytic acid (3.18) and polysaccharides such as pectins and alginates was discussed in Chapter 3.

Phosphorus

It seems inevitable that discussion of phosphorus will follow closely that of calcium. In the form of inorganic phosphates and organic esters, phosphorus is a component of all living organisms. In animals it is of course concentrated in the skeleton, but it should not be forgotten that the intermediary metabolism of sugars takes place in the form of phosphate esters. Sugar phosphates are essential components of the nucleotide coenzymes and nucleic acids and all biological membranes contain phospholipids.

The metabolic role of phosphate is reflected in its distribution in foodstuffs. For example, lean meat (muscle) contains around 180 mg of phosphorus per 100 g, while the corresponding figure for liver is 370 mg. In foodstuffs of plant origin, those derived from seeds, such as legumes, cereals and nuts, also have high levels of phosphorus, between 100 and 400 mg per 100 g. Other vegetables, including potatoes, spinach and fruit such as apples, have levels of phosphorus below 100 mg per 100 g. The phosphorus in the wheat grain is concentrated in the bran fraction, which means that white flour has only about one-third of the level in wholemeal flour, 130 mg compared with 340 mg per 100 g. Most of the phosphorus in bran is in the form of phytic acid, whose role is simply that of a compact phosphate reserve for the germinating seed. Nuts are similar, in that 70–85% of the phosphorus content is also in the form of phytic acid. The human digestive system is normally equipped with low levels of the enzyme phytase, which liberates the phosphate from the phytic acid, but there is evidence to suggest that more of the enzyme is synthesised by people whose diet is high in phytic acid over a long period. The high phosphorus content of milk has already been mentioned, and the presence of phosphorus in many types of cheese at around 500 mg per 100 g is therefore to be expected. Eggs are another good source of phosphorus, with around 220 mg per 100 g. The widespread distribution of phosphorus in foodstuffs means that phosphorus deficiency is not encountered in otherwise satisfactory diets.

Not all the phosphorus in food is entirely natural in origin. Polyphosphates are popular additives in many meat products, especially ready-sliced ham, luncheon meat and pre-packed frozen poultry. Polyphosphates such as tetrasodium diphosphate (11.1) are manufactured by heating orthophosphates:

$$2x\ Na-O-\overset{\overset{\displaystyle O}{\|}}{\underset{\underset{\displaystyle Na}{|}}{P}}-OH \longrightarrow Na-O-\overset{\overset{\displaystyle O}{\|}}{\underset{\underset{\displaystyle Na}{|}}{P}}-O-\overset{\overset{\displaystyle O}{\|}}{\underset{\underset{\displaystyle Na}{|}}{P}}-O-Na\ +\ H_2O \qquad (11.1)$$

Those most commonly used have two or three phosphorus atoms per molecule, but polymeric forms with more than twenty phosphorus atoms have some applications. Those with two phosphorus atoms are known either as pyrophosphates or diphosphates. Polyphosphates enhance the water-binding properties of muscle protein. Apart from the obvious advantage (to the processor) of improving the yield of meat products from a given weight of raw meat, the greater retention of muscle water and its solutes during cooking enhances the juiciness and flavour of products such as burgers, ham and bacon. In spite of the addition of water entailed by the use of polyphosphates, and the apparent wetness of many pre-packaged ham products, their final water content can be lower than that of the uncured meat. This results in the curious statement occasionally seen on labels of cured meats that the product contains 110% meat! This is the unavoidable consequence of the meat content being calculated on the basis of the Kjeldahl nitrogen protein determination and the factors used to convert the protein content to meat content being based on the protein content of fresh, *i.e.* uncured, meat. The phosphates and polyphosphates permitted for use as food additives in Europe are listed in Table 11.1.

An incidental advantage of polyphosphates is that they appear to delay the onset of rancidity in some products. As they are powerful chelating agents, it is assumed that this effect arises through their sequestration of iron and copper, which would otherwise promote the autoxidation of meat lipids (see Chapter 4). They are normally used at levels around 0.1–0.3%. At these levels they pose no health

Table 11.1 Phosphates and polyphosphates permitted as food additives in Europe.

	E number		*E number*
Sodium phosphates	E339	Sodium, potassium or calcium diphosphates	E450
Potassium phosphates	E340	Sodium triphosphates	E451
Calcium phosphates	E341	Sodium, potassium or calcium polyphosphates	E452

problems, since they appear to be hydrolysed to orthophosphates by the pyrophosphatases present in most animal and plant tissues.

THE TRACE MINERALS

Iron

Of all the metals, iron is probably the one which the layman is most aware of as a nutrient, and also the one which is perceived to be potentially in short supply in the diet. One reason for this is that most people are familiar with the need for iron in the blood, even though they may not know what it is doing there, in the myoglobin of muscle, or in respiratory enzyme systems generally. Iron is the most important transition metal in the animal body, where it occurs almost entirely in elaborate coordination compounds based on the porphyrin nucleus, notably the haem pigments (*see* Figure 5.5), which are responsible for carrying oxygen. Alternation between its two oxidation states, Fe^{II} (ferrous) and Fe^{III} (ferric), is an essential feature of its contribution to the oxidation–reduction reactions involving the other iron porphyrin proteins involved in respiration, the cytochromes. A deficiency of iron becomes apparent as anaemia, *i.e.* an abnormally low blood haemoglobin level.

Iron is abundant in most foodstuffs, of plant as well as animal origin. Lean meat contains between 2 and 4 mg per 100 g, mostly as myoglobin, which means that the relative redness of different cuts is a guide to the abundance of the metal. Chicken, lamb and ox livers contain rather more iron, around 9 mg per 100 g, and pig's liver about twice as much. The iron in liver is not present as myoglobin or haemoglobin but is bound by specialist iron-binding storage proteins, notably ferritin. The core of the ferritin molecule is able to contain up to around 4000 iron atoms in a crystalline structure resembling the mineral ferrihydrate, which has the overall formula $[FeO(OH)]_8[FeO(H_2PO_4)]$. Leafy green vegetables, legumes, nuts and whole cereal grains all contain between 2 and 4 mg of iron per 100 g. Most fruit, potatoes, and white fish such as cod have between 0.3 and 1.2 mg per 100 g. Although spinach is one of the vegetables at the upper end of the range for iron content (~ 4 mg per 100 g) sadly there is no scientific basis for Popeye's enthusiasm for it.[v]

---

[v] There are at least two versions of how this error arose. One suggests that in 1890 a German scientist, Gustav von Bunge, found the iron content of spinach to be 35 mg per 100 g, but it was overlooked that he had analysed powdered dried spinach, and this figure came to be associated in error with the fresh vegetable. An alternative possibility is that some 20 years earlier another scientist, Dr E. von Wolf, had performed a similar analysis—but had misplaced the decimal point in the result. Wherever the truth lies, the error was not picked up until 1937 and the myth has managed to survive the subsequent 70 years.

Much more remarkable is the figure of 26 mg per 100 g reported for the humble cockle, (*Cardium* sp.), but whether this observation is relevant to anyone's regular diet is rather dubious. The iron content of cow's milk is remarkably low, 0.1–0.4 mg per 100 g, but in dairy products such as butter and cheese the iron is concentrated to around 0.16 and 0.4 mg per 100 g, respectively. Eggs contain around 2 mg per 100 g.

It is tempting to deduce from these data that iron is abundant in the diet and that iron deficiency should be rare. It is true that most people's diet contains between 10 and 14 mg per day, and the absence of any specific mechanism for iron secretion from the body would appear to support this. Losses during menstruation can reach 15–20 mg of iron per month, but other types of loss are much smaller. A total of up to 1 mg daily may be lost *via* the urine, sweat and the epithelial cells shed from the skin and the lining of the intestines. The difficulty is that iron is poorly absorbed from the intestine, and the degree of absorption is highly dependent upon the nature of the iron compounds present in the diet. Only about 10% is absorbed from a typical European diet. The different behaviour of ferric and ferrous ions in aqueous solution is a key factor.

In the acidic environment of the stomach ferrous and ferric ions occur as the soluble, hydrated ions $Fe(H_2O)_6^{2+}$ and $Fe(H_2O)^{3+}$, but as the pH rises to 8 in the duodenum the much less soluble hydroxides are formed, $Fe(OH)_2$ and $Fe(OH)_3$. Ferrous hydroxide is soluble up to 10 mmol dm^{-3}, but ferric hydroxide is essentially insoluble, the maximum concentration in solution being $10\text{–}18\ mol\,dm^{-3}$. Clearly, there is little prospect of absorbing useful quantities of ferric iron at pH 8. Studies with radioactive isotopes of iron, [55]Fe or ^{59}Fe, have shown that the presence of ascorbic acid in the duodenum greatly enhances the degree of iron uptake. This is partly due to its action in reducing Fe^{III} to Fe^{II}. The acidic conditions in the stomach will also ensure that any insoluble ferric iron in the food becomes solubilised. The other role played by ascorbic acid is shared by many other substances, notably sugars, citric acid and the amino acids. All of these are effective chelating agents and form soluble complexes with iron, making it available for absorption. Home-made beer, with a significant residual sugar level and brewed in iron vessels, has been blamed for the excessive iron intake which is one cause of the cirrhosis of the liver common among the urban African population of South Africa. Whenever iron vessels are extensively used for food preparation there is a likelihood that iron overload will occur.

Although there is no suggestion that toxicity has resulted, it is worth noting that some canned fruits and vegetables may have high levels of dissolved iron if the cans are inadequately tinned. An instance has been recorded in which canned blackcurrants were found to contain 130 mg per 100 g of iron. The high nitrate level in the tropical pawpaw (*Carica papaya*) solubilises so much iron that this fruit is not regarded as suitable for canning. It is sometimes observed that the discoloration of canned foods only appears if the contents are kept in the can after opening. When air is admitted to the can any ferrous ions dissolved from the can may be oxidised to ferric. The latter are able to form a black coordination compound with rutin (11.2), a flavonoid commonly found in fruit and vegetables. This is a glycoside of the flavonol quercitin.

(11.2)

Some natural drinking water contains quite high levels of iron; if water is stored in iron tanks or conducted through cast-iron pipes objectionably high concentrations of iron may result. Iron gives a bitter or astringent taste to drinking water and beverages.

The substances complexing the iron in the diet have a considerable influence on the proportion absorbed by the body. When haem iron is the dietary source, *e.g.* the myoglobin of meat or fish skeletal muscle tissue, 15–25% is absorbed. On the other hand, no more than 8% of the iron in plant foods, including cereals, legumes and leafy vegetables, is absorbed. The figure for spinach has been reported to be around 2.5%. The iron in plant foods is mainly present as insoluble complexes of Fe^{III} with phytic acid, oxalates, phosphates and carbonates. One reassuring aspect of iron absorption from the intestine is that its efficiency rises when the body's iron reserves are low, and *vice versa*. Since there is no mechanism for the excretion of excess iron, such regulation of intake seems essential, even though it may not always be wholly effective.

The low level of iron in both human and cow's milk has already been mentioned. The iron in milk is carried by a specialised protein, lactoferrin. This is a similar protein to transferrin, the non-haem iron protein which transports iron in the bloodstream. The most curious aspect of this protein is that in milk its full iron-binding capacity is not taken up, even when there is no suggestion of iron deficiency in the lactating animal. For their first few months of life, humans and many other mammals have to rely on reserves built up in the foetus before birth. The most likely explanation for this phenomenon is that having low levels of free iron in the milk has other benefits. Lactoferrin undersaturated with iron maintains a low concentration of free iron in the milk, and in consequence in the upper reaches of the gastrointestinal tract. This is believed to provide the new-born mammal with protection against undesirable bacteria that have a strict requirement for iron. Among these are members of the *Enterobacteriaceae*, including pathogenic strains of *Escherichia coli*, which can cause severe diarrhoea in infants. More welcome bacteria, such as the lactobacilli, do not have the same demand for iron. It is worth noting that the concentration of lactoferrin in human milk is approximately 100 times greater than that in cow's milk, even though the total iron content is only marginally greater than in cow's milk.

Copper

Copper nowadays poses few problems. It is widely distributed as a component of a variety of enzymes in foodstuffs of all kinds, at levels between 0.1 and 0.5 mg per 100 g. Milk is notably low in copper at about 0.2 mg per 100 g, and mammalian liver is exceptionally high, at around 8 mg per 100 g. The daily intake from normal adult diets is between 1 and 3 mg, which is roughly in line with the level recommended by many authorities. As with iron, it appears that nature expects newborn infants to employ the reserves built up before birth until they are weaned. The occasional cases of copper deficiency in infants, seen as a form of anaemia, are invariably associated with other physiological or nutritional disorders.

Copper is not as toxic as one might imagine. Large doses of copper sulfate are emetic, and a dose of 100 g is reported to have resulted in liver and kidney damage. Long-term intake of moderately high amounts of copper can arise from regular drinking of water from the hot tap in houses with copper plumbing, but the probable outcome is no more than a gastrointestinal upset. This fairly optimistic attitude to the toxicity of copper salts has not always been the rule. Elizabeth Raffald, an 18th century forerunner of Mrs Beeton, warned against the practice of using copper salts when pickling vegetables:

'... *for nothing is more common than to green pickles in a brass pan for the sake of having them a good green, when at the same time they will green as well by heating the liquor without the help of brass, or verdegrease of any kind, for it is a poison to a great degree ...*'[vi]

[vi] E. Raffald, *The Experienced English Housekeeper*, Baldwin, London, 1782; 8th edn published in facsimile by E. W. Books, London, 1970.

It was not until 1855 that public pressure began to force the commercial manufacturers of pickles to abandon the use of copper, but for a time sales actually went down until consumers got used to the more natural brown colour. It is ironic that nowadays caramel has to be added to enhance the brown colour that was once looked upon as undesirable.

Zinc

The distribution of zinc in foodstuffs has much in common with copper. It is an essential component of the active sites of many enzymes, and it is therefore not surprising to find it at high levels in animal tissues such as lean meat and liver, at around 4 mg per 100 g. Wholemeal flour, dried peas and nuts contain similar levels, although white flour has only one-third as much. Other fruit and vegetables contain about 0.5 mg per 100 g or less. Of the dairy products, only eggs, with 1.5 mg per 100 g, can be regarded as a good source of zinc. Cow's milk has around 0.35 mg per 100 cm^3, but human milk less than a third of this. As with so many minerals, the new-born infant has to rely on reserves laid down in the later stages of foetal development. Typical diets have been calculated to supply between 7 and 17 mg per day, of which about 20% is absorbed.

The potential for zinc absorption to be reduced by the presence of large quantities of fibre in the diet does not seem to be borne out in the case of ordinary European diets. Although the occurrence of zinc deficiency and its symptoms are well recognised in veterinary practice, there is little to indicate that it is a significant problem in human diets. There is some evidence that zinc deficiency is common among some poor Middle Eastern peasant communities, a major part of whose diet consists of unleavened bread made from wholemeal flour. When this poor diet is supplemented with zinc a number of symptoms tend to disappear, including delayed puberty and reduced stature.

Selenium

Selenium[vii] was regarded as toxic decades before it was recognised in the 1950s to be an essential nutrient. The toxic effects of selenium were identified in farm animals in a number of areas of the world in which there are unusually high levels of selenium in the soil. There are a few species of pasture plants able to accumulate extraordinary amounts, over 1 g kg^{-1} (dry weight), when growing on selenium-rich soil. Wide variations are found in the selenium content of human diets. At one

---

[vii] Chemically selenium is defined as a metalloid, *i.e.* a substance having the appearance and other physical properties of a metal but the chemical properties of a non-metal.

extreme is New Zealand, where much of the agricultural land is deficient in this element and the average *per capita* daily intake has been calculated to be only 25 μg. The corresponding figure for the UK, taken from the 1995 Total Diet Survey, is 34 μg. This is also low by international standards and also when compared with the Reference Nutrient Intake (*see* Appendix I) figures of 60 and 75 μg per day for adult females and males, respectively. In different areas of the North American continent diets range from 60 to 220 μg. Even within the British Isles a wide range of selenium contents can be found in the same commodity produced in different regions. In the UK the selenium content of cereals lies in the range 0.03–0.23 mg kg^{-1}, meat and fish 0.09–0.38 mg kg^{-1}, and fruit and vegetables and milk ~ 0.01 mg kg^{-1}. Nuts are generally very rich in selenium; some samples of Brazil nuts have been found with 53 mg kg^{-1}. Just one of these nuts would supply the daily intake of selenium recommended by some authorities. Care must always be exercised in the pursuit of a healthy selenium intake, as it is believed that prolonged intake of in excess of 3 mg daily can give symptoms of selenium poisoning. Proposals for maximum permitted limits for selenium in foods (1.0 mg kg^{-1} for most foods) have been developed in Australia.

The biological function of selenium was mentioned in Chapter 8 during discussion of the antioxidant role of vitamin E. Evidence is beginning to accumulate that selenium intake towards the top of the normal range is beneficial; for example, the incidence of certain cancers in the USA is being tentatively linked to the selenium levels in the local diet. Human diseases that can definitely be blamed on selenium deficiency are quite rare. One example is Keshan disease, a cardiomyopathy which occurs in children in parts of China where selenium intake is particularly low, and there are also indications of thyroid function disorders associated with deficiency. However, the relatively narrow safety margin between beneficial and harmful intake makes wide-scale dietary supplementation difficult—and self-medication decidedly hazardous. One alternative approach has been adopted in Finland since the early 1980s, where supplementation of agricultural fertilisers has raised intake from 30–40 to 85 μg per day.

Iodine

Iodine is another nutrient whose concentration in the diet can vary dramatically from one locality to another. Unlike selenium, however, the variation is sufficient to result in iodine deficiency being widespread throughout the world. Some years ago the World Health Organisation estimated that at least 200 million people were affected. Iodine has only one role in the body, as a component of the pair of thyroid hormones thyroxine (11.3) and tri-iodothyronine (which differs from thyroxine only in lacking the arrowed iodine atom). The release of these two hormones enables the thyroid gland in the neck to regulate metabolic activity and to promote growth and development. Inadequate levels of the hormones are associated with a lowered metabolic rate and what might be regarded as the corresponding physical symptoms, listlessness, constipation, a slow pulse, and a tendency to feel cold. The visible manifestation of iodine deficiency is goitre, a massive enlargement of the thyroid gland as the body attempts to synthesise more thyroxine and tri-iodothyronine. More severe symptoms occur if iodine intake is very low during the critical stages of growth. The mental and physical abnorm-alities described collectively as cretinism are found in the children of mothers who have suffered severe iodine deficiency during or immedi-ately before pregnancy. Children with low iodine intake during their growing years suffer a related disorder known as myxoedema.

(11.3)

The incidence of goitre is closely related to low levels of iodine in the local soil and water supply and is most prevalent in mountainous regions, including the Alps, Pyrenees, Himalayas, parts of the Andes and the Rocky Mountains. In Britain, where goitre has been known as "Derbyshire neck", it is particularly associated with areas of the

Pennines. The flatter regions also potentially vulnerable to endemic goitre include the Great Lakes of North America and the Thames Valley in Britain.

The wide variations in the iodine content of foodstuffs (in which it occurs as the iodide) mean that lists of typical values are relatively meaningless. For the record, the 1995 UK Total Diet Study found average intake to be in the range 151–209 μg per day. What is certain is that most foods are poor sources of iodine. Fruit and vegetables, meat, freshwater fish, cereals and dairy products usually contain between 20 and 50 μg kg^{-1}. Seafood is the only consistently useful source, some popular fish species being recorded as containing several thousand μg kg^{-1}. Goitre is almost unknown in Japan, where seaweed, containing 4–5 mg of iodine per 100 g, is regularly consumed. Some countries have regions free from goitre adjoining others where it is endemic, and studies have clearly shown the influence of local food supplies. For example, a study of two such regions in Greece showed differences in iodine content across a wide range of foodstuffs ranging from 1.66 times higher for cow's milk to 7.05 times higher in eggs. Superimposed on these factors was the proximity to the sea of the low goitre region.

A number of different methods have been employed to supply iodine to populations in deficient areas. The only consistently successful technique has been to include it in domestic salt. A good example of what can be achieved is provided by data from Michigan, in the USA. Before iodised salt became available (at 0.02% w/w) in 1924, nearly 40% of Michigan schoolchildren showed signs of goitre. By 1928 the figure had fallen to 9% and by 1951 to 1.4%. Sadly, by 1970 incidence had risen to 6%, due to complacency on the part of the public and less intensive health education. That non-iodised salt should be on sale at all in such regions is hard to understand.

In Britain it is generally agreed that the average diet contains adequate iodine. One reason for this is the popularity of fish—and there is no evidence that fish fingers lose iodine during manufacture. The increasing degree of industrialisation of our food supply also means that we are much less dependent on the produce of our immediate neighbourhood than in past years. Nevertheless, endemic goitre is still a problem in some parts of Derbyshire, although not on the scale of the past.

Other Trace Minerals

There are a number of other essential trace elements, including boron, silicon, vanadium, chromium, manganese, cobalt, nickel and molybdenum. Our need for these has been inferred from their occurrence in various enzymes isolated from animal sources, rather than by identifying human diseases caused by their deficiency. For the foreseeable future food chemists are likely to have more important matters to occupy their attention.

FURTHER READING

Mineral Components in Foods, ed. P. Szefer and J. O. Nriagu, CRC Press, Boca Raton, Florida, 2007.
Reducing Salt in Foods: Practical Strategies, ed. D. Kilcast and F. Angus, CRC Press, Boca Raton, Florida, 2007.
L. R. McDowell, *Minerals in Animal and Human Nutrition*, Academic Press, San Diego, California, 1992.

RECENT REVIEWS

M. Zijp, O. Korver and L. B. Tijburg, *Effect of Tea and Other Dietary Factors on Iron Absorption; Crit. Rev. Food Sci. Nutr.*, 2000, **40**, p. 371.
E. C. Theil, *Iron, Ferritin, and Nutrition; Annu. Rev. Nutr.*, 2004, **24**, p. 327.
J. R. Hunt, *Bioavailability of Iron, Zinc, and Other Trace Minerals from Vegetarian Diets; Am. J. Clin. Nutr.*, 2003, **78**, p. 633.

CHAPTER 12

Water

The final chapter of this book is devoted to the most abundant and most frequently overlooked of all the components of food, water. Although we are quick to recognise our biological need for drinking water, we tend to overlook the presence of the vast quantities of water in our solid foods. Table 12.1 lists the water content of a range of foodstuffs and a few drinks, and it is not difficult to see from these data the contribution that solid foods can make to one's total water intake.

Table 12.1 Typical water content of food and drinks.

Food	Water (% w/w)	Food	Water (% w/w)
Lettuce, tomatoes	95	Dried fruit (currants, sultanas, *etc.*)	18
Cabbage, broccoli	92	Butter, margarine	16
Carrots, potatoes	90	Low fat spread	50
Citrus fruit	87	Wheat flour	12
Apples, cherries	85	Dried pasta	12
Raw poultry	72	Milk powder	4
Egg white (yolk)	88 (51)		
Raw lean meat	60	*Drink*	
Cheese	37	Beer	
White bread	35	Fruit juices	90
Salami	30	Milk	87
Jam (and other preserves)	28	Scotch whisky	87
Honey	20		60

Food: The Chemistry of its Components, Fifth edition
By T.P. Coultate
© T.P. Coultate, 2009
Published by the Royal Society of Chemistry, www.rsc.org

WATER STRUCTURE

The chemical structure of water, H_2O, seems at first sight very simple, but all is not as it seems. The first point is that water cannot be automatically regarded as a single pure compound. Present in all natural water are a number of variants caused by the natural occurrence of isotopes such as ^{17}O, ^{18}O, ^{2}H and ^{3}H. Fortunately for food chemists, their quantitative and qualitative contributions are small enough to be ignored. Furthermore, the ionisation of water to hydroxyl (OH^-) and hydrogen ions (as H_3O^+), while obviously important to the behaviour of other ionising molecules, does not move the composition of water very far from 100% H_2O, except at extremes of pH irrelevant to food materials.

The problems arise from the structure and behaviour of H_2O itself. Figure 12.1 shows the dimensions of the water molecule. If one visualises the oxygen atom being at the centre of a tetrahedron, the σ-bonds point towards two of its corners. The orbitals of the non-bonding electron pairs of the oxygen atom point towards the other two corners.

By comparison with similar molecules involving neighbouring atoms in the Periodic Table (e.g. H_2S), one would expect water to melt at $-90\,^{\circ}C$ and boil at $-80\,^{\circ}C$. As these are clearly not the figures observed,

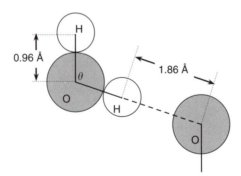

Figure 12.1   The dimensions of the water molecule and its hydrogen bonds ($1\,\text{Å} = 0.1\,\text{nm}$). The angle θ formed by the H–O–H σ–bonds depends on the state of the water:
In the gaseous state, $\theta = 104^{\circ}\ 52'$
In the liquid state, $\theta = 105^{\circ}\ 0'$
In the solid state (ice), $\theta = 109^{\circ}\ 47'$
In a perfect tetrahedron $\theta = 109^{\circ}\ 28'$. The other preferred angles are:
O–H———O: $\sim 180^{\circ}$ (*i.e.* a straight line)
H———O———O and H———O–H: $\sim 109^{\circ}$.

we must conclude that in the liquid and solid states water molecules in some way associate together so that their effective molecular weight becomes much higher. The large forces of association between the water molecules can be traced to the polarised nature of the O–H bond (40% partial ionic character) and the V-shape of the molecule. This facilitates hydrogen bonding between the oxygen atom, which carries a small negative charge, and the corresponding positively charged hydrogen atoms of two other water molecules. It is this ability of water molecules to form links with four others which allows the formation of three-dimensional structures. An extended three-dimensional lattice occurs in ice, as shown in Figure 12.2. Layers of oxygen atoms are arranged in an array of chair structures resembling those of pyranose sugars. The links between the layers create rings in the boat conformation. There is continual oscillation of the hydrogen atoms between their two possible

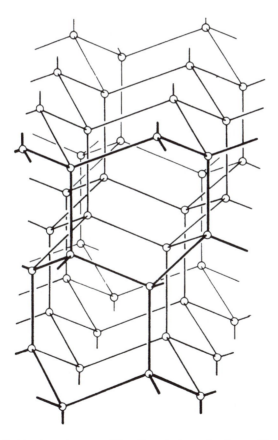

Figure 12.2 The lattice structure of ice. For clarity only the positions of the oxygen atoms at the junctions of the tetrahedrally arranged bonds are shown.

positions, bound covalently to neighbouring oxygen atoms:

$$O–H---O \leftrightarrow O---H–O$$

The structure of ice is by no means static. Even at the lowest temperatures normally encountered in food preservation (around $-20\,°C$) the hydrogen atoms oscillate as shown above and a considerable fraction of the water molecules will not be fully integrated into the lattice structure. There is plenty of evidence to show that water molecules can diffuse slowly through ice. The OH^- and H_3O^+ ions will also be joined by entrapped solute molecules to disrupt the lattice. As the temperature is raised, the proportion of the hydroxyl groups not involved in hydrogen bonding slowly rises. When liquid at $0\,°C$, more than 90% of the hydroxyl groups in water are still involved in hydrogen bonding. This figure falls only a little below 80% when the normal boiling point is reached. There is considerable evidence to suggest that the hydroxyl groups not involved in hydrogen bonding are not evenly, or randomly, distributed. Water appears instead to consist of irregular clusters or aggregates of molecules arranged in a lattice structure, although not necessarily identical to that in ice. The surface of these clusters is marked by the orientation defects caused by non-hydrogen bonded hydroxyl groups. At room temperature the average number of molecules in a cluster is estimated to be between 200 and 300.

The structure of the liquid water clusters has attracted considerable attention in recent years. Although attractive, the idea that the clusters have the same lattice structure as ice seems unlikely. One proposal, made by the author's colleague, Professor Martin Chaplin (Chaplin, M. F., *Biophys. Chem.,* 1999, **83**, p. 211), is that the clusters contain water molecules arranged in rings, as in ice (*see* Figure 12.2). However, whereas the lattice in ice consists only of hexamers in chair and boat conformations, in liquid water the cluster also contains a substantial proportion of pentamers of water molecules. (The conformations of pyranose and furanose monosaccharides provide a useful model, *see* Chapter 2.) The H–O–H angle in a five-membered ring is 108°, and is closer to the supposed ideal (in the gaseous state) of 104° 52′ than that in ice, which is 109° 47′. In their idealised almost spherical state, Chaplin's clusters contain 280 water molecules with 70 pentamers and 116 hexamers (80 chairs and 36 boats). Figure 12.3 shows 14 water molecules linked together, forming in the example shown two planar pentamers and three hexamers in boat conformation.

It should never be overlooked that the lattice structure of liquid water is essentially dynamic. The lifetime of an individual hydrogen bond is

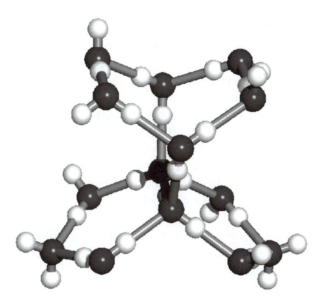

Figure 12.3   Clusters of hydrogen bonded water molecules containing two pentamers, and three hexamers in a boat conformation. Oxygen and hydrogen atoms are represented by black and white spheres, respectively. No attempt has been made to discriminate between covalent and hydrogen bonds, and the location of the hydrogen atoms closer to one or the other of the two oxygen atoms is arbitrary. (Based on a computer-generated drawing kindly provided by Professor Chaplin.)

measured in picoseconds, as hydrogen atoms oscillate from one oxygen atom to another. Studies of the "dielectric relaxation time" suggest that a molecule of liquid water at $0\,°C$ changes its orientation with respect to its neighbours about 10^{11} times per second. In spite of these apparently brief periods of continuous stability for individual molecules the lifetimes of individual clusters are much longer, up to 10^{-8} seconds. This is the result of cooperation between nearby hydrogen bonds in maintaining the lattice structure. Although the bonds holding one molecule in position may be broken, the structure as a whole is maintained by the bonding between surrounding molecules.

Although the details of the lattice structure of water is unresolved, there is general agreement that in the liquid state two types of lattice are always present. One, typified by that described above, is referred to as *expanded*, being very open and offering the possibility of non-hydrogen bonded water molecules penetrating into its pores. The other is a *collapsed* structure in which the rigid geometry imposed by the hydrogen bonds is to some extent lost when they weaken at higher temperature, and other forces draw the water molecules closer together. The hexamer and pentamer rings then become distorted, with values of θ closer to 90°.

As the temperature rises, the proportion of water in the collapsed state also rises. The coordination number, *i.e.* the average number of neighbours nearest to individual molecules, also rises, from 4.0 in ice to 4.4 in water at 1.5 °C, and 4.9 at 83 °C. A water molecule having more than the four nearest neighbours expected from the formal lattice structure would still not be expected to be hydrogen bonded to more than four others. As would be expected, temperature also affects the average distance between neighbouring water molecules. In ice at 0 °C, and in water at 1.5 °C and 83 °C, the $O \leftrightarrow O$ distances are 2.76 Å, 2.9 Å, and 3.05 Å, respectively.

There are several consequences of this behaviour which influence the properties of food materials, but one of the most obvious is the change in density. The difference in density between ice at 0 °C, 0.9168 g cm^{-3}, and liquid water at the same temperature, 0.9998 g cm^{-3}, is readily accounted for by the more compact packing shown by the rise in coordination number. Up to 3.98 °C the increase in coordination number is the dominating influence on the density, but above this temperature the nearest neighbour distance takes over. These data correctly predict that water will increase in volume by 9% as it freezes. This density change is of course of profound significance to the evolution of life in aquatic environments. However, food scientists have to be content with its more mundane influences, for example in the design of moulds for ice lollies. The expansion taking place on freezing is also partially responsible for the destructive effect caused by freezing and thawing on the structure of soft fruit such as strawberries.

The interrelationships between water in its liquid, solid and gaseous states are summarised in the phase diagram shown in Figure 12.4. The diagram shows that if the pressure is low enough ice can sublime directly into the vapour state. This is the basis of the technique of food dehydration known as freeze drying. Delicate food extracts, as for example in the manufacture of instant coffee, are first frozen and then subjected to high vacuum. The latent heat of sublimation of ice is 2813 J g^{-1}. This high figure means that when the vacuum is first drawn on the frozen material, sublimation of water vapour abstracts so much heat that the temperature falls, to the point at which water is no longer removed at a satisfactory rate. In the modified commercial process, known as accelerated freeze drying, this is counteracted by applying radiant heat to the frozen material until almost all the water has been removed. In spite of the heating, the temperature remains well below 0 °C throughout the process.

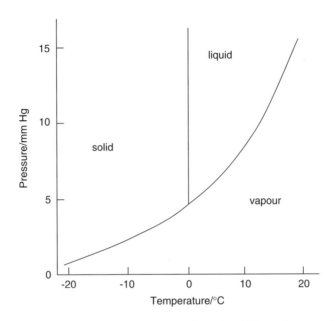

Figure 12.4 The phase diagram for water. At least five different ice crystal structures have been found to occur at particular combinations of pressure and temperature, but since the pressure required for these are very high they are irrelevant to food studies and have been ignored.

INTERACTIONS OF WATER WITH FOOD COMPONENTS

The interactions between water and other food components lie at the root of many problems in food systems. Interactions at the molecular level will first be considered, with particular attention to carbohydrates, lipids and proteins.

Some features of the interaction of sugars with water were discussed in Chapter 2, where it was pointed out that in aqueous solution some water molecules are tightly bound to sugar molecules. The value of 3.7 for the number of water molecules bound by glucose is illustrated in the structural diagram given in Figure 12.5. The distance between pairs of hydroxyl oxygen atoms on the same face of the pyranose ring of β-D-glucose (*i.e.* those on carbons 1 and 3 and those on carbons 2 and 4) is 4.86 Å. This is remarkably close to the distance of 4.9 Å. separating the oxygen atoms in the same plane of the water lattice. Not only are the dimensions therefore correct, but for equatorially disposed hydroxyl groups the geometry is also correct. The figure of 3.7 is close to the average number of equatorial hydroxyl groups of D-glucose in aqueous solution (4 on the more abundant β-anomer and 3 on the α-anomer).

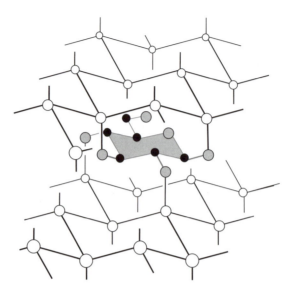

Figure 12.5   The insertion of a β-D-glucopyranose molecule into a water lattice. For clarity only the oxygen and carbon atoms are shown (open circles: water oxygen; grey circles: glucose oxygen; and black circles: glucose carbon). Comparison with Figure 12.2 shows that, at least in an ice-type lattice, the insertion of the glucose molecule has not disrupted the relationships of water in the layers above and below. The possibility of water molecules binding to the oxygen atoms in the ring and the –CH$_2$OH groups of the glucose molecule remains unconfirmed.

It has been suggested that the tendency of glucose to retard the formation of ice crystals in frozen dessert products is a result of this close fit with liquid water, but not with ice. In ice at ordinary pressures neighbouring oxygen atoms are 4.5 Å. apart. The anti-freeze properties of substances such as glycerol, CH$_2$OH.CHOH.CH$_2$OH, and ethylene glycol, (CH$_2$OH)$_2$, which have α-glycol oxygen atoms capable of similar orientation, may have a similar basis.

In solution isolated glucose molecules will clearly have a strong stabilising effect on the clusters of water molecules around them, and it is tempting to extend this concept to the properties of polysaccharide-based systems. The discussions of gel structure in Chapter 3 should have left the reader in no doubt as to the underlying strength of the polysaccharide–water interaction. For example, a good quality agar for microbiological use will give a firm gel at a polysaccharide concentration of 1.5% w/v. The average molecular weight of the monosaccharide units of agarose (the gel-forming and major component of agar, *see* Figure 3.9) is about 150. This implies that each monosaccharide unit is responsible for the immobilisation of at least 550 water molecules! The lifetime of an individual water cluster is far too brief to envisage the stability of the gel being the result of water clusters becoming ensnared in the polymer network. Capillary attraction is a far more plausible suggestion, with the gel adopting the character of a wet sponge. As was suggested in Chapter 3 (*see* page 62), it will not be difficult to modify our perception of this model of gel structure to account for the behaviour of gums as well as the less soluble fibre polysaccharides.

It comes as little surprise that the interaction of water with non-polar molecules is a more aloof affair. It seems axiomatic that water and oil do not mix, but what is not so obvious is why this should be so. What is the advantage, in free energy terms, for two small oil droplets suspended in water to coalesce and become one larger one?

The answer lies in the collective strength of the hydrogen bonds of the water. If a single small non-polar molecule is suspended in water, it will have to occupy a cavity in a mass of associating water molecules. To form this cavity numerous hydrogen bonds will need to be broken, requiring a large amount of free energy. The free energy of the system is then partially reduced (by compensating entropy changes) as the water molecules reorganise around the cavity to maximise the number of hydrogen bonds between them. There is evidence that this reorganisation can involve water structures based on strained five-membered rings. At room temperature the transfer of a small hydrocarbon molecule from an aqueous solution to a non-polar solvent is accompanied by a change in free energy, ΔG^0, between -10 and $-15\,\mathrm{kJ\,mol^{-1}}$. The much larger molecules of typical food lipids can be expected to require correspondingly larger free energy changes.

When an isolated non-polar solute molecule is considered, it is clear that the larger the surface area of the solute molecule the less soluble it will be in water; in other words, larger non-polar molecules tend to be more hydrophobic. However, when two non-polar molecules come to share the same cavity in the water, the total disruption of hydrogen bonding will be less than when they were apart, and the free energy of the system will be lower. This principle readily extends to the coalescence of oil droplets suspended in water.

The interaction of proteins with water has received a great deal of attention from biochemists for many years. However, this is mostly because of the importance of water to the structure and behaviour of proteins (*in vitro* as well as *in vivo*) rather than the influence of proteins on the behaviour of water. A protein molecule presents three kinds of surface group to its aqueous environment depending on the nature of the amino acid side-chain: polar charged groups such as those of glutamic acid ($-CH_2CH_2COO^-$) or lysine [$-CH_2)_4NH_3^+$], polar neutral groups such as those of serine ($-CH_2OH$) or glutamic acid at low pH ($-CH_2CH_2COOH$), and non-polar groups such as those of valine [$-CH(CH_3)_2$] or methionine ($-CH_2CH_2SCH_3$). For the reasons discussed above, the non-polar side-chains are usually found buried in the interior of the protein and need not concern us further. The pH value will obviously be a major influence on the properties of ionisable side-chains.

Studies using sophisticated physical methods have probed the water-binding properties of amino acid side-chains and have obtained results comparable with those for sugars. When ionised, the acidic side-chains, aspartate and glutamate, bind with about six molecules of water; on protonation this figure drops to two, a similar figure to that of the polar but un-ionised side-chains of asparagine, serine, *etc.* The positively charged side-chains of lysine and histidine each bind about four molecules of water. These figures refer to the situation when the polypeptide chain is surrounded by an abundance of water. The high values for the charged side-chains probably result from the tendency of these to orientate multiple layers of polarised water molecules. The effect of pH on the water-binding capacity of amino acids is reflected in the behaviour of entire proteins, a phenomenon referred to in Chapter 5 in connection with the loss of water from fresh meat. A homogenised muscle preparation may double its water-binding capacity if the pH is lowered from 5.0 (approximately the mean isoelectric point of the muscle proteins) to 3.5.

As a general rule, only a small proportion of the polar groups of a native (*i.e.* undenatured) protein will be unable to bind water due to their being buried in the largely hydrophobic interior of the molecule. Thermal denaturation is assumed to lead to unfolding of the polypeptide chain, which can cause extra water-binding as the last few polar groups are exposed. The newly exposed hydrophobic regions of the polypeptide chain also show limited water-binding capacity, partly due to the hydrophilic nature of the peptide bond regions of the backbone. Thus many proteins show a modest increase in water-binding on denaturation, from 30% to around 45% w/w. However, when denaturation leads to aggregation, as unfolded polypeptide chains become entangled around each other, there is sometimes a tendency for protein–water interactions to be usurped by protein–protein interactions, leading to an actual loss of water-binding capacity. The release of bound water which accompanies the roasting of meat is a good example of this.

The behaviour of protein solutions is markedly affected by the presence of ions of low molecular weight, notably the anions and cations of inorganic salts. Up to a concentration of about $1 \, mol \, dm^{-3}$ salts such as sodium or potassium chloride enhance the solubility of many proteins. Indeed, the old-fashioned division of proteins into albumins and globulins was based on the ability of the former to dissolve in pure water, whereas the latter would only dissolve in a dilute salt solution. The mechanism of this "salting-in" effect is straightforward. Charged groups on the surface of the protein attract and bind anions and cations more strongly than they do water. However, these ions will still bring with them an ordered cluster of their own solvating water molecules, which will then function to maintain the protein molecule in solution.

The solubility enhancing effect of increasing salt concentrations does not continue indefinitely. At concentrations above 1 or $2\,mol\,dm^{-3}$ many salts cause proteins to precipitate out of solution, a phenomenon known as "salting-out". For biochemists this is the basis of the time-honoured procedure for the fractionation of crude protein mixtures known as ammonium sulfate precipitation. The use of ammonium sulfate stems from its extreme solubility ($1\,dm^3$ of water can dissolve 760 g!) rather than from any special aspect of its interaction with proteins. The explanation is simple. As these salts reach high concentrations, the proportion of the total water that they bind starts to become more significant. The ions of the salt compete for the water necessary for the protein to remain in solution, and as the salt concentration is raised the various proteins precipitate out.

Salting-out occurs only at salt concentrations of little relevance to food systems, but salting-in is a very important phenomenon in food processing. When pork is cured by the Wiltshire method (see Chapter 5), sodium and chloride ions are injected as brine into the muscle tissues. These bind to the muscle proteins, particularly the abundant actin and myosin of the myofibrils, and enhance their water-binding capacity. This gives the ham its attractive moistness, as well as incidentally increasing its weight. Many modern ham-type products, prepared for off-the-bone machine slicing for pre-packaged sale, are subjected to the additional process of "tumbling", in which large pieces of ham are literally massaged for periods of up to half an hour in rotating vessels resembling a concrete mixer. This has the effect of bringing dissolved myofibrillar proteins out of the cell structure so that they are able to give the moulded or pressed finished product a firmer texture, based on gels formed by these proteins.

In products such as luncheon meat and sausages, which include a major proportion of use homogenised or macerated muscle tissue, the solubilisation of proteins by inorganic salts acquires a special importance. While water binding in these products is not unimportant, the binding of fat is of paramount significance. The ability of the hydrophobic regions of proteins to bind fat molecules makes proteins excellent emulsifying agents. Getting myofibrillar proteins into solution and into action as emulsifiers is the task of added "emulsifying salts". In addition to the ubiquitous sodium chloride, we find sodium and potassium phosphates and polyphosphates, citrates and tartrates in this role; the phosphates and polyphosphates were considered in greater detail in the previous chapter. We also find emulsifying salts used in processed cheese, where they release casein molecules from their micelles to assist in binding the added fat.

INTERACTIONS OF WATER WITH FOOD MATERIALS

When we come to consider the interaction between water and food materials we must leave molecular niceties behind and look at the behaviour of water on the large scale. Food scientists need to understand how water behaves, since so many of the important properties of food are related to its presence, in particular the susceptibility of food to microbial growth. It is clear from data such as those in Table 12.1 that the water content of a foodstuff is not the major influence on stability. What is more important is the availability of the water to micro-organisms, rather than its abundance. The concept of "water activity" is nowadays universally adopted by food scientists to quantify availability.

Water activity, a_W, is defined as follows:

$$a_W = p/p_0$$

where p is the partial pressure of water vapour above the sample (which can be solid or liquid) and p_0 is the partial pressure of water vapour above pure water at the same specified temperature. At equilibrium, which may take hours or even days to be achieved, there is a close relationship between the a_W of a food material and the equilibrium relative humidity (ERH) of air above it:

$$a_W = ERH/100$$

This relationship enables us to predict which foods will gain or lose water when exposed to air at a particular relative humidity.

While the concept of water activity is sometimes regarded as lacking the full rigour of classical physical chemistry, it can provide a practical basis for describing the water relations of foodstuffs. Table 12.2 gives typical examples of water activity for a range of foods. An important

Table 12.2 The water activity of various foodstuffs.

Foodstuff	Typical water activity (a_W)
Fresh meat	0.98
Cheese	0.97
Preserves	0.88
Salami	0.83
Dried fruit	0.76
Honey	0.75
Dried pasta	0.50

feature of the data is the lack of correlation between water content and water activity, seen by comparison with Table 12.1.

The data in Table 12.3 illustrate the importance of water activity in controlling microbial growth. The halophilic (*i.e.* salt-loving) bacteria are those normally associated with marine environments, including the fish obtained from them. The xerophilic moulds and osmophilic yeasts are particularly adapted to environments of low water activity. Their cells contain high concentrations of particular solutes, which means that they do not lose water by osmosis to their surroundings. In bacteria these "protective solutes" are usually the amino acids proline or glutamic acid, and in fungi they are polyols such as glycerol or ribitol. These fungi and yeasts can pose serious problems in the storage of some foods which are normally regarded as stable, notably dried fruit. Fortunately they have not so far been associated with mycotoxin production, only the more obvious physical deterioration.

Microbial growth is not the only important phenomenon affected by water activity. The rates of both enzymic and non-enzymic reactions are also markedly affected. However, to gain a clearer understanding of the mechanisms of these influences we need to examine the issue of water binding.

Table 12.3 Minimum water activity for the growth of micro-organisms.

Micro-organism	Minimum water activity (a_W)
Normal bacteria	0.91
Normal yeasts	0.88
Normal moulds	0.80
Halophilic bacteria	0.75
Xerophilic moulds	0.65
Osmophilic yeasts	0.60

WATER BINDING

The term "bound water" is frequently used in discussions of the properties of food materials, but unhappily it rarely means exactly the same thing to any two different scientists. Linked to this is another term, "water-holding capacity", which refers with equal ambiguity to the maximum quantity of bound water a material is able to contain. Fennema has described a sequence of four categories for the water present in a solid food material to cover the range of possible definitions of bound water. The percentage figures given for each category indicate its relative contribution to the total water in a typical high-moisture food, *i.e.* one with a water content of about 90%.

- Vicinal water ($0.5 \pm 0.4\%$). This is water held at specific hydrophilic sites on molecules in the material, either by hydrogen bonding or by the attraction between charged groups and the water dipole. Vicinal water may give an uninterrupted monolayer surrounding a hydrophilic molecule. Vicinal water is regarded as unfreezable down to at least $-40\,°C$, and to be unable to act as a solvent.

- Multilayer water ($3 \pm 2\%$). This is the water forming additional layers around the hydrophilic groups in the material. Water–solute hydrogen bonds can occur, but water–water bonds predominate. Most multilayer water does not freeze at $-40\,°C$, and the remainder still has a very low freezing point. Multilayer water does have limited solvent capacity.

- Entrapped water (up to $\sim 96\%$). This water has properties similar to that in dilute salt solutions, with only a slight reduction in freezing point and virtually normal solvent capability. However, it is prevented from flowing freely by the matrix of gel structures, or due to capillary attraction in tissues.

- Free water (up to $\sim 96\%$). This has properties essentially similar to pure water. While it is held within the structure of a food material as a consequence of capillary action, or trapped in layers of fatty material, it does not require much pressure to squeeze it out.

Starting with a completely dry food material, one can visualise the gradual addition of water filling each of these categories in turn. The layer of vicinal water is considered to be complete at an a_W value of approximately 0.25. The boundary between multilayer water and either entrapped or free water lies at an a_W of approximately 0.8. This sequence of events is normally expressed in the form of a sorption

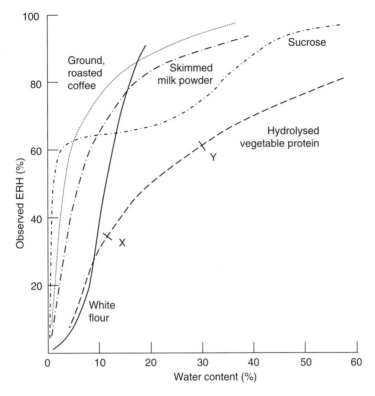

Figure 12.6 Sorption isotherms. These are representative curves at room temperature for the food materials shown. The points marked X and Y on the hydrolysed vegetable protein isotherm represent the transition, firstly from a free-flowing powder, to an increasingly sticky mass at X, and then to a viscous solution at Y.

isotherm, a plot of water content against water activity. A number of plots are shown in Figure 12.6.

It is tempting to offer speculative explanations for the differences between the shapes of these curves in terms of the molecular properties of the different materials present. However, the reader's guesses will probably be no better than the author's, so none will be offered. The value of plots like these lies more in their practical application rather than in their theoretical interpretation. Whenever two or more ingredients which start with different a_W values are confined together in an airtight space, water will move from the high a_W ingredient to the lower until they become equal. Their biological, chemical and physical properties are thus liable to change in a manner which was not to be expected unless their respective sorption isotherms had been studied beforehand. An obvious problem area is dehydrated instant meals, which may have pieces of meat, vegetables, *etc.* inside the same compartment. To make matters even more complicated, the sorption isotherm is by definition recorded at one particular temperature. Low temperature storage of a product could well change all the relationships which had been carefully established at the temperature of the production line.

Another complication is that the isotherm obtained when a material is dried (the desorption isotherm) may be different from that produced when it regains water (the resorption isotherm). As a general rule the a_W for a given water content will be higher when a material gains water than as it loses it. There are numerous explanations for this, including the time dependency of the process, capillary collapse and swelling phenomena, and all may eventually be found to be involved to some extent.

The growth of micro-organisms is by no means the only "process" affected by the water activity of a foodstuff: A number of important enzyme-catalysed reactions are also affected. For example, the rate of enzymic hydrolysis of lipids in foods such as meat becomes negligible at a_W values below 0.4. Chemical reactions can show other effects. The rate of the Maillard reaction (*see* Chapter 2) rises sharply as the water activity falls to ~ 0.8, and only decreases again below ~ 0.3. The usual explanation for this is that enhancement of the reaction rate initially occurs when removal of water concentrates the reactants. The subsequent fall is the result of loss of reactant mobility in the residual multilayer water. The loss of reactant mobility appears to be the dominant influence on reaction rates at low water activity. An interesting exception to the general trend is the autoxidation of unsaturated lipids in food materials (*see* Chapter 4). Although this process shows a steady decline in rate down to $a_W \sim 0.4$, it then rises again at a_W values approaching zero to within about 10% of its maximum rate. It is assumed that the removal of vicinal water from hydrophilic sites in the material causes the exposure of vulnerable lipid molecules.

Knowledge of the sorption isotherm of a material thus enables food processors to identify the water content at which the effect of degradative reactions is minimised, usually the point at which multilayer water has been removed but the monolayer of vicinal water is still intact.

WATER DETERMINATION

It will now be evident that there are two distinct approaches to the determination of water in food materials. The first is to determine the amount of water the material actually contains. The second approach is to determine the activity of the water in the material. As we have seen, this gives useful information about the properties and behaviour of the material. In many cases the water content is of course sufficient to predict the behaviour of the material. For example, we knew how far the water content of cereal grains should be reduced to prevent mould growth or sprouting long before the concept of water activity was introduced.

There are three basic methods for the determination of overall water content of a food material: gravimetric, volumetric and chemical. Gravimetric methods simply depend on the difference in weight of a sample before and after drying. The actual drying method must be chosen with care. High temperatures will obviously hasten water loss but at the risk of causing decomposition of some food components such as sugars; other volatiles such as short-chain fatty acids may also be driven off. Before drying it is often necessary to reduce the particle size of the material. Cereal grains need to be milled and wet materials such as jam must be dispersed on to a particulate matrix such as sand. Without these steps the drying period can be inordinately long. Whichever drying method is adopted it is essential that it is carried out to constant weight, *i.e.* until the sample shows no further weight loss. For many purposes, especially where accounting for the final 1% of the moisture content is not critical, an ordinary laboratory fan oven is sufficient; for precise work a vacuum oven is preferable. Using temperatures around 100 °C and pressures below 50 mm of mercury, the drying time for cereals can be brought to within 16 hours. It must never be forgotten that, having been dried at elevated temperatures, the sample must be cooled in a dry environment such as a desiccator before being weighed.

Volumetric methods depend on refluxing with a water-immiscible solvent such as toluene, followed by distilling off the solvent and estimating the amount of water present. The modified Dean and Stark apparatus shown in Figure 12.7 employs this principle.

The only chemical method for water determination in regular use is the Karl Fischer titration. This can be applied to most foodstuffs, apart from those with heterogeneous structures such as fruit, vegetables and meat, which cannot be dispersed in the solvent. It is especially effective with heat-sensitive materials, those rich in volatiles, and sugary materials in which the water is particularly tightly bound. The method depends on the reaction between sulfur dioxide and iodine in the presence of water, pyridine and methanol:

$$SO_2 + I_2 + 2H_2O \rightarrow H_2SO_4 + 2HI$$

Although not shown in the equation, the pyridine appears to form loose compounds with the actual reactants, sulfur dioxide and iodine.

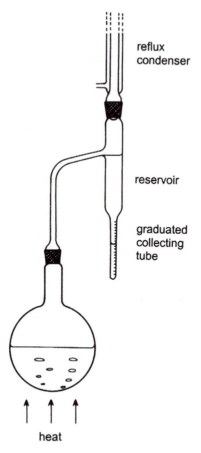

reflux
condenser

reservoir

graduated
collecting
tube

heat

Figure 12.7   The modified Dean and Stark apparatus. As distillation proceeds the
water collects in the graduated tube while the toluene, being less dense,
overflows back into the flask containing the sample.

All the reagents must be carefully prepared to be totally free from water,
and the titration apparatus is designed so that water vapour from the
atmosphere is excluded. The food sample is dispersed or dissolved in a
combination of methanol, sulfur dioxide and pyridine and titrated with
standardised iodine in methanol. The end-point (the appearance of
excess iodine) can be detected either visually or using an electrometric
method.

  Although the range of materials is limited, there are two additional
instrumental methods of water determination which deserve a mention.
One of these is near-infrared (NIR) spectrophotometry, mentioned in
Chapter 5 in connection with protein determination. Water has a
number of absorbance bands in the near infrared, of which the one at

1.93 μm is most often used. Comparison of the absorbance at this wavelength (or at 1.94 μm where interference is minimal) with that at 2.08 μm (outside the major absorbance bands) provides a rapid and sensitive estimate of the water content of cereal grains, flour and similar materials. Low-resolution nuclear magnetic resonance (NMR) can also be used for water determination. Although the instrumentation is expensive and the theory intimidating, NMR is proving increasingly popular within the cereal industry, since it is both rapid (1 min in total) and non-destructive. Even whole grains can be examined.

The determination of the water activity of food materials is less easy. All the methods available depend on the relationship between water activity and equilibrium relative humidity (ERH), and measure the tendency of the test material to gain or lose water when exposed to air at a specific temperature and relative humidity. The usual technique is to place the sample in a small open vessel alongside another in a sealed container containing a solution of known a_W, the two vessels being kept at constant temperature for a few days. At intervals the container is opened and the weight of the vessels compared, to determine whether the sample is gaining or losing moisture. If a range of different standard solutions is available, it is possible to arrive at the a_W value of the sample by interpolation. Table 12.4 shows the a_W values of a number of salt solutions. Other values may be obtained using standard solutions of concentrated sulfuric acid or glycerol, but of course the water activity of these changes as they absorb water from the sample under test.

The instruments marketed for the determination of a_W use a variation of this approach. The sample is allowed to equilibrate with a small known volume of air whose relative humidity can be monitored by a special sensor built into the sample container. One common type of sensor contains a tiny crystal of lithium chloride. This rapidly gains or loses water of crystallisation as the relative humidity changes and shows a corresponding change in electrical conductivity. Although it may take several hours for even a small sample to reach equilibrium, modern instruments record several readings over the first few minutes of

Table 12.4 Water activity of saturated salt solutions as 25 °C.

Salt	a_W	Salt	a_W
Potassium dichromate	0.980	Potassium chloride	0.843
Potassium nitrate	0.925	Ammonium sulfate	0.790
Barium chloride	0.902	Sodium chloride	0.753
Sodium benzoate	0.880	Ammonium nitrate	0.618

equilibration and use computerised mathematical modelling to extrapolate, with considerable accuracy, the final result at equilibrium.

FURTHER READING

F. Franks, *Water, a Matrix of Life*, Royal Society of Chemistry, London, 2000.

Properties of Water in Relation to Food Quality and Stability, ed. L. B. Rockland and G. F. Stewart, Academic Press, London, 1981.

Water Activity in Foods: Fundamentals and Applications, ed. G. V. Barbosa-Canovas, A. J. Fontana, S. J. Schmidt and T. P. Labuza, Blackwell, Oxford, 2007.

Water and Food Quality, ed. T. M. Hardman, Elsevier Applied Science, London, 1989.

RECENT REVIEWS

J. L. Finney, *Water? What's so special about it?; Philos. Trans. R. Soc. London, Ser. B*, 2004, **359**, p. 1145.

M. J. Blandamer, J. B. F. N. Engberts, P. T. Gleeson and J. C. R. Reis, *Activity of Water in Aqueous Systems; A Frequently Neglected Property; Chem. Soc. Rev.*, 2005, **34**, p. 440.

M. Mathlouthi, *Water Content, Water Activity, Water Structure and the Stability of Foodstuffs; Food Control*, 2001, **12**, p. 409.

H.-D. Isengard, *Rapid Water Determination in Foodstuffs; Trends Food Sci. Technol.*, 1995, **6**, p. 155.

L. I. Leake, *Water Activity and Food Quality; Food Technol.*, 2006, **60**, p. 62.

Appendix

NUTRITIONAL REQUIREMENTS AND DIETARY SOURCES

This book does not set out to be a textbook of nutrition. Extensive consideration of the amounts of different food components as nutrients required in a healthy diet is beyond its scope. Nevertheless it is obviously valuable to be able to place the quantity, stability and availability of the nutrients discussed in this book within the context of what nutritionists would regard as desirable. Such data were at one time presented as Recommended Daily Amounts (RDAs). Unfortunately RDA values were frequently misused, and interpreted as absolute values to be applied without qualification in decisions about an individual's diet. In 1991 the UK Government Committee on Medical Aspects of Food (COMA) introduced Dietary Reference Values (DRVs). For any given food ingredient (*e.g.* a particular vitamin), and within a defined section of the population (*e.g.* males aged 11–14 years), there are normally three DRVs:

  (i) **The Estimated Average Requirement (EAR)**. Statistically, roughly half the population will require less than the EAR and the other half more.[i]
 (ii) **The Reference Nutrient Intake (RNI)**. This is the nutrient intake adequate for almost all the population. This is not dissimilar from the old RDA and is arbitrarily set at two standard deviations

---

[i] Readers familiar with statistics will appreciate that the average (or mean) is not the same as the median, which would precisely divide the population into two numerically equal halves.

Food: The Chemistry of its Components, Fifth edition
By T.P. Coultate
© T.P. Coultate, 2009
Published by the Royal Society of Chemistry, www.rsc.org

above the EAR. This assumes that the spread of actual requirements between individual members of the population follows
normal distribution, and implies that "almost all" means 97.5%.

(iii) **The Lower Reference Nutrient Intake (LRNI).** This is the nutrient
intake which is sufficient for only a small proportion of the
population. This is arbitrarily set at two standard deviations
below the EAR, in which case "a small proportion" means 2.5%.

The DRVs for most vitamins and minerals are presented in Table A1.
An important feature is that DRVs for three of the vitamins are quoted
in terms of the intake of other nutrients. This is a reflection of the close
involvement of these vitamins in particular aspects of metabolism—
thiamin and niacin with energy metabolism, and pyridoxine with protein
metabolism. Complete DRVs have not been established for some vitamins and minerals, since the absence of deficiency symptoms leads to the
conclusion that even poor diets are able to supply adequate quantities.

Table A1 Dietary Reference Values (DRVs) for daily intake of minerals and
vitamins. The values shown refer to males aged 11–14, and are
intended merely to illustrate the principles involved.

	LRNI	EAR	RNI	Units
Thiamin (per 1000 kcal)	230	300	400	µg
—or assuming the EAR for energy of 2220 kcal per day is met:	–	–	900	µg
Riboflavin	0.8	1.0	1.2	
Pyridoxine (per g protein)	11	13	15	µg
—or assuming the EAR for high quality protein of 33.8 g per day is met:	–	–	51	µg
Niacin (per 1000 kcal)	4.4	5.5	6.6	mg
—or assuming the EAR for energy of 2220 kcal per day is met:	–	–	15	mg
Folic acid	100	150	200	µg
Cobalamin	0.8	1.0	1.2	µg
Ascorbic acid	9	22	35	mg
Vitamin A (retinol equivalents)	250	400	600	µg
Calcium	0.45	0.75	1	g
Phosphorus (intake should match calcium on an equimolar basis)	350	580	770	g
Magnesium	180	230	280	mg
Sodium (no EAR set)	0.46	–	1.6	g
Potassium (no EAR set)	1.6	–	3.1	g
Zinc	5.3	7.0	9.0	mg
Iron	6.1	8.7	11.3	mg
Iodine (no EAR set)	65	–	130	µg
Copper (no LRNI or EAR set)	–	–	800	µg
Selenium (no EAR set)	25	–	45	µg

Another problem is that for some nutrients (particularly iron) the statistical distribution of requirements is not "normal" in the statistical sense, but skewed, which makes these simple statistical concepts invalid.

In the case of pantothenic acid the normal intake rate of 3–7 mg per day is assumed reasonably to be both safe and adequate. A similar conclusion is drawn for a biotin intake of 10–200 µg per day. Most individuals synthesise their own vitamin D, as discussed in Chapter 8, but for individuals confined indoors the RNI is set at 10 µg per day. The

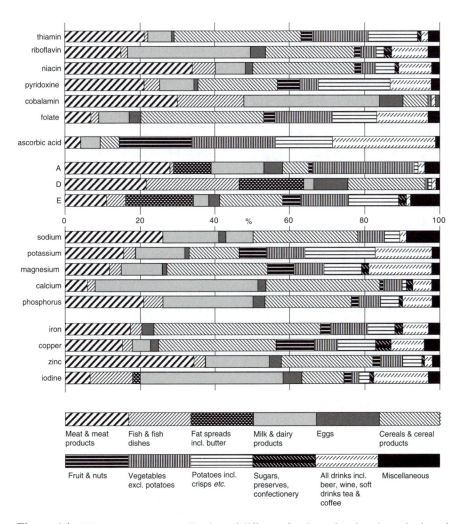

Figure A1 The percentage contribution of different foods to the vitamin and mineral content of the British diet in 2000–01. The data are the average of a representative sample of 1724 UK adults aged 19–64, omitting women who were pregnant or breastfeeding.

special needs of infants for vitamin D are reflected in RNI values of 8.5 μg per day for the first 6 months and 7 μg per day for the following 18 months. The requirement for vitamin E is highly dependent on the polyunsaturated fatty acid (PUFA) content of the diet, but the indications are that at a reasonable PUFA intake 3–4 mg per day is adequate. New-born infants are particularly prone to vitamin K deficiency and in this case an RNI of 10 μg per day is suggested. For adults, a figure related to body weight (1 μg kg^{-1} per day) is recommended as adequate and safe.

Although the chapters on vitamins and minerals give information on the amounts of these nutrients in specific foods, to appreciate such data fully one needs to understand the contribution of different foods to the average diet. The survey designed to provide this information conducted by the UK Ministry of Agriculture, Fisheries and Food (MAFF) in 1986–87 was superseded by the National Diet and Nutrition Survey carried out for the Food Standards Agency and published in 2003. It includes data on the influence of region, household income and other factors on diet and health. Figure A1 summarises data from the survey on the contribution of various types of foodstuff to the supply of vitamins and minerals. Readers are recommended to consult the original data for further information and in order to understand the changes since the earlier survey.

In addition to data of the type shown the survey records details of the contribution of particular foods to the supply of specific nutrients, usually in relation to particular sections of the population. For example, cooked carrots supply 30% of total carotene, but adult males obtain 26% of their vitamin A from liver. Beer and lager make a significant contribution to the vitamin B intake of adult males (*e.g.* riboflavin 7%, niacin 9%, pyridoxine 13% and folic acid 11%), but the corresponding figures for females are 3% or less.

FURTHER READING

Department of Health, *Dietary Reference Values for Food Energy and Nutrients for the United Kingdom*, HMSO, London, 1991.

National Academy of Sciences, *Dietary Reference Intakes for Thiamine, Riboflavin, Niacin, Vitamin B$_6$, Folate, Vitamin B$_{12}$, Pantothenic Acid, Biotin and Choline*, Institute of Medicine Standing Committee on the Scientific Evaluation of Dietary Reference Intakes, Washington DC, 1998.

Dietary Reference Intakes for Calcium, Phosphorus, Magnesium, Vitamin D, and Fluoride, ibid., 1997.

Dietary Reference Intakes for Vitamin C, Vitamin E, Selenium, and Carotenoids, ibid., 2000.

Dietary Reference Intakes for Water, Potassium, Sodium, Chloride and Sulfate, ibid., 2005.

Dietary Reference Intakes for Energy, Carbohydrate, Fiber, Fat, Fatty Acids, Cholesterol, Protein and Amino Acids, ibid., 2005.

Food Standards Agency, *National Diet and Nutrition Survey*, HMSO, 2002–03, vol. 1, 2 and 3.

Subject Index

Numerical prefixes to chemical names have been omitted, *e.g.* 3,4-Benzpyrene is listed as Benzpyrene. Other prefixes (*e.g. N-, α-, epi, p*) are ignored in the ordering of the index. Page references to structural formulae are printed in **bold**.

Acacia gum (gum arabic), 85
Accelerated freeze drying, 460
Acceptable Daily Intake, *see* ADI
Aceric acid, **93,** 94
Acesulfame K, 272, **274**–276
Acetaldehyde, 307, 368
Acetic acid, *see* ethanoic acid
Acetol (hydroxypropanone), **36,** 295
Acetone, 367
Acetylformoin, **33,** 34
N-Acetylglucosamine, 49, **50,** 185
Acrolein, **33**
Acrylamide, 44–47, **45**
Adenosine triphosphate (ATP), 188–191, 292
ADI values, 416–417
Adjuncts, brewing, 24
Aflatoxins, **400**–402
Agar, 70–72, 85, 462–463
Agave, blue, 42
Age-related macular degeneration, 227
Ajoene, **304**
Alanine, **162,** 169
Aldehydes, 298–299
Aldoses, 9
Alfalfa, 217
Algal toxins, 395–397
Alginates, 67–69
Alitame, 272, **274**

Alkaloids, 279, 382–385 , 398–399
Allergens and food allergy, 250, 407–412
Allicin, **303**
Alliinase, 303
Allose, **12,** 44
Allura Red, 246, 251
Allyl isothiocyanate, **289**
Almonds, 161, 299, 411
Altrose, **12,** 44
Aluminium, 249, 427–428, 436
Amaranth, 246, 250
AMD, *see* age-related macular degeneration
Ames test, 40
Amidated pectins, 66
p-Aminobenzoic acid, 355–**356**
Amnesic shellfish poisoning, 396
Amygdalin, **20,** 387
Amyl acetate, butyrate and valerate, 307
Amylase, 24, 56, 88
Amyloglucosidase, 89
Amylopectin, 51–57, **52**
Amylose, 51–57, **52**
Anaemia, 237, 330, 393, 435
Aneurine, *see* thiamin
Annatto, 224, 225, 250
Anomers, 14–15, 44
Antheraxanthin, **220,** 221
Anthocyanidins, 227–**229,** 256

Anthocyanins, 227–233, 250
Antibiotic residues, 420
Antioxidants, natural, 111, 241, 258, 337, 342
Antioxidants and rancidity, 123–127
Apiose, **93,** 94
Apocarotenoids, 221, 223
Apoptosis, 78
Apple juice, 18, 403, 422
Apples
  aroma, 298–300
  ascorbic acid content, 333
  browning, 236–238
  calcium content, 442
  fibre content, 75
  flavonoid content, 258
  glycaemic index, 91
  pectin, 60, 92
  pesticide residues, 414, 415
  protein content, 161
  starch content, 50
  sugar content, 8
  water content, 455
Apricots, 65, 67, 69
Arabinan, 60
Arabinose, **12,**
  anthocyanins, 228
  pectins, 63, 64
  xylans, **77**
Arachidic acid, **101,** 103
Arachidonic acid, **101,**103
  essential fatty acid, 106, 109
  chicken flavour, 297
Arginine, 38, **162,** 169, 172–173
Arsenic, 24, 426, 436
Arterial disease, atherosclerosis, 79, 111, 115, 82, 257
  and see heart or cardiovascular disease
Ascorbic acid, 320, 332–338, **333**
  anthocyanins interaction, 233
  antioxidant, 127, 353–355
  benzoic acid, 371
  bread-making, 205, 206–208
  collagen formation, 195
  curing, 365–366
  Dietary Reference Values, 476

enzymic browning, 238
enzyme cofactor, 353–355
sources in UK diet, 477
Ascorbyl palmitate, **124,** 127
Asparagine, 46–47, **162,** 169
Asparagus, 305
Aspartame, 272, **274**–276, 281
Aspartic acid, **162,** 169
Astaxanthin, **220,** 225
Asthma, 369
Astringency, 284–285
ATP, *see* adenosine triphosphate
Atropine, 279
Attention Deficit Hyperactivity Disorder, 251
Aubergines, 230, 382
Autoxidation of fatty acids, 117–123
Azo dyes, 245–246
Azodicarbonamide, **205**

Baby foods, 414
Bacon, 161, 363–364, 372, 445
Baking powder, 57
Banana vasopressor amines, 386–387
Bananas
  ascorbic acid content, 333
  browing, 236
  flavour, 298, 307
  glycaemic index, 91
  protein content, 161
  resistant starch, 58
  starch content, 50
  sugar content, 8
BaP, *see* benzpyrene
Barbecues, 428–429
Barley, 198
Basil, 306
Bean sprouts, 161, 389
Beans
  glycaemic index, 91
  protease inhibitors, 388
  protein content, 161
  resistant starch, 58
  starch content, 50
  sugar content, 7
  and see specific varieties

Beef
  amino acid composition, 169
  essential amino acid content, 170
  flavour, 297, 293
  protein content, 161
  texture, 196
Beef fat, 98, 102, 129, 132
Beef myoglobin content, 192
Beer
  caramel in, 41
  food colour, 250
  hops, 280
  iron content, 447
  magnesium content, 442
  nitrosamines, 365
  protein content, 161
  sugar content, 7
  water content, 455
Beet sugar, 42–43
Beetroot, 50, 234–236, 419
Behenic acid, **101**
Benedict's solution, 41
Benzaldehyde, 300
Benzene, 371
Benzoic acid and benzoates
  intolerance 250
  preservative, 362, 370–372, **371**
Benzoyl peroxide, 204, **205**
Benzpyrene, **367**
Benzyl propionate, 307
Beriberi, 313–314
Betanin, 234, 236, 250
Betel nuts, 257, 318
BHA, *see* butylated hydroxyanisole
BHT, *see* butylated hydroxytoluene
Bifidobacteria, 80
Bile salts, 79, 82, 142–**143**
Biltong, 366
Biogenic amines, 398
Biotin, 330–332, **331**
Biscuits
  aroma, 36
  colour, 224
  emulsifiers, 149
  fat content, 98
  flour, 199
  sugar content, 7

Bitter peptides, 280–281
Bitterness, 278–281
Bixin, **221**–223, 225
Black PN, 246–248, **247**
Black pudding, 402
Blackberries, 229, 283
Blackcurrants, 229, 300, 333
Blanching, 166, 216, 238, 302
Blancmange, 54
Bloom, on chocolate, 134
Boar taint, 309
Bonito, 293
Borage, 109
Boron, 453
Bottling, home, 405
Botulism and botulinum toxin, 365–366,
  404–406
Bran (wheat), 76, 79
Brassicas and glucosinolates, 287–289
Brazil nuts, 98, 411
Bread
  acrylamide, 46
  aroma, 36
  crust, 37
  emulsifiers, 149
  fat, 98, 132
  glycaemic index, 91
  making, 198–208
  niacin content, 325
  pesticide residues, 414
  protein content, 161
  starch, 50, 56–57, 59
  sugar content, 7
  water content, 455
Breakfast cereals, 46, 250, 345
Breast milk, *see* human milk
Brewing sugar in brewing, 24
Brilliant Blue, 246
Broad beans, 394
Broccoli
  ascorbic acid content, 333
  calcium content, 442
  fibre content, 75
  glucosinolates and colon cancer,
    287–289
  starch content, 50
  water content, 455

Bromates in flour, 205
Brown FK, 246, 250
Brussels sprouts, 287–289
Burger buns, 205
Burgers
  acrylamide, 46
  carcinogens, 40
  food poisoning, 407
  polyphosphates, 445
  protein content, 161
  sulfite, 369
  fat content, 98, 145
  protein content, 161
Butanedione, *see* diacetyl
Butanoic acid, *see* butyric acid
Butter, 148
  cholecalciferol content, 345
  colour, 250
  fat content, 98
  retinol content, 340
  water content, 455
Butyl acetate, 307
Butylated hydroxyanisole, **124,** 127
Butylated hydroxytoluene, **124,** 127
Butyric acid (butanoic acid) , **101**
  colon 58, 73, 76, 78, 80

Cabbage
  fibre content, 75
  glucosinolates, 287–288
  protein content, 161
  (red) anthocyanins, 233
  sugar content, 8
  vitamin content, 333, 348, 350
  water content, 455
Cadaverine, **398**
Cadmium, 426–427, 436
Caffeic acid, 228–**229**
Caffeine, **383–384**
Cake
  aroma, 36
  colour, 224
  emulsifiers, 144, 145, 149
  fat, 98, 132
  flour, 204
  sorbic acid, 372
  sugars, 7, 24

Calciferol, *see* cholecalciferol
Calcifetriol, 345–**346**
Calcium, 442–444, 477
  and vitamin D, 345
  (phosphate) in milk, 180–182
  muscle contraction, 189–191
  pectin gels, 65
Calcium
Camphor, **306**
Camposterol, **150**
Cancer
  glucosinolates, 289
  tea, 242
  colo-rectal, 58, 77–78, 365
  larynx, 227
  liver, 401
  lung, 227
  mouth, 227
  oesophagus, 227, 257
  prostate, 225
  stomach, 227, 257, 367, 419
  vaginal, 418
Cane sugar, 42–43
Canned fruit, 365, 404, 447–448
Canthaxanthin, **224,** 225
Capons, 418
Capric acid, **101,** 103
Caprylic acid, **101,** 103
Capsaicin, **285**
Capsanthin, **220,** 223, 225
Carageenans, 70–72
Caramel food colour, 41, , 249 250
Caramelisation, 30–33
Caraway, 306
Carbon black, 249
Carcinogens in cured meats, 40, 364, 367
Cardiovascular disease
  fatty acids, 108–112, 122, 125
  flavonoid consumption, 257
  phytosteroids, 151
  sodium chloride, 441
  tea, 242
  vitamin E, 348
  vitamin K, 351
  glycaemic index, 89
Cardomom, 306
Carmine (carminic acid), 243–**244**

Carmoisine, 246–248, **247,** 251
Carnesol, **126**
Carnitine, **355**
Carnosine, **293**
β-apo-8'-Carotenal, **224,** 225
α-, β- and γ-Carotenes, 218–**219,** 222, 250
ζ-Carotene, **254**
Carotenoids, 123, 204, 218–227, 339
Carrageenan, 85
Carrots
  carotenoids, 222, 223
  myristicin, 394
  pesticide residues, 415
  petroselenic acid, 104
  starch content, 50
  sugar content, 8
  vitamin content, 333, 339
  water content, 455
Carvacrol, **306**
Carvone, **306**
Casein (micelle), 176–183
Cashews, 411
Cassava, 50, 386–388
Casseroles, 36, 196
Caster sugar, 26
Castor oil, 104
Catalase, 210
Catechin and epi-catechin, **240,** 242
Catechins, 125–126, 236, 256
Cats (and other carnivores), 26, 107
Cauliflower, 333
Celery, 104, 385, 394, 415, 419
Cellobiose, **21,** 22
Cellulose, 73–76, 75, 199
Cephalin, *see* phosphatidyl ethanolamine
Cereals, 58, 319, 334, 350,446, 451
Character impact substances, 299–301
CHD (coronary heart disease), *see* cardiovascular disease
Cheese
  biogenic amines, 398
  umami substances, 293
  colour, 224, 250
  fat content, 98
  flavour, 115–116

  making, 182–184
  mineral content, 443, 444
  protein content, 161
  sugar content, 7
  vasopressor amines, 386
  vitamin content, 319, 322, 324, 340
  water activity, 467
  water content, 455
Cheese, processed and sorbic acid, 372
Chemesthesis, 285, 290
Cherries, 69, 232, 250, 299, 456
Chestnuts, 50
Chicken, 46, 161, 169, 293
Chicken fat, 98, 129
Chickpeas, 59, 91
Chicory, 79, 80
Chilli peppers, 285
Chips acrylamide levels, 46
Chips (USA), *see* crisps
Chitin, 49
Chlorine, 205
Chlorine dioxide (in flour), 204
Chlorogenic acid, **237**–238
Chlorophyll, 117, 214–217, **215,** 250
Chlorophyllase, 216
Chocolate
  aroma, 36
  caffeine content, 383–384
  emulsifier (lecithin), 144, 147, 148
  fat content, 98
  flavour, 308
  glycaemic index, 91
  making, 133–135, 156–157
  protein content, 161
  sugars, 7, 26–27
  vasopressor amines, 386–387
Chocolate Brown HT, 246
Cholecalciferol, **343**–346
Cholesterol, **141**
  garlic consumption
  levels in blood, 79, 112, 151, 305
  occurence, 140, 143
Cholesterol oxides, 122
Chorleywood Bread Process, 205, 207–208
Chromium, 453

Chutney, 217
Chymosin, 183–184
Cider, 67, 250, 422
Cineole, **306**
Cinnamon, 306
Citral, 301
Citrates (emulsifying salts), 146–147, 466
β-Citraurin, **221**–222
Citric acid, 64, **238**
Citrus essential oils, 301
Citrus fruit
  aroma, 298, 300–301,
  ascorbic acid, 333
  water, 455
Citrus peel, 60, 67
Clams, 395
Cling film, 430–431
Clostridium botulinum, 365–366
Cloudberries, 370
Cloves, 126, 286, 306
Clupanodonic acid, **101**
Cobalamin, **326**–328, 388, 476, 477
Cobalt, 326, 454
Cochineal, 243–244, 250
Cockles, 395
Cocoa powder, 383
Cocoa butter, 102, 128–135, 137–139, 156
Coconut milk, 22
Coconut oil, 102, 103, 112, 132
Cod, 98, 161, 169, 293
Codex Alimentarius Commission, 74
Coeliac disease, 410
Coffee, 46, 324, 384, 460
Cola drinks, 7, 41, 91, 371, 383–384
Collagen, 195–197, 334
Colour Index numbers, 245–246
Comfrey, 385
Conalbumin, 185, 186
Condensed (sweetened) milk, 27
Confectionery, 23–24, 246, 290
Conjugated linoleic acids, 105–106
Connective tissue, 195
Constipation, 77
Convicine, **393**
Cookies (USA), *see* biscuits
Cooling, taste, 289–290

Copper
  food colours, 217, 245, 449–450, 477
  phenolase, 238
  dietary sources, 450, 477
  Dietary Reference Values, 476
Corn (USA), *see* maize
Corn chips, 325
Cornflour, *see* maize flour
Corn oil, 102, 103, 132
  phytosterols, 151
  tocopherols, 348
Corn syrup, 23–24
Cornflakes, 7, 50, 59, 91, 161,
Cottonseed oil, 104, 132, 135, 348
p-Coumaric acid, 228–229
p-Cresol, **310**
Coumarin, 351
Coumestrol, **390**
Cranberries, 370
Cream, 98, 147–148
Cream of tartar, 187
Creatine, **33**
Crispbread (rye), 46, 59, 91
Crisps (potato), 46, 91, 161, 284
Crocetin, **221**–223, 225
Cryptoxanthins, 221–222
Cucumber, 301
Curcumin, **243,** 250
Curing, 363–366, 372, 465
Curry powder, 224, 243
Custard, 407
Cyancobalamin, *see* cobalamin
Cyanide, 279, 386–388
Cyanidin, **229**
Cyanogenic glycosides, 279, 387–388
Cyclamate, 272, **274**–276
Cyclohexyl propionate, 307
Cysteine, **162,** 169,
  breakdown, 172–173
  in ovalbumin, 185
  in gluten proteins, 201
  in meat flavour, 295–296
Cystine, 164

Daidzien, **390–391**
DATEM esters, 145–**146,** 149

DDT, **412–414,** 416
Decadienal, 119–**120**
Decaffeination, 384
γ-Decalactone, **300**
Decanoic acid, *see* capric acid
Decenal, **297**
Decontamination, 376
DEHA, *see* dioctyl adipate
Dehydroascorbic acid, DHAA, **333,** 337
Delphinidin, **229**–230
Denaturation of proteins, 166, 186
Deoxyribose, **19**
Dextrins, 86, 87
Dextrose 11, 24, 86
   and see glucose
DHA, *see* docosahexaenoic acid
DHAA, *see* dehydroascorbic acid
Diabetes, 19, 39–40, 89–90, 275
Diacetyl (butanedione), **36,** 121, 148,
   295, 367
Diacetylated tartaric acid esters of
   monoglycerides, 145–**146,** 149
Diarrhetic shellfish poisoning, 396
Diarrhoea, 19, 275
Dibenzofurans, **432**–433
Dichlorodiphenyltrichloroethane, *see*
   DDT
Dicoumarol, **351**
Dieldrin, 413, 416
Dietary fibre, 73–79
Dietary Reference Values, 475–478
Diethylstilboestrol, **418**
Dihydroxyacetone, 8–**9**
Dihydroxypropanal, *see* glyceraldehyde
Dihydroxypropanone, *see*
   dihydroxyacetone
Diketogulonic acid, **333,** 337
Dimethyl sulfide, 308
Dimethylpyrazine, 308
Dinitrosalicylic acid, **42**
Dinoflagellates, 395
Dioctyl adipate, **430**–431
Dioxins, **432**–434
Disaccharides, 20–23
Disinfestation, 376
Disulfite, **368**

Diverticulosis, 77
Divicine, **393,** 435
DNA, 159
Docosahexaenoic acid formation, **101,** 110
Docosanoic acid, *see* behenic acid
Dodecanoic acid, *see* lauric acid
Domoic acid, 396–**397**
Dopamine, **334,** 387
Dough formation and properties,
   204–208
Dried fruit (raisins, sultanas *etc.*),
   369, 456, 467
Durum wheat, 210

Eczema and food colours, 250
Egg
   allergy, 410
   amino acids, 169, 170, 171
   fat content, 98
   iodine content, 453
   phosphorus content, 444
   powder, 37
   protein content, 161
   vitamin content, 322, 324, 348
   water content, 455
Egg white proteins, 166, 184–187
Egg yolk lipids, 140, 143–144, 224
Eggplants (USA), *see* aubergines
Eicosanoic acid, *see* arachidic acid
Eicosanoids, 108, 111
Eicosapentaenoic acid, **101,** 110
Einkorn wheat, 210
Elaidic acid, 103
Elastin, 202
Elderberries, 229, 233
Emetine, 279
Emmer wheat, 210
Emulsifiers, 142–150
Emulsifying salts, 146–147, 466
Emulsions, 81, 141–146
Enantiomers, 10
Endosperm, wheat, 198–199
Energy drinks, 384
Englyst (and Cummings), 73, 75
Enzymic browning, 236
EPA, *see* eicosapentaenoic acid

Ergocalciferol, **343**
Ergosterol, **343**
Ergot(ism), 398–399
Ergotamine, **399**
Erucic acid, 104
Erythorbic acid, 127
  and see isoascorbic acid
Erythrose, **12**
Erythrosine, 246–248, **247,** 250
Essential amino acids, 168–172
Essential fatty acids, 106–112
Essential oils, 301
Esters, 298–299
Estragole, **306**
Ethanal, *see* acetaldehyde
Ethanedial, *see* glyoxal
Ethanoic acid (acetic acid)
  colon, 73, 78, 80,
  sourness, 283
  preservative, 362, 373
Ethyl butyrate, 307
Ethyl caproate, 307
Ethylvanillin, 307
Eugenol, **126,** 286, **287,** 299
Evaporated milk, 27
Evening primrose, 109

FAD and FMN, *see* riboflavin
Fat soluble vitamins, 314, 338–351
Fatty fish, 108, 110, 111
Fava beans and favism, 393–394, 434–435
FD&C numbers, 246
Fehling's solution, 21, 41, 42
Ferritin, 446
Ferulic acid, **229**
Fibre, 73–79
Figs and aflatoxins, 8, 401
Fish
  toxic metals, 425, 426
  colouring, 248
  nutrient metals, 442, 446
  iodine, 451, 453
  and see named species
Fish oils
  margarine, 113
  vitamins, 344, 346, 348
  and see named species

Fish sauce, 291
Flatulence, 26, 73
Flavanoids, *see* flavonoids
Flavanols, **256**
Flavanones, **256**
Flavonoids, 227, 255–258
Flavour enhancers, 291
Flour
  acrylamide, 46
  amino acid composition, 169
  calcium content, 442
  fat content, 98
  fibre content, 75
  pesticide residues, 414
  phosphorus content, 444
  resistant starch, 59
  starch content, 50
  water content, 455
Foams, 141, 186
Folacin, *see* folic acid
Folic acid, 328–330, **329,** 476, 477
Food Green S, 217, 246–248, **247**
Formaldehyde, **367**
Formic acid, **367**
Fractionation, of fats and oils, 135–137
Frankfurters, 364
Free radicals
  in meat, 194, 209
  in rancidity, 116–120, 122–125
Freeze drying, 460
French fries, 114, 122 (USA)
  and see chips
Fried food and carcinogens, 40, 46
Frozen desserts, 462
Fructans, 79–80
Fructose, 8, 12, 13, **14,** 29
Fruit
  aroma, 297–301
  juices, 371, 422, 456
  mineral content, 446, 451, 453
  vitamin content, 324, 350
  and see named species
Frying oils, 121–122, 136
Fucose, **81,** 93, 94
Fumonisins, 402–**403**
Furaneol, **299**
Furanose rings, 14–15

Furans, 432
Furfural in smoke, **367**
Furosine, **39**
Fusel oils, 299

Galactan, 60
Galactose, 8, 12, 44, 63, 64
Galacturonan, 60
Galacturonic acid, **18**, 60, 61, 81
Gallic acid, **240**
Gallocatechin and epi- gallocatechin,
    239–**240, 242**
Game (pheasants, rabbits etc.), 423
Garlic, 303–305
Garum (liquamen), 291
Gel structure, 60–63
Gelatin, 197–198
Gelatinisation of starch, 54–57
Gellan gum, 83–85, **84**
Genetic engineering, 184
Genistein, **391**
Gentiobiose, **21**
Geosmin, 235–**236**, 267
Geranial, **301**
Geranyl propionate, 307
Germ (of wheat), 198–199
Ghee, 122
Ginger, 286
Gingerols, **286**
Gliadin, 200–206, 211–212
β-Glucans, **78**–79
Glucobrassicin, **288**
Glucomannans, 76
Gluconasturtiin, **288**
Gluconic acid, 17, 18
Glucoraphanin, **288**
Glucoraphasatin, **288**
Glucose
    analysis, 17
    anthocyanins, 228
    content of foods, 8
    pectins, 63
    relative sweetness, 269
    structure, **11**, 12, 29, 44
Glucose isomerase, 24, 89
Glucose oxidase, 17

Glucose syrup, 23–24
Glucosinolates, **287**–289
Glucotropaeolin, 287, 288
Glucuronic acid, **77**, 83, 93, 94
Glutamic acid, **162**, 169
Glutamine, **162**, 169,
    breakdown, 172–173
    cereal proteins, 201, 203
Glutathione, **207**, 334, 435–436
Gluten, 199–208
Glutenin, 200–206, 211–212
Glycaemic Index, 89–91
Glycaemic Load, 90–91
Glyceraldehyde, 8–**10**, 12
Glycerine, *see* glycerol
Glycerol, **100**
Glycerol oleate(s) HLB values, 145
Glycetein, **391**
Glycine, **162**, 169
    collagen, 195–196
    cereal proteins, 201
    taste, 272
Glycogen, 49, 190–191
Glycosides, 19–20
Glycyrrhizin, 272, **274**, 277
Glyerophospholipids, 140
Glyoxal, **33**, 36
Goat's milk, 409
Goats milk fat, 103
Goitrins and goitre, 288–289, 388, 452
Gold, 249
Golden Syrup, 32
Goose fat, 102
Gooseberries, 65, 75
Granulated sugar, 26
Grape juice, 43
Grapefruit, 277, 279, 283, 300
Grapes
    anthocyanins, 228–230
    organic acids, 283
    pesticide residues, 414
    sugar content, 7, 8
Gravy, 54, 197
Green S, *see* Food Green S
Green tea, 242
Grilled food and carcinogens, 40

Groundnut oil, 129, 132
Groundnuts, *see* peanuts
Guanosine monophosphate, 291–293, **292**
Guar gum, 80–82, 81, 85
Guarana, 384
Gulose, **12,** 44
Guluronic acid, **18,** 67–68
Gums, 80–85

Haem, 117, 167
Haemagglutinins, 389
Haemoglobin, 193
Halophiles, 467
Ham
    food poisoning, 407
    polyphosphates in, 445
    water binding, 465
    and see curing
Hamburgers, *see* burgers
Haricot beans, 59
Haworth convention, 16, 27–28
Heart disease, *see* cardiovascular
    disease
Heat, *see* pungency, 285
Heliotropin in synthetic flavourings, 307
Hemiacetals, **13**
Hemicelluloses, 73–76, 199
Hepatitis, 401
Herbs, 305–306
Herring (oil), 102, 130, 132, 345
Herring oil triglycerides, 130
Heterocyclic amines, 40
Hexachlorocyclohexane, 413
Hexadecanoic acid, *see* palmitic acid
Hexanoic acid, *see* caproic acid
Hexanal, **299**
Hexenal, 121, 299, 307
Hexenol, **299**
Hexoestrol, 418
Hexoses, 9
High- flour ratio, 205
High fructose corn syrups, 89
Histamine, 397–**398,** 408
Histidine, **162,** 169
    Maillard reaction, 38, 39
    histamine poisoning, 397–398

HLB system, 144–145
Hollandaise sauce, 144
Homogalacturonan, 92–94
Homogenised milk, 148
Honey, 7, 32, 91, 248, 456
Hops, 280, 390
Hormones, 417–418
Horseradish, 287, 288, 333
Human milk and MSG, 293 414
Human milk fat, 102, 130, 139,
Humulones, **280**
Hydoxybenzalehyde, 76–**77**
p-Hydroxybenzoate esters, 362
Hydrogen sulfide, 172, 296
Hydrogenated sugars, 273–274
Hydrogenation, 113–115, 133, 152–153
Hydroperoxides, 117–119, 204
Hydroxyacetyl furan, **31**
Hydroxylysine, **162**
Hydroxymethyl furfural (HMF), 31–32, 34, 35, 295
Hydroxyproline, **162,** 169, 195
Hydroxypropanone, *see* acetol
Hygienisation, 376
Hyperactivity, 250–251
Hypertension, 386, 441
Hypocholesterolaemia, 151

Ice structure, 457–460
Ice cream
    and emulsifiers, 147
    and gums, 82
    and oat β-glucan, 79
    colour, 224
    colours for, 243
    fat content, 98
    glycaemic index, 91
    protein content, 161
    sugar content, 7
Icing sugar, 26
Idose, **12,** 13, 44
Immunoglobulins, 177, 181–182
IMP, *see* inosine monophosphate
Improvers, flour, 204
Indigo Carmine, 246–248, **247**
Indigoid dyes, 246
Indigotine, *see* Indigo carmine

Infant foods
  added colours, 248
  lead contamination, 423
Infant formula feeds
  aluminium content, 428
  phytoestrogens, 391
Inosine monophosphate, 290–293,
  291
myo-Inositol, 355–**356**
INS numbers, 86
Insoluble fibre definition, 74
Instant desserts, 56, 147
Interesterification, 133, 134, 137–139
International Units of vitamins, 347
Inulin, 79–**80**
Invert sugar, syrup, 23, 25, 32
Invertase, 23, 26–27
Iodates in flour, 205
Iodine, 451–453
  Dietary Reference Values, 476
  and goitrinogens, 288
  reaction with starch, 55
  sources in UK diet, 477
Iodine number, 112
Ionone (α- and β-), **218**, 222, **223**, 307
Irish moss, 70
Iron, 446–449
  absorption, 447–448
  and ascorbic acid, **335**
Iron oxides and hydroxides food
  colours, 249
Irradiation, 375–380
Isoamyl acetate, 307
Isoamyl alcohol, 307
Isoamyl butyrate, 307
Isoascorbic acid, **333**
Isobutyraldehyde, **37**
Isocitric acid sourness, **283**
Isoflavones, 390
Isoleucine, **162,** 169, 280
Isomalt, 269, **273**
Isomaltol, 34, 35
Isomaltose, **21,** 22
Isopentyl acetate, 299
Isosyrups, 24
Isothiocyanates, **287**

Isouramil, 393, 435
Isovaleraldehyde in synthetic
  flavourings, 307

Jam
  and patulin, 403
  fruit content, 441–442
  pectin, 64–65
  protein content, 161
  sugars, 7, 25, 64–65
  water content, 455
Jelly (dessert), 197
Jerusalem artichokes, 79
Junction zones in gels, 61–63, 69,
  71–72

Karaya gum, 85, 147
Katemfe, 278
Ketchup, 83
α-Ketoglutaric acid, **316**
Ketones in fruit aromas, 298–299
Ketoses, 9, 12, 13
Kidney, 187, 327
Kidney beans, 59, 230
Kidney disease and ochratoxin, 402
Kippers, 245, 250
Kjeldahl method, 174–175
Kohlrabi pesticide residues, 415
Konjac, 85

α-Lactalbumin, 177, 181
Lactic acid
  in cheese, 182–**183**, 284
  in muscle, 190–191
  preservative, 362
Lactitol, 269, **273**–276
Lactoferrin, 449
β-Lactoglobulin, 160, 177, 181
Lactose, **22**
  crystals, 27
  Maillard reaction, 39
  intolerance, 26
  relative sweetness, 269
Lactulose, **39**
Laetrile, 375
Laevulose, 11

Lager, 7
Lamb
  fat content, 98
  flavour, 293, 297
  protein content, 161
Lane and Eynon method, 42
Lard fat content, 98
  fatty acid composition, 102
  melting, 131
  triglycerides, 128–130, 132
Lauric acid, **101,** 103, 108
Lead, 421–424, 436
Lead chromate, 245
Lead oxide, 245
Lecithin, 140
Lectins, 389
Legumes
  lectins, 389
  protease inhibitors, 388
  essential amino acids, 171
  niacin content, 324
  oligosaccharides, 25
  thiamin content, 316
  and see peas, beans *etc.*
Lemons, 283, 414
Lemon oil in synthetic flavourings, 307
Lentils, 91, 161
Lettuce, 414, 419, 455
Leucine, **162,** 169
  bitter taste, 280
  in fruit aromas, 298
  sweetness, 272
Leucoanthocyanidins, **256**
Leukotrienes, 108, 111
Lignan, **390**
Lignin, 73, 75–77
Lima beans, 388
Lime juice, 386
Limonene, 300, **301**
Limonin, **279**
Linalool, 307
Linoleic acid, **101**
  conjugated, 105–106
  essential fatty acid, 106–109
  fruit aroma, 298
  melting point, 103

α-Linolenic acid, **101,** 103, 106, 110
γ-Linolenic acid, **101,** 109
Lipases
  cheese, 184
  rancidity, 116
  interesterification, 138, 139
Lipoxidase, *see* lipoxygenase
Lipoxygenase
  blanching, 166
  vegetable flavour, 298–299, 301–302
Liquid glucose, 24
Liquid smoke, 367
Liquorice root, 277–278
Liver
  cholecalciferol content, 345
  cobalamin content, 327
  copper content, 449
  folic acid content, 328
  iron content, 446
  phosphorus content, 444
  protein content, 161
  retinol content, 340
  riboflavin content, 319
  vitamin E content, 348
Lobster colour, 224, 225
Locust bean gum, 81, 85
Low fat spread, 98
Luncheon meat, 465–466
Lutein, **220**–222, 224–227
Lychees, 300
Lycopene, 123, **219,** 225
Lye, 238
Lysergic acid, 398–399
Lysine, **162,** 169
  and collagen, 197
  breakdown, 173
  essential amino acid, 170–172
  Maillard reaction losses, 38, 39
Lysozyme, 185
Lyxose, **12**

Mackerel
  cholecalciferol, 345
  fat content, 98
  histamine, 397
Macroglobulins, 181

Magnesium, 442
  bitterness, 442
  chlorophyll, 215
  Dietary Reference Values, 476
  sources in UK diet, 477
Maillard reaction, 33–41
  bread-making, 208
  meat flavour, 295
  process flavours, 308
Maize (corn)
  mycotoxins, 402
  coeliac disease, 410
  niacin content, 325
Maize flour, 50
Maize oil, *see* corn oil
Maize starch, 24, 88–89
Malaria
  cyanogenic glycosides, 279
  DDT *etc.*, 412–413
  favism, 393–394, 435
Malathion, 415
Malic acid, 64, 238, **283**
Malonaldehyde, 120, **121**
Maltitol, 269, **273**–276
Maltodextrin, 86
Maltol, 34, 35
Maltose, 8, **21,** 22, 269
Malvalic acid, **104**
Malvinidin, **229**
Malvones, **233**
Mandarin oranges, 301
Manganese, 453
Mango, 218, 222
Manioc, *see* cassava
Mannans, 76
Mannitol, 269, **273**–276
Mannose, **12,** 44, 82
Mannuronic acid, 67–**68**
Maragaritas, 386
Margarine
  cholecalciferol, 345
  colour, 224
  emulsifiers, 148
  fat content, 98
  food colour, 250
  low fat, 87
  manufacture, 104, 105, 113–115, 152

retinol content, 340
  sorbic acid, 372
  trans fatty acids, 137
  water content, 455
Marmalade, 65
Maximum Residue Level, *see* MRL
Mayonnaise, 87, 144
Meat, 187–198
  aroma, 36
  calcium content, 442
  cholecalciferol content, 345
  cobalamin content, 327
  colour, 192–195, 248
  conditioning, 191
  folic acid content, 328
  iodine content, 453
  iron content, 446
  niacin content, 324
  odour, 295–297
  phosphorus content, 444
  pyridoxine content, 322
  retinol content, 340
  riboflavin content, 319
  selenium content, 451
  thiamin content, 316
  water activity, 467
  water binding, 465
  water content, 455
Meatiness (umami), 290–293
Mediterranean diet, 225
Melanoidins (brown pigments), 37–38
Membranes, 140
Menadione, *see* vitamin K
Menaquinone, *see* vitamin K
Menthofuran, **306**
Menthol, **290, 306**
Menthone, **306**
Mercuric sulfide, 245
Mercury, 424–425, 436
Meringue, 186–187
Metabisulfite, 362, 369–370
Metamerism, 262
Methanal, *see* formaldehyde
Methanoic acid, *see* formic acid
Methanol in smoke, 367
Methionine, **162,** 169
Methyl butyrate, **300**

Methyl furanolone, **295**
Methyl heptanone, 307
Methyl mercaptan, **305**
Methylbutanal, **298**
Methylchavicol, **306**
Methylnonanoic acid, **309**
Methyloctanoic acid, **309**
Methylthiophenone, **295**
Methylxanthines, 383–384
Metmyoglobin, 193–194, 208–209
Mevalonic acid, **218**
Microbiological assay, 321
Migraine, 386
Milk
  aflatoxins, 401
  allergy, 409–410
  colours, 248
  ascorbic acid content, 334
  calcium content, 443
  cholecalciferol content, 345
  cobalamin content, 327
  copper content, 449
  amino acids, 169–172
  fat content, 98
  human *versus* cow's, 182
  iodine content, 453
  iron content, 447, 448–449
  niacin content, 324
  pesticide residues, 414, 415
  powder, 37
  proteins, 161, 176–184
  pyridoxine content, 322
  retinol content, 340
  riboflavin content, 319–320
  selenium content, 451
  sugar content, 7, 22
  tea, 285
  vitamin E content, 348
  vitamin K content, 350
  water content, 455
  umami substances, 293
Milk fat
  fatty acid composition, 102
  globules, 147–148
  melting point, 103
  triglycerides, 129, 130,, 132, 139

Milk powder, 455
Mint, 306
Miso, 391
Modified starch, 85–88
Molybdenum, 453
Monellin, 271
Monochloropropanediol, **392**
Monoglycerides as emulsifiers, 149
Monosaccharides, 7
Monosodium glutamate, 290–293, **291**
MRL values, 415
MSG, *see* monosodium glutamate
Muffins (USA), 205
Munson and Walker method, 42
Muscle structure and contraction, 187–190
Mushrooms, 161, 293, 301
Mussels, 395
Mustard oils, 287
Mustard, 287
Mutarotation, 13, 21
Mutton, 309
Mycoprotein, 200, 360
Mycotoxins, 398–404
Myoglobin, 166–167
  nitrosamines, 429
  cured meats, 364
  free-radical formation, 208–210
  muscle, 190–195
Myristic acid, **101**, 103, 108
Myristicin, **394**
Myrosinase, 287

NAD and NADP, 323–**324**
Naringin, 277, **279**
Natamycin, 362, 373–**375**
Near-infrared spectrophotometry, 175, 472–473
Neocarotenes, 225
Neoglucobrassicin, **288**
Neohesperidine dehydrochalcone, 272, **274**–277
Neohesperidose, **274**
Neoxanthin, **220**–222
Neral, **301,** 302

Neurosporene, **254**

Niacin, **323**–325, 476, 477

Nickel, 453

Nicotinic acid and nicotinamide, *see*
   niacin

Ninhydrin, 176

Nisin, 362, 373–**374**

Nitrates
  beetroot, 236
  nitrosamine formation, 419
  preservatives, 362–364
  vegetables, 418–420

Nitriles, **287,** 288

Nitrites, 362–366

Nitrogen factors, 174–175

Nitrogen oxides, 364

Nitrosamines, 364, 429

No Observable Effect Level, *see*
   NO(A)EL

NO(A)EL, 417

Nonenal, 119–120, **297**, 301–**302**

Non-starch polysaccharide (NSP), 73

Non-sugar sweeteners, 271–278

Nootkatone, 300, **301**

Noradrenaline, **334**
  and see norepinephrine

Norepinephrine, **387**

NSP, *see* Non-starch polysaccharide

Tetradecanoic acid, *see* myristic acid

Nut allergy, 411

Nutmeg, 306, 394

Nutraceuticals, 389

Nuts, 444, 446, 451
  and see specific types

Oatmeal fibre content, 75

Oats β-glucan, 78, 198

Obesity, 89

Ochratoxin, **402**

Octadecanoic acid, *see* stearic acid

Octanoic acid, *see* caprylic acid

Octenol, **302**

Oestradiol, **390**

Oestrogens, 418

Offal, 187

Off-flavours, 202, 309

Okadaic acid, 396–**397**

Oleic acid, **101**, 103, 107, 145

Olein, 135, 136

Oligosaccharides, 20–26

Olive oil
  fatty acid composition, 102
  melting point, 103
  triglycerides, 130, 132
  vitamin E content, 348

Omega-3, -6 and -9 fatty acids, 99,
   107–111

Onion
  flavonoids, 258
  flavour, 302–304
  phenolics, 237–238
  sugar content, 8

Optical activity and isomerism, 10–15

Oranges
  carotenoids, 222
  flavour, 283, 301
  sugar content, 7, 8
  and see Seville orange

Orange juice, 91

Orange oil, 307

Oregano, 126, 306

Orotic acid, **356**–357

Osmophiles, 467

Ovalbumin, 185

Ovoglobulins, 185, 186

Ovomucin and ovomucoid, 185, 186

Oxalic acid, **283**

Oxazoles, 295–**296**

Oxidising agents in flour, 204

Oxoglutaric acid, *see* α-ketoglutaric acid

Oxopropanol, 295
  and see pyruvaldehyde

Oxymyoglobin, 193–194, 208–209

PABA, *see* p-aminobenzoic acid

Packaging residues, 430–431

PAHs, *see* polynuclear aromatic
   hydrocarbons

Palm kernel oil, 132, 134

Palm oil
  carotenoids, 224
  fatty acid composition, 102

for margarine making, 113
fractionation, 136–137
retinol content, 339, 340
triglyceride composition, 129
triglyceride polymorphism, 132
vitamin E content, 348
Palmitic acid, **101** 103, 108, 110
Palmitoleic acid, **101**
Pangamic acid, 357
Panthenol, **332**
Pantothenic acid, 330–**332**
Paprika, 218, 250
Parabens, **372**
Paralytic shellfish poisoning, 396
Parsley, 104, 333
Pasta
    durum wheat, 210
    protein content, 161
    sauces, 225
    water activity, 467
    water content, 455
Pasteurisation of milk, 38
Pastry, 98, 124, 132, 199
Pâté, 250
Patent (household) flour, 200
Patent Blue, 246–248, **247**
Patulin, **403**
PCBs, *see* polychlorinated biphenyls
Peach aroma, 300
Peaches, 67, 333
Peanut butter, 401
Peanuts
    acrylamide levels, 46
    allergy, 411–412
    fat content, 98
    glycaemic index, 91
    protease inhibitors, 388
    protein content, 161
    starch content, 50
    sugar content, 7
    and see groundnut oil
Pear juice, 18
Peas
    amino acid composition, 169
    ascorbic acid content, 333
    calcium content, 442

canned, 217
essential amino acid content, 170, 171
fibre content, 75
glycaemic index, 91
off-flavour, 302
oligosaccharides, 25
protease inhibitors, 388
resistant starch, 59
starch content, 50
sugar content, 7, 8
vitamin K content, 350
Pectin, 60–67
Pectin lyase, 66
Pectin methyl esterase, 66
Pectinases, 66–67
Pelargonidin, **229**
Pellagra, 313, 325
Pentadienal, 121
Pentadione, 121
Pentoses, 7
Peonidin, **229**
Pepper, black and white, 286, 394
Pepper, red, 223
Peppermint (oil), 290
Peppers, sweet, 333
Peptide bonds, 163
Perillartine, 272, **274**
Pernicious anaemia, 327
Peroxidase, 17
Persimmon, 222, 415
Persulphates, 204, 205
Pesticides, 412–417
Petroselenic acid, **104**
Petunidin, **229**
Phenolase, 235, 236–239
Phenolphthalein, 245
Phenylalanine, **162**, 169
    bitter taste, 280
    in gluten proteins, 201
Phenylethylamine, 386–**387**
Pheophytin, 216–217
PhIP, 40
Phosphate, calcium in milk, 180–182
Phosphates as emulsifying salts, 146–147
Phosphatidic acid, **141**
Phosphatidyl choline, **140**

Phosphatidyl ethanolamine, **141**
Phosphatidyl inositol, **141**
Phosphatidyl serine, **141**
Phosphorus, 444–446
    Dietary Reference Values, 476
    sources in UK diet, 477
Phosphoserine, 179–181, **180,** 185
Phthalates, **431**
Phylloquinone, *see* vitamin K
Phytic acid, **79**, 444
Phytoene, **254**
Phytoestrogens, 389–392
Phytofluene, **254**
Phytostanols, **150**
Phytosteroids, 150
Piccalilli, 243
Pickles
    cabbage and cucumber, 283–284
    copper salts, , 217, 449–450
    lead contamination, 422
    sourness, 238
    turmeric, 243
    colours, 250
Pie filling, 69
Pimaricin, *see* natamycin
Pineapple
    aroma in wine, 300
    ascorbic acid content, 333
    carotenoids, 225
    sugars, 8, 42–43
Piperine, **286**
Plantain, 387–388
Plums, 333
Polarised light, 11
Polyaromatic hydrocarbons, 428–429
Polychlorinated biphenyls (PCBs),
    432–434
Polygalacturonase, 66
Polyglyceryl ricinoleate, 148–**149**
Polymorphism of triglycerides, 131–135
Polynuclear aromatic hydrocarbons, 367
Polyoxyethylene sorbitan
    monopalmitate, 145–**146**
Polyphenol oxidase, *see* phenolase
Polyphenolics, 126, 236–242
Polyphosphates, 445, 466

Pomace, apple, 65
Pomegranates, 415
Ponceau, 246–248, **247**
Popcorn, 46
Pork
    boar taint, 309
    flavour, 293, 297
    myoglobin content, 192
    umami substances, 293
Porridge (oats)
    β-glucan, 78
    glycaemic index, 91
    resistant starch, 59
Port , 7
Potassium, 441–442, 476, 477
Potatoes
    acrylamide, 46
    ascorbic acid content, 333
    calcium content, 442
    enzymic browning, 236, 238
    fibre, 75
    glycaemic index and load, 91
    iron content, 446
    nitrate content, 419
    pesticide residues, 414, 415
    protein content, 161
    resistant starch, 58, 59
    solanine, 382–383
    starch, 24, 50
    sugar content, 7
    vitamin K content, 350
    water content, 455
Poultry, 192, 248
Prawns, 293
Preserves and water activity, 467
Proanthocyanidins, 241–**242, 256**–257
Process flavours, 308
Processed cheese, 466
Procyanidins, 233
Progoitrin, **288**
    Proline **162,** 169
    gluten proteins, 201, 203
    collagen, 195–196
    nitrosamines, 364
Propanone, *see* acetone
Propenal, *see* acrolein

Propionic acid
  in the colon, 73, 78, 80
  preservative, 362, **373**
Propyl gallate, **124,** 127
Prostaglandins, 108, 111
Protease inhibitors, 388–389
Proteases, 116, 184
Proteins, 160–213
  thermal denaturation, 464
  water binding, 463–464
Protein hydrolysates, 280–281
Protocatechuic acid, **237**–238
Protopectin, 60
Provitamin A, 226
Provitamins, 321, 338–339
Proximate analysis, 2
Prunes, 370
Psoralens, 385–**386**
Puffer fish, 394–395
Pullulanase, 89
Putrescine, **398**
Pyranose rings, 14–15
Pyrazines, 36, 37, **296,** 308
Pyridines, 36
Pyridoxal *etc., see* pyridoxine
Pyridoxine, 321–323, **322**
  coenzyme functions, 352–353
  Dietary Reference Values, 476
  sources in UK diet, 477
Pyrroles, 36
Pyrrolizidine alkaloids, 385
Pyruvaldehyde, **33,** 36
Pyruvic acid, 82, **190, 316**

Quercitin, **258**
Quinine, **278**
Quinoline Yellow, 246–248, **247,** 251
Quinones, **236**–238, 241
Quorn™, *see* mycoprotein

Radapperisation, 376
Radicidation, 376
Radish, 230, 287, 288
Radurisation, 376
Raffinose content of foods, 8, **25,** 26
Raisins protein content, 7, 161

Rancidity
  lipolytic, 115–116
  oxidative, 116–123, 121, 195
Rapeseed meal, 288–289
Rapeseed oil, 104, 132
Raspberries, 229, 300
Raspberry ketone, **300**
Recommended Daily Amounts, 475
Red 2G, 246, 248
Red cabbage anthocyanins, 233
Reducing sugars, 21
Reference Nutrient Intake, 475
Rennet, 183
Rennin, *see* chymosin
Resveratrol, **258**
Retinal and retinol formation, **226**
Retinol, **338**–342
Reversion, of soya bean oil, 121
Rhamnogalacturonan, 92–94
Rhamnose, **19**
  anthocyanins, 228
  gums, 81, 83
  pectins, 63, 64, 94
Rhubarb, 283
Riboflavin, 318–321, **319**
  Dietary Reference Values, 476
  rancidity, 117
  sources in UK diet, 477
Ribonucleotides in meat flavour, 295
Ribose, **12,** 269
Rice
  coeliac disease, 410
  aflatoxins, 401
  glycaemic index and load, 91
  niacin content, 324–324
  protein content, 161, 198
  starch content, 50
  thiamin content, 318
  vitamin A (golden rice), 341
Ricinoleic acid, **104,** 148, 149
Rickets, 345
Rigor mortis, 191
Ripening of fruit, 66, 298
RNA, 159
Root beer, 286
Rope (in bread), 373

Rose hips, 333
cis-Rose oxide, 300
Rosemary, 126, 306
Rosmarinic acid, **126**
Roughage, 73, 75
Roux, 57
Ruminants, 76, 103, 105
Rutin, **448**
Rutinose, **228**
Rye, 198, 398

Sabinene, **306**
Saccharin, 272, **274**–276
Safflower oil, 132
Saffron, 223
Safrole, **286**
Sage, 126, 306
Salad dressing, 83, 87, 104
Salad oils, 136
Salami, 455, 467
Salami, 364
Salicylates, 250
Salmon
   cholecalciferol, 345
   colour, 224, 225
   fat content, 98
   umami substances, 293
Salt, *see* sodium chloride
Saltpetre, 363
Sardines, 293, 397
Sauerkraut, 283, 287
Sausages
   fat, 79, 98
   colour, 224, 250
   protein content, 161
   sulphur dioxide, 369
   sugar content, 7
   water binding, 465–466, 145
   and see salami
Saxitoxin, **396**
Scallops, 395, 396
SCP, *see* single-cell protein
Scurvy, 313, 334, 355
Seaweed, 67–72, 291–293
Selenium, 349, 450–451, 476
Serine, **162,** 169, 172–173
Serotinin, 386–**387**

Seville orange, 277, 279
Shellfish, 224, 395, 423
Shogaols, **286**
Shortbread, 57, 59, 98
Shortening, 132, 137, 152
Silicon, 453
Silver, 249
Sinalbin, **288**
Sinensal, **301**
Single-cell protein, 360
Singlet oxygen, 117, 123, 153–154
Sinigrin, 287, **288**
β-Sitostanol, **150**
β-Sitosterol, **150**
Skim milk, 176
Smoke, 366–367, 428–429
Snack foods, 308
Sodium (chloride), 440–441, 476, 477
Sodium chloride
   taste, 282
   preservative, 361–362
Sodium copper chlorophyllin, 217
Sodium erythorbate, 127
Sodium stearoyl-2-lactylate, 145–**146**
Soft drinks, 247
Solanine, **382**–**383**
Soluble fibre, 74, 75
Sophorose, **228**, 274
Sorbic acid, 362, **372**–373
Sorbitan monooleate, 145
Sorbitan tristearate, 145–**146**
Sorbitol, **18**–19
   ascorbic acid manufacture, 336
   sweetener, 269, 272, 273–276
Sorghum, 198
Sorption isotherms, 468–470
Sotolon, 34, **35**, 36
Southgate method, 73, 75
Soy sauce, 391–392
Soya essential amino acid content, 170, 171
Soya bean oil
   phytosteroids, 151
   lecithin, 144, 145
   margarine making, 113
   rancidity, 121
   production, 137
   vitamin E content, 348

Soya beans
  lectins, 389
  oligosaccharides, 25, 26
  phytoestrogens, 390–392
  protease inhibitors, 388
  starch content, 50
  sugar content, 8
Soya milk, 98, 161, 410
Spaghetti, 50, 59, 91
Spans(emulsifiers), 145, 146
Spearmint, 306
Sphingomyelin, **143**
Spices, 126, 305–306
Spina bifida (and neural tube defects), 330
Spinach
  calcium content, 442
  chlorophyll source, 217
  iron content, 446
  nitrate content, 419
  vitamin E content, 348
  vitamin K content, 350
β-Spiral in gluten, 202–203
Spirits (distilled), 250, 299
Stabilisers, 81
Stachyose, 8, **25,** 26, 73
Staling, 57, 149
Staphylococcal food poisoning, 406–407
Starch, 50–60
  granules, 51, 53–55
  modified, 85–88
  resistant, 57–59, 75
  retrogradation, 56–58, 149
  structure, 51–53
  syrups, 88–89
Stearic acid, 103, **101,** 108
Stearin, 135, 136, 138
Sterculic acid, **104**
Stevioside, 272, **274,** 277
Stigmasterol, **150**
Strawberries, 60, 229, 333, 299
Strecker degradation, 36–37, 46–47, 295–296
Sucralose, 272, **274**–277, 281
Sucrose, **22**
  content of foods, 8
  crystals, 26–27

glycaemic index, 91
inversion, 23
jam, 25
sweetness, 269
Sucrose monolaurate, 145–**146**
Sugar alcohols, 18–19
Sulfhydryl groups in gluten, 206
Sulfides in chocolate flavour, 308
Sulfites, 238, 362, 368–370
Sulforophane, **289**
Sulfur dioxide
  preservative, 362, 367–370
  anthocyanins, 232–233
  enzymic browning, 238
Sunset Yellow, 246–248, **247,** 251
Superoxide (dismutase), 194–195, 209–210
Supplements, dietary, 105
Surfactants, 142
Swedes, 415
Sweetcorn, 8, 161
Sweets (boiled), 23
Symphytine, **385**
Syringaldehyde, 76–**77**
Syringic acid, **367**

Talin^{TM}, 278
Tall oil, 151
Tallow, *see* beef fat
Talose, **12,** 44
Tannins, 233, 256–257
Tara gum, 85
Taramasalata, 98
Tartaric acid, 146–147, **283**
Tartrates, 466
Tartrazine, 217, 246–248, **247,** 250–251
Taurocholate, sodium, 142–**143**
TDI values, 417
Tea
  aluminium content, 427
  antioxidant function, 126
  astringency, 284
  caffeine, 383
  oesophagal cancer, 257
  polyphenolics, 239–242
Tecnazene, 415, 416

Tempeh, 391–392
Tequila, 42–43
Terpenoids, 218
a-Terpinyl acetate, **306**
Tetrodotoxin, 394–**395**
Tetroses, 7
Texture, of meat, 195–197
Thaumatin, 271, 272, 278
Theaflavins, 126, **241**–242
Thearubigens, 241
Theobromine, **383**
Thiamin, 315–318, **316**
  coenzyme functions, 352–353
  Dietary Reference Values, 476
  in meat flavour, 295
  sources in UK diet, 477
Thiazoles, 295–**296**
Thiocyanates, **287**, 388
Thixotropic fluids, 83
Threonine, **162,** 169
Threose, **12**
Thromboxanes, 108, 111
a-Thujone, **306**
Thymol, **306**
Thyroxine, 452
Tin, 427–428, 436
Titanium dioxide, 249
Toast, 46
Tocopherols
  antioxidant, 125–127
  and see vitamin E
Tocotrienols, **347**–348
Tofu protein content, 161, 391–392
Tolerable Daily Intake, *see* TDI values
Tomato juice, 67
Tomatoes
  ascorbic acid content, 333
  carotenoids, 222, 225
  nitrate content, 419
  sugar content, 8
  umami substances, 293
  vitamin K content, 350
  water content, 455
Tortillas, 325
Tragacanth, gum, 80–81, 85
Trans acid reduction

Trans fatty acids, 113–115, 135, 152 153
Trehalose, **21**, 269
Triarylmethane dyes, 246
Tricaprin, 132
Trichloroanisole, **310**
Triglyceride polymorphism, 155–157
Trigonelline, **324**
Trilaurin, 132
Trimyristin, 132
Triolein, 127, 132
Trioses, 7
Tripalmitin, 127, 132
Triplet oxygen, 153–154
Tristearin, 127, 132
Tristimulus values, 259
Tropocollagen, 195–197
Tryptophan, **162,** 169
  bitter taste, 280
  Maillard reaction, 38, 39
Tuna, 161, 293, 397, 425
Turkey, 98
Turmeric, 243
Turnips, 289
Tweens (emulsifiers) 145,146
Tyramine, **387**, 398
Tyrosine, **162,** 169, 203, 280

UHT (ultra high temperature), 38
Umami, 290–293
Vaccenic acid, **105**
Valine, **162,** 169, 280
Vanadium, 453
Vanillic acid, **367**
Vanillin, 76–**77**, , **266** 307
Vasopressor amines, 386–387
Veal, 192, 293
Vegan diets, 327
Vegetable oils, 98
Vegetables
  carotenoids, 222
  colours, 248
  vitamins, 324, 329, 339, 350
  iodine content, 453
  iron content, 446
  phosphorus content, 444
  selenium content, 451

Vegetarian diets and products, 107, 144, 184

Verbascose, 8, **25,** 26

Verbenone, **306**

Vermouth, 250

Vicine, **393**

Vinegar, 250, 283, 422

Vinyl chloride, 430–431

Violaxanthin, **220,** 222

Vitamin toxicity, 342, 346

Vitamin A, *see* retinol

Vitamin B_1, *see* thiamin

Vitamin B_2, *see* riboflavin

Vitamin B_3, *see* niacin

Vitamin B_6, *see* pyridoxine

Vitamin B_{12}, *see* cobalamin

Vitamin B_{13}, B_{15}, and B_{17}, 357

Vitamin C, *see* ascorbic acid

Vitamin D, *see* cholecalciferol

Vitamin E, 346–349, **347,** 477

Vitamin K, **350**–351

Vitamin supplementation, 330, 340–341

Walnuts, 50

Warfarin, **351**

Water activity, 466–470, 473–474

Water binding, 30, 468–470

Water soluble vitamins, 314, 315–338

Watercress, 288

Waxy starch (incl. maize), 51, 54, 56

Whale oil, 132

Wheat, 198–208
    essential amino acids, 170, 171
    coeliac disease, 410
    milling, 199–200
    niacin content, 324

pesticide residues, 415

proteins, 199–208

pyridoxine content, 322

Wheatgerm oil, 348

Whey (proteins), 177, 183

Whisky lead contamination, 41, 267, 422, 455

Wine and
    added colours, , 245 248
    anthocyanins, 232–233
    aroma, 300
    astringency, 285
    flavonoid content, 257–258
    lead contamination, 422
    sugar content, 7
    sulfur dioxide, 368–370
    vasopressor amines, 386

Wine lactone, 300

Winterization, 135, 136

Xanthan gum, 82–**83,** 85

Xanthene dyes, 246

Xanthophylls, 220–224

Xerophiles, 467

Xylans, 76–77

Xylitol, **18,** 269, 273–276, **274**

Xylose, **12**
    gums, 81
    pectins, 63, 64, 93, 94

Xylulose, 12, **13**

Yeast, 22, 319, 368, 371

Yoghurt, 7, 79, 161, 183

Yolk, *see* egg

Zeaxanthin, **220,** 221, 224, 226–227

Zinc, 79, 450, 476, 477